D0845250

Complete Wireless Design

Cotter W. Sayre

McGraw-Hill

New York Chicago San Francisco Lisbon London
Madrid Mexico City Milan New Delhi San Juan Seoul
Singapore Sydney Toronto

Cataloging-in-Publication Data is on file with the Library of Congress

McGraw-Hill

A Division of The ***McGraw·Hill*** *Companies*

4 5 6 7 8 9 DOC/DOC 0 6 5 4 3

P/N 137017-X
PART OF
ISBN 0-07-137016-1

The sponsoring editor for this book was Steve Chapman, the editing supervisor was Carol Levine, and the production supervisor was Sherri Souffrance. This book was set in Century Schoolbook per the MHT design by Joanne Morbit of McGraw-Hill's Professional Book Group composition unit, Hightstown, N.J.

Printed and bound by R. R. Donnelley & Sons Company.

This book is printed on recycled, acid-free paper containing a minimum of 50% recycled, de-inked fiber.

To my lovely wife Linda, without whom this book would not have been possible.

Contents

Preface

Complete Wireless Design gives the reader a solid grounding in the latest radio-frequency (RF) design methods and communication circuits employed in today's wireless equipment and systems, and will assist any engineer, technician, or ham, to design—down to the circuit level—anything from a basic two-way radio to the wireless receivers and transmitters of a digital communications system.

Included with the book is a free and complete copy of Caltech's Puff RF/microwave circuit simulation software, along with Sonnet Lite's electromagnetic simulator, Agilent's AppCad RF design software, and National's PLL design programs.

Unlike many wireless books, *Complete Wireless Design* does not simply present predesigned circuits and expect readers to modify them in some haphazard fashion for their own wireless applications, nor does this book present overly complex equations for the design of wireless circuits and systems, which most readers, even engineers, would have difficulty understanding, much less applying. Instead, *Complete Wireless Design* allows the reader, using simple algebra, to design cutting-edge oscillators, amplifiers, mixers, filters, phase-locked loops (PLLs), frequency multipliers, RF switches, microstrip elements, automatic gain control (AGC) loops, power splitters, attenuators, and diplexers easily and quickly. This book will also explain the practical aspects of designing with radio-frequency integrated circuits (RFICs) and monolithic microwave integrated circuits (MMICs); and how to perform all the necessary calculations for impedance matching, perform wireless link analyses, complete a frequency plan, and integrate a complete communications system. The book covers vital high speed and circuit design issues as well.

Cotter W. Sayre

Acknowledgments

The author wishes to thank the following individuals for their assistance and contributions in giving permission to reproduce the enclosed RF software, as well as the software documentation:

David Rutledge of Caltech University, along with Richard Compton, Scott Wedge, and Andreas Gerstlauer, for the linear circuit simulator *PUFF*.

Dr. James C. Rautio, president of Sonnet Software, for the EM simulator *Sonnet Lite em*.

Christopher A. Schell, senior engineer at National Semiconductor, for *Codeloader* and *EasyPLL*.

Robert L. Myers, AppCad Program Manager, for Agilent's *AppCad*.

Chapter

1

Wireless Essentials

A firm understanding of how passive and active components function at high frequencies, as well as a strong grasp of the fundamental concepts of lumped and distributed transmission lines, S-parameters, and radio-frequency (RF) propagation, is essential to successful circuit design.

1.1 Passive Components at RF

1.1.1 Introduction

At radio frequencies, lumped (physical) resistors, capacitors, and inductors are not the "pure" components they are assumed to be at lower frequencies. As shown in Fig. 1.1, their true nature at higher frequencies has undesirable resistances, capacitances, and inductances—which must be taken into account during design, simulation, and layout of any wireless circuit.

At microwave frequencies the lengths of all component leads have to be minimized in order to decrease losses due to lead inductance, while even the board traces that connect these passive components must be converted to transmission line structures. *Surface mount devices* (SMDs) are perfect for decreasing this lead length, and thus the series inductance, of any component (Fig. 1.2), while the most common transmission line structure is *microstrip,* which maintains a 50-ohm constant impedance throughout its length—and without adding inductance or capacitance.

As the frequency of operation of any wireless circuit begins to increase, so does the requirement that the actual physical structure of all of the lumped components themselves be as small as possible, since the part's effective frequency of operation increases as it shrinks in size: the smaller package lowers the harmful distributed reactances and series or parallel resonances.

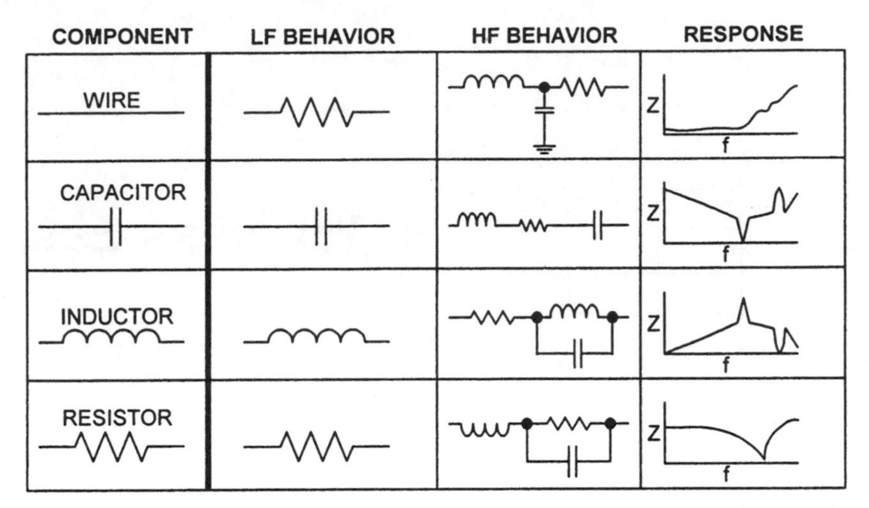

Figure 1.1 A component's real-life behavior at high frequencies (HF) and low frequencies (LF).

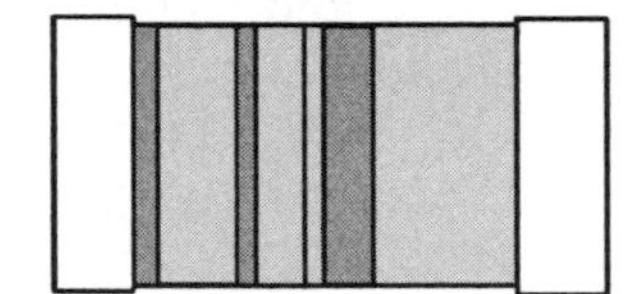

Figure 1.2 A surface mount resistor.

1.1.2 Resistors

As shown in Fig. 1.3, a resistor's actual value will begin to decrease as the frequency of operation is increased. This is caused by the distributed capacitance that is always effectively in parallel with the resistor, shunting the signal around the component; thus lowering its effective value of resistance. As shown in the figure, this distributed capacitance is especially problematic not only as the frequency increases, but also as the resistance values increase. If the resistor is not of the high-frequency, thin-film type, a high-value resistor can lose much of its marked resistance to this capacitive effect at relatively low microwave frequencies. And since the series inductance of the leads of the surface-mount technology resistor are typically quite low, the added reactive effect is negligible in assisting the resistor in maintaining its marked resistance value.

1.1.3 Capacitors

Capacitors at RF and microwave frequencies must be chosen not only for their cost and temperature stability, but also for their ability to properly function at these high frequencies. As shown in Fig. 1.1, a capacitor has an undesired lead inductance that begins to adversely change the capacitor's characteristics as

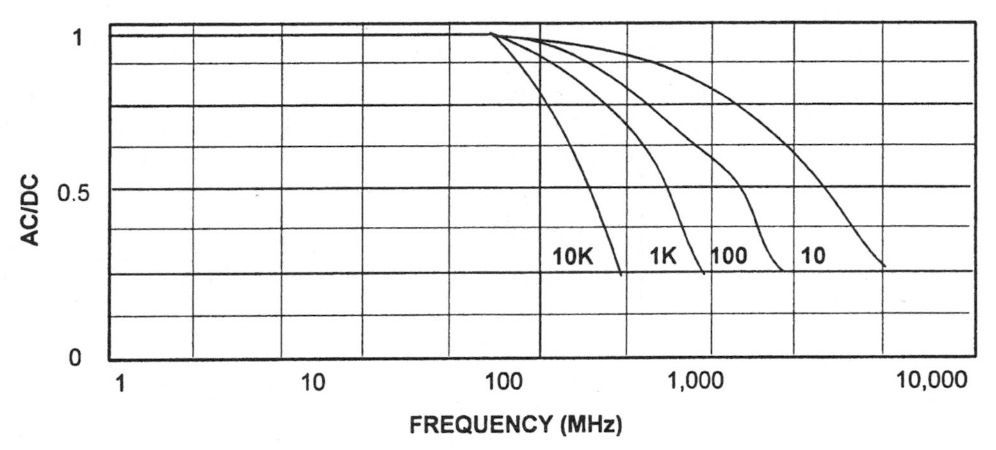

Figure 1.3 Ratio of an SMD resistor's resistance at DC to its resistance at AC for increasing frequencies.

the frequency is increased. This effect is most pronounced if the lead inductance resonates with the capacitance of the physical capacitor, resulting in a series resonance—or a total reactance of nearly *zero* ohms (resonating a capacitor can also be purposeful: a *j0 capacitor* is the type that becomes series resonant at the frequency of interest by resonating its own parasitic inductance with its own small value of marked capacitance, which creates a very low series impedance, perfect for coupling and decoupling at very high frequencies). Above this series resonant frequency the capacitor itself will actually become more inductive than capacitive, making it quite important to confirm that the circuit's design frequency will not be over the series resonance of the capacitor. This is vital for coupling and decoupling functions, while a capacitor for tuned circuits should have a series resonance comfortably *well above* the design frequency. The higher the value of the capacitor, the lower the frequency of this series resonance—and thus the closer the capacitor is to its inductive region. Consequently, a higher-value capacitor will demonstrate a higher inductance, on average, than a smaller value capacitor. This makes it necessary to compromise between the capacitive reactance of the capacitor in coupling applications and its series resonance. In other words, a coupling capacitor that is expected to have a capacitive reactance at the frequency of interest of 0.1 ohm may actually be a much poorer choice than one that has a capacitive reactance of 5 ohms—unless the capacitor is chosen to operate as a *j*0 type.

Only certain capacitor classifications are able to function at both higher frequencies *and* over real-life temperature ranges while maintaining their capacitance tolerance to within manageable levels. The following paragraphs discuss the various capacitor types and their uses in wireless circuits:

Electrolytic capacitors, both aluminum and tantalum, are utilized for very low frequency coupling and decoupling tasks. They have poor *equivalent series resistance* (ESR) and high DC leakage through the dielectric, and most are polarized. However, they possess a very large amount of capacitance per unit volume, with this value ranging from greater than 22,000 μF down to 1 μF for the aluminum types. Aluminum electrolytics have a limited life span of between 5 to 20 years while tantalums, with their dry internal electrolyte, have a much longer lifetime—and less DC dielectric leakage. Unfortunately, tantalums have less of a range of values (between 0.047 μF and 330 μF) and a lower maximum working voltage rating.

Metallized film capacitors are commonly good up to about 6 MHz and are adopted for low-frequency decoupling. These capacitors are available in capacitance ranges from 10 pF to 10 μF, and include the polystyrene, metallized paper, polycarbonate, and Mylar™ (polyester) families. Metallized film capacitors can be constructed by thinly metallizing the dielectric layers.

Silver mica capacitors are an older, less used type of high-frequency capacitor. They have a low ESR and good temperature stability, with a capacitance range available between 2 and 1500 pF.

Ceramic leaded capacitors are found in all parts of RF circuits up to a maximum of 600 MHz. They come as a single-layer type (ceramic disk) and as a stacked ceramic (monolithic) structure. Capacitance values range from 1.5 pF to 0.047 μF, with the dielectric available in three different grades: *COG* (*NPO*) for critical temperature-stable applications with tight capacitance tolerance values of 5 percent or better (with a capacitance range of 10 to 10,000 pF); *X7R* types, with less temperature stability and a poorer tolerance (±10 percent) than COG (with available values of 270 pF to 0.33 μF); and *Z5U* types, which are typically utilized only for bypass and coupling because of extremely poor capacitance tolerances (±20 percent) and bad temperature stability (with a range of values from 0.001 to 2.2 μF). However, the dominant microwave frequency capacitors today are the SMD *ceramic* and *porcelain chip* capacitors, which are used in all parts of RF circuits up to about 15 GHz. Nonetheless, even for these ultra-high-quality RF and microwave chip capacitors, the capacitance values must be quite small in order for them to function properly at elevated frequencies. Depending on the frequency, a maximum value of 10 pF or less may be all that we can use in our circuit because of the increasing internal inductance of the capacitor as its own capacitance value is raised. These leadless microwave chip capacitors are also available in multilayer and single-layer configurations, with the multilayer types normally coming in a basic SMD package, while single-layer capacitors are more difficult to mount on a board because of their nonstandard SMD cases. Nonetheless, single-layer capacitors can operate at much higher frequencies—up to tens of GHz—than multilayer; but they will also have a much lower capacitance range. In addition, some ceramic and porcelain microwave SMD capacitors will have a microstrip ribbon as part of their structure for easier bonding to the microstrip transmission lines of the printed circuit board.

1.1.4 Inductors

A significant, real-world high-frequency effect in an inductor is undesired distributed capacitance—which is a capacitance that is in parallel with the actual desired inductance of the coil (Fig. 1.1). This also means that there must be some frequency that will allow the coil's inductance to be in parallel resonance with the distributed capacitance, causing a high impedance peak to form at that frequency. In fact, the impedance created by this parallel resonance would be infinite if not for the small value of wire resistance found in series with the inductor's structure. The point of resonance is called the *self-resonant frequency* (SRF) of the inductor and must be much higher than the circuit's actual frequency of operation if the inductor is to be used in a tuned resonant circuit (to maintain the tank's proper impedance). RF inductors for use at the higher frequencies are built with small form factors in order to decrease this distributed capacitance effect, and thus increase their SRF (this technique will also lower the maximum inductance available, however).

An inductor parameter that is especially important for tuned circuits is the Q, or quality factor, of the inductor. The Q indicates the quality of the inductor at a certain test frequency; Q equals the inductive reactance divided by the combined DC series resistance, core losses, and skin effect of the coil. At low frequencies Q will increase, but at high frequencies the Q of an inductor will begin to decrease as a result of the skin effect raising the resistance of the wire. (Even while this is occurring, the distributed capacitance is also decreasing the desired inductance of the coil. Thus, the Q will soon reach zero, which is the value at its SRF). The coil's DC series resistance is the amount of physical resistance, measured by a standard ohmmeter, that is due to the innate resistance within the inductor's own wire. The DC series resistance affects not only the Q of a coil as mentioned above (and can reach relatively high levels in physically small, high-value, high-frequency inductors), but will also drop a significant amount of DC bias voltage. This is important in choosing a coil for a circuit that demands that the inductor must not have an excessive DC voltage drop across it, which can cause erratic circuit operation because of decreased bias voltages available to the active device. The last major loss effect that can create problems in high-inductance coils at high frequencies is created by coil-form losses, which can become substantial because of hysteresis, eddy currents, and residual losses, so much so that the only acceptable type of inductor core material is typically that of the air-core type.

Inductor coil design. There are times when the proper value or type of inductor is just not available for a small project or prototype, and one must be designed and constructed.

For a high-frequency, single-layer air-core coil (a *helix*), we can calculate the number of turns required to obtain a desired inductance with the following formula.

$$n = \frac{\sqrt{L\,[(18d) + (40l)]}}{d}$$

where n = number of single layer turns required to meet the desired inductance (L)

L = desired inductance of the air coil, μh

d= diameter, in inches, of the inside of the coil (the same diameter as the form used to wind the coil)

l= length, in inches, of the coil (if this length is not met after winding the turns, then spread the individual coils outward until this value is reached)

But this should be kept in mind: The formula is only accurate for coils with a length that is at least half the coil's diameter or longer, while accuracy also suffers as the frequency is increased into the very high frequency (VHF) region and above. This is a result of the excessive growth of conductor thickness with coil diameter. Only varnished ("magnet") wire should be used in coil construction to prevent turn-to-turn shorts.

Toroids. Inductors that are constructed from doughnut-shaped powdered iron or ferrite cores are called toroids (Fig. 1.4). Ferrite toroidal cores can function from as low as 1 kHz all the way up to 1 GHz, but the maximum frequency attainable with a particular toroid will depend on the kind of ferrite material employed in its construction. Toroids are mainly found in low- to medium-power, lower-frequency designs.

Toroidal inductors are valuable components because they will exhibit only small amounts of flux leakage and are thus far less sensitive to coupling effects between other coils and the toroid inductor itself. This circular construction keeps the toroid from radiating RF into the surrounding circuits, unlike air-core inductors (and transformers), which may require some type of shielding and/or an alteration in their physical positioning on the printed circuit board (PCB). And since almost every magnetic field line that is created by the primary makes it to the secondary, toroids are also very efficient. Air-core transformers do not share these abilities.

At low frequencies, toroids are also used to prevent hum from reaching the receiver from the mains and any transmitter-generated interference from entering the power lines. This is accomplished by placing toroid inductors in series with the supply power, choking out most of the undesired "hash."

Toroids are identified by their outer diameter and their core material. For instance, an FT-23-61 core designation would indicate that the core is a ferrite toroid (FT) with an outer diameter of 0.23 inches and composed of a 61-mix

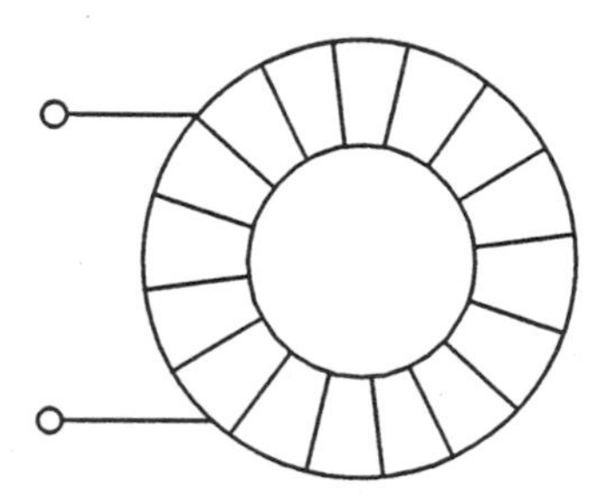

Figure 1.4 A toroid core inductor.

type of ferrite material. A T designation (instead of FT) would indicate a powdered iron core as opposed to a ferrite core.

Toroid coil design. As mentioned above, powdered-iron toroidal inductor cores are available up to 1 GHz. To design and wind an iron toroidal inductor or choke the A_L must be found on the core's data sheet. A_L symbolizes the value of the inductance in microhenrys (μH) when the core is wrapped with 100 turns of single-layer wire. All the inductor designer is required to do in order to design a powdered-iron toroidal coil is to choose the core size that is *just* large enough to hold the number of turns:

$$N = 100 \sqrt{\frac{L}{A_L}}$$

where N = number of single-layer turns for the desired value of L
L = inductance desired for the coil, μH
A_L = value, as read on the core's data sheet, of the chosen size and powdered-iron mix of the core, μH per 100 turns

Alternatively, if designing a *ferrite* toroidal core, the designer would use the formula

$$N = 1000 \sqrt{\frac{L}{A_L}}$$

where N = number of single-layer turns
L = inductance desired, mH
A_L = value, as read on the core's data sheet, of the chosen core size and ferrite mix, mH per 1000 turns

Notes

A_L values have a tolerance of typically ±20 percent.

The core material must never become saturated by excess power levels, either DC or AC.

Wind a single-layer toroid inductor or transformer with a 30-degree spacing between ends 1 and 2, as shown in the inductor of Fig. 1.5, to minimize distributed capacitance, and thus to maximize inductor Q.

The chosen mix for the core determines the core's maximum operating frequency.

1.1.5 Transformers

RF transformers are typically purchased as a complete component, but can also be constructed in toroidal form (Fig. 1.6). Toroids have replaced most air cores as interstage transformers in low-frequency radio designs (Fig. 1.7).

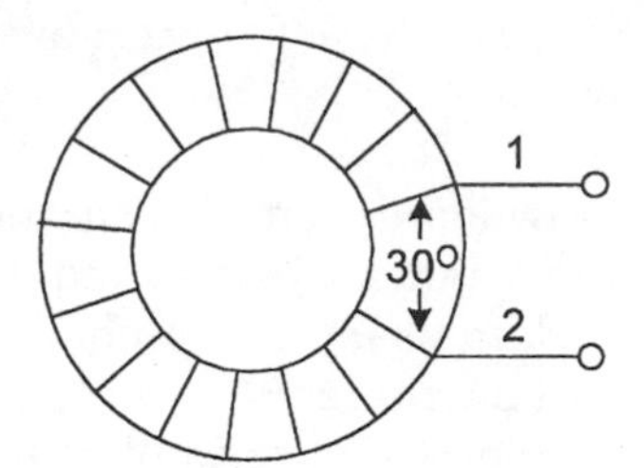

Figure 1.5 Proper winding of a toroidal inductor.

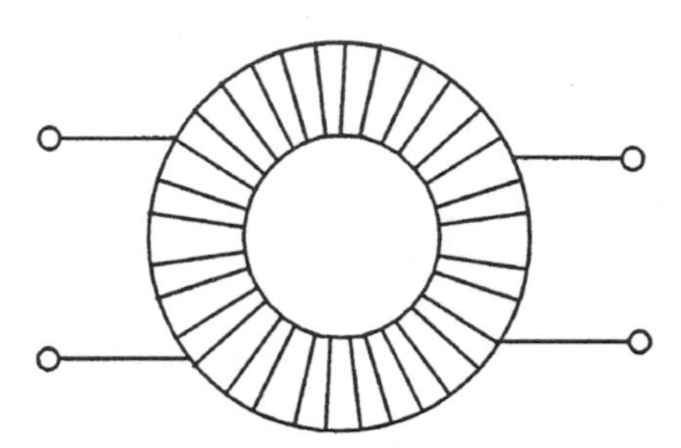

Figure 1.6 A toroid used to form a transformer.

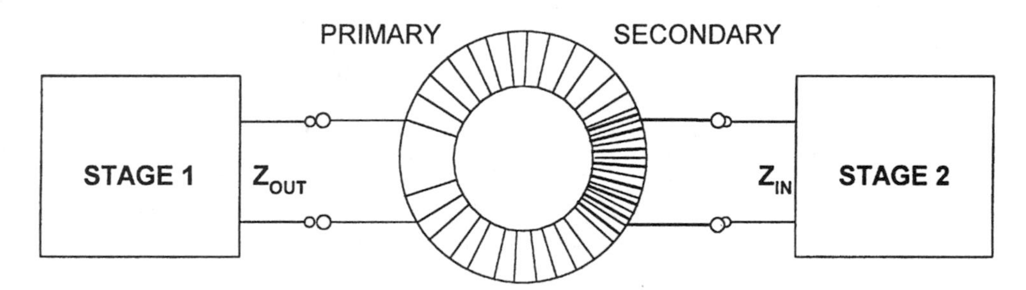

Figure 1.7 Impedance matching with a toroidal transformer.

Toroidal transformers, with the proper core material, are quite effective up to 1 GHz as broadband transformers. As the broadband transformer increases in frequency, however, the capacitance between the transformer's windings becomes more of a limiting factor. This internal capacitance will decrease the transformer's maximum operating frequency, since the signal to be transformed will now simply pass through the transformer. However, this effect can be minimized by choosing a high-permeability core, which will allow fewer turns for the very same reactance, and thus permit less distributed capacitance for higher-frequency operation.

Toroidal transformer design. For proper toroidal transformer operation, the reactances of the primary and secondary windings must be 4 or more times greater than the source and loads of the transformer at the lowest frequency of operation. As an example, if a 1:1 transformer's primary had a 50-ohm amplifier attached to its input, and the secondary had a 50-ohm antenna at its output, then the primary winding's reactance (X_P) should be at least 200 ohms, while the secondary winding's reactance (X_S) should also be 200 ohms at its lowest frequency of operation.

To design a toroidal transformer, follow these steps:

1. Calculate the required reactances of both the primary and the secondary of the transformer at its lowest frequency:

$$X_P = 4 \times Z_{OUT} \quad \text{and} \quad X_S = 4 \times Z_{IN}$$

where X_P = required primary reactance at the lowest frequency of transformer operation
Z_{OUT} = output impedance of the prior stage
X_S = required secondary reactance at its lowest frequency
Z_{IN} = input impedance of the next stage

2. Now, calculate the inductance of the primary and secondary windings:

$$L_P = \frac{X_P}{2\pi f_{LOW}} \quad \text{and} \quad L_S = \frac{X_S}{2\pi f_{LOW}}$$

3. Choose a core that can operate at the desired frequency, with a high permeability and as small a size as practical, and then calculate the number of primary and secondary turns required*

$$N_S = 100\sqrt{\frac{L_S}{A_L}} \quad \text{or} \quad N_S = 1000\sqrt{\frac{L_S}{A_L}}$$

$$N_P = N_S\sqrt{\frac{L_P}{L_S}}$$

4. Now wind the primary as a single layer around the entire toroid. Wind the secondary over the top of the primary winding at one end (Fig. 1.8). Reverse the windings for a step-up transformer.

1.2 Semiconductors

1.2.1 Introduction

Semiconductors, as opposed to the vacuum tubes of the past, are small, dependable, rugged, and need only low bias voltages. These devices are utilized not

*The formula for N_S will depend on how A_L is given in data sheet: 100 for µH, 1000 for mH.

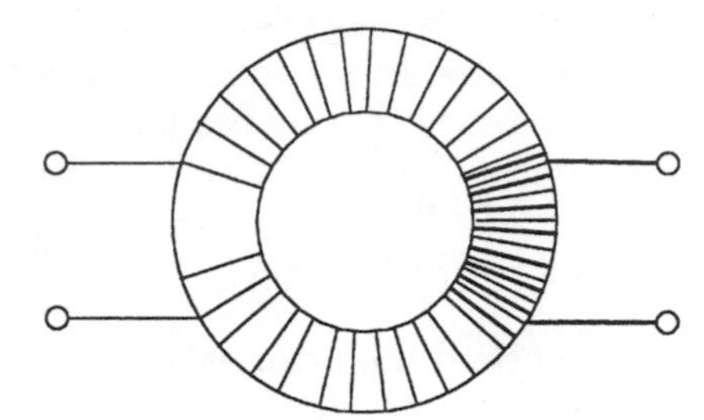

Figure 1.8 Proper winding for a toroidal transformer.

only to amplify signals, but also to mix and detect such signals, as well as create RF by oscillation. Indeed, integrated circuits, and thus most modern wireless devices, would not be possible without semiconductors. The following is a quick overview of the dominant semiconductor components.

1.2.2 DIODES

PN junction diodes. A *PN junction diode* (Fig. 1.9) is composed of both N- and P-type semiconductor materials that have been fused together. The N-type material will contain a surplus of electrons, called the *majority carriers,* and only a small number of holes, the *minority carriers.* The reason for this overabundance of electrons and lack of holes is the insertion of impurities, called doping, to the pure (or *intrinsic*) semiconductor material. This is accomplished by adding atoms that have five outer shell, or *valence,* electrons, compared to the four valence electrons of intrinsic silicon. The P-type material will have a surplus of holes and a deficiency of electrons within its crystal lattice structure due to the doping of the intrinsic semiconductor material with atoms that contain three valence electrons, in contrast to the four valence electrons of pure silicon. Thus, P-type semiconductor current is considered to be by hole flow through the crystal lattice, while the N-type semiconductor's current is caused by electron flow.

In a diode with no bias voltage (Fig. 1.10), electrons are drawn toward the P side, while the holes are attracted to the N side. At the fused PN junction a *depletion region* is created by the joining of these electrons and holes, generating neutral electron-hole pairs at the junction itself; while the depletion region area on either side of the PN junction is composed of charged ions. If the semiconductor material is silicon, then the depletion region will have a barrier potential of 0.7 V, with this region not increasing above this 0.7 value since any attempted increase in majority carriers will now be repulsed by this barrier voltage.

However, when a voltage of sufficient strength and of the suitable polarity is applied to the PN junction, then the semiconductor diode junction will be forward biased (Fig. 1.11). This will cause the barrier voltage to be neutral-

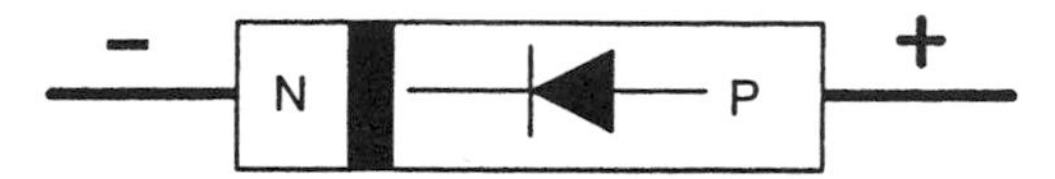

Figure 1.9 The semiconductor diode.

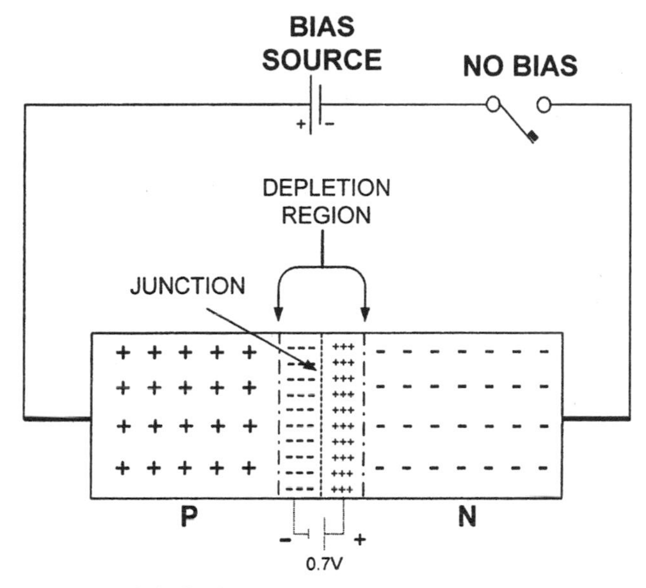

Figure 1.10 A diode shown with zero bias and its formed depletion region.

ized, and electrons will then be able to flow. The bias, consisting of the battery, has a positive terminal, which repulses the holes but attracts the electrons, while the negative battery terminal repels the electrons into the positive terminal. This action produces a current through the diode.

If a reverse bias is applied to a diode's terminals, as shown in Fig. 1.12, the depletion region will begin to enlarge. This is caused by the holes being attracted to the battery's negative terminal, while the positive terminal draws in the electrons, forcing the diode to function as a very high resistance. Except for some small leakage current, very little current will now flow through the diode. The depletion region will continue to expand until the barrier potential equals that of the bias potential or breakdown occurs, causing unchecked reverse current flow, which will damage or destroy the diode.

As shown in the characteristic curves for a typical silicon diode (Fig. 1.13), roughly 0.7 V will invariably be dropped across a forward-biased silicon diode, no matter how much its forward current increases. This is because of the small value of dynamic internal resistance inherent in the diode's semiconductor materials.

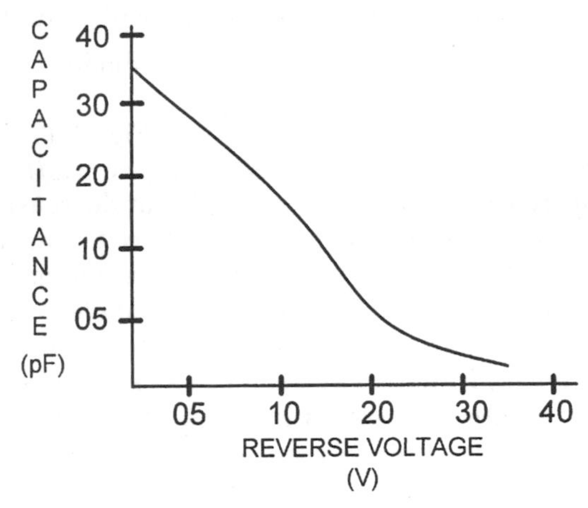

Figure 1.19 Capacitance versus the applied reverse voltage for a varactor diode.

Varactors are found in circuits that require a voltage-variable capacitance, such as tunable resonant filters and voltage-controlled oscillators (VCOs). They are available in many diverse capacitance values for almost any RF application.

PIN diodes. PIN diodes are constructed of a thin intrinsic layer sandwiched between a positive and a negative doped layer. They can be operated as RF switches and attenuators. PIN diodes, above certain frequencies (greater than 50 MHz), do not act as normal PN junction rectifier diodes, but as current-controlled resistors (the *carrier lifetime* rating will decide the diode's low frequency limit, under which the PIN begins to function as a normal PN junction diode). PINs will also have a much lower on resistance than do normal PN junction rectifier diodes, which can be changed over a range of $^1/_2$ ohm to over 10,000 ohms with the application of a DC control current. When employed as a switch, this control current is switched on or off, thus going from a very low resistance (on), to a very high resistance (off). When used as an attenuator, this control current is changed continuously, normally in nondiscrete steps, allowing the PIN to alter its resistance from anywhere between its lowest to its highest resistance values. Figure 1.20 displays a typical PIN diode's forward-bias current and resultant RF resistance.

Schottky diode. The Schottky diode is constructed of a metal that is deposited on a semiconductor material, creating an electrostatic boundary between the resulting Schottky barrier. These diodes can be found in microwave detectors, double-balanced modulators, harmonic generators, rectifiers, and mixers. Some Schottky diodes can function up to 100 GHz, have a low forward barrier voltage, and are mechanically sturdy.

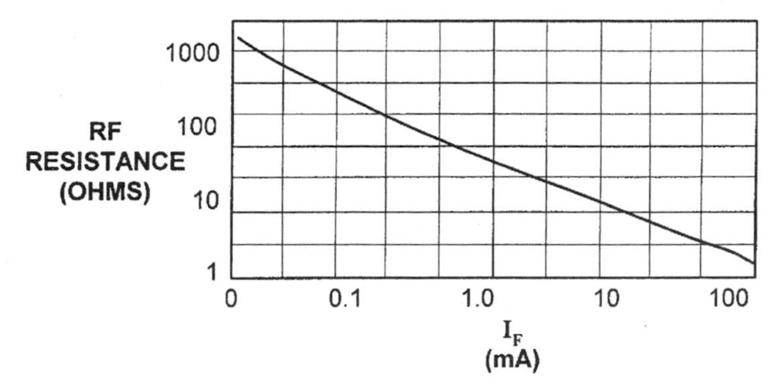

Figure 1.20 PIN diode forward-bias current and RF resistance.

Zero-bias Schottkys are a type of diode with a very low forward voltage. Figure 1.21 displays their *I-V* curves, showing their low forward voltage and the resultant forward current.

Gunn diodes. Gunn diodes can function as an oscillator at microwave frequencies. The transit time of an electron through the Gunn diode determines the actual frequency of oscillation and, when the diode is inserted into a suitable resonant cavity, the Gunn device can oscillate at frequencies of up to 100 GHz. However, the higher the frequency of the Gunn, the thinner it must be, which lowers its power dissipation abilities.

Step-recovery diodes. A step-recovery diode (SRD) is a special diode employed in some microwave frequency-multiplication circuits. The SRD functions in this role by switching between two impedance conditions: low and high. This change of state may occur in only 200 ps or less, thus discharging a very narrow pulse of energy. An SRD can best be visualized as a capacitor that stores a charge, then discharges it at a very rapid rate, causing a pulse that is plentiful in harmonics.

1.2.3 Transistors

Bipolar junction transistor (BJT). A bipolar transistor is constructed of NPN or PNP doped regions, with the NPN being by far the most common. The *emitter* provides the charges, while the *base* controls these charges. The charges that have not entered the base are gathered by the *collector.*

Figure 1.22 reveals a silicon NPN transistor that has its emitter and base forward biased, with the collector reversed biased, to form a simple amplifier. The negative terminal of the emitter-base battery repels the emitter's electrons, forcing them into the thin base. But the thin base structure, because of the small amount of holes available for recombination, cannot support the large number of electrons coming from the emitter. This is why base current

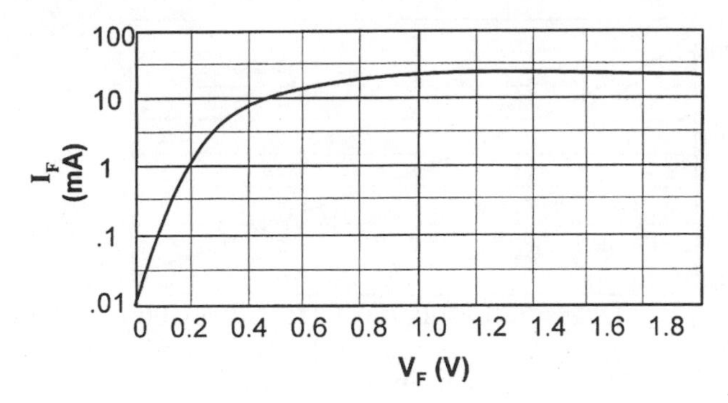

Figure 1.21 Zero-bias Schottky diode *I-V* curves showing forward voltage and the resultant forward current.

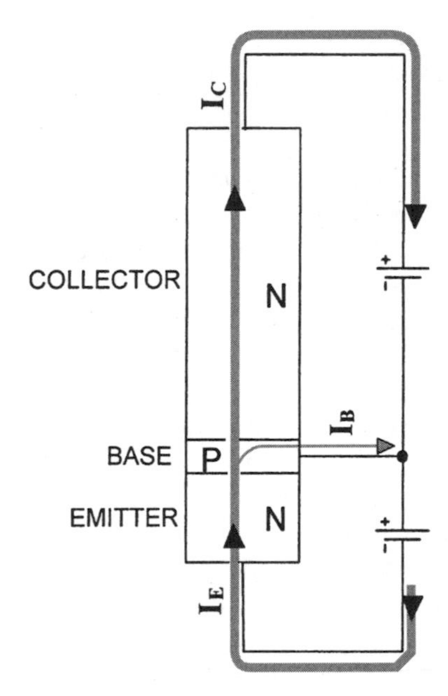

Figure 1.22 The current flow through the emitter, base, and collector of a bipolar NPN transistor.

is always a small value, since the majority of the electrons—over 99 percent—are attracted by the positive potential on the much larger collector, where they continue to flood into the collector's positive bias supply. This action is what forms the transistor's output current.

From the foregoing explanation, we see that $I_E = I_B + I_C$ and $I_B = I_E - I_C$, meaning that the currents through a transistor are completely proportional. Thus, if the emitter current doubles, then so will the currents in the base and the collector. But more important, this also means that if a small external bias or signal should increase this small base current, then a proportional—but far greater—emitter and collector current will flow through the transistor. This will produce voltage amplification if the collector current is sent through a high output resistance.

The input of a common-emitter transistor has a low resistance because of its forward bias, so any signal inserted into the base-emitter junction will be across this low input resistance, thus causing the bipolar transistor to be current controlled by both the DC bias and any external signal voltages. This is shown in the BJT's characteristic curves of Fig. 1.23. The input signal, such as an RF or audio signal, will then add to or subtract from the DC bias voltage that is across the transistor.

Before significant collector current can flow, the transistor's emitter-base barrier voltage V_{BE} of approximately 0.6 V (for silicon) must be overcome. This task is performed by the base bias circuit. In a linear amplifier, the initial transistor's operating point is set by the bias circuits to be around 0.7 V in order to allow any incoming signal to be able to swing above and below this amount. The region of active amplification of a BJT is only about 0.2 V wide, so any voltage between saturation (0.8 V) and cutoff (0.6 V) is the only range that a semiconductor is capable of amplifying in a linear manner. Between these two V_{BE} values of 0.6 and 0.8 V, the I_B, and thus the I_C, is controlled.

A BJT can be thought of as a current-controlled resistance, with a tiny base current controlling the transistor's resistance, which influences the much larger emitter-to-collector current. This collector current is then made to run through a high load resistance, generating an amplified output voltage.

Some high-frequency power transistors may be internally impedance matched to increase their normally very low input and output impedances (as low as 0.5 ohm), while some metal-can transistors may be found with four leads; with one lead attached to the metal can itself, which is then grounded to provide an RF shield.

A few of the more common transistor specifications found in BJT data sheets are:

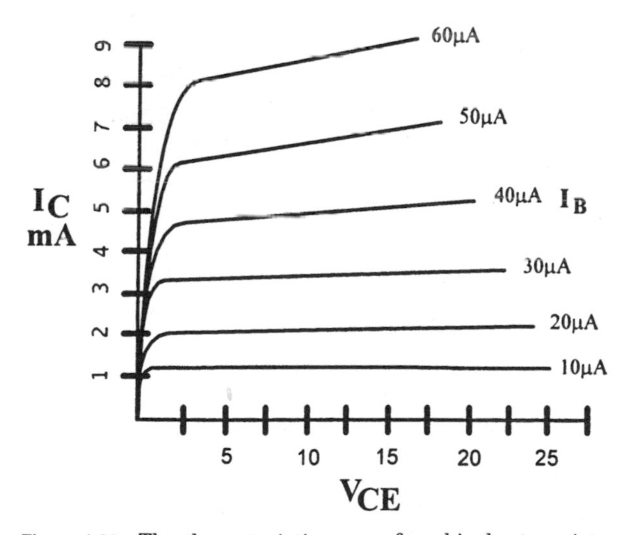

Figure 1.23 The characteristic curves for a bipolar transistor.

BV_{CBO}, the collector-to-base breakdown voltage, the amplitude of collector voltage that will normally break down the collector junction

$P_{D(MAX)}$, the maximum total power dissipation a transistor is capable of in an ambient air temperature of 25°C

$T_{J(MAX)}$, the maximum internal junction temperature before the semiconductor material breaks down

$I_{C(MAX)}$, the maximum collector current of the BJT

f_T, the current gain–bandwidth product, the frequency that a common-emitter transistor will be at a beta of unity

f_{ae}, the beta cutoff frequency, the frequency that the BJT's beta decreases to 70.7 percent of its low frequency value

I_{CEO}, the temperature-dependent leakage current that occurs from the emitter to the collector with the base open

Junction field-effect transistor (JFET). Since a JFET's input gates are always reversed biased, they will have a very high low-frequency input impedance, and are thus voltage controlled. Junction field-effect transistors are also quite capable of receiving an input of up to several volts (compared to the bipolar transistor's few tenths of a volt), and create less internal noise than a BJT, but they display less voltage gain and more signal distortion.

As shown in Fig. 1.24, the structure of a JFET is composed of a *gate,* a *source,* and a *drain.* The JFET's terminals are voltage biased in such a way that the drain-to-source voltage (V_{DS}) causes the source to be more negative than the drain. This lets the drain current (I_D) flow from the source to the drain through the N channel.

The JFET characteristic curves of Fig. 1.25 readily indicate that a JFET is a normally on device when there is no bias voltage present at the gate. This permits the maximum JFET current (I_{DSS}) to flow from the source to the drain. When the gate and source are presented with a negative voltage ($-V_{GS}$), an area lacking charge carriers (the *depletion region*) starts to form within the JFET's N channel. This N channel depletion region functions as an insulator; therefore, as the JFET becomes increasingly reverse biased and increasingly exhausted of any charge carriers, the N channel continues to be narrowed by this developing depletion region. The channel's resistance rises, decreasing the JFET's current output into its load resistor, which lowers the device's output voltage across this resistor. As the negative gate voltage of $-V_{GS}$ is increased, the depletion region continues to widen, decreasing current flow even further—but a point is ultimately reached where the channel is totally depleted of all majority carriers, and no more current flow is possible. The voltage at which the current flow stops is referred to as $V_{GS(OFF)}$. In short, the V_{GS} successfully controls the JFET's channel resistance, and thus its drain current. However, it is important that the drain-to-source voltage V_{DS} should be of a high enough amplitude to allow the JFET to operate within its linear region,

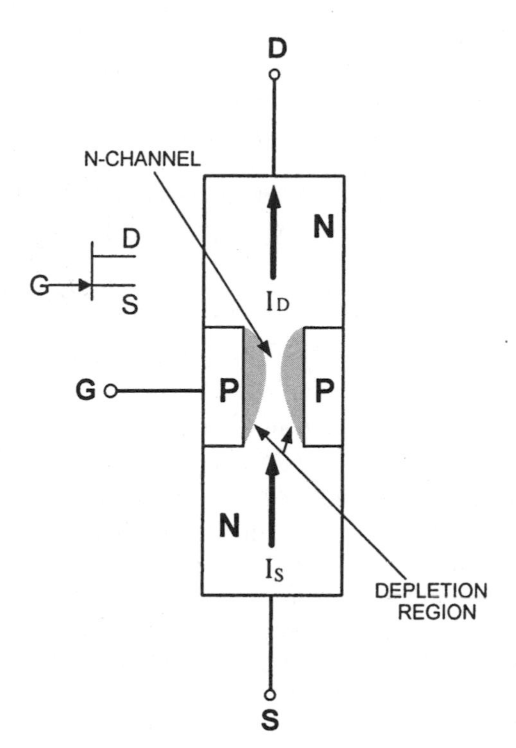

Figure 1.24 The internal structure of, and current flow through, a JFET.

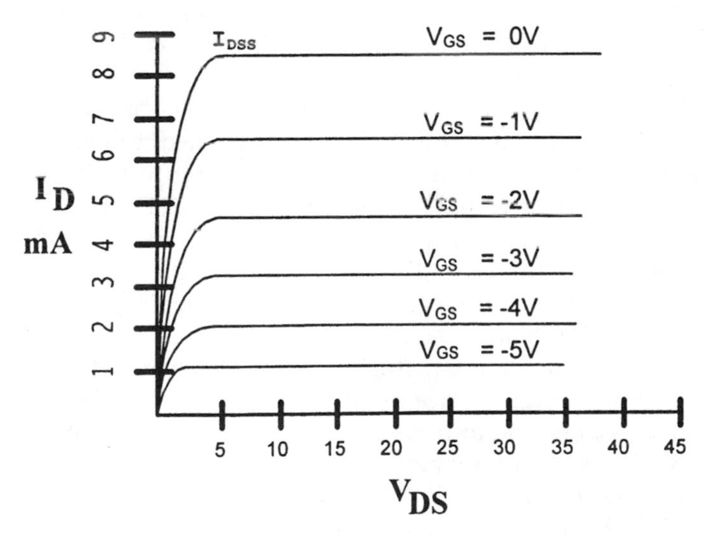

Figure 1.25 A JFET's characteristic curves.

or above *pinch-off* (V_P). Pinch-off is simply an area where the drain current will stay constant even if the drain-to-source voltage is increased; now only the gate-to-source voltage can affect the drain current.

A few of the more common JFET parameters are:

I_{DSS}, the maximum JFET drain current possible (with a V_{GS} at 0 V)

g_m or g_{fs}, the transconductance gain (or $\Delta I_D/\Delta V_{GS}$), measured in *siemens* or *mhos*

$V_{DS(MAX)}$, the maximum safe drain-to-source voltage

V_P, the pinch-off voltage, the minimum V_{DS} required for the JFET's linear operation

P_D, the JFET's maximum power dissipation rating

Metal-oxide-silicon field-effect transistor (MOSFET). Metal-oxide-silicon field-effect transistors use a gate structure that is well insulated from the *source, drain,* and *channel.* This produces an active device with an almost infinite DC input resistance. However, this high input resistance is significantly decreased by its bias components, as well as by high-frequency operation. In fact, as the frequency of operation is increased, the MOSFET's input impedance *approaches* that of a BJT.

MOSFETs are available that can operate in one of two modes: the *depletion mode,* as a normally on device, and the *enhancement mode,* as a normally off device.

Drain current in a depletion-mode N-channel MOSFET (Fig. 1.26) is controlled through the application of a negative *and* positive gate voltage (Fig. 1.27). By raising the negative voltage at the MOSFET's gate we would soon reach a point where, as a result of the channel being depleted of all majority carriers, no significant drain current can flow. But as the gate-to-source voltage V_{GS} becomes less negative, more current will start to move. Even as we pass 0 V for V_{GS}, the drain current will still continue to rise, since at zero V_{GS} the MOSFET—unlike the JFET—has not reached the maximum current. Nonetheless, the drain current is still quite substantial, since many majority carriers are present within the depletion MOSFET's N channel. The V_{GS} increases until it reaches some maximum positive value; now the maximum number of electrons has been drawn into the N channel, and the maximum current is flowing through the channel and into the drain.

Depletion MOSFETs are used extensively in wireless circuits because of their low-noise-producing characteristics. A similar structure, but employing two gates within a single device, is the dual-gate MOSFET (Fig. 1.28). These are utilized in mixers and automatic gain control (AGC) amplifiers, with each of the MOSFET's gate inputs having an equal control over the drain current.

The other type of MOSFET, the enhancement-mode type, or E-MOSFET (Fig. 1.29) is, as mentioned above, a normally off transistor. So, almost no source-to-drain current flows when there is no bias across the E-MOSFET's gate, as shown in the characteristic curves of Fig. 1.30. However, almost any

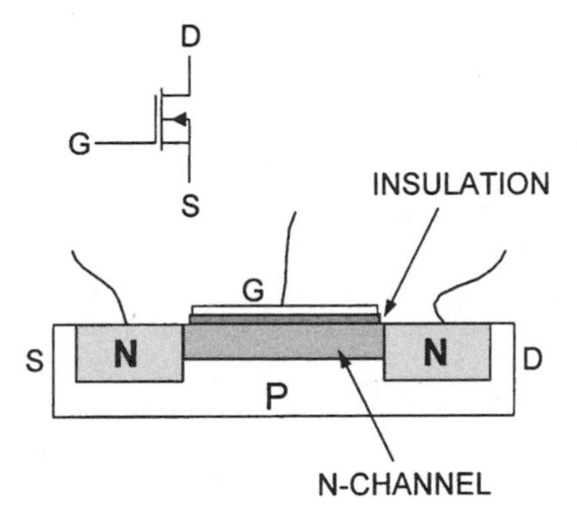

Figure 1.26 The internal structure of an N-channel depletion-mode MOSFET.

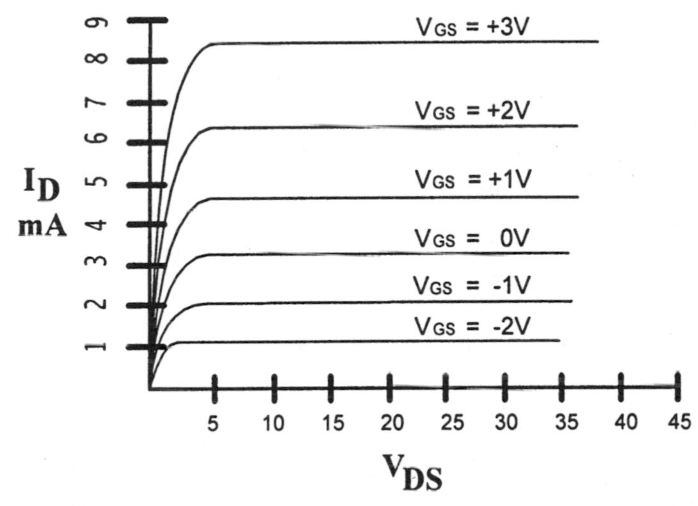

Figure 1.27 The characteristic curves of an N-channel depletion-mode MOSFET.

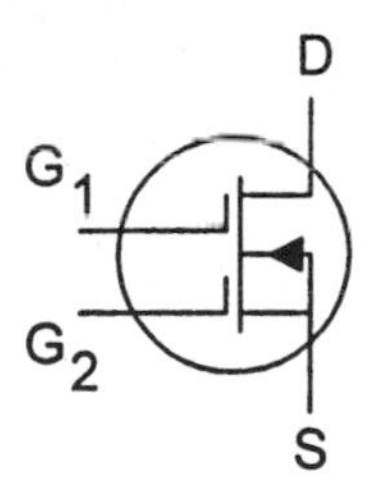

Figure 1.28 A dual-gate MOSFET's schematic symbol.

positive voltage that is placed across the gate will produce a channel between the device's source and drain (Fig. 1.31). Thus, as electrons are pulled to the gate, an N-channel is created within the P-type substrate. This action permits electrons to flow steadily toward the positively charged drain, creating a continuous current flow.

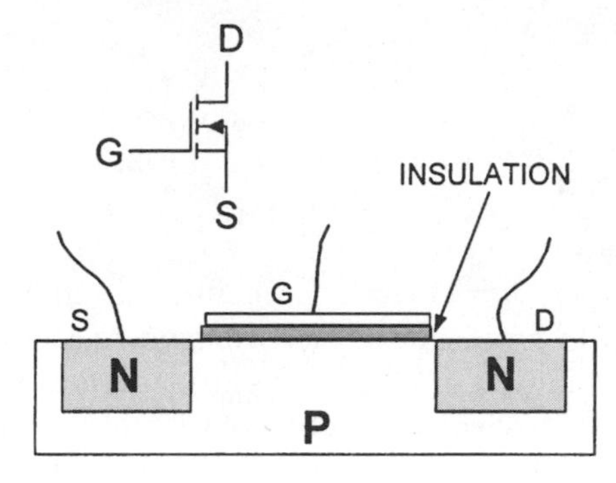

Figure 1.29 The internal structure of an enhancement-mode MOSFET.

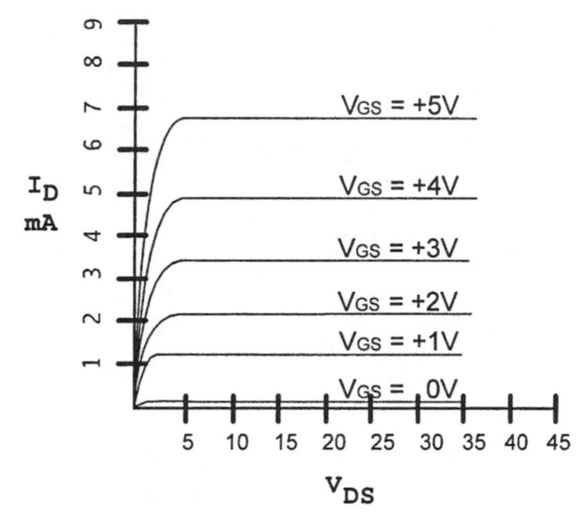

Figure 1.30 The characteristic curves of an enhancement-mode MOSFET.

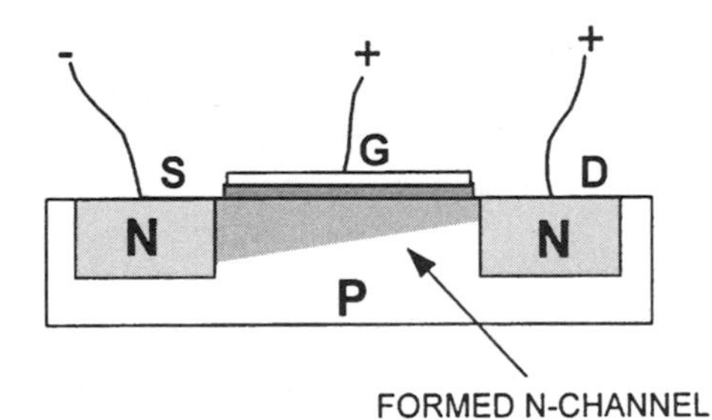

Figure 1.31 The formation of the N channel in an E-MOSFET's substrate by a positive gate voltage.

Nonetheless, enhancement-mode MOSFETs will have a 1-V gate threshold voltage before any significant drain current will flow. In fact, enhancement-mode MOSFET power devices must use a positive gate bias to overcome this gate threshold voltage in order to optimize gain and output power. This bias requirement means that, unlike a BJT, an E-MOSFET cannot simply employ a zero gate bias at its input to run in Class C power amplifier operation.

E-MOSFETs are popular in digital ICs as voltage-controlled switches, and are found as the active element in high-frequency, very high frequency, and ultrahigh frequency (HF, VHF, and UHF) power amplifiers and drivers. This is a result of the E-MOSFET parameters' superiority to those of a typical power BJT, such as higher input impedance and gain, increased thermal stability, lower noise, and a higher tolerance for load mismatches. Another advantage that any MOSFET enjoys over a BJT is the impossibility of thermal runaway, since MOSFETs are designed to have a *positive temperature coefficient* at high drain currents. This means that, as the temperature increases, a MOSFET will actually *decrease* its source-to-drain current, instead of increasing its current output as a BJT will (see "Thermal Runaway"). This makes thermal runaway impossible and temperature stabilization components less necessary, except to stabilize the MOSFET's Q point. In addition, a MOSFET's input and output impedances will change much less for different input drive levels than a BJT's, and a MOSFET offers better single-stage stability and 20 percent more gain. MOSFETs also can survive a very high voltage standing wave ratio (VSWR), second only to BJTs with an emitter ballast resistor in this respect. On the negative side, however, MOSFETs are very sensitive to destruction by static electricity, with almost any electrical spark possibly causing damage to the gate insulation. And N-channel enhancement MOSFETs, the most common in RF power applications, as well as depletion-mode power MOSFETs in Class C and B operation, can begin to exhibit low-frequency oscillations if they are directly paralleled for increased output. MOSFETs also have inferior low-order intermodulation distortion (IMD) performance to that of BJTs.

SiGe BiCMOS. SiGe stands for *silicon-germanium,* while CMOS stands for *complementary metal-oxide-semiconductor.* SiGe BiCMOS comprises these two major technologies: SiGe and the integration of SiGe with CMOS.

SiGe devices, also called SiGe HBTs (*silicon-germanium heterojunction bipolar transistors*), is a mixture of silicon (Si) and germanium (Ge) within a single transistor structure. This produces much higher cutoff frequencies (60 Ghz for SiGe; 20 GHz for standard silicon), a reduction in noise, and greatly decreased power dissipation, with the added benefit of increased gain over that of standard silicon. A primary limitation of current HBTs is that the breakdown voltage of the device is rather low, decreasing dependability somewhat. This will be rapidly improved, however.

Current SiGe technology allows the high-frequency performance of GaAs at much lower costs (equivalent to VLSI silicon, or about a quarter of the cost of GaAs). SiGe also employs much simpler manufacturing techniques (GaAs manufacturing is intensive, complex, and has lower chip yields than SiGe). In fact, many companies are claiming that SiGe will eventually completely obsolesce GaAs in all frequencies below 60 GHz.

The recent ability to economically combine CMOS with SiGe will permit the integration of microwave RF front ends with the intermediate-frequency (IF) and baseband circuitry—as well as the necessary control logic—on a single chip.

This will significantly decrease the number of components, and thus the cost, of high-volume items such as mobile phones, direct-conversion (zero IF) data radio receivers, cheap Global Positioning System (GPS) receivers, wireless local-area networks (LANs), pagers, and other (high-volume) systems-on-a-chip.

Many major companies, such as Harris Semiconductor, Hughes Networking, National Semiconductor, Northern Telecom, and Tektronix, find this technology important enough that they have obtained expensive foundry licenses for the IBM SiGe BiCMOS technology.

The first products are just becoming available, and are small-density components meant to supersede GaAs products. Inevitably, higher-integration devices will be introduced that will lower cost and increase performance in many high-volume wireless systems, with cellular phones being the primary market.

1.3 Microstrip

1.3.1 Introduction

At microwave frequencies, microstrip (Fig. 1.32) is employed as transmission lines, as equivalent passive components, and as tuned circuits and high-Q microwave filters on printed circuit boards. Microstrip is used for these functions for its low loss and ease of implementation, since high-frequency components, such as surface mount capacitors, resistors, and transistors, can be mounted directly onto the PCB's microstrip metallization layer. The metallization layer can be formed of copper or gold.

Microstrip itself is unbalanced transmission line and, because of its unshielded nature, can radiate RF. However, radiation from properly terminated microstrip is quite small. Stripline (Fig. 1.33) is similar to microstrip, but is placed between the metallization layers of a PCB and, because of the balanced twin ground planes, does not radiate. Both microstrip and stripline normally have a printed circuit board substrate constructed of fiberglass, polystyrene, or Teflon. Since microstrip can utilize standard PCB manufacturing techniques, it is easier and cheaper to fabricate than stripline.

The characteristic impedance of microstrip is governed by the width of the conductor, the thickness of the dielectric, and the dielectric constant; low-

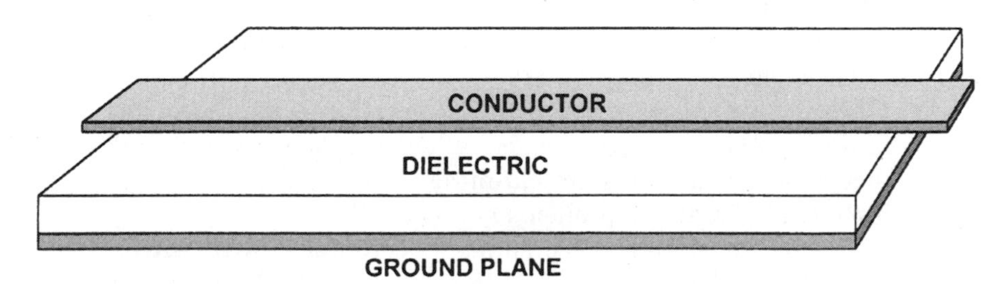

Figure 1.32 Microstrip, showing the dielectric and conductive layers.

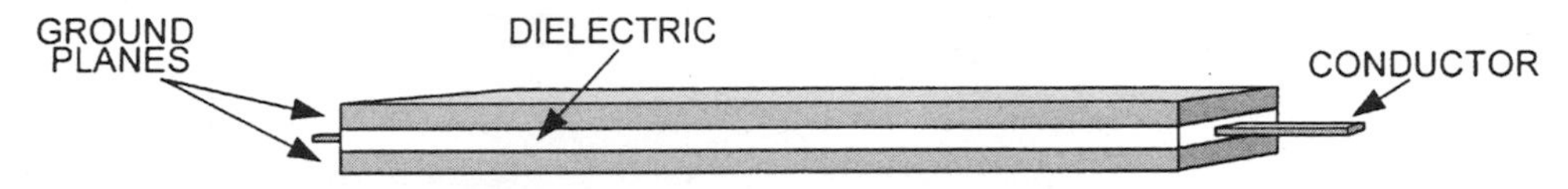

Figure 1.33 Stripline, showing the dielectric and conductive layers.

impedance microstrip lines are wide, and high-impedance microstrip lines are narrow. But the most important attribute of microstrip is that its impedance does not change with frequency or with line length. The characteristic impedances of microstrip and stripline are designed to be anywhere between 10 to 100 ohms, with 50 ohms being the norm for transmission line use. Microstrip is very common in frequencies of operation at 150 MHz and above.

1.3.2 Microstrip as transmission line

Fifty-ohm microstrip is utilized in microwave circuits to prevent reflections and mismatch losses between physically separated components, with a calculated nominal width that will prevent the line from being either inductive or capacitive at any point along its length. In fact, with a source's output impedance matched to the microstrip, and the microstrip matched to the input impedance of the load, no standing or reflected waves will result; thus there will be no power dissipated as heat, except in the actual resistance of the copper as I^2R losses.

In microstrip the dielectric constant (E_r) of the dielectric material will not be exactly what the microstrip transmission line itself "sees." This is due to the flux leakage into the air above the board, combined with the flux penetrating into the dielectric material. This means that the actual *effective dielectric constant* (E_{eff}), which is the true dielectric constant that the microstrip will now see, will be some value between that of the surrounding air and the true dielectric constant of the PCB.

Because of the small RF field leakage that emanates from all microstrip, these types of transmission lines should be isolated by at least two or more line widths to decrease any mutual coupling effects when run side by side on a PCB. To lower the chances of cross talk even further, a ground trace may be necessary between the two microstrip transmission lines. Microstrip should also always be run as short and straight as possible, with any angle using a mitered or slow round bend (Fig. 1.34) to decrease any impedance bumps—and the ensuing radiation output [electromagnetic interference (EMI)] caused by a sharp or unmitered bend.

Another issue to watch for in designing microwave circuits with microstrip transmission lines is the *waveguide effect:* Any metal enclosure used to shield the microstrip—or its source or load circuit—can act as a waveguide, and drastically alter circuit behavior. This effect can be eliminated by changing the width of the shield to cover a smaller area or by inserting special microwave foam attenuator material within the top of the enclosure.

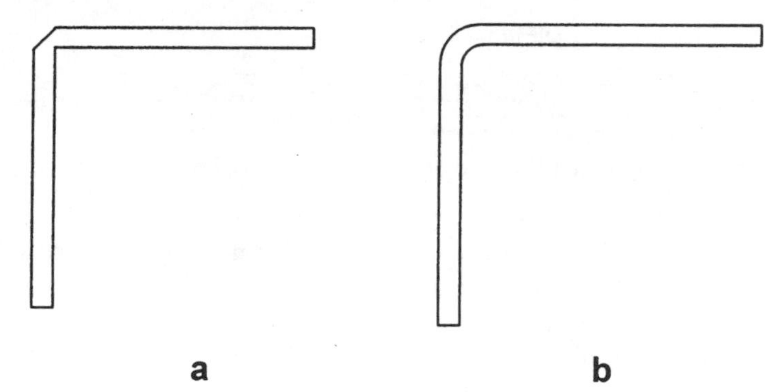

Figure 1.34 Proper way to work with bends in microstrip lines: (*a*) miter; (*b*) curve.

Microstrip transmission line design. Use the following equation to plug in different microstrip widths to obtain the desired impedance:

$$Z_0 = \frac{377}{\left(\frac{W}{h} + 1\right)\sqrt{E_r + \sqrt{E_r}}}$$

where Z_0 = characteristic impedance of the microstrip, ohms
W = width of the microstrip conductor (use same units as h)
h = thickness of the substrate between the ground plane and the microstrip conductor (use same units as W)
E_r = dielectric constant of the board material

1.3.3 Microstrip as equivalent components

Distributed components such as inductors, transformers, and capacitors can be formed from microstrip transmission line sections on PCBs at microwave frequencies. A series or shunt inductor can be formed from a thin trace (Fig. 1.35), a shunt capacitor can be formed by a wide trace (Fig. 1.36), and a transformer can be formed by varying the width of the microstrip (Fig. 1.37).

Distributed equivalent component design. It is important to never make a distributed component longer than 30 degrees out of the 360 degrees of an entire wavelength or the *equivalent component effect* will depart more and more from that of an ideal lumped component. To calculate how long 30 degrees is out of 360 degrees, simply divide 30 by 360, then multiply this value by the actual wavelength of the signal on the PCB, keeping in mind that the signal's wavelength in the substrate will not be the same as if it were traveling through a vacuum.

To find the actual wavelength of the signal, which is being slowed down by the substrate material, calculate the microstrip's *velocity of propagation* (V_p). First, find the *effective dielectric constant* (E_{EFF}) of the microstrip, since, as

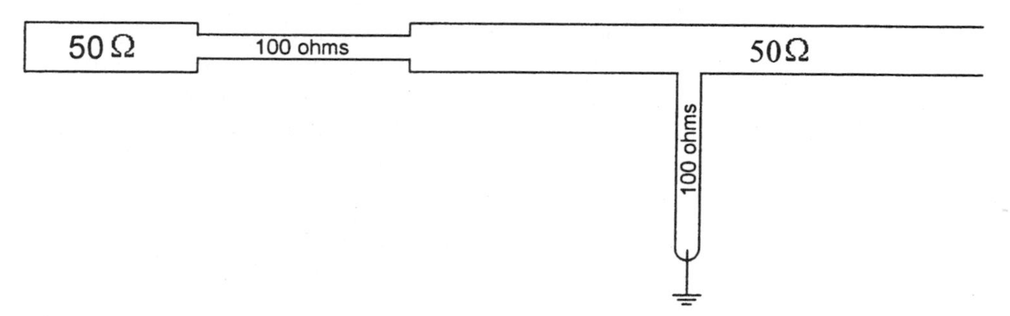

Figure 1.35 A distributed inductor.

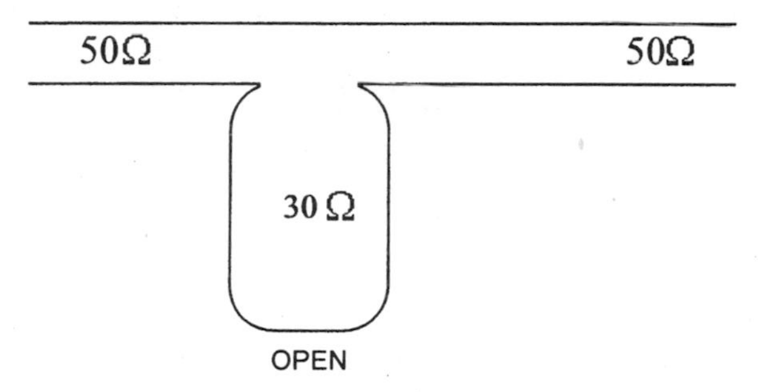

Figure 1.36 A distributed capacitor.

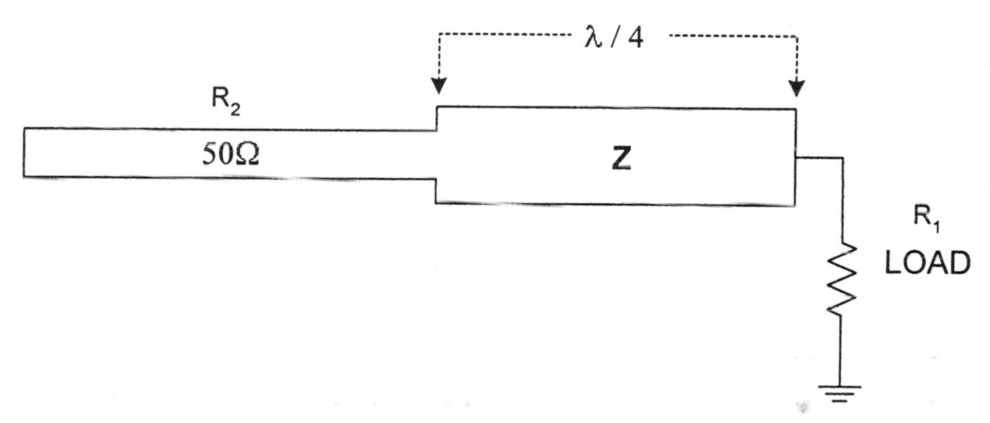

Figure 1.37 Using a distributed transformer for resistive matching.

stated above, the signal will be partly in the dielectric and partly in the air above the microstrip, which will affect the propagation velocity through this combination of the two dielectric mediums:

$$E_{\mathrm{EFF}} = \frac{E_r + 1}{2} + \left(\frac{E_r - 1}{2} \cdot \frac{1}{\sqrt{1 + \frac{12h}{W}}} \right)$$

where E_{EFF} = effective dielectric constant that the microstrip sees
E_r = actual dielectric constant of the PCB's substrate material
h = thickness of the substrate material between the top conductor and the bottom ground plane of the microstrip
W = width of the top conductor of the microstrip

Then:

$$V_P = \frac{1}{\sqrt{E_{EFF}}}$$

where V_P = fraction of the speed of light compared to light in a vacuum.

Then calculate the wavelength of the signal of interest in a perfect vacuum:

$$\lambda_{VAC} = \frac{11{,}800}{f}$$

where λ_{VAC} = wavelength of the frequency of interest (f), mils, in a true vacuum
11,800 = speed of light value to obtain a λ_{VAC} in mils while using an f in GHz
f = frequency of the signal of interest, GHz

Then multiply the velocity of propagation (V_P) times the wavelength (λ_{VAC}) of the signal as calculated above in order to arrive at the wavelength of the signal of interest (λ), in mils, when the signal is placed into the microstrip:

$$\lambda = V_P \times \lambda_{VAC}$$

Distributed parallel (shunt) capacitor. First, knowing the capacitance of the desired component for your circuit, calculate the reactance of the shunt capacitor required, at the frequency of interest, by the common formula

$$X_c = \frac{1}{2\pi f C}$$

Second, utilize 30-ohm microstrip (Z_L = 30 ohms) for the substrate's dielectric. Find the microstrip width required for this 30-ohm value by using one of the many microstrip calculation programs available free on the Web (such as HP's AppCad, or AWR's TXLine, or Daniel Swanson's MWTLC), or use the formula above. As shown in Fig. 1.38, the microstrip of the equivalent shunt capacitor is open, and not grounded, at its end. The capacitor section is also attached to the 50-ohm microstrip transmission line by a small tapered section to improve the transition. A further improvement is possible by splitting the capacitor in two and placing it on both sides of the transmission line.

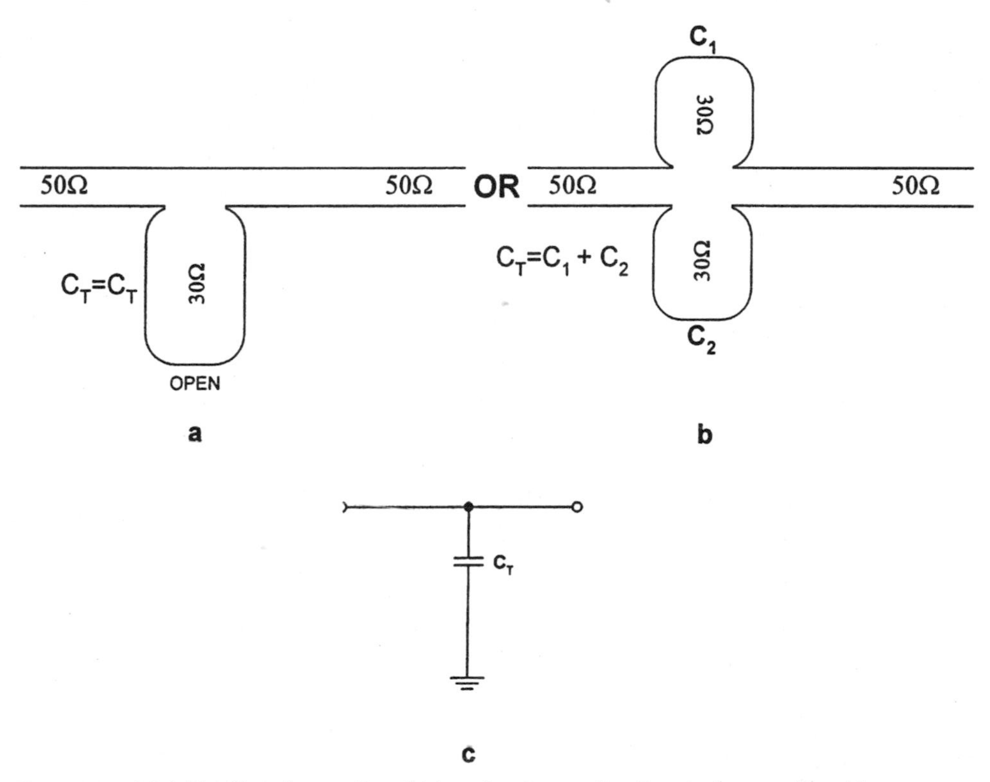

Figure 1.38 (*a*) A distributed capacitor; (*b*) two shunts equaling the single capacitor; (*c*) equivalent lumped shunt capacitor.

Third, calculate the required microstrip's length to become a capacitor of value X_c, as calculated above, with this formula:

$$\frac{\left(\text{Arctan}\,\dfrac{30}{X_c}\right)}{360} \times \lambda = \text{length}$$

where X_c = capacitive reactance required in the distributed circuit, ohms
length = length of the microstrip required to imitate a lumped component of value X_c in mils (which should never be longer than 30 degrees, or 12 percent, of λ)
λ = wavelength of the frequency of interest *for the substrate of interest* (or $V_P \times \lambda$; see above formulas) , mils

Series inductor. As shown in Fig. 1.39, the equivalent series inductor is placed in series with the 50-ohm microstrip transmission line, or placed between other distributed or lumped components.

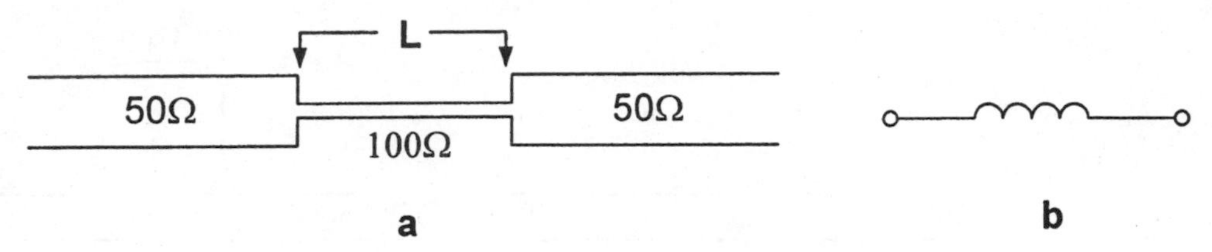

Figure 1.39 (*a*) A series distributed inductor; (*b*) equivalent lumped circuit.

First, knowing the inductance required of the distributed inductor, calculate the reactance, at the frequency of interest, by the common formula

$$X_L = 2\,\pi f L$$

Second, utilize 100-ohm microstrip (Z_L = 100 ohms) for the substrate's dielectric in use. Find the microstrip width required for this 100-ohm value by either working with one of the many microstrip calculation programs available free on the Web (such as HP's AppCad, or AWR's TXLine, or Daniel Swanson's MWTLC) or by employing the microstrip formula above.

Third, calculate the microstrip's required length to become an inductor of value X_L:

$$\frac{\left(\text{Artcan}\,\dfrac{X_L}{100}\right)}{360} \times \lambda = \text{length}$$

where X_L = inductive reactance needed in the distributed circuit, ohms
length = length of the microstrip required to imitate a lumped component of value X_L (should never be longer than 30 degrees, or 12 percent, of λ) , mils
λ = wavelength of the frequency of interest for the substrate of interest (or $V_P \times \lambda$; see wavelength calculations above) , mils.

Parallel (shunt) inductor. As shown in Fig. 1.40, the equivalent shunt inductor is grounded at one end (a grounded stub) through a via to the ground plane of the PCB. Alternatively, as will be shown, it can also be RF grounded through a distributed equivalent capacitor to ground.

First, knowing the inductance required within the circuit, calculate the reactance of the shunt inductor, at the frequency of interest, by the common formula:

$$X_L = 2\,\pi f L$$

Second, use 100-ohm microstrip (Z_L = 100 ohms) for the substrate's dielectric. Find the microstrip width required for this 100-ohm value either by using one of the many microstrip calculation programs available free on the Web

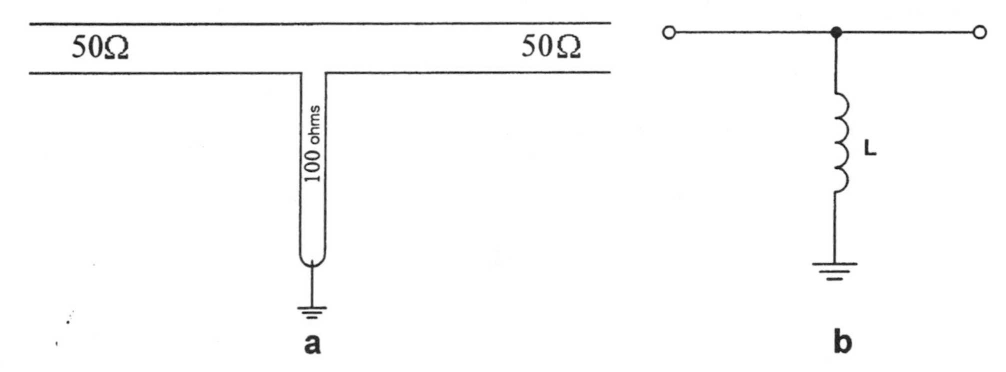

Figure 1.40 (*a*) A shunt distributed inductor; (*b*) equivalent lumped circuit.

(such as HP's AppCad, or AWR's TXLine, or Daniel Swanson's MWTLC) or by calculating with the microstrip formula above.

Third, calculate the microstrip's required length to become an inductor of value X_L:

$$\frac{\left(\text{Artcan}\,\frac{X_L}{100}\right)}{360} \times \lambda = \text{length}$$

where X_L = inductive reactance needed in the distributed circuit, ohms
length = length of the microstrip required to imitate a lumped component of value X_L (should never be longer than 30 degrees, or 12 percent, of λ) , mils
λ = wavelength of the frequency of interest for the substrate of interest (or $V_P \times \lambda$; see the above wavelength calculations), mils.

Choke. The distributed choke is RF grounded (a grounded stub) through a distributed or lumped capacitor (Fig. 1.41); or by a direct connection through a via to the ground plane (Fig. 1.42). The width of a distributed choke is that of 100-ohm microstrip for the substrate's dielectric (Z_L = 100 ohms, 100 ohms is the impedance of the microstrip only, and not that of the equivalent choke). Find the microstrip width required for this 100-ohm value either by using one of the many microstrip calculation programs available free on the Web (such as HP's AppCad, or AWR's TXLine, or Daniel Swanson's MWTLC) or by calculating with the microstrip formulas above. The length of the choke will be exactly $V_P \times \lambda/4$, or 90 degrees electrical. The distributed choke is theoretically now a complete open circuit because the distributed circuit is at precisely $\lambda/4$.

The equivalent choke can be used in the bias decoupling circuit of Fig. 1.43. L acts as a shorted quarter-wave stub because of the RF ground provided by C; R_{BIAS} and C_1 function as low-frequency decoupling [R_{BIAS} can also act as a

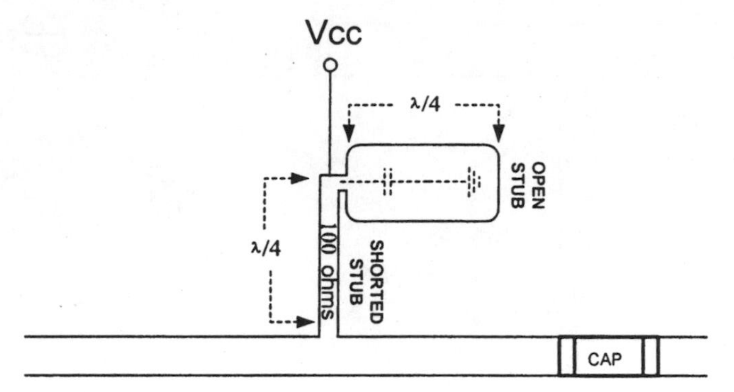

Figure 1.41 Distributed DC bias decoupling.

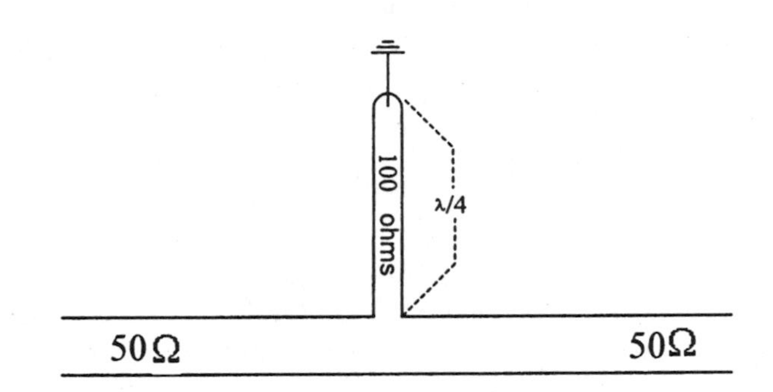

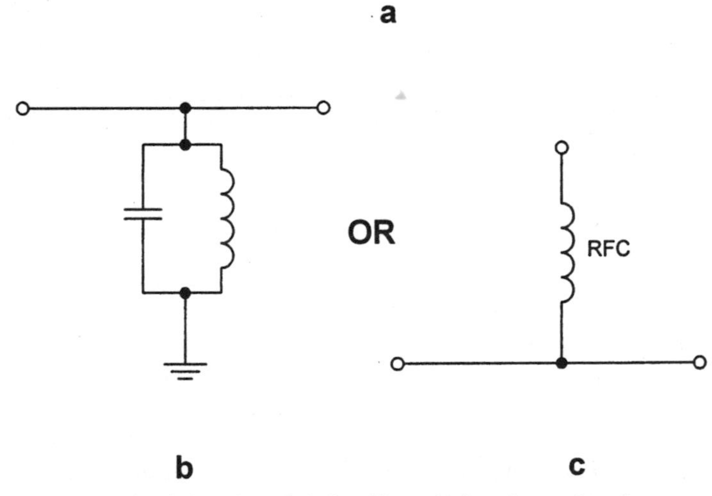

Figure 1.42 (*a*) A distributed choke; (*b*) equivalent lumped tank circuit; (*c*) a lumped choke.

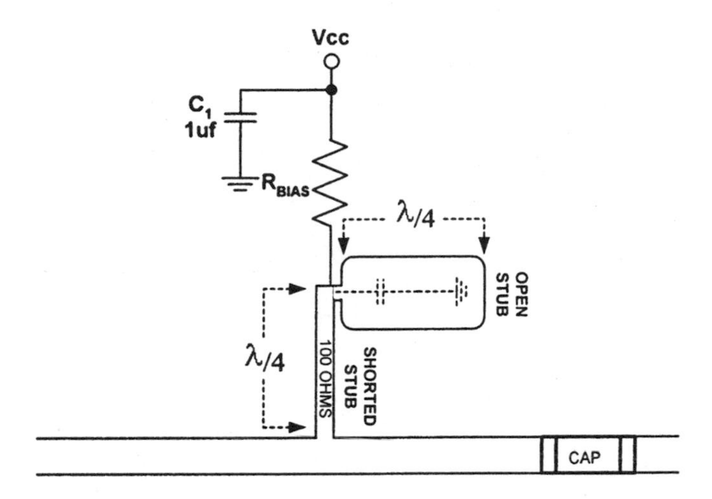

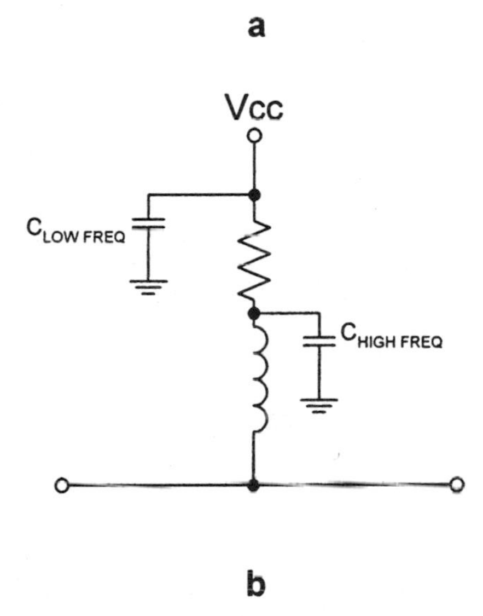

Figure 1.43 (*a*) Distributed DC bias decoupling, (*b*) equivalent lumped circuit.

bias resistor for a monolithic microwave integrated circuit (MMIC)]; *C* behaves as an open stub to work as an RF short circuit (by being exactly $\lambda/4$), and the fact that it is wide lowers its impedance even further.

Transformer. The narrowband transformer of Fig. 1.44 is employed for resistive terminations only, such as those between different values of microstrip, between two resistive stages, or between two reactive stages with the reactances tuned out by a capacitor or inductor. The transformer's

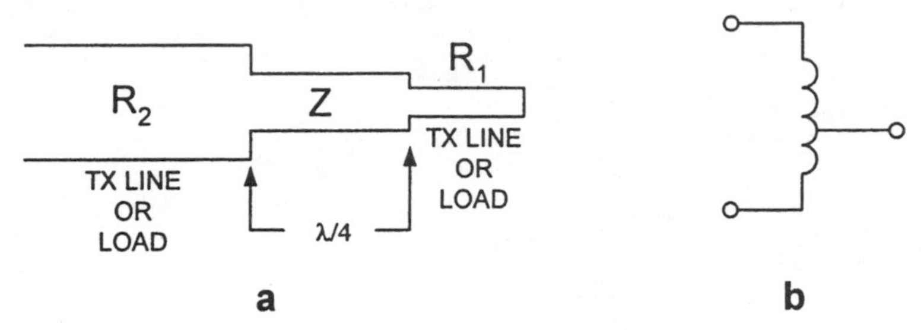

Figure 1.44 (*a*) A distributed transformer for resistive transformations; (*b*) equivalent lumped circuit.

length is exactly $V_P \times \lambda/4$, or 90 degrees electrical, and its impedance can be calculated by

$$Z = \sqrt{R_1 R_2}$$

After the impedance is found, calculate the *Z* section's required width either by employing one of the many microstrip calculation programs available free on the Web (such as HP's AppCad, or AWR's TXLine, or Daniel Swanson's MWTLC), or by calculating with the microstrip formula above.

Microstrip component equivalency issues. Inductors, transformers, capacitors, and series and parallel tank circuits will function only for the particular dielectric constant, board thickness, and frequency used in the original equivalency calculations.

As stated above, the length of the equivalent inductor and capacitor elements should not be longer than 12 percent (30 degrees) of λ, or they will begin to lose their lumped component equivalence effect. In calculating the wavelength of the frequency of interest, the velocity factor of the substrate must be considered, since this changes the actual wavelength of the signal over that of free air. And inasmuch as the wavelength of the signal varies with the propagation velocity of the substrate, and the dielectric constant varies the V_P, then all distributed components are frequency and dielectric constant dependent.

In shielding microstrip distributed equivalent capacitors and inductors, as well as microstrip transmission lines, the shield should be kept at least 10 substrate thicknesses away from the microstrip because of the field leakage above the etched copper—which causes a disruption within this field—and subsequent impedance variations.

The calculations for a frequency's velocity of propagation (V_P) will change slightly with the *width* of the microstrip conductor. This is due to the electric field that is created by the signal being bounded not by the dielectric and ground plane but by air, on one side of the microstrip.

Figure 1.45 displays proper and improper methods to construct a distributed inductor and capacitor equivalent circuit, which in this case is being used as a

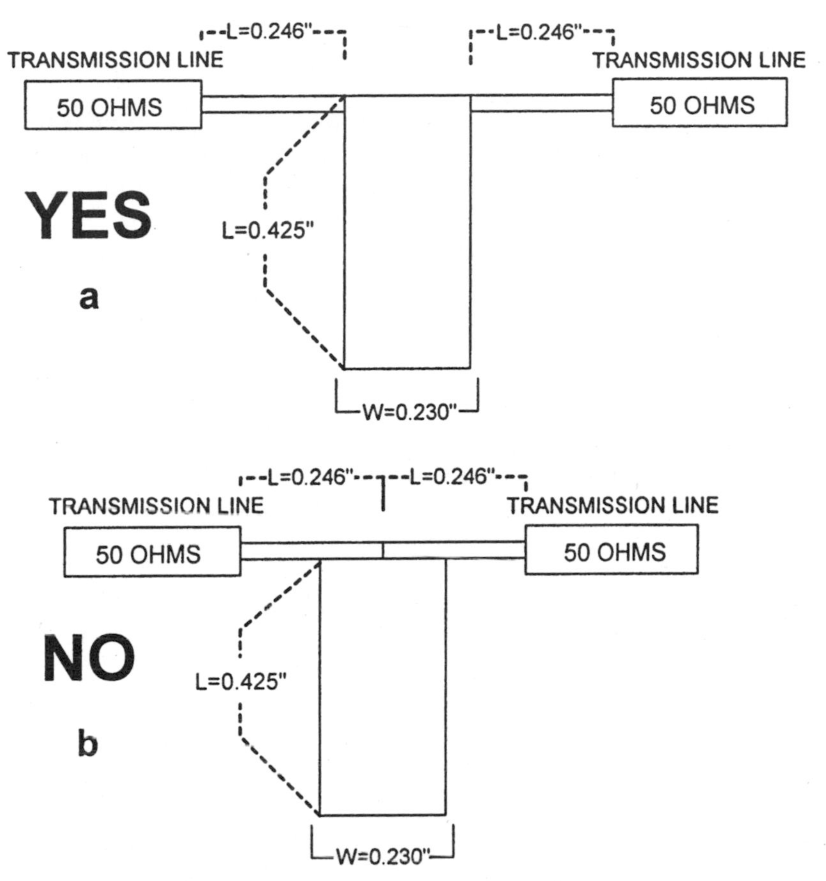

Figure 1.45 Proper (*a*) and improper (*b*) layouts of equivalent microstrip components.

3-pole low-pass filter. The proper way to position the microstrip distributed inductors and capacitors is shown in Fig. 1.45*a*, where the length has been calculated to be 0.246 inch for the series inductors and 0.425 inch for the shunt capacitor. This layout clearly allows the microstrip lengths and widths, as calculated, to function as desired. However, in Fig. 1.45*b* the length of each of the distributed inductors is now far less than calculated, and even the length of the capacitor is a little longer than desired. The figure demonstrates that each equivalent distributed component must be laid out properly—with no length ambiguities—or improper circuit operation will result.

The capacitive *end effect* of open stubs should be taken into account in all distributed designs. This effect creates an open stub that is approximately 5 percent longer electrically than the microstrip actually is physically on the PCB, causing the stub to resonate at a frequency lower than expected. This can be corrected by removing 5 percent from the calculated length of the open stub.

1.4 Transmission Lines

1.4.1 Introduction

Transmission lines are conductors intended to move current from one location to another not only without radiating, but also at a selected impedance. There are two kinds of lumped RF transmission lines: *unbalanced,* normally in the form of coaxial cable, and *balanced,* such as twin-lead.

Waveguide, a type of transmission line, can still be found in high-powered microwave transmitters, but is normally more expensive than coaxial cable and is much harder to work with.

1.4.2 Transmission line types

Balanced lines are characteristically 300-ohm twin-lead (Fig. 1.46), and are distinctly different from unbalanced coaxial line, since there is no conductor in balanced line that is at a ground potential. In fact, each conductor has an equal-in-amplitude but opposite-in-phase signal present on each of its two conductors.

Commonly operated as a feedline to a television or FM receiver antenna or, more infrequently, as a balanced feed to a dipole transmitting/receiving antenna, twin-lead has very little line losses and is able to survive high line voltages. However, twin-lead is not found in the impedance required for most transmitters and receivers (50 ohms), and matching networks must be used.

By far the most popular line is unbalanced, which comes in the form of *coaxial* cable (Fig. 1.47) and is shielded with varying degrees of copper braid (or aluminum foil) to prevent the coax from receiving or radiating any signal. The inner conductor carries the RF current, while the outer shield is at ground potential.

Coax cable comes in many diameters, qualities, and losses per foot. It is commonly the flexible type, which is covered with a protective rubber sleeve, but the semirigid type, with solid copper outer conductor, is also used. Flexible coax is available, at a high cost, that can function with low losses up to frequencies as high as 50 GHz.

Now that coax cables can work in the microwave region, *waveguide* (Fig. 1.48) has become a little less widespread. Whenever possible, modern microwave designs have removed waveguides in favor of low-loss, semirigid

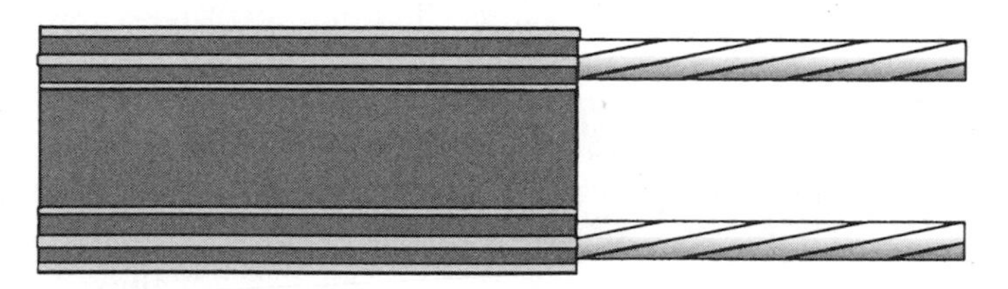

Figure 1.46 Twin-lead transmission line.

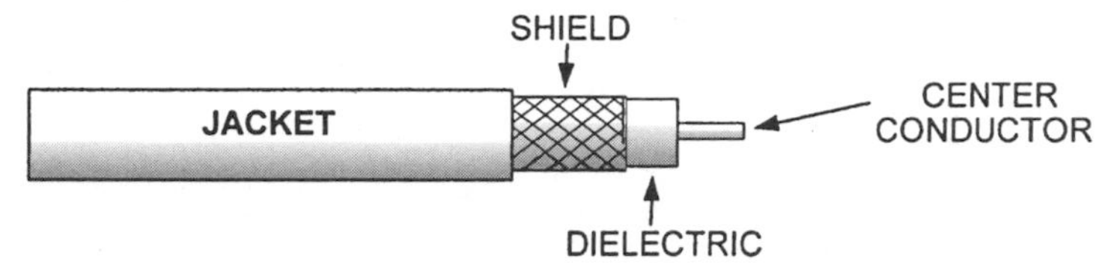

Figure 1.47 Flexible coaxial cable.

coax cables to transmit and receive high-frequency signals. However, waveguide is still favored as the transmission line of choice in certain demanding microwave high-power applications. Waveguide can be a round or a rectangular hollow metal channel made to transport microwave radiation from one point to another, with minimal signal loss, for very long distances. The actual size of the waveguide itself will govern its working frequency (Fig. 1.49), with one-quarter wavelength straight or loop probes adopted to inject or remove the microwave energy from the waveguide structure.

Waveguides perform as a type of high-pass filter, since they will propagate microwave radiation above their working frequency but not below their cutoff frequency. However, mode shifts that arise within the waveguide structure will limit the highest frequencies they are capable of propagating, thus making a waveguide more of a very wide bandpass filter.

1.4.3 Transmission line issues

With a frequency source's output and its transmission line at the same impedance, and with the transmission line also equal to the load's input impedance, no standing or reflected waves will exist on the transmission line. Thus, no power will be dissipated as heat—apart from that generated by the transmission line center conductor's natural resistance—and the line will seem infinitely long, with no standing waves reflected back into the source, while sending the maximum power to the load. The transmission line is now considered to be *flat line* (Fig. 1.50). However, if there were high standing waves (high VSWR) existing on the transmission line (Fig. 1.51), the line's dielectric and/or the wireless transmitter's final amplifier can be damaged by the reflections.

Generally, the larger the diameter of the coaxial cable, the higher the operating frequency and the smaller the losses. This is not true at the higher microwave frequencies, where the diameter of the cable can approach a certain fraction of the signal's wavelength, causing high transverse electric mode (TEM) losses due to the coax transitioning to an undesired waveguide mode.

1.5 *S* Parameters

1.5.1 Introduction

S parameters characterize any RF device's complicated behavior at different bias points and/or frequencies, and give the circuit designer the ability to

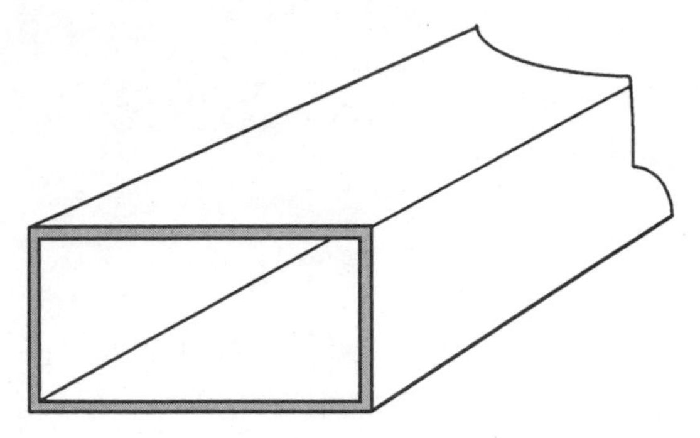

Figure 1.48 Waveguide microwave transmission line.

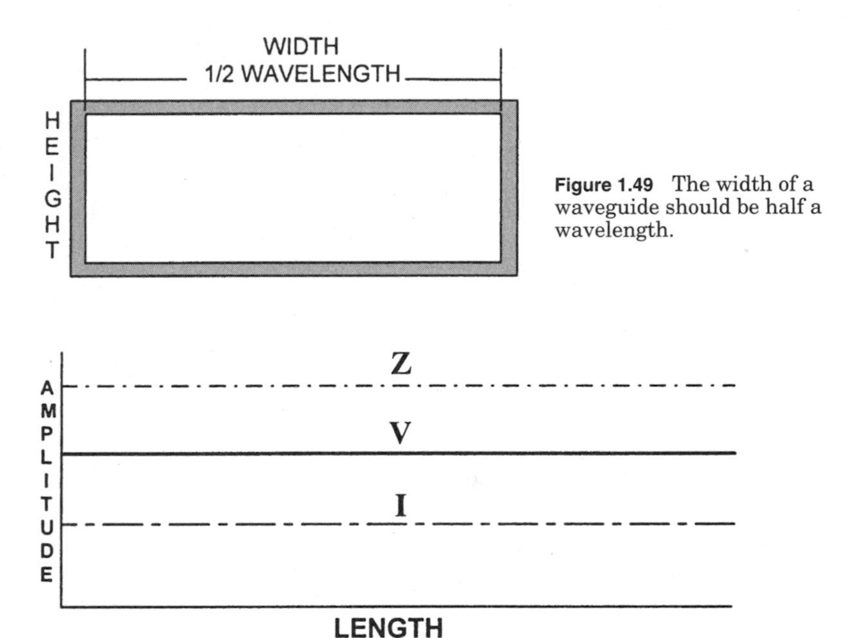

Figure 1.49 The width of a waveguide should be half a wavelength.

Figure 1.50 Voltage, current, and impedance distribution on a matched line, with no standing waves.

easily calculate a wireless device's gain, return loss, stability, reverse isolation, matching networks, and other vital parameters.

S parameters, or *scattering parameters,* are effective for small-signal design in linear, Class A amplifiers. Typically practical only in amplifiers running under 1 watt, they are not considered useful in most power amplifier designs (RF power amplifiers operate at 1 watt and above). As intimated above, power is only one of the aspects that determine whether an RF amplifier can be designed and described with *S* parameters: The amplifier must also be operated within its linear region. This would leave out any amplifier, even under

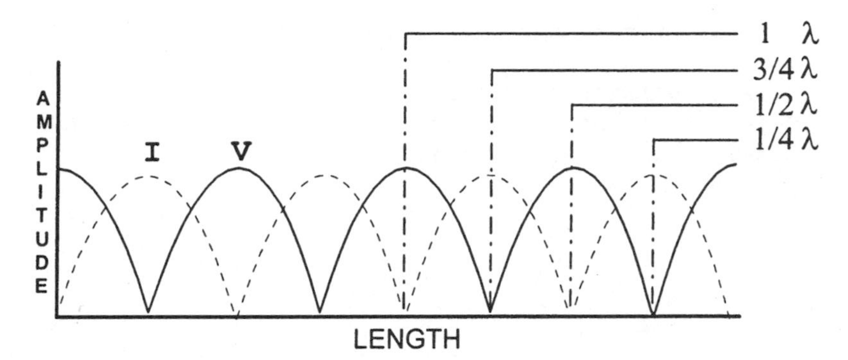

Figure 1.51 Voltage and current distribution on a nonterminated transmission line, with maximum standing waves.

nonpower conditions, that is biased at anything other than Class A, such as Class AB, B, or C. However, Class AB is normally accepted as performing acceptably when designed with S parameters, while even a few Class B designs have at least been started with S-parameter techniques.

What are S parameters? For small-signal transistors meant to operate at frequencies of over 50 MHz, S parameters are typically utilized to design the transistor's input and output matching networks for maximum power output, and to define the forward and reverse gain as well as the input/output *reflection coefficients* of a linear amplifier (or any linear black box) that is terminated at both of its ports with $50 + j0$. [Reflection coefficients are the ratio of the reflected wave to the forward wave, and are a measure of the quality of the match between one impedance and another, or $V_{\text{REFLECTED}}/V_{\text{FORWARD}}$—with a perfect match equaling zero, worst match equaling 1—and can be expressed in rectangular ($\Gamma = R \pm jX$) or polar ($\Gamma = P \angle \pm 0$) forms.] It is extremely important to remember that all S parameters are collected, and are valid, for only one V_{CE}, I_C, and f_r. However, this is not as limiting as it may seem, as multiple frequencies will always be given in the device's own S parameter file (*.S2P file; Fig. 1.52), and many microwave transistor manufacturers will also supply multiple S-parameter files with a few different selections of V_{CE} and/or I_C for each active device. This allows the engineer more flexibility in common-emitter amplifier bias design. There may also be other S-parameter files available, depending on whether the active device is to be part of a common-source or a common-base amplifier.

S parameters can be taken for any device, whether active or passive, not only to be used in calculating matching circuit elements, but also to simulate a complete circuit in a computer at high frequencies for gain, stability, and return loss. These S-parameter measurements are required in high-frequency design, since at elevated frequencies most Spice simulation models will completely break down. This is due to the lack of proper Spice modeling

```
! MANUFACTURER'S NAME (Small Signal Semiconductors)
! MODEL OF TRANSISTOR
! TYPE OF TRANSISTOR AND PACKAGE
! S-PARAMETERS at Vce=3.0 V  Ic=50 mA.    LAST UPDATED 09-06-00
! Common Emitter S-Parameters:
# GHz  S   MA  R  50
!  f          S11             S21              S12              S22
! GHz     MAG    ANG     MAG    ANG      MAG     ANG      MAG     ANG
 0.010 0.1968    0.2   8.019  178.9  0.0777     0.6  0.3276    -0.5
 0.020 0.1967    0.2   8.070  178.3  0.0777     1.0  0.3266    -1.2
 0.030 0.1975   -1.4   8.032  177.5  0.0777     1.4  0.3270    -2.4
 0.050 0.1926   -2.3   8.027  175.8  0.0777     2.5  0.3251    -4.3
 0.100 0.1938   -4.8   8.002  171.6  0.0781     4.9  0.3245    -8.6
 0.150 0.1953   -6.2   7.960  167.6  0.0789     7.3  0.3231   -12.7
 0.200 0.1920   -9.6   7.924  163.4  0.0799     9.5  0.3188   -16.8
 0.250 0.1923  -12.0   7.824  159.4  0.0809    11.9  0.3177   -21.2
 0.300 0.1947  -14.4   7.753  155.4  0.0823    13.8  0.3126   -25.0
 0.400 0.1931  -19.5   7.528  147.5  0.0859    17.8  0.3047   -33.3
 0.500 0.1912  -25.9   7.284  139.9  0.0898    21.1  0.2948   -41.2
 0.600 0.1895  -31.9   7.019  132.7  0.0945    23.9  0.2828   -48.6
 0.700 0.1885  -37.7   6.707  125.8  0.0999    26.0  0.2719   -55.9
 0.800 0.1850  -45.0   6.437  119.1  0.1054    27.9  0.2603   -62.9
 0.900 0.1787  -52.4   6.124  112.8  0.1114    29.2  0.2491   -69.8
 1.000 0.1753  -60.3   5.836  106.8  0.1175    30.2  0.2389   -76.2
```

Figure 1.52 A *.S2P *S*-parameter file at a set bias condition and at assorted frequencies. Lines preceded by ! are comments for the user and are ignored by the simulation programs.

of a device's natural internal parasitic distributed capacitances and inductances and the problems that occur as the wavelength of the RF signal becomes a significant portion of the size of the physical components. Both will have a significant effect on the device's true RF behavior.

S parameters are described by S_{21}, S_{12}, S_{11}, and S_{22}. S_{21} is the forward transmission coefficient (representing the stage gain); S_{12} is the reverse transmission coefficient (representing the reverse gain, or isolation); S_{11} is the input reflection coefficient (representing the input return loss); and S_{22} is the output reflection coefficient (representing the output return loss). Figure 1.53 combines all of these *S* parameters into the gains and reflections of a single black box for a simplified graphical description.

Basically, *S* parameters are voltage ratios referenced to 50 ohms, since a voltage comparison is all that need be obtained to find any *S* parameter at the same impedance. But a complete *S* parameter is a vector quantity, taking into account not only the magnitude, but also the phase. The vector quantity allows the analysis of stability and complex impedances while both ports of the device are terminated in 50 ohms. Considering that *S* parameters are the voltage ratio of a port input potential to the reflected potential, or of the input potential to the output potential, it can be seen that any device can easily be accurately characterized for any RF circuit.

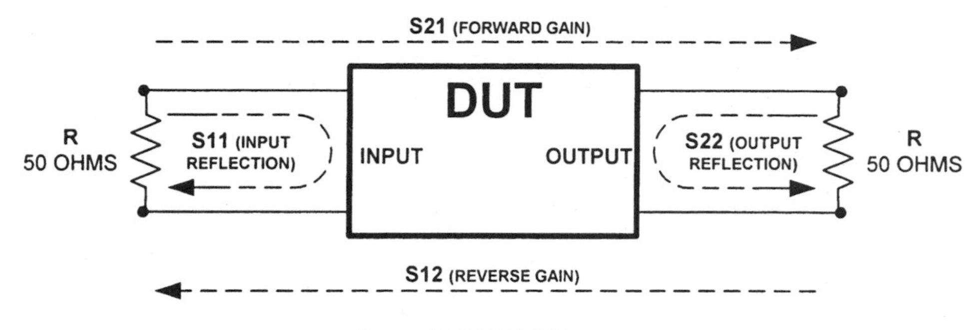

Figure 1.53 A 2-port network showing transmission and reflection parameters. DUT stands for device under test.

As it is far easier for the designer to deal with decibels than with voltage levels, the majority of the S_{21}, S_{11}, S_{12}, S_{22} values are in dB which, when compared to the measured voltages in a 50-ohm system is

$$S_{XX}\,(dB) = 20 \log_{10} |S_{XX}|$$

where XX is 21, 12, 11, or 22.

Even though S parameters are frequently associated with 2-port devices, they work equally well with 3, 4, or more ports by the addition of the suitable subscripts, such as S_{31} for the forward gain through one branch of a splitter, as shown in Fig. 1.54.

1.5.2 S parameter measurement

To clarify just what S parameters are, the measurement of an active device is shown in Fig. 1.55, which demonstrates a simple test setup for a method of taking the S parameters in the forward direction.

To obtain the S parameters for a BJT, the bias voltage is injected at bias voltage 1 (with L and C acting as decoupling components), which will control the base current, and thus the collector current, of the BJT. The emitter is grounded, while the V_{CE} for the BJT's collector is supplied by bias voltage 2. Bias voltage 2 furnishes the Class A static DC bias conditions for the transistor. This hints as to why all S-parameter files are supplied only for a specific frequency, I_C, and V_{CE}—and why other frequencies and I_C's or V_{CE}'s will change the device's parameters (more on this later). An AC signal is now injected into port 1, through the dual directional coupler, by the 50-ohm RF signal generator (S parameters are taken only when terminated at the input and output by 50 ohms, or $50 + j0$). The vector voltmeter M_1, a device that is able to measure not only the voltage of a signal, but also its phase, reads the amplitude and phase of V_A of the signal into port 1 of the BJT. Meter M_2 reads the amplitude and phase of V_B of the signal that is reflected back from

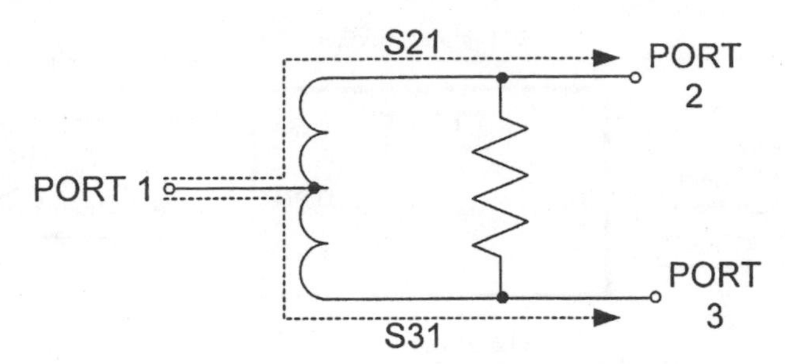

Figure 1.54 Three-port splitter demonstrating forward S parameters.

the input of the BJT (due to any impedance mismatch from the system impedance of 50 ohms). The ratio V_B/V_A of the amplitude of the reflected signal, V_B, to the amplitude of the V_A signal from the generator gives the magnitude of S_{11}, the *input reflection coefficient.* This value will invariably be less than unity.

The phase angle difference between V_B and V_A depicts the phase angle of S_{11}:

$$\angle \theta_B - \theta_A$$

So the S parameter S_{11} will be

$$S_{11} = \frac{V_B}{V_A} \angle \theta_B - \theta_A$$

S_{21}, the *forward transducer gain,* equals the voltage measured at V_C by M_3, and will be some value greater than unity since it is the amplified value of V_A, or V_C/V_A.

The phase difference between V_C and V_A is measured as

$$\angle \theta_C - \theta_A$$

So S_{21} will be

$$S_{21} = \frac{V_C}{V_A} \angle \theta_C - \theta_A$$

Figure 1.56 shows one technique for measuring the *reverse* S parameters of an active device. Basically, the setup of Fig. 1.55 is reversed, but the bias voltages and the DUT orientation remain the same, with the input now terminated with 50 ohms and the active device's output now fed by the 50-ohm signal generator. A signal is injected into port 2, through the dual directional coupler, by the 50-ohm signal generator. The vector voltmeter M_4 reads the amplitude and phase of V_D of the signal at port 2 of the DUT. Meter M_3 reads the amplitude and phase of V_C of the signal reflected back from the output of the DUT (due to any impedance mismatch from the system's impedance of 50 ohms).

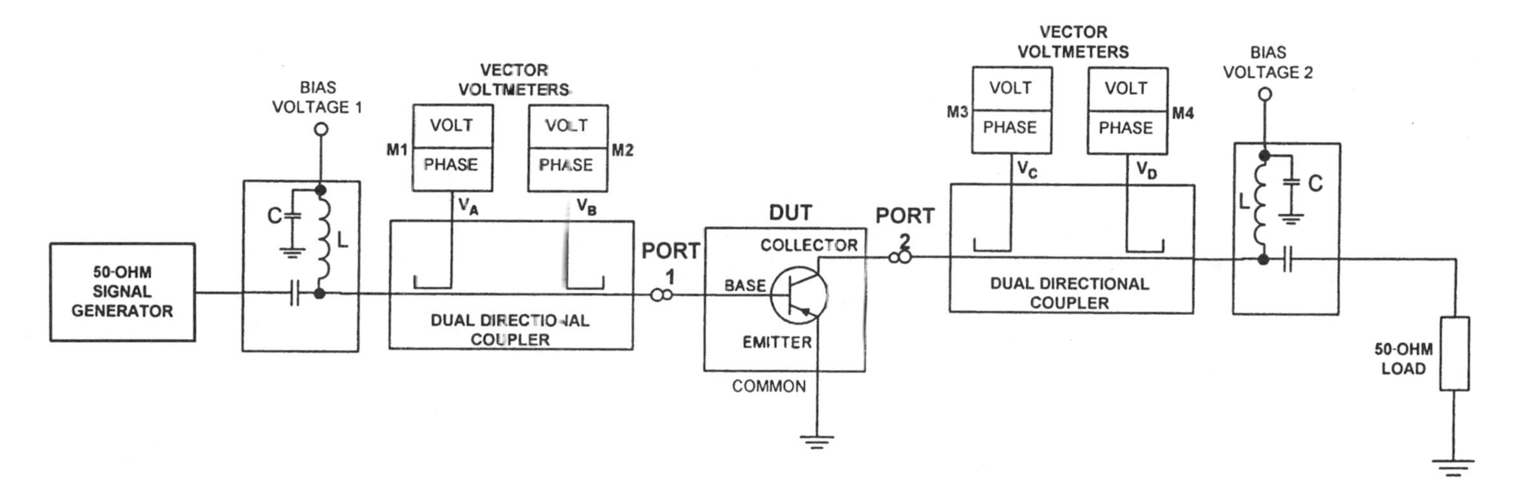

Figure 1.55 Test setup for forward S-parameter measurement.

1.6 Propagation

1.6.1 Introduction

An electromagnetic wave propagates through a vacuum at approximately 300,000 meters/second. This speed decreases as it passes through any type of dielectric material—even air. The *E* (electric) field and the *H* (magnetic) field of the electromagnetic wave are not only at 90 degree angles to each other (orthogonal), but they also increase and decrease in amplitude together over time, with one field regenerating the other as they travel through space. This is referred to as TEM (*transverse electric mode*) propagation.

Radio wave propagation is conditional on the frequency of the RF carrier, and is by three main modes:

Ground waves, which travel on top of and through the surface of the earth at frequencies below 1 MHz

Surface waves—also called *space* or *direct waves*—which propagate through the atmosphere in an almost straight line from the transmitter to the receiver, and are the primary form of propagation for RF signals at 30 MHz and above (surface waves reach to slightly more than line-of-sight (LOS) distances because of atmospheric bending of the signal)

Sky waves, which are RF signals at less than 30 MHz that refract and reflect off of the atmosphere's ionosphere, and are the chief means of low- and high-powered simplex long-range RF communications.

Even when direct line-of-site communications paths are used between the transmitter and the receiver, natural signal losses will begin to decrease the transmitted signal power to a level that comes closer and closer to the noise floor of the receiver's output. This will obviously decrease signal-to-noise ratio (SNR), which increases the bit error rate (BER) of a digital system or increases the noise level of an analog system.

1.6.2 Multipath

Multipath fading effects, especially problematic at microwave frequencies, occur when a transmitted RF signal of interest bounces off a conductive object—such as building pylons, light poles, or even the earth itself—and reaches the receiver at a slightly different time than the direct RF signal. This produces an out-of-phase reception condition, or phase cancellation, causing fading of the received signal. The severity of microwave multipath effects depends on antenna height, frequency, gain, and sidelobe suppression. The fading effect is also a huge problem in HF communications, since it also creates an intermittent or continuous decrease in the received signal's amplitude. However, it is produced by the changing conditions of another reflective surface that is not so close to the surface of the earth: the ionosphere. HF multipath can also be caused by multiple-path reception, such as when an HF receiver is accepting both ground waves and sky waves, or sky waves and surface waves, or waves arriving from one-hop and two-hop paths. The two signals are, as above, alternately in phase and out of phase with each other, causing severe fading conditions.

Multipath can create high intersymbol interference (ISI) and BER in digital communications systems, and is magnified by misalignment of the transmit or receive antenna, by a receive location that is near an RF reflective site (such as a building or a mountain behind the receiver), or by the surface movement of trucks or automobiles sending back a secondary signal to the receive antenna. This multipath can be examined as a frequency-selective amplitude notch and/or a tilting and misshaping of the received waveform on the screen of a spectrum analyzer (Fig. 1.57) in digital communications systems. Without an *equalizer* built into the receiver's baseband circuits to fill-in for these multipath effects, small tilts and notches of 1 dB (or less) would quickly render

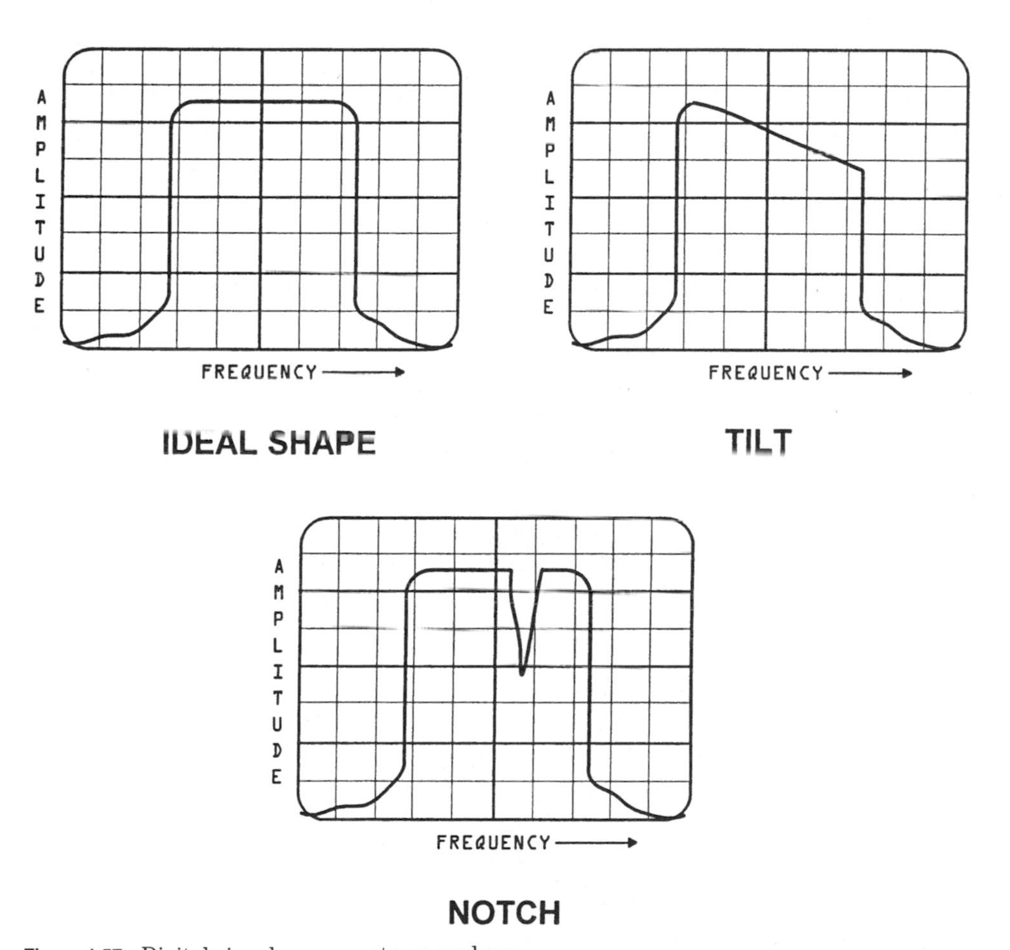

Figure 1.57 Digital signals on a spectrum analyzer.

many digital signals unreadable. *Space diversity* is another common method to mitigate multipath effects, but requires two antennas and the associated diversity circuitry.

An important term to remember in discussing multipath transmission is *delay spread.* Since multipath signals take longer to reach the receiver's antenna (the time is dependent on how far they have traveled compared to the direct wave), the spread between the time the direct signal arrives and the time the last multipath signal arrives is called the delay spread. Delay spread causes increased ISI due to the received data actually overlapping. As stated above, equalizers are typically employed in an attempt to mitigate this problem; but if the delay spread is excessive, then equalizers may not be able to properly handle the multipath effects.

Chapter

2

Modulation

Different modulation schemes have been adopted for radio services, such as broadband AM and FM for broadcast, narrowband FM for line-of-site two-way voice communications, single sideband (SSB) for long-distance voice communications via the ionosphere, and digital modulation for high-speed point-to-point and multipoint microwave radio communications links.

2.1 Amplitude Modulation

2.1.1 Introduction

Amplitude modulation (AM) is the earliest modulation method for wireless voice communications. It is very simple and cheap to work with from a hardware standpoint, and it is still extensively used today for commercial and shortwave broadcast, as well as in certain citizens band and limited ham radio systems.

2.1.2 Fundamentals

Modulation is the way we insert baseband information on an RF carrier wave. The baseband information can be voice, digital data, analog video, etc. Demodulation is the procedure of extracting this baseband information, which is then sent to a speaker for voice and music, or on to digital circuits for processing or storage.

The most basic way we have of imprinting voice, data, or music on an RF carrier is by modulating the amplitude of the carrier (Fig. 2.1). The unmodulated carrier, which is produced by an oscillator, functions as the RF that will transport the baseband modulation through space to a receiver. The baseband is the intelligence—always at a much lower frequency than the RF carrier—and is inserted onto the carrier by nonlinear mixing of these two signals. As seen in the time domain, the amplitude of the RF carrier is modified at the rate of the baseband's own amplitude and frequency variations. In fact, if the amplitude of the baseband signal increases, then so will the amplitude of the RF carrier (Fig. 2.2), while decreasing the baseband's amplitude decreases the amplitude of the carrier (Fig. 2.3).

The baseband modulation travels with the RF carrier to the receiver. The receiver then takes these amplitude variations that are riding on the carrier and removes them, thus converting them back into the original audio amplitude variations that were inserted at the transmitter. The recovered baseband is then amplified and fed into a speaker, or some other appropriate transducer. The percent of modulation controls the final amplitude of the detected signal, and the higher the amplitude of the baseband signal, the higher the volume at the receiver's speaker, as shown in Fig. 2.1 above.

When the baseband signal is modulated at the transmitter, both the positive and negative alternations of the RF carrier will be influenced symmetrically. This means that the missing negative alternation lost by the Class C collector modulation circuit will be recreated again by the tuned output tank of the transmitter's final amplifier, forming a mirror image of the positive alternation (Fig. 2.4).

Sidebands formed by the modulation between a carrier and its baseband signal are viewable in the frequency domain as shown in Fig. 2.5. The sidebands are created by the modulator's nonlinear mixing circuit, which produces sum

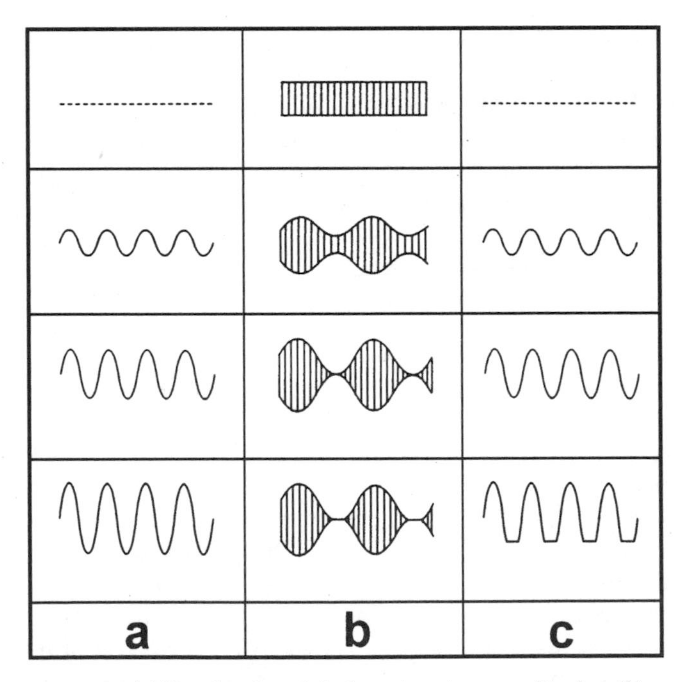

Figure 2.1 (*a*) Baseband modulation at various amplitudes; (*b*) a carrier unmodulated and at various modulation percentages; (*c*) the demodulated waveform at baseband.

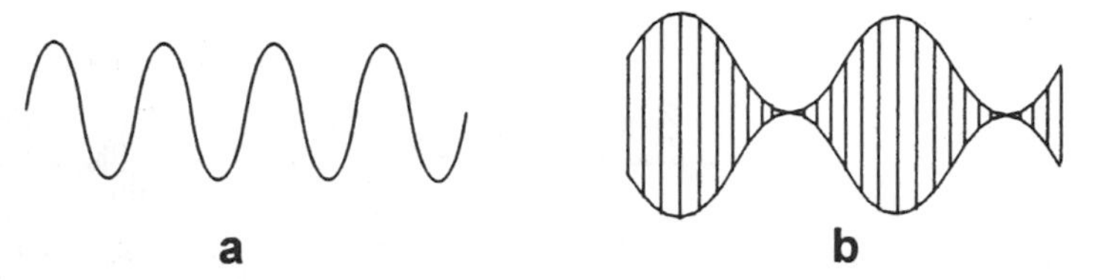

Figure 2.2 (*a*) The baseband audio modulation; (*b*) the 100 percent modulated RF waveform.

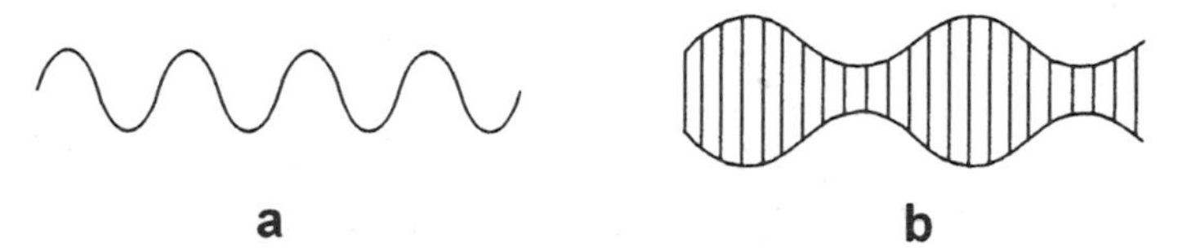

Figure 2.3 (*a*) The baseband audio modulation; (*b*) the 50 percent modulated RF waveform.

(the upper sideband) and difference (the lower sideband) frequencies. But it is the phase relationships between the RF carrier and the upper and lower sidebands that actually create a new waveform that will deviate in amplitude in the time domain (Fig. 2.6). This effect occurs when the two sidebands and the carrier are in phase, causing the amplitude of the carrier waveform to be double that of the carrier when unmodulated; when the carrier and the two sidebands are completely out of phase, the amplitude of this new carrier waveform will be virtually zero. The new waveform will therefore have high peaks and low valleys (Fig. 2.7). In the time domain, the percent of modulation of an AM signal can be found on an oscilloscope display by this formula:

$$\%\ \text{MOD} = \frac{V_{\text{PEAK}} - V_{\text{MIN}}}{V_{\text{PEAK}} + V_{\text{MIN}}} \times 100$$

However, when such a modulated signal is observed in the frequency domain, the RF carrier's frequency and amplitude will not actually change, whether it is modulated or not (Fig. 2.8). This confirms that the carrier itself holds no information that can be demodulated, but that the information is in fact embodied within the two sidebands only. Indeed, when an AM signal is inspected in the frequency domain, we clearly see that when the transmitter's baseband modulation is varied both in frequency and amplitude, the carrier will stay at its original frequency and amplitude, while only the sidebands themselves will change in frequency and amplitude (Fig. 2.9). This distinctly verifies that there is no actual information contained within the RF carrier, but only within its sidebands, each sideband holding the same information and power as the other.

These sidebands, both the upper (USB) and the lower (LSB), can be found at the sum and difference frequencies of the carrier and modulating frequencies:

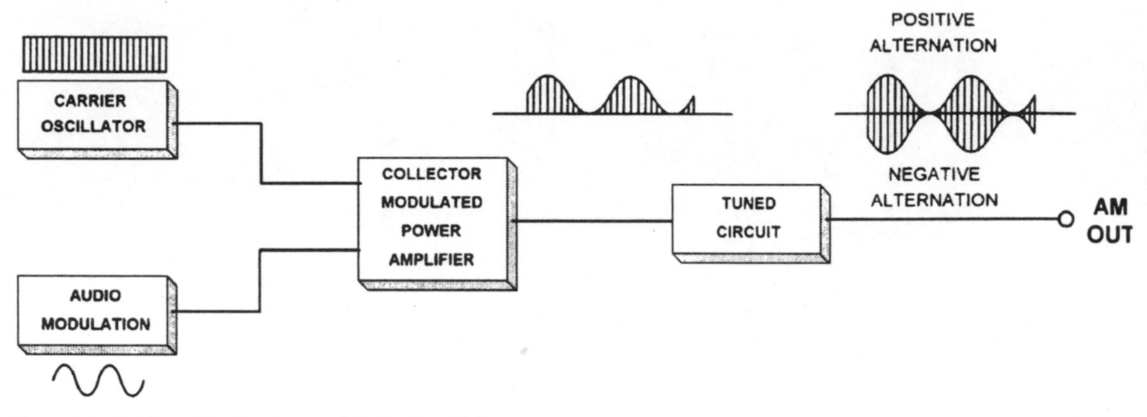

Figure 2.4 A Class C collector modulator for AM.

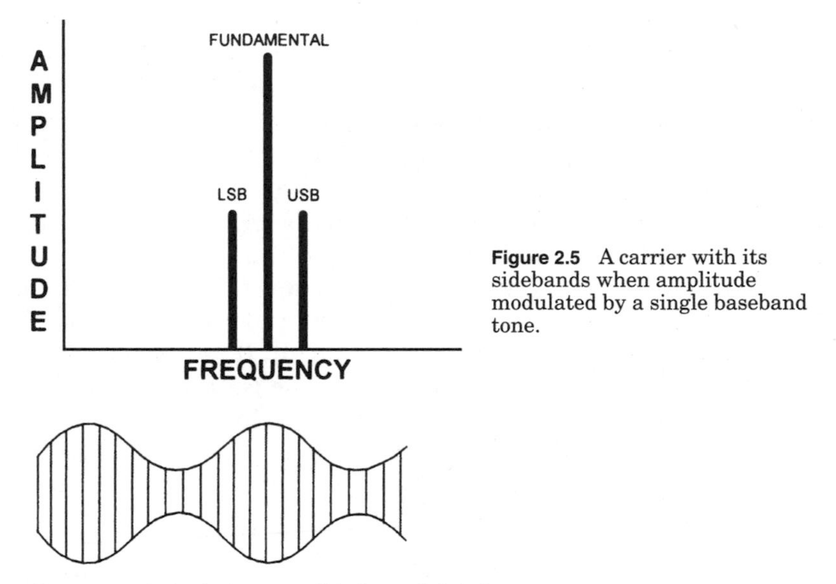

Figure 2.5 A carrier with its sidebands when amplitude modulated by a single baseband tone.

Figure 2.6 A single-tone amplitude-modulated RF carrier in the time domain.

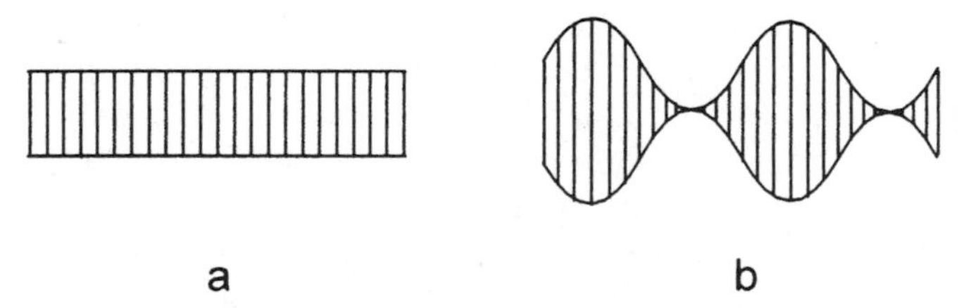

Figure 2.7 (*a*) An unmodulated carrier and (*b*) a 100 percent amplitude-modulated carrier.

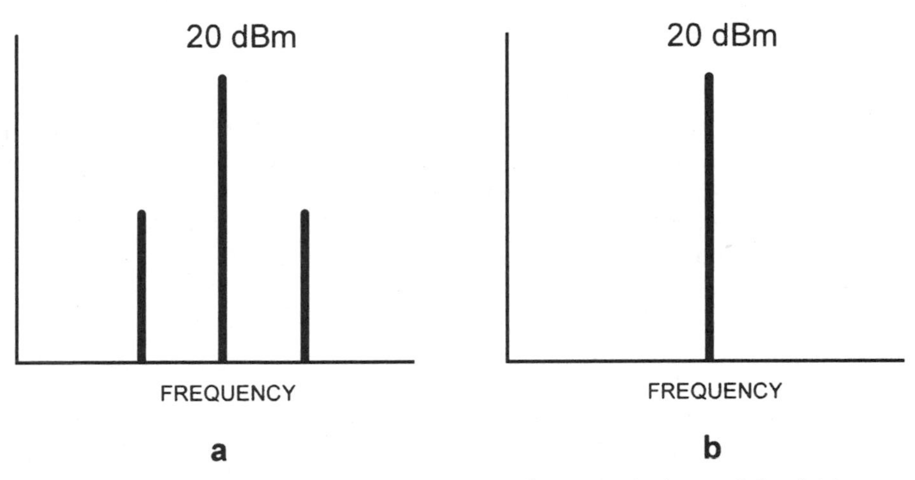

Figure 2.8 The amplitude of an AM carrier remains the same whether modulated (*a*) or not (*b*).

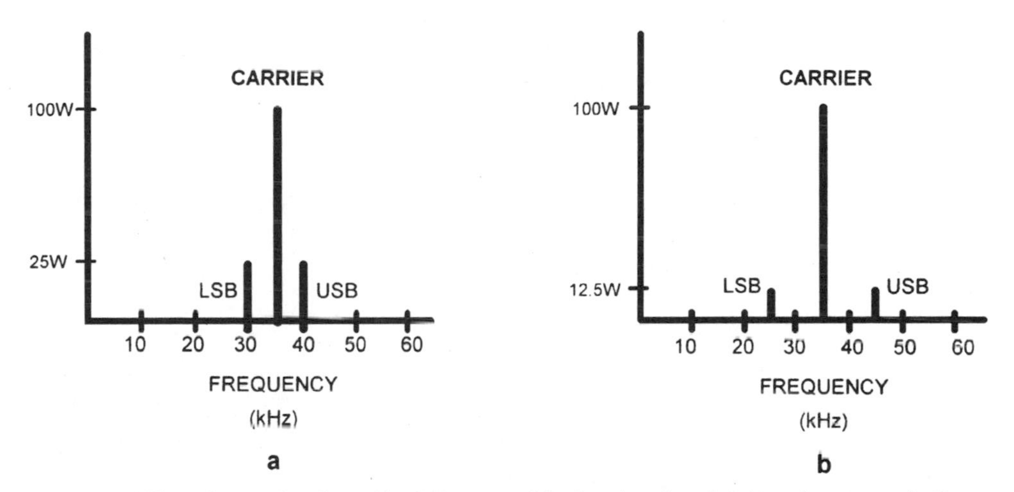

Figure 2.9 When the amplitude and/or frequency of the baseband modulation changes, only the sidebands are affected: (*a*) 5-kHz high-amplitude and (*b*) 10-kHz low-amplitude modulation.

$f_{\text{CARRIER}} - f_{\text{AUDIO}}$ = LSB and $f_{\text{CARRIER}} + f_{\text{AUDIO}}$ = USB. As an example, if a 1000-kHz carrier is modulated with a 10-kHz baseband audio tone, then the sideband frequencies can be located at 1000 kHz + 10 kHz = 1010 kHz, and at 1000 kHz − 10 kHz = 990 kHz. The bandwidth of this signal will thus be twice the baseband frequency, or the upper sideband minus the lower sideband. In the above example, the bandwidth is 2 × 10 kHz, or 20 kHz.

Only the total power changes in an AM signal, or $P_T = P_C + P_{\text{LSB}} + P_{\text{USB}}$. This is because, as discussed above, the carrier power remains unchanged no matter what baseband modulation amplitude we use, and only the sideband amplitudes will vary. Thus, the total power of an AM signal will equal the sum of the carrier and sideband powers. In fact, when the carrier is amplitude modulated by the baseband, the antenna current will rise because of the power added by the increasing sidebands.

An enormous amount of sidebands are created during normal voice modulation, and they are located at many different frequencies and amplitudes. As shown in Fig. 2.10, a spectrum analyzer display of a voice amplitude-modulated signal can become quite complex. This is why we must employ single- or dual-tone baseband input signals for testing purposes.

A majority of AM voice transmitters confine their modulation frequencies to between 300 and 3000 Hz to limit transmitted bandwidth. Limiting the baseband frequencies is easily accomplished by the use of a bandpass filter located just after the first audio (microphone) amplifier. An amplitude limiter circuit can also be employed in order to limit the maximum audio baseband amplitude to prevent AM overmodulation, which causes an unwelcome increase in transmitted bandwidth due to spectral "splatter," as well as distortion. Splatter is the harmonic production in the original baseband frequencies created by clipping of the signal's modulation envelope. This action further modulates the RF carrier, producing adjacent channel interference (ACI). The distortion level is increased because now part of the AM signal is not actually present at the demodulator (see Fig. 2.1), so intelligibility of the received signal is degraded.

2.1.3 Power measurement

The power of an AM signal can be measured as the *peak envelope power* (PEP), which is utilized to gauge the average peak power, with 100 percent modulation applied, of the transmitted signal:

$$\text{PEP} = V^2_{\text{RMS}}/R = V_{\text{RMS}} \times I_{\text{RMS}} = I^2_{\text{RMS}} \times R$$

The carrier power can also be calculated with these same formulas, but with zero transmitter modulation.

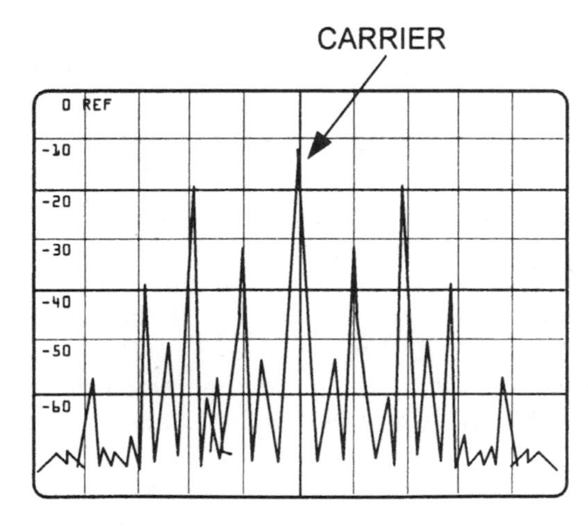

Figure 2.10 A voice signal, as viewed in the frequency domain, is composed of many sidebands at various frequencies and amplitudes.

As an important aside to AM power measurement, increasing the transmitter's range by increasing its power is not linear. In other words, to extend a transmitter's distance by 2, multiply the transmitter's power, in watts, by 2^2 (or 4 × power). To increase the distance by 3, multiply the power by 3^2 (or 9 × power).

2.1.4 Disadvantages

The disadvantages of AM are many: The bandwidth of an AM signal is twice that actually required for the reception of the intelligence being sent, since only one sideband is absolutely necessary to convey the baseband information; a significant amount of power is in the carrier, which is not even required to furnish the intelligence; the phase relationship between the carrier and the sidebands must be precise, or severe fading will result within the demodulated signal, and this is quite difficult to maintain under most atmospheric and multipath conditions.

2.2 Frequency Modulation

2.2.1 Introduction

Frequency modulation (FM) was originally invented as an answer to the many deficiencies inherent in AM, primarily that of excessive noise sensitivity. Since noise is normally produced by undesired amplitude variations in a signal, this is removed in frequency-modulated receivers by amplitude limiters.

Two techniques can be employed to generate an FM signal. The first, *direct FM,* directly alters the frequency of the carrier in step with the baseband's amplitude variations; the second method, *indirect FM,* changes the phase of the carrier, which creates phase modulation. However, both of these techniques produce the end effect of frequency modulation of the RF carrier. Both methods are classified under the designation of *angle modulation.*

2.2.2 Fundamentals

Modulation is the method we use to insert baseband information on an RF carrier wave. The baseband information can be voice, digital data, analog video, etc. Demodulation is the procedure of extracting this baseband information, which is then sent to a speaker to reproduce the original voice and music, or on to digital circuits for processing or storage.

FM accomplishes this modulation process by altering the carrier's frequency in step with the baseband signal's changes in amplitude. When this frequency-modulated RF carrier arrives at the receiver, the frequency variations created by the original baseband modulations are changed back into amplitude variations. The baseband is then amplified and inserted into an appropriate transducer. As stated, in FM the baseband's amplitude alters the frequency of the RF carrier, and not the amplitude as it does with AM, while the amount of the actual *frequency deviation* of the FM carrier depends on the increase or decrease in the baseband's amplitude. Frequency deviation is the amount the RF carrier deviates from its center frequency in *one* direction during modulation. Without any baseband modulation present, however, the frequency of the RF carrier will stay at the transmitter's predetermined *center frequency,* which is the frequency of the master oscillator after any multiplication. Thus, as the baseband modulation occurs, the carrier will increase and decrease in frequency; as the baseband swings positive in amplitude, the carrier will increase in frequency, but as the baseband modulation swings negative in amplitude, the frequency of the carrier will fall below its rest frequency (Fig. 2.11).

The *frequency* of the baseband signal will change the rate that the frequency-modulated RF carrier intersects its own rest frequency, and will vary at this same baseband rate. As an example; if a baseband audio tone is inserted at 2 kHz, the FM carrier will actually swing past its own rest frequency 2000 times in 1 second.

Unlike AM, the percent of modulation for FM is directed by government rules and regulations, and not by any natural limitations. For instance, for narrowband voice communications, 5 kHz is the maximum allowed deviation for 100 percent frequency modulation, while for wideband FM broadcast, the maximum allowed deviation is 75 kHz. If the baseband signal's amplitude should induce the FM deviation to go above the 100 percent limit, then more frequency sidebands will be created, broadening the bandwidth and conceivably causing interference to any adjacent channels.

As shown in Fig. 2.12, when FM is observed on an oscilloscope in the time domain, the modulated RF carrier will not change in amplitude, but only in

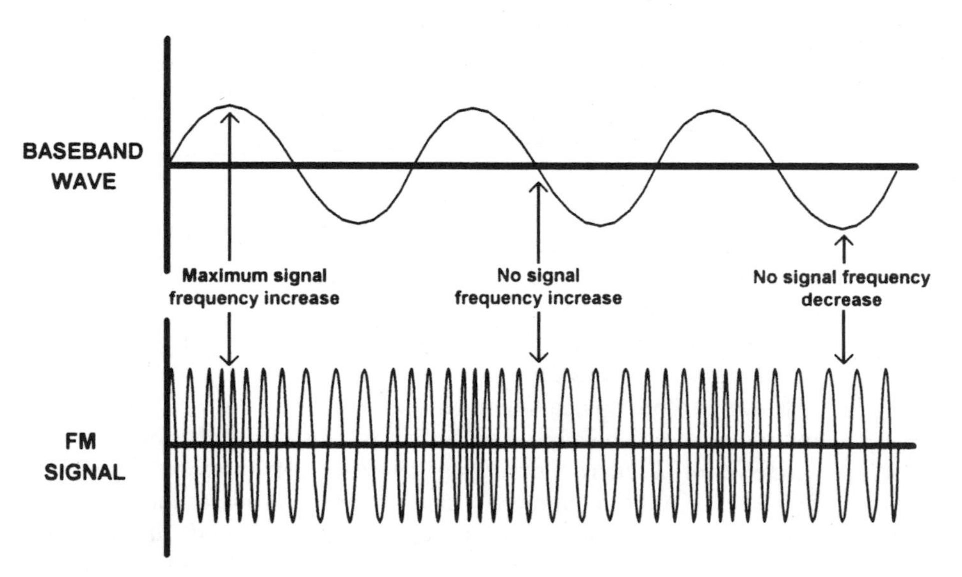

Figure 2.11 A baseband modulating signal's effect on a carrier after frequency modulation.

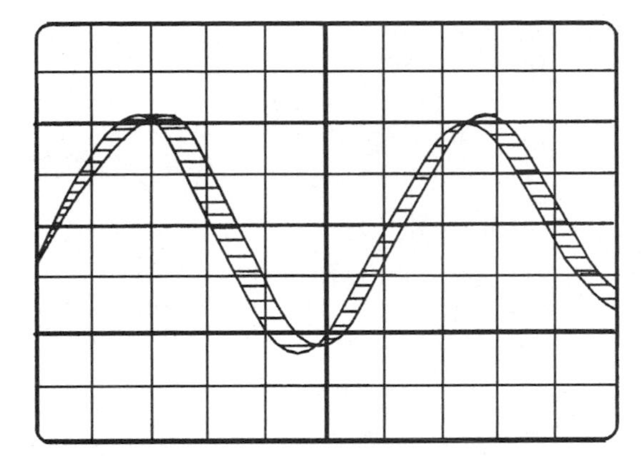

Figure 2.12 A time-domain view of frequency modulation.

frequency (exaggerated here for clarity). These rapid FM frequency fluctuations are evidenced by the shortening and lengthening of the carrier's wavelength on the scope's screen, creating a blurring of the signal. And since wavelength equals the speed of light divided by the frequency, we can readily see that any shift in wavelength corresponds to a change in frequency.

The total FM transmitter power will always stay constant during baseband modulation, so the combined power or voltage in an FM signal will not vary whether it is modulated or unmodulated. However, any sidebands formed by the modulation must gain their power from the carrier itself. This carrier must then sacrifice some of its own power in the creation of the FM sidebands. For instance, let us assume that an FM transmitter is sending out an unmodulated carrier at 100 watts. When the RF carrier is modulated by the baseband signal it must give some—or even all—of its power to these sidebands. Thus, the carrier and its *significant sidebands* must all total up to the original 100 watts that was present in the unmodulated carrier. Indeed, at certain *modulation indexes* (see below), the carrier itself will actually vanish, while the sidebands will now contain all of the power.

An infinite number of sidebands will be created during the modulation process, since the carrier is sent through an infinite number of frequency or phase values by the continuously changing baseband frequencies. This action produces an infinite amount of sideband frequencies; even a single test-tone, changing in a sinusoidal manner, has an infinite number of discrete amplitudes within a single cycle.

Because of the difficulties inherent in "infinite," the concept of the *significant sideband* was created. Significant sidebands are any sidebands with an amplitude that is 1 percent or more of the amplitude of the unmodulated carrier. When a sideband is below this level it can be ignored, while the higher the amplitude of the baseband modulation, the higher the number of these significant sideband frequencies that will be produced.

However, unlike amplitude modulation, more than one pair of sidebands will be created for each single-tone modulation (Fig. 2.13). The sidebands are also separated on each side of the carrier and from each other by an amount that is equal to the frequency of the single-tone baseband signal.

The ratio between the FM carrier's instantaneous frequency deviation (f_{DEV}) divided by the instantaneous frequency of the modulation (f_{MOD}) is an important FM specification, and is referred to as the *modulation index*. We can find the number and amplitudes of all significant sidebands generated during FM modulation from the modulation index by simply reading the chart of Table 2.1. To use this table, first calculate the FM signal's modulation index by f_{DEV}/f_{MOD}; take this number and find its value under the *Modulation index* column; now read across. The relative amplitude of the carrier, and each sideband with its number of significant sidebands, will be shown.

We can also find the bandwidth of the modulated RF signal by multiplying the number of significant sidebands by two, then multiplying by the maximum modulating frequency, or BW = $2N \cdot f_{mod(max)}$.

The following is a basic example of the modulation index and its effect on what we might see in the frequency domain. With a modulation index of zero, we would be generating no sidebands at all (Fig. 2.14), since this would be just a simple continuous-wave (CW) carrier with no baseband modulation. But as the modulation index increases to 1.5, we see in Fig. 2.15 that the sidebands will start to consume more bandwidth. This is a good example of why the frequency of the baseband modulation, and its amplitude, must be controlled so that we may lower FM bandwidth demands and adjacent channel interference (ACI).

The two-way, narrowband FM radio modulation index is normally maintained at 2 or less, since the maximum allowed frequency deviation would be approximately 5 kHz with a maximum baseband audio frequency of about 2.5 kHz. Thus, a bandwidth of between 12 and 20 kHz is customarily considered

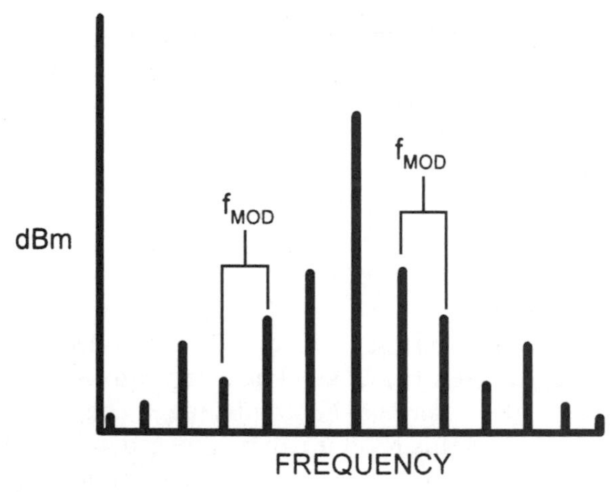

Figure 2.13 A single-tone baseband signal creating multiple sidebands in FM.

TABLE 2.1 Number and Amplitudes of Sidebands

		Pairs of sidebands to the last significant sideband							
Modulation index	Carrier	1st	2d	3d	4th	5th	6th	7th	8th
0.00	1.00								
0.25	0.98	0.12							
0.5	0.94	0.24	0.03						
1.0	0.77	0.44	0.11	0.02					
1.5	0.51	0.56	0.23	0.06	0.01				
2.0	0.22	0.58	0.35	0.13	0.03				
2.5	−0.05	0.50	0.45	0.22	0.07	0.02			
3.0	−0.26	0.34	0.49	0.31	0.13	0.04	0.01		
4.0	−0.40	−0.07	0.36	0.43	0.28	0.13	0.05	0.02	
5.0	−0.18	−0.33	0.05	0.36	0.39	0.26	0.13	0.05	0.02

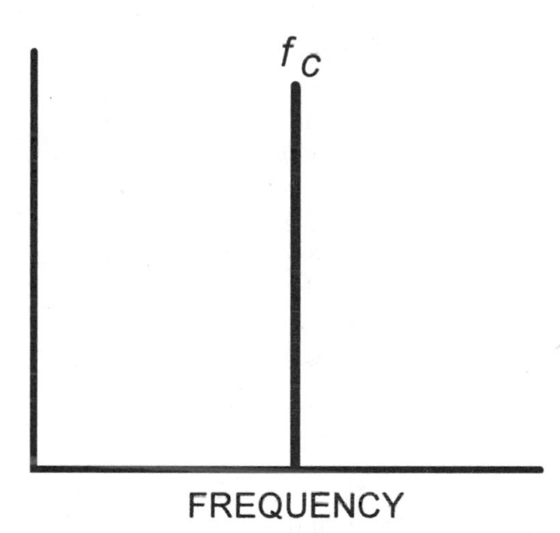

Figure 2.14 An FM signal with a modulation index of 0 (an unmodulated FM carrier).

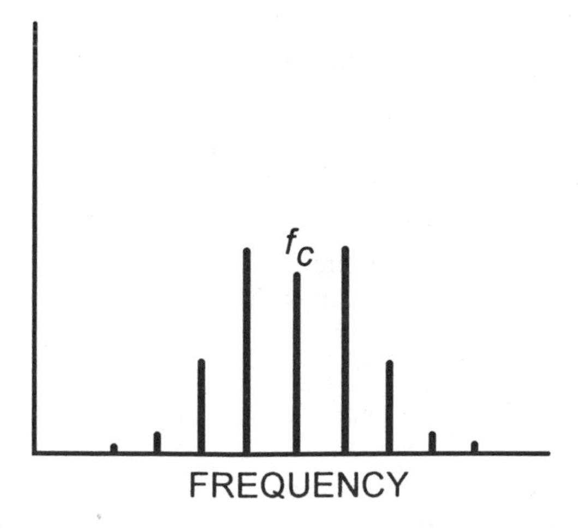

Figure 2.15 An FM signal with a modulation index of 1.5, showing multiple sidebands.

to be adequate to pass enough sideband power for ample intelligibility of voice signals. However, as mentioned above, for FM broadcast use the maximum deviation is set at 75 kHz, with the maximum audio frequency being 15 kHz and the deviation ratio at 5, or 75 kHz/15 kHz. This allows for high-fidelity music to pass without significant loss of quality.

2.2.3 FM and AM comparisons

Frequency modulation holds many benefits over amplitude modulation: superior noise immunity, helped by amplitude limiting to eliminate AM noise; a decrease in high-frequency noise constituents, due to preemphasis circuits, which boost the higher frequencies at the transmitter, and deemphasis circuits, which attenuate the now overemphasized frequencies at the receiver; FM's *capture effect,* which forces any undesired signal that is near, or at the same, frequency as the desired signal to be rejected. And, since FM does not have a delicate modulation envelope, as does AM, it does not require Class A linear amplifiers, but instead can utilize the far more efficient Class C types in both its RF and IF sections. Also, transmitter efficiency in FM is quite high, since the transmitter itself can be modulated by low-level techniques, needing little baseband modulation power.

Frequency modulation does have its drawbacks: Increased bandwidth is necessary because of the additional sideband production over AM; the FM transmitter and receiver are more expensive to design and construct because of their higher frequencies of operation, along with higher stability requirements; bouncing the FM signal off the atmosphere's ionosphere creates distortion of the FM wave, so it's normally (unless repeaters are used) a line-of-sight communications medium.

A reprise of some of the more important FM terms:

Center frequency, sometimes referred to as the *rest frequency,* is the FM transmitter's carrier frequency with zero percent modulation.

Frequency deviation is the amount the RF carrier shifts from its center frequency in a single direction when modulated.

Frequency swing is the movement of the modulated carrier on *both* sides of the center frequency, or twice the *frequency deviation.*

Modulation index, which is employed when one tone, at a steady deviation, is transmitted, is the ratio of the carrier's instantaneous frequency deviation to the instantaneous frequency of the modulation.

Deviation ratio is the ratio between the *maximum* frequency deviation—with 100 percent modulation—to the *maximum* audio modulation frequency.

2.3 Single-Sideband Modulation

2.3.1 Introduction

As a result of the highly stable oscillators and phase-locked loops available today, along with advances in mass production and RF integrated circuits, low-cost transmitters and receivers for single-sideband (SSB) communications have become quite popular, and have completely replaced the older AM long-range voice communication systems.

2.3.2 Fundamentals

Single-sideband suppressed carrier is a form of AM, but transmits only a single sideband, rather than the two sidebands and the complete carrier of amplitude modulation. In SSB the carrier, which holds no information, and the other sideband, which duplicates the information present in the transmitted sideband, are strongly attenuated (Fig. 2.16). Since one of the sidebands is attenuated, SSB requires only half the bandwidth that AM consumes for its transmissions, which also translates into less noise received (Fig. 2.17). Although amplitude modulation's fading characteristics are quite poor, in SSB fading is much less of a problem. This is because the multiple phase dependencies between all of AM's transmitted elements—both sidebands and the carrier—need not be sustained in SSB, inasmuch as only a single sideband is actually being transmitted. Power efficiency is also much higher in SSB than in AM because of the power savings in transmitting only a single sideband, while further power is conserved since a transmitted signal is produced only when the baseband is actually present at the modulator.

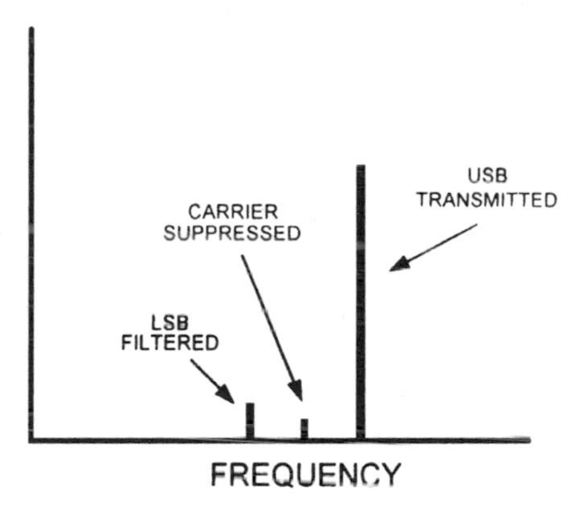

Figure 2.16 A single-sideband signal.

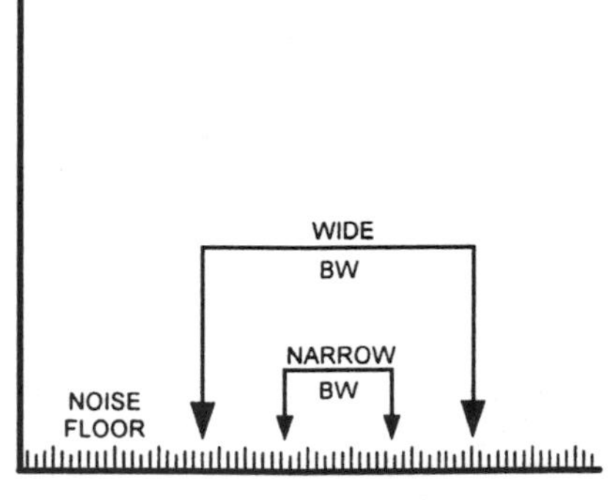

Figure 2.17 A narrower received bandwidth means less noise.

2.3.3 Modulation

The SSB transmitter (Fig. 2.18) creates a single-sideband signal by inserting both the oscillator (OSC) generated carrier and a modulating audio signal from the AUDIO AMP into the *balanced modulator.* The balanced modulator nonlinearly combines, or mixes, the carrier and baseband inputs, producing both lower and upper sidebands. The modulator will also severely attenuate the carrier from the OSC stage by phase cancellation or common-mode rejection methods. The ensuing *double-sideband (DSB) suppressed carrier* signal is then injected into the next stage, which is an upper sideband/lower sideband (USB/LSB) filter.

These filter stages of the SSB transmitter consist of very selective bandpass filters that have a center frequency to pass either the upper or lower sidebands. There are nonfilter phase-cancellation methods that can be utilized to reject the undesired sideband by twin balanced modulators and phase-shifter circuits. Either way, the SSB signal is then upconverted, amplified, and sent out through the antenna. However, since the modulated signal will contain a nonconstant amplitude modulation envelope that can easily become distorted, linear amplifiers must be utilized throughout an SSB system's signal path.

The RF signal is then picked up at the SSB receiver's antenna, filtered, amplified, and downconverted (Fig. 2.19). The signal is inserted into a type of nonlinear mixer called a *product detector,* along with the *carrier oscillator* [or beat frequency oscillator (BFO)] frequency to supply the missing carrier. The

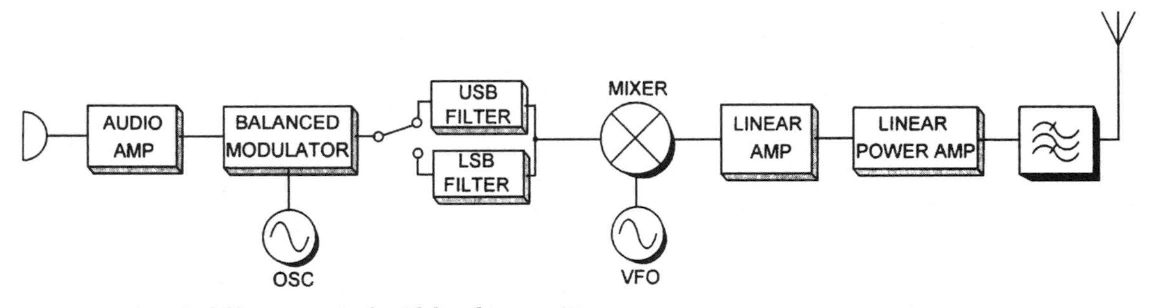

Figure 2.18 A typical filter-type single-sideband transmitter.

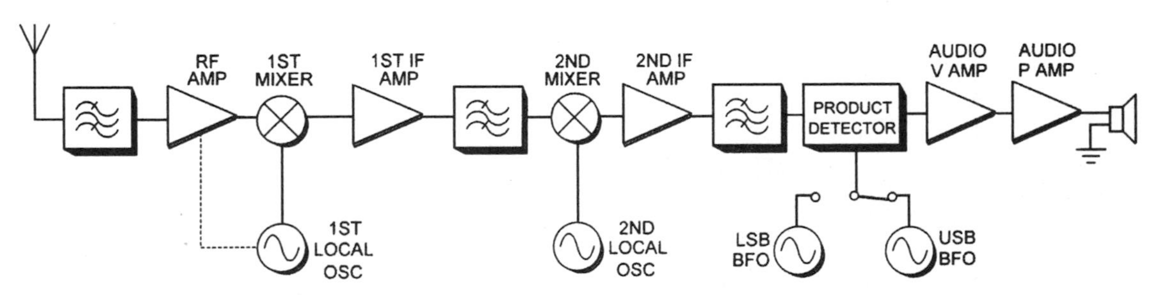

Figure 2.19 A typical dual-conversion single-sideband receiver.

output is the baseband signal, which is amplified and sent on to headphones or speakers.

Depending on the number of modulating tones and their amplitude, different time-domain outputs can be viewed on an oscilloscope. If a single baseband tone is injected into an SSB modulator, a steady RF signal in both amplitude and frequency will be created, as in Fig. 2.20. This is simply a CW signal. However, a two-tone baseband signal will generate the consummate SSB modulation envelope display of Fig. 2.21, with the amplitude of the modulation envelope dependent on the baseband modulation level. The two-tone RF signal will start to "flat-top" (Fig. 2.22) if overmodulation occurs, causing extreme distortion and spurious outputs.

2.3.4 Output power

The measurement of output power in SSB is the same as in AM, the *peak envelope power* (PEP) being the measurement of the average peak power of the transmitted signal with 100 percent modulation. PEP can be calculated by V^2_{RMS}/R, or $V_{RMS} \times I_{RMS}$, or $I^2_{RMS} \times R$, where V and I are those of the maximum modulated peak.

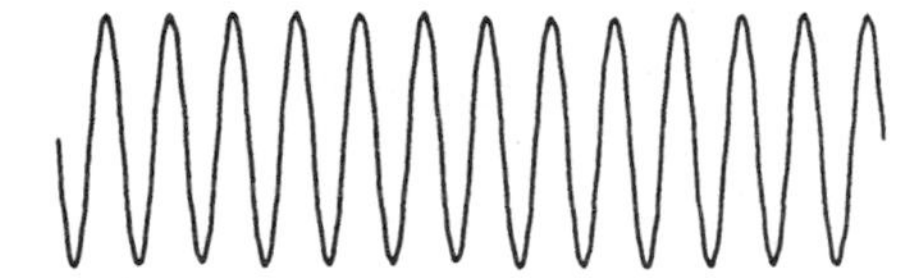

Figure 2.20 A single-tone SSB signal in the time domain.

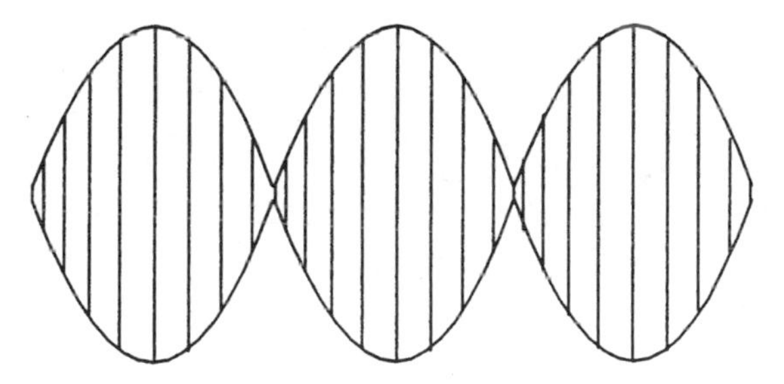

Figure 2.21 A two-tone SSB signal showing its modulation envelope.

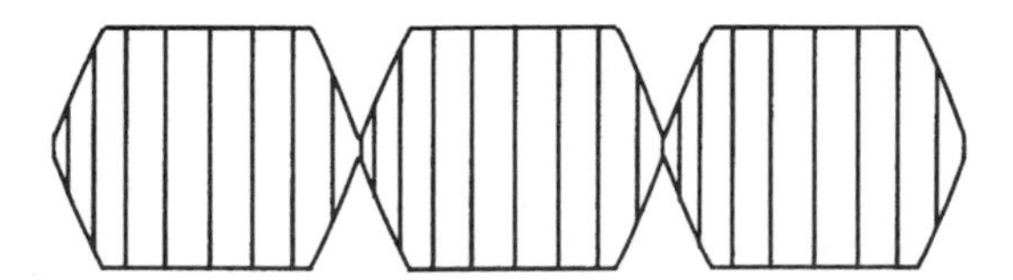

Figure 2.22 An overmodulated two-tone SSB signal.

2.4 Digital Modulation

2.4.1 Introduction

With the advent of digital modulation techniques, far higher data rates are possible within constrained bandwidths, and at higher reliability and noise immunity levels, than the older analog modulation methods of FM, AM, frequency shift keying (FSK), on-off keying (OOK), pulse width modulation (PWM), pulse position modulation (PPM), pulse amplitude modulation (PAM), etc.

The newer digital modulation has much in common with some of the older discrete digital/analog modulation methods, such as OOK and FSK, in that it has discrete states at discrete times—whether these states are amplitude, phase, or amplitude/phase—and these states define the information being transmitted, while the number of states possible governs the amount of data that can be transmitted across the link. However, digital modulation may be considered to be only the QAM, QPSK, and BPSK modulations (defined below), and their many variants.

2.4.2 Types of digital modulation

Modulation methods in general can most easily be viewed with a phasor diagram (Fig. 2.23). I is the in-phase (0 degree) reference plane, while Q is the quadrature (90 degree) reference plane. In between these two I and Q states is the signal (S), which can vary in phase (θ) and amplitude (A). Since any modulation will alter either the phase or the amplitude (or both) of a carrier, this is an effective way to visualize various modulation methods.

For comparison to digital modulation, Fig. 2.24 displays the phasor diagram of analog phase modulation which, since it has no information in its carrier's amplitude, shows a full circle as the carrier rotates from 0 to 360 degrees. On the other hand, AM contains no phase information in its carrier, so its phasor will vary only in amplitude. This is indicated in Fig. 2.25 by its carrier amplitude varying from 0 to 100 percent in magnitude.

More efficient digital-like methods of analog modulation are possible by using only discrete states in the phasor diagrams. The simplest is *on-off keying* (OOK), a type of *amplitude shift keying* (ASK) modulation, Fig. 2.26, which

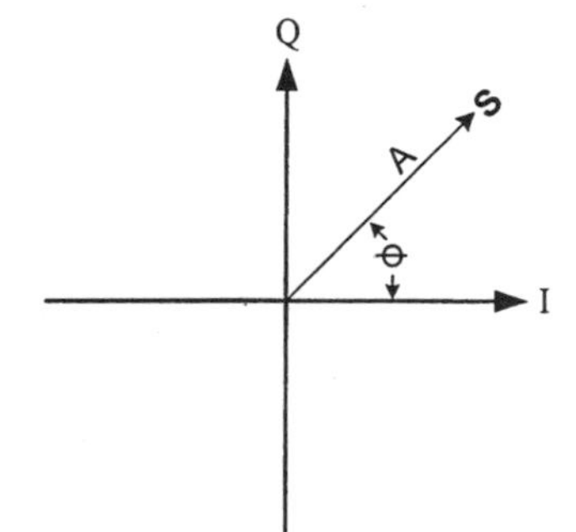

Figure 2.23 A phasor diagram.

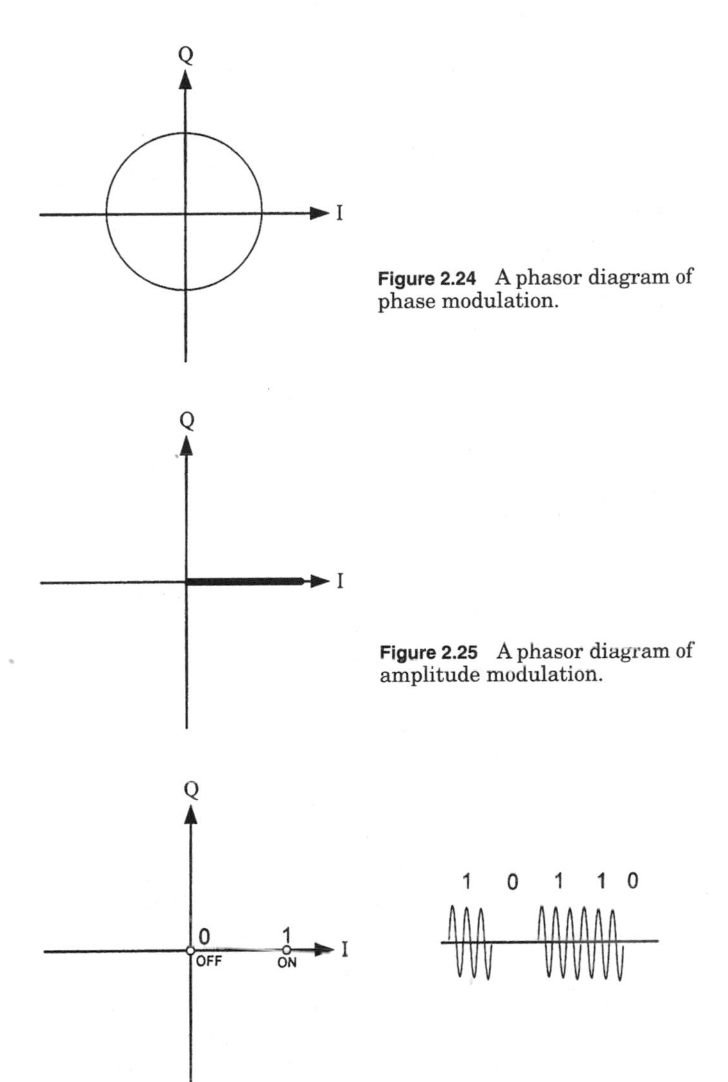

Figure 2.24 A phasor diagram of phase modulation.

Figure 2.25 A phasor diagram of amplitude modulation.

Figure 2.26 A phasor diagram of OOK modulation with accompanying time-domain sine waveforms.

is adopted to send Morse code or to send 1s and 0s by turning on and off the RF carrier frequency. This allows one bit of data to be sent between each discrete amplitude transition.

Instead of varying the amplitude in discrete states—while maintaining the phase—we can maintain the amplitude of the carrier while changing the phase of the signal to two discrete states; such as 0 and 180 degrees, as shown in Fig. 2.27. This type of digital modulation is the most basic, and is referred to as *binary phase shift keying* (BPSK) with a 0 degree reference phase indicating a 1, and a 180 degree discrete state indicating a binary 0.

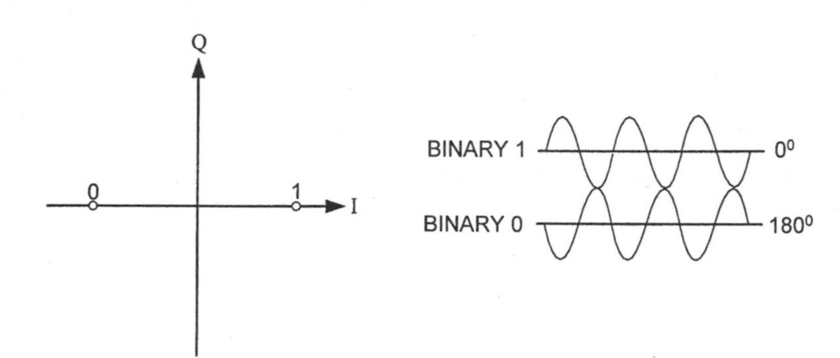

Figure 2.27 A phasor diagram of BPSK modulation with accompanying time-domain waveforms.

When we modulate the phase of a carrier into four discrete states, we have *quadrature phase shift keying* (QPSK). As shown in Fig. 2.28, four discrete phase states have been chosen to convey information, unlike analog phase modulation, which has infinite phase points as it rotates from 0 to 360 degrees. The four discrete states for QPSK are 45, 135, 225, and 315 degrees and are located on a constant-amplitude carrier. The four states supply 2 bits for each shift of phase (00, 01, 10, 11), instead of 1 bit (1, 0) as in the BPSK system. This technique would clearly contribute double the information within the identical bandwidth and time period.

However, when we say that the carrier of a QPSK signal is of a "constant amplitude" during modulation, this is not quite true. Amplitude variations may play no role in actually transferring the information across a QPSK-modulated wireless link, but amplitude variations of the carrier *do* occur. We will go into this in further detail below.

Quadrature amplitude modulation (QAM) is the most widespread method today for sending data at very high bit rates across terrestrial microwave links. It employs a blend of amplitude and phase modulation. QAM utilizes various phase shifts to the carrier, each of these phase shifts being able to also have two or more discrete amplitudes. In this way, every amplitude-phase combination can symbolize a different and distinct binary value. As an example, in QAM-8, a digital value of 111 could be represented by a carrier that displays a phase shift of 180 degrees and an amplitude of +2; or 010 can be symbolized if the phase is shifted to 90 degrees with an amplitude of −1. Indeed, QAM-8 exploits four phase shifts and two carrier amplitudes for a total of eight possible states of 3 bits each: 000, 001, 010, 011, 100, 101, 110, and 111. Another example of quadrature amplitude modulation, QAM-16, shown in Fig. 2.29, provides for 4 bits per AM/PM change.

More data can be transmitted within an allocated bandwidth or time period as the number of AM/PM states are increased, since more bits per change can now be encoded. But as the number of the AM/PM states is increased, the states become closer together, so noise will begin to become more of a problem for the signal's BER. This means that the higher the QAM state, the more it

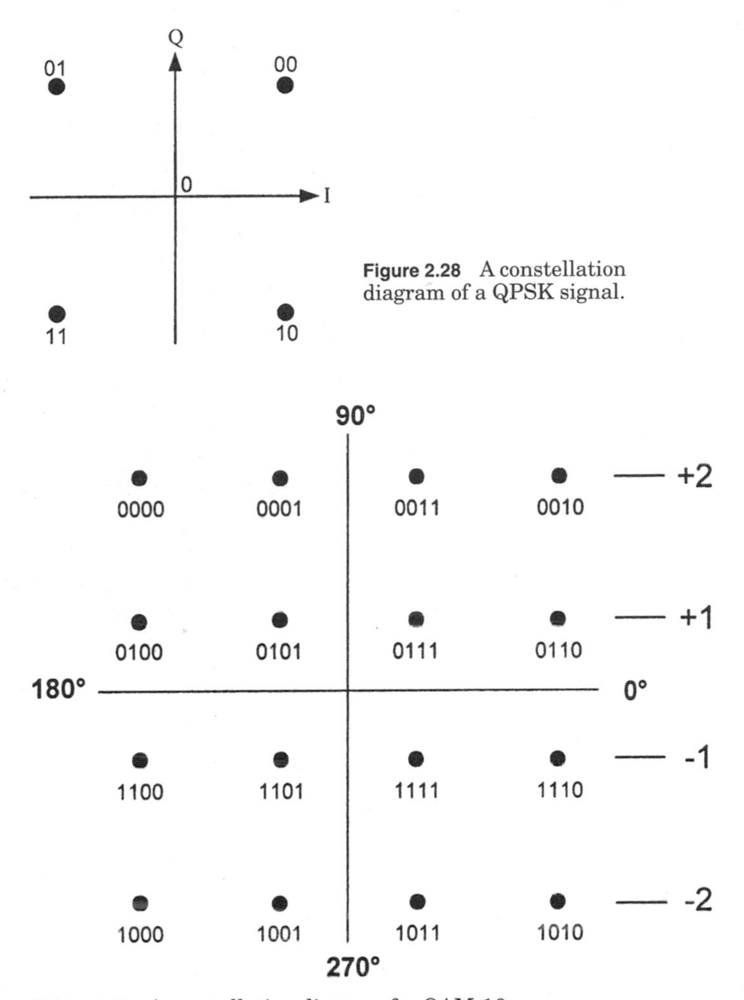

Figure 2.28 A constellation diagram of a QPSK signal.

Figure 2.29 A constellation diagram for QAM-16.

can be subjected to damaging interference under comparatively low noise conditions, since the signal's SNR must be increased for each increase in the number of states in order to maintain the BER. For this reason satellite communication, which is subject to a high-noise environment, will normally use the simpler modulation schemes, such as BPSK or QPSK. However, low-noise terrestrial radio links can use 32, 64, 128, and 256 QAM to send data at much higher rates than possible with the older digital systems within the confined bandwidth restraints of most wireless information channels.

Because of the noiselike nature of all digitally modulated signals, viewing such signals in the frequency domain will usually not tell us much about the complexity of the modulation—such as whether it is QPSK, QAM-16, or QAM-256—but only the signal's amplitude, frequency, flatness, spectral regrowth, etc. In fact, a digital signal, when transmitting, will normally appear the same whether or not it is transporting any data; this is due to the coding and encryption added to the RF signal.

2.4.3 Power and digital signals

In describing these digital modulation formats, we display either a static constellation diagram or a phasor diagram at a discrete point in time and completely ignore how these transitions from symbol to symbol occur. These transitions are especially important in the common four-phase state QPSK modulation format and its variants, because this governs whether the modulation will have a constant modulation envelope, or one that varies in amplitude. A constant-amplitude modulation envelope will allow the use of an efficient, near-saturated power amplifier that does not need to be backed off in its power output, while a QPSK modulation with a nonconstant amplitude requires a highly inefficient linear amplifier that must be heavily backed off from its maximum power output in order to avoid massive spectral regrowth. (Too much spectral regrowth, a form of IMD, will cause an otherwise legal signal to cause interference to channels on either side of its own channel, and a bandpass that no longer fits within the mandated FCC spectral mask limits.) The problem of a non-constant-amplitude RF carrier during QPSK modulation occurs because the carrier of regular QPSK will sometimes pass through *zero* amplitude on its way from one phase state to another, as shown in Fig. 2.30, thus causing the QPSK modulation envelope to vary in amplitude, disallowing the use of a nonlinear amplifier because of the resultant IMD production. This is not quite the same thing as QAM, in which much of the information to be conveyed is actually *contained* within the amplitude variations of the signal; in QPSK, amplitude variations are but an annoying side effect of the digital phase modulation.

Using amplifiers that do not have to be as extensively backed off in power to prevent spectral splatter problems can be accomplished by employing a less bitrate efficient modulation, such as offset QPSK (OQPSK). This type of modulation allows only changes in phase states that do *not* pass through the zero amplitude origin. Referring to Fig. 2.30, this would mandate symbol changes between 00 and 01, or between 10 and 11, but not between 11 and 00, or 01 and

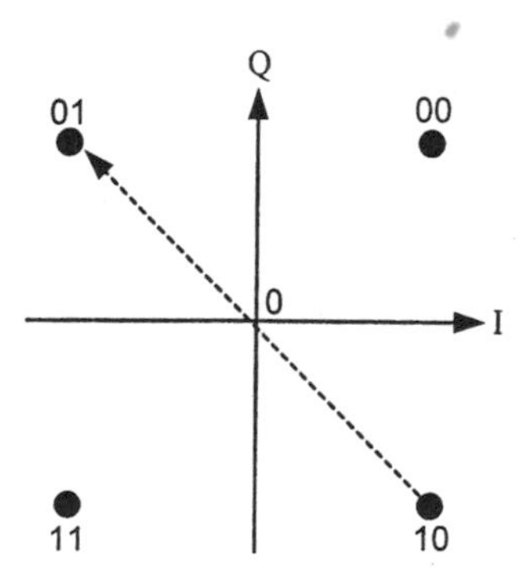

Figure 2.30 A QPSK signal passing through zero amplitude, quenching the carrier.

10. Thus OQPSK does not allow the carrier to be snuffed, and will have carrier amplitude variations of, at most, only 3 dB. As stated above, this permits the use of more efficient nonlinear amplifiers, with higher output powers possible, and with less spectral regrowth. Another similar QPSK modulation format, referred to as $\pi/4$ DQPSK, will also not let the carrier be completely snuffed out, while also allowing for easier clock recovery during demodulation. Other modulation schemes have a modulation envelope that has a completely constant envelope, and can thus employ very efficient nonlinear Class C amplifiers without any spectral regrowth. Two of these modulations are *minimum shift keying* (MSK) and its derivative *gaussian minimum shift keying* (GMSK).

Digital power measurement. The *peak* amplitude of a digital signal is completely unpredictable, and will vary dramatically over time, because of its noiselike nature. The average power output of an analog amplitude-modulated transmitter will vary depending on the baseband waveform, and its peak envelope power can easily be measured over a single cycle of the waveform (and need not be averaged over time) by commonly available, low-cost test equipment. The situation is not quite the same with digitally modulated signals. To find the peak amplitude of these erratic digital signals, the power measurement must be taken over time to obtain a *statistical* peak amplitude, since there is only a statistical chance that a much higher peak will come along at any time. This peak reading is then compared to the digital signal's *average* power (the average power is the same power that a DC signal would require to heat up a resistance element to the same temperature as the RF signal). With this type of average power measurement, any wave shape can be gauged.

If an RF signal did not vary its power over time (as with DC signals, with $P_{DC(PEAK)} = P_{DC(AVG)}$), then its average power would be delivered constantly to the load, with its peak power equaling its average power. But since RF signals *do* vary (as any AC signal will), their peak power will be different from their average power. This ratio between the peak and the average values of a modulated signal is referred to as the *peak-to-average* ratio. The lower this ratio, the closer to the P1dB level an amplifier can be driven without producing excessive intermodulation distortion (IMD) products, because the occasional power peaks will be lower in amplitude with a modulation format that has a lower peak-to-average ratio (see "P1dB compression point test" in Sec. 2.6.2). Thus the amplifier will not have to have as much of a power margin to gracefully accept these higher-amplitude power spikes without creating inordinate IMD. A signal with very high absolute peaks, but a low average power, will have a very poor peak-to-average ratio, forcing any amplifier that is working with this type of signal to have a large amount of reserve power to amplify these occasional peaks without excessive distortion. However, while the peaks may be erratic over time—since the peaks are caused by the modulation shifting from certain constellation points to other constellation points—the average power in a digital signal will be constant. This is due to the digital signal's encoding.

As stated above, this peak-to-average power ratio, if high, can cause excessive inefficiency since the amplifier must be backed off by this amount to allow for the occasional peaks. This ratio, however, will vary with the symbol patterns and clock speeds, as well as the channel filter and bandwidth. General peak-to-average ratios, nevertheless, are at least 5 dB for QPSK, 8 dB for 64-QAM and *orthogonal frequency division multiplexing* (OFDM), and up to 15 dB for *code division multiple access* (CDMA). This means that for some QPSK modulations, the intermittent peaks will rise above the RMS power by 5 dB.

To measure this peak-to-average ratio, first measure the power peaks with a *fast-acting,* digital-modulation-capable, peak-reading power meter (such as the Boonton 4400) over a time of at least 10 seconds. This will give a reasonably accurate indication of the signal's peak power. To measure the average power, see "Measuring digital signal power" in Sec. 2.6.2, or use a special digital-modulation-compliant *average* power meter. Now subtract the average from the peak, in dBm. This equals the peak-to-average ratio in dB.

To recap this very important concept of digital signal power: Digital signals, because of their nonrepetitive, random nature, as well as the fact that all of their power is spread over frequency rather than condensed as it is in analog signals on all sides of the carrier, have no predictable, recurrent peak power points to measure. Since it is difficult to measure these peaks in a nonstatistical manner, we are forced to take an average measurement of the digital signal's power over its *entire* bandwidth. But how much bandwidth does the normal digital signal consume? It is considered to be the −30-dB bandwidth where most of the power of the digital signal resides, rather than just the −3-dB bandwidth used in most analog signal measurements.

Any digital baseband signal we work with will also have been filtered, since any unfiltered (ideal) digital signal would theoretically take up infinite bandwidth. But filtering a digital signal will cause the signal to go from a square wave to a more rounded signal. This allows the signal to be placed into a narrower bandwidth; but will also increase the required power that the *power amplifier* (PA) of the digital transmitter must occasionally transmit. In fact, this filtering is mainly responsible for the peak-to-average problem we just discussed (along with the signal passing through the origin), so the more filtering of the digital square wave we employ to decrease the bandwidth, the more we create a higher peak-to-average ratio. Thus, the PA must be backed off in power to allow these occasional peaks in power (Fig. 2.31) to be sent through the PA without causing nonlinear (IMD) performance and a widening of the bandwidth. More detail on filtering of digital signals is in Sec. 2.4.4, "Digital modulation issues."

2.4.4 Digital modulation issues

In digital communications, everything is geared to not only supplying reliable communications at the lowest transmitted power and bandwidth practical, but also to maximizing the data rate. In fact, bandwidth, power, noise, and information capacity are all interrelated by *Shannon's information theorem,* which

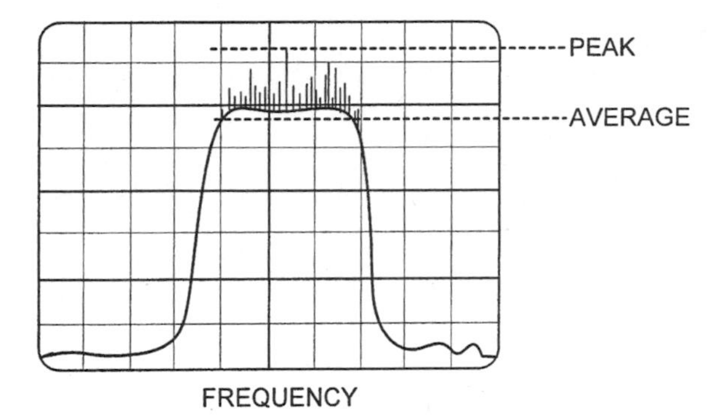

Figure 2.31 Peak and average amplitudes of a digital signal.

states that the speed of information transfer is limited by the bandwidth and the signal-to-noise ratio of a communications channel:

$$C = W \log_2 (1 + \text{SNR})$$

where C = capacity of the data link, bits per second (bps)
W = bandwidth of the channel, Hz
SNR = signal-to-noise ratio, dB

Table 2.2 shows the theoretical maximum amount of bits per symbol that can be transmitted for the most common modulation schemes. In a real radio, these maximum rates are never attained, because of hardware and transmission impairments. And, depending on error correction techniques, Table 2.3 lists the different SNRs required of the various modulation formats to sustain a desired BER.

Since the *symbol rate* equals the *bit rate* divided by number of bits represented by each symbol, then a modulation format such as BPSK would transmit at a symbol rate that is equal to its bit rate, or bit rate = symbol rate. Since symbol rate and baud rate have the same meaning, BPSK's baud rate equals its bit rate. However, in modulation schemes that encode more than one bit per symbol, such as QPSK (2 bits/baud), the baud rate will be less than the bit rate (in this case, half). This, as discussed above, allows more data to be transmitted within a narrower bandwidth.

The *modulation index* (*h,* bits/symbol), also referred to as *bandwidth efficiency,* is measured in bits per second per hertz (bits/s/Hz). The higher the modulation efficiency, the higher the data rate that can be sent through a certain fixed bandwidth. For instance, BPSK has an h of 1, while QAM-64 has an h of 6. However, a higher h comes at the expense of higher equipment cost, complexity, and linearity, and an increased SNR to maintain the same BER as the lower h systems. Table 2.4 displays the various common modulation schemes and their h values, number of states, amplitudes, and phases.

TABLE 2.2 Maximum Bits/Symbol for Common Modulation Schemes

Modulation	Bits/symbol
MSK	1
BPSK	1
QPSK	2
QAM-16	4
QAM-32	5
QAM-64	6
QAM-256	8

TABLE 2.3 SNR for Various Modulation Formats

	Signal-to-noise ratio, dB					
BER	QPSK	QAM-16	QAM-32	QAM-64	QAM-128	QAM-256
10^{-4}	8	13	15	17	19	21.5
10^{-5}	10	14	16	18	20.5	23
10^{-5}	11	15	17	19	21.5	24
10^{-7}	12	15.75	18	20	22.5	25
10^{-8}	12.5	16.25	18.5	21	23.5	25.75
10^{-9}	13	16.5	19	21.5	24	26.25
10^{-10}	13.25	16.75	19.25	21.25	24.25	26.5

TABLE 2.4 Common Modulation Schemes and Their Properties

type	Bits/symbol (h)	States	Amplitudes	Phases
BPSK	1	2	1	2
QPSK	2	4	1	4
PSK-8	3	8	1	8
QAM-16	4	16	3	12
QAM-32	5	32	5	28
QAM-64	6	64	9	52

Adaptive equalization will correct certain signal impairments in real time, such as group delay variations (GDV), amplitude tilt, ripple, and notches. Adaptive equalization, however, will not improve impairments created by a nonlinear amplifier, noise, or interference, but it will mitigate the sometimes massive multipath effects that would normally render a digitally modulated signal unreadable because of the high BER caused by the resultant amplitude variations.

Adaptive equalization basically uses a dynamically varying adaptive filter that corrects the received signal in amplitude, phase, and delay, making high-density modulations possible. Virtually all terrestrial microwave communication systems employ some form of adaptive equalization, located right after the receiver's demodulator.

Because of the nature of digital signals, they can maintain a relatively high quality at the receiver—even when close to becoming unreadable as a result of impairments. This makes the testing of a digital signal for its merits at the receiver of little use, since the digital signal may actually be only a few dB in signal strength from crashing the entire link. This is referred to as the *cliff* (or *waterfall*) *effect,* due to the rapid degradation, or complete elimination, of the digital signal. BER will lessen to unacceptably high levels quite rapidly (Fig. 2.32). But digital communication systems can be examined for proper operation by sending and receiving certain digital test patterns that incorporate a recurring succession of logical 1's and 0's. The test then compares the impaired received pattern to the perfect transmitted pattern. The BER can then be established by contrasting the bits received that were incorrect with the *total* number of bits received.

This degradation in digital signal quality can be caused by many things: reflections off metallic surfaces (multipath), producing amplitude ripple within the signal's passband; inadequate signal strength at the receiver creating decreased SNR and a corresponding blurring of the symbol points (poor SNR can be due to transmitter power levels being too low, high receiver noise figure (NF), or path attenuation caused by trees, weather, or Fresnel zone clearance problems); group delay variations and amplitude ripple produced by improper analog filtering; strong phase noise components in the frequency synthesizers of the conversion stages; or noise and cochannel interference levels induced by interferers of all types.

Since many communication systems live or die by their bit-error rate figures, it is therefore worthwhile to not only recapitulate what the dominant causes of BER degradation are in a digital communications system, but also to dig a little deeper into the reasons behind this increase in BER. Decreased signal-to-noise ratio is the main mechanism for poor BER, since noise will smudge the symbol points, making their exact location hard to distinguish by the receiver's demodulator. Phase noise, another important contributor, will cause an input signal into a radio's frequency converter stage to be slightly changed at its output; this phase noise is introduced by the real-world local oscillators (LOs) of a communication system, since the LOs are not perfect sin-

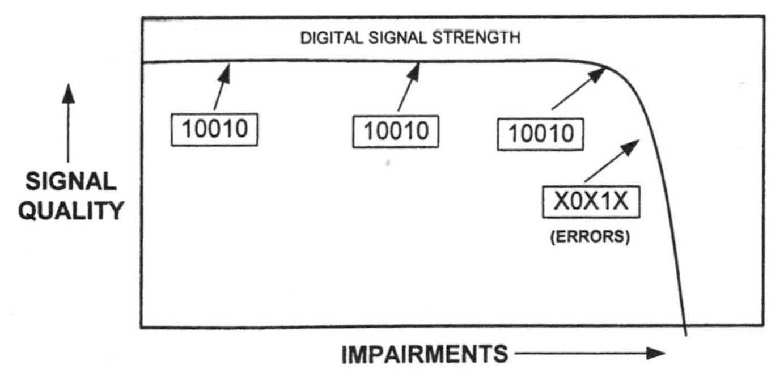

Figure 2.32 A digital signal and the cliff effect.

gle CW frequency sources, but possess phase noise. Since digital signals carry their information in the phase of the signal, this insertion of phase variances will create increased BER, with the density of the modulation affecting the severity of the BER degradation: The higher-order QAM constellations, such as QAM-256, can be severely degraded at relatively small levels of phase noise. This is because their constellation points are so densely packed, causing their phase/amplitude points to bisect digital decision boundaries.

Another major impairment, intermodulation distortion (IMD), will induce noiselike sidebands in a digital system, increasing distortion and decreasing the SNR—which degrades the BER—as well as creating adjacent channel interference (Fig. 2.33). And analog filters, especially at their band edges, can add significant group delay variations, which force the digital signal to arrive at different times at the filter's output, sometimes causing catastrophic BER problems. Analog filter–related amplitude variations that are located within the passband, called *ripple,* can produce a high BER in many digital systems. Ripple is caused by poorly designed or implemented *LC* filters. Multipath itself will also create both amplitude impediments (ripple and notches) and

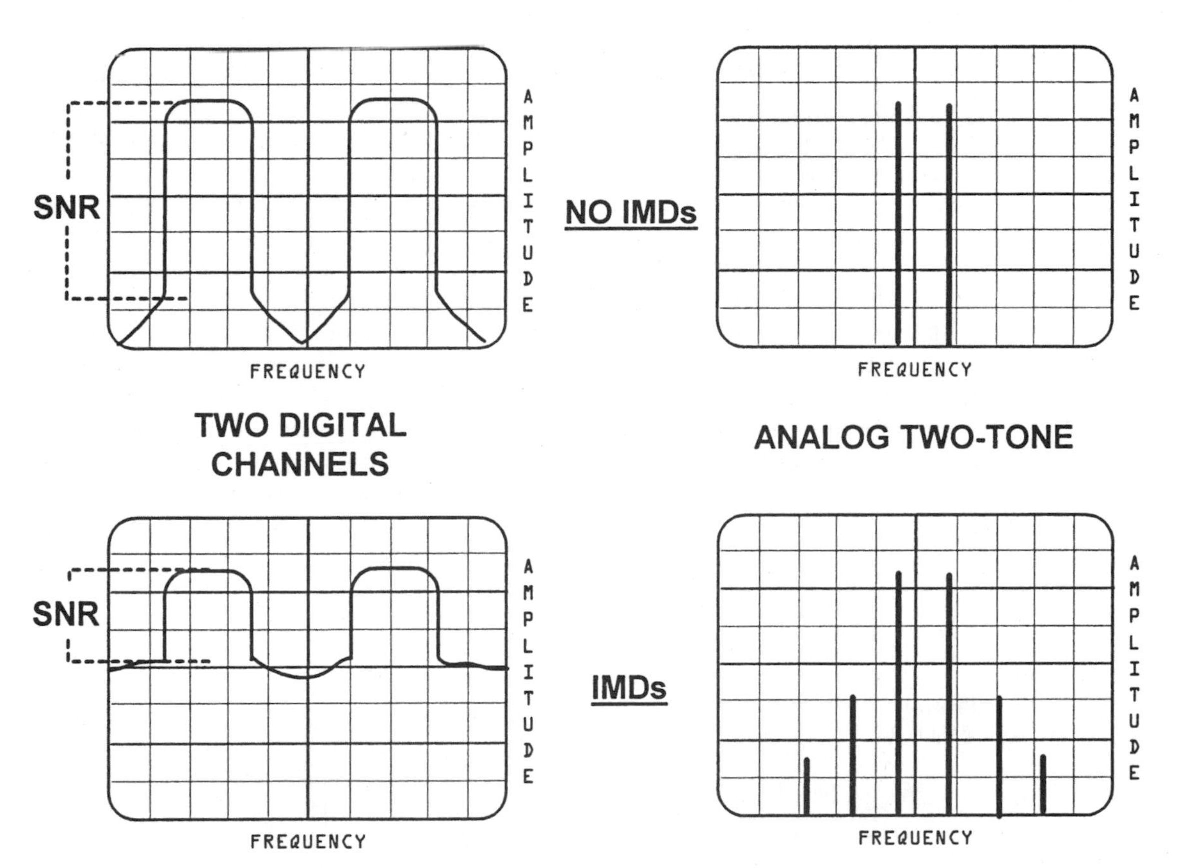

Figure 2.33 Digital signals as affected by IMD, along with a standard CW two-tone display for comparison.

phase distortions of the signal through phase cancellation, decreasing the received signal strength—which decreases SNR, and thus increases the BER.

Therefore, digitally modulated radio systems must be designed for low levels of phase noise, group delay variations, IMD, amplitude ripple and shape, frequency variations, and multipath and high levels of SNR so as not to adversely influence the BER of phase/amplitude-modulated digital signals.

Another very important issue in digital modulation that has, as yet, only been touched on is the effect that the filtering within the transmitter's modulator has on digital signals. This filtering, as stated above, is employed to limit transmitted bandwidth to reasonable or legal levels. Our example for the following discussion will be with *filtered QPSK*.

As shown in Fig. 2.34*a*, a quadrature modulator adopted for QPSK transmitters receives a data bit stream, which is then inserted into the *bit splitter*. The bit splitter sends the *odd* bits to the I input of the quadrature modulator chip, and the *even* bits to the Q input. However, before exiting the modulator, these bits must first pass through a low-pass filter, which rounds off the bits' sharp rise and fall times. This shaping of the digital signal before it actually enters the I/Q modulator chip helps to avoid interference to the important central lobe of the RF or IF digital signal—and to specifically reduce the bandwidth that will exit the modulator chip. Notwithstanding, bandlimiting can also be added through a bandpass filter at the modulator's output (*after* the I and Q are linearly added in the combiner), along with the low-pass filters in the I and Q legs.

Even at the receiver's demodulator, filtering is taking place. In fact, the filtering and bandshaping is typically shared among the transmitter and receiver. The transmit filter reduces interference of *adjacent channel power* (ACP) in other channels, while the receive filter reduces the effect of ACP and noise on the received signal. This scheme allows an almost zero group delay variation from the input of the transmitter to the output of the receiver so as to obtain low intersymbol interference (ISI) and BER. At the receive end, one method for demodulating an incoming received QPSK input is displayed in Fig. 2.34*b*. The IF or RF enters the demodulator's input, where the signal is split into two paths and enters the respective mixers. Each mixer's LO input is fed by the *carrier recovery circuit*, which strips the carrier from the incoming signal at the exact frequency that the transmitted signal would be after going through the receiver's conversion stages (if conversion stages are present). The outputs of these mixers are fed into the low-pass filters (LPFs), which eliminate the now undesired IF signals. Some of this output from the LPFs is tapped and placed into the *symbol timing recovery* and the *threshold comparison loop* to judge whether a 1 or a 0 is present. This also reshapes the digital data into a recognizable bit stream. The bits from both mixers are then combined in the shift register as a replica of the originally transmitted binary signal—if the SNR is high enough to assure a low BER, that is.

The LPFs in the modulator and demodulator sections just discussed are not just any breed of filter. The low-pass filters must be of a very special type that

will limit excessive intersymbol interference, since the demodulator would have great difficulty in deciding whether an input signal was a 1 or a 0 if high ISI were present. A *raised cosine filter* (a type of *Nyquist filter*) is commonly employed for this purpose. Raised cosine filters are utilized to slow the transitions of a digitally modulated signal from high to low, or from low to high, in order to decrease the bandwidth needed to transmit the desired information, without degrading the ISI and the BER at the symbol decision times, as discussed above. These filters are usually matched, with one placed between the incoming data and the digital-to-analog converter (DAC) in the transmitter and the other half placed in the demodulator of the receiver. This replicates the response of a full Nyquist filter.

To compute the required bandwidth needed for a wide cosine filtered symbol rate, the formula is BW = symbol rate $\cdot$ $(1 + \alpha)$, with α between 0 and 1. It would be very bandwidth efficient if the BW could be equal to the symbol rate (this is not quite practical), which is the same as $\alpha = 0$ for a raised cosine filter. Anything over this value of zero for α is referred to as the *excess bandwidth factor,* because it is this bandwidth that is necessary beyond the symbol rate = BW value. We will always require an excess bandwidth greater than the symbol rate; or an α at some value that is over zero. If α equaled 1, the bandwidth necessary to transmit a signal would be twice the symbol rate. In other words, twice the bandwidth is required than the almost ideal situation of $\alpha = 0$. A contemporary digitally modulated radio, however, will usually filter the baseband signal to a value of α between 0.2 and 0.5, with a corresponding decrease in bandwidth and increase in the required output power headroom compared to an $\alpha = 1$ device. Figure 2.35 demonstrates the effect on the digital input signal's rise and fall times, the channel's bandwidth, and the received constellations, as α is varied.

Values of α lower than 0.2 are very uncommon because of the increased cost and complexity of building sustainably accurate filters (with high clock precision) in mass production environments. Any attempts at lower α will also increase ISI to unacceptable levels, along with the added expense of producing amplifiers that must be capable of greater peak output powers without excessive distortion products. Power back-off is required of these amplifiers because of the elevated power overshoots (see Fig. 2.35) created by the increased filtering of the digital signals by the Nyquist-type filtering, which limits the transmitted bandwidth. For heavily filtered QPSK, the excess peak power requires the solid-state power amplifier (SSPA) to have a P1dB that is at least 5 dB over what would normally be required for an unfiltered signal. This is to allow the power overshoots of the signal enough headroom so as not to place the SSPA into limiting, which would create spectral splatter into adjacent channels. All signals that have a modulation envelope—even if not used to carry information—will be affected by this Nyquist filtering, including QPSK, DQPSK, and QAM signals.

Gaussian filters are another method of slowing the transitions of the signal in order to decrease occupied bandwidth in the modulation scheme GMSK. Unlike raised cosine filters, however, these filters create a certain amount of

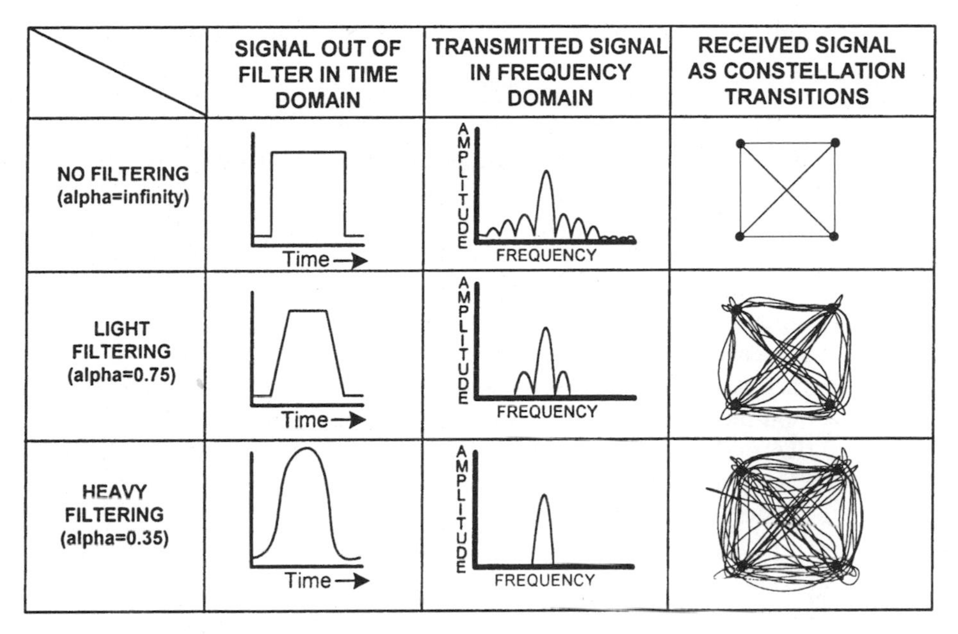

Figure 2.35 Baseband filtering effects on a digital QPSK signal.

unavoidable ISI. However, there are no power overshoots so these filters allow the use of more efficient amplifiers, with less power back-off, than the raised cosine filters discussed above.

It is important to note the difference between the filtering that takes place in each modulator/demodulator leg, as opposed to the rest of the analog radio sections. Since the I and Q signals at the input to the quadrature modulator are filtered separately in each I and Q input leg, then each of the I and Q legs of the modulator will low-pass-filter at BW = symbol rate $\cdot$ $(0.5 + \alpha)$. But when the I/Q stream is modulated onto the IF—thus creating a double-sideband signal—this will cause BW = symbol rate $\cdot$ $(1 + \alpha)$. So the actual shaping of the digital QPSK (or QAM) signal will take place in these modulator sections (usually composed of a modem), with the rest of the radio design merely used to maintain the modulator-generated spectral shape, while adding as little distortion as possible to the already predefined signal. Thus, the actual analog filtration that occurs within the IF and RF sections of the analog transmitter and receiver units may be significantly wider than BW = symbol rate $\cdot$ $(1 + \alpha)$, especially in block up- and down-converter designs.

2.5 Designing with Modulator/Demodulator ICs

2.5.1 Introduction

Quadrature (I/Q) modulators and demodulators are the most popular method today to perform modulation and demodulation of digital, as well as analog, signals. Quadrature modulators have only recently become popular, after they were integrated onto a single low-cost chip. These devices solve the problem of imparting complex amplitude/phase information onto an RF or IF carrier.

Any part of a signal's parameters can be modified by the quadrature modulator—phase, frequency, and/or amplitude—thus can add information to an unmodulated carrier. Simply employing a single mixer for this role would be unacceptable, since only one parameter (such as phase for a BPSK signal) could be modified at a time, making an efficient digital modulation scheme infeasible.

Figure 2.36 shows a quadrature modular for digital signals that is capable of varying two of the three modulation parameters; typically, phase and/or amplitude are chosen to generate BPSK, QPSK, or QAM. Many quadrature modulators are also proficient at generating AM, FM, CDMA, and SSB. The *I/Q* modulator shown will accept data at its *I/Q* inputs, modulate it, and then upconvert the baseband to hundreds of megahertz. There are some specialized *I/Q* modulators that are actually capable of functioning into the gigahertz range. Many will also be fed by DACs into their *I/Q* inputs (Fig. 2.37). The digital data is placed at the input to the DAC, which outputs in-phase (*I*) and quadrature (*Q*) baseband signals into the *I/Q* modulator inputs. The *I* modulating signal enters the *I* input, where it is mixed with the LO, which converts it to RF or IF. The *Q* modulating signal enters the *Q* input, where it is mixed with the 90 degree phase shifted LO signal, which converts it to RF or IF. Both of these signals are then linearly added in the combiner, with each mixer outputting a two-phase state BPSK, which (depending on the bits entering the modulator) will be in any one of four phase states. This combining of the two BPSK signals produces QPSK, which is shown in the time domain in Fig. 2.38. Since each mixer's output is 90 degree phase shifted from the other, the algebraic summing of the combiner creates a single phase out of four possible phase states. In other words, the incoming baseband signals to be modulated are mixed with orthogonal carriers (90 degrees), and thus will not interfere with each other. And when the *I* and *Q* signals are summed in the combiner, they become a complex signal, with both signals independent and distinct from each other. This complex signal is later effortlessly separated at the receiver into its individual *I* and *Q* components—all without the amplitude and phase constituents causing cointerference.

Now, the quadrature *demodulator* will take the incoming RF or IF signal, demodulate it, and then down-convert the signal's *I/Q* outputs into baseband for further processing by digital logic circuits. An *I/Q* demodulator (Fig. 2.39) performs the reverse of operation of the *I/Q* modulator above. It accepts the amplified and filtered RF or IF modulated signal—in this case QPSK—from the receiver's front end or IF section. The demodulator then recovers the signal's carrier (which can be employed as the LO to maintain the original phase information from the transmitter), splits it, and inserts it in phase into mixer 1 and out of phase into mixer 2. A baseband signal in *I/Q* format is then output at I_{OUT} and Q_{OUT} of the demodulator for processing.

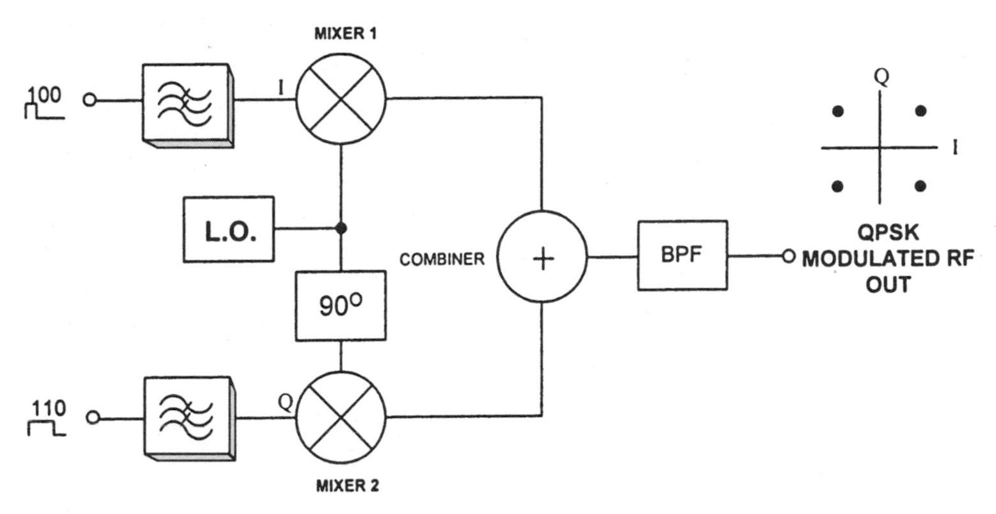

Figure 2.36 Simplified internal structure of a QPSK quadrature modulator.

Virtually all high-speed digital communication systems will employ a pre-designed modem (modulator-demodulator) for the task of modulating and demodulating the digital signal stream through the wireless system. Still, a methodology for the design of a modulator/demodulator is included in this section to assist in the construction of lower-speed systems that may not have a separate modem.

2.5.2 Designing with the RFMD RF2703

A popular chip that can perform both modulation and demodulation is the RFMD RF2703; with a few component changes, it can be adopted as a modulator (Fig. 2.40) or as a demodulator (Fig. 2.41). The RF2703 is a monolithic IC that can operate with an IF from 100 kHz to 250 MHz, and with a V_{CC} of 3 to 6 V.

In the *modulator* configuration, pins 1 and 3 are the single-ended I and Q inputs (they can be driven differentially, while 2 and 4 are at RF ground). Since pins 1 and 3 are at a high input impedance, 51-ohm resistors are added for 50-ohm matching, while the capacitors, at less than 1-ohm X_c, supply DC blocking. Pin 5 is at RF ground through a 0.1-μF capacitor. In normal modulator operation, pins 8 and 9 are left floating, while pins 10, 11, and 12 are connected directly to the ground plane. Pin 13 is a high-impedance input, so a 51-ohm resistor can be placed in shunt to match to the 50-ohm LO signal input.

A low power LO with medium-voltage outputs (0.1 to 1 V_{PP}) is required, as is a LO that is at twice the desired frequency of the carrier (this is because an internal divide-by-2 frequency divider is used for the 90 degree splitter; or LO = 2 × IF). Pin 14 supplies DC power to the chip and must be adequately bypassed. Pin 6 should be tied to pin 7, while V_{CC} is connected through a 1200-ohm resistor, or an inductor, to bias the internal active mixers. However, since

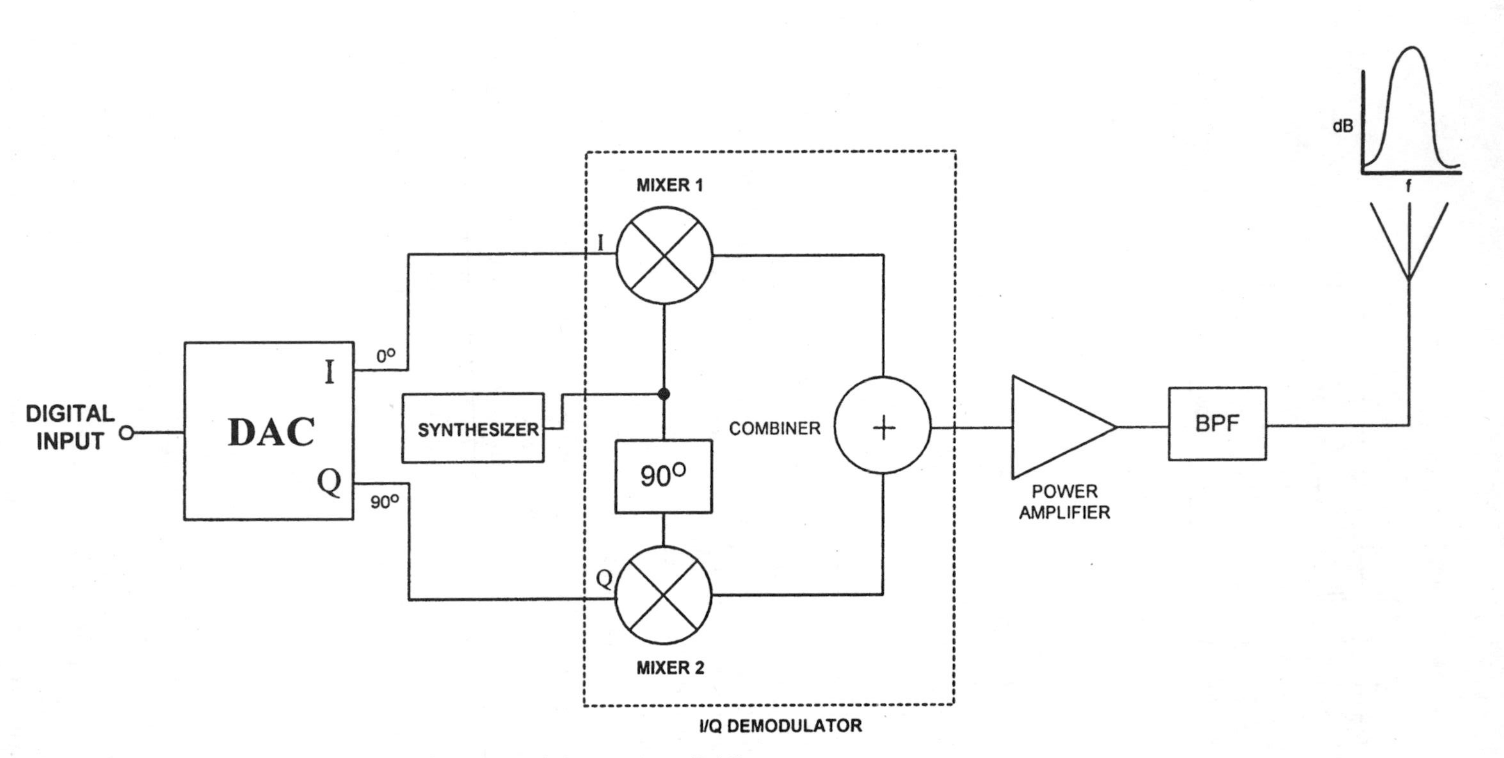

Figure 2.37 Simplified structure of a quadrature modulator using a DAC.

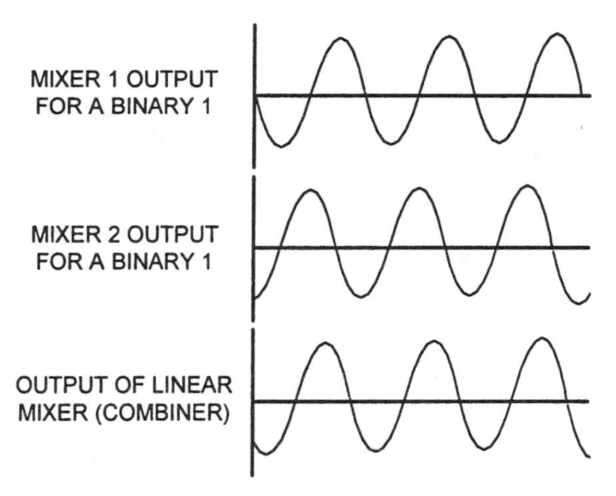

Figure 2.38 A signal's phase states through a quadrature modulator in the time domain.

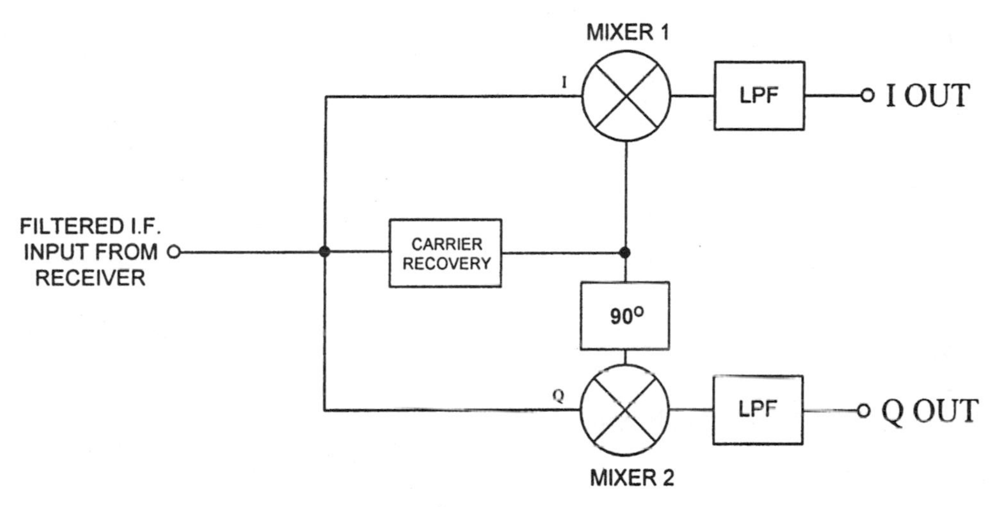

Figure 2.39 An I/Q demodulator for QPSK.

the signal output at pin 6/7 is at 1200 ohms, either a high input impedance filter should be used, or *LC* matching should be performed to obtain an impedance match for maximum gain.

When the RF2703 is in *demodulator* configuration, pin 1—with pin 3 tied to it—has the IF injected into the high input impedance, which is shunted by 51 ohms for matching. Pins 2 and 4 are connected together and placed at RF ground. Pin 5 is at RF ground through the 0.1-μF capacitor, while pins 6 and 7 are tied to V_{CC} to bias the internal active mixers. Pins 8 and 9 are Q and I out, with 50-ohm outputs (but can drive only a high-impedance load and are not internally DC blocked). Pins 10, 11, and 12 are sent directly to ground. Pin

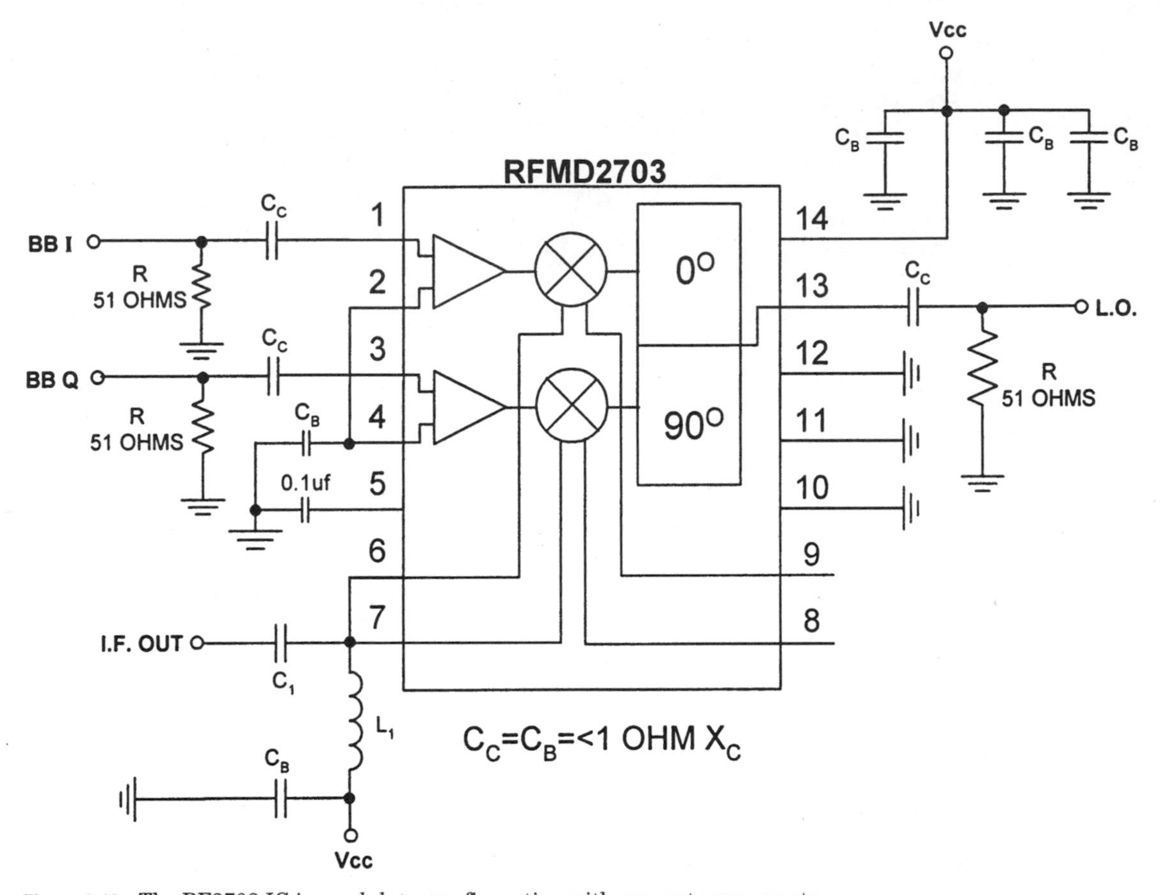

Figure 2.40 The RF2703 IC in *modulator* configuration with support components.

13 is the high-impedance LO input, with 51 ohms shunted to ground for a 50-ohm oscillator. Pin 14 is V_{CC}, which must be adequately bypassed.

2.6 Digital Test and Measurement

2.6.1 Introduction

This section describes new tests created to confirm proper radio operation with digital modulations by measuring BER and generating constellation and eye diagrams, as well as the more rigorous testing methods necessary to detect the smaller levels of error-producing phase noise, frequency instabilities, group delay variations, etc., found in today's high-end digital radios. And since digital power measurement is not the same as for the narrowband analog modulations, techniques for its accurate measurement will be presented, as will a few of the more important testing procedures for QAM and QPSK wireless devices.

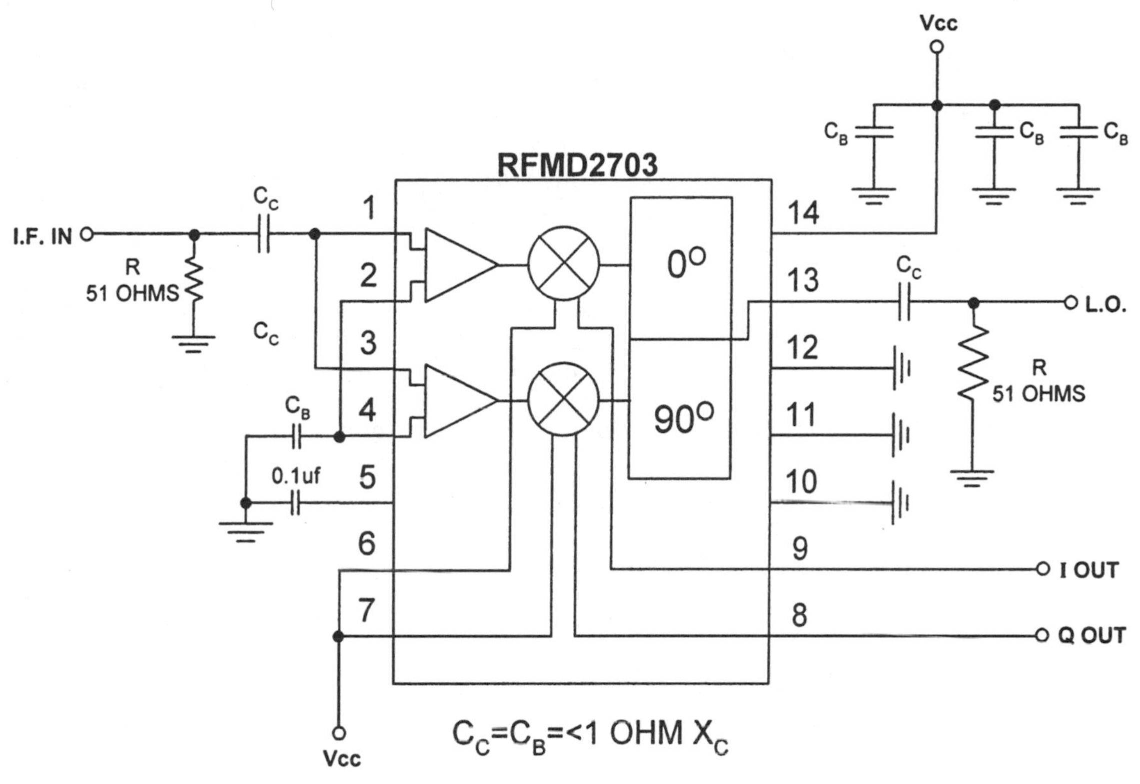

Figure 2.41 The RF2703 IC in *demodulator* configuration with support components.

2.6.2 Common digital tests and measurements

Measuring digital signal power. Accurately measuring the power of wideband digital signals can be much more difficult than measuring the output power of most analog signals.

Normal analog CW wattmeters can be very inaccurate when utilized to measure the power of digitally modulated signals, which can have a peak-to-average power ratio of 10 dB or more. Most analog modulation power meters were not calibrated to measure these types of wideband signals.

As the symbol rate of a digital signal increases, the bandwidth also increases, so the power in a digital signal is stretched across a very wide frequency range, and not localized around the center frequency of the carrier as in AM or FM modulation. This means that a single power measurement, at a single location within the digital channel, taken with a spectrum analyzer will give a deceptively low power output measurement. This is because the spectrum analyzer's maximum resolution bandwidth (RBW), its internal IF filtering, is actually narrower than the digital signal's own total bandwidth. Any accurate method of measuring the output power on a spectrum analyzer would involve taking many discrete average power measurements of the digital signal over its entire bandwidth, and then summing them together. However, this is not necessary, as you will see below.

To precisely measure a digital signal's power, the bandwidth must first be found. In order to measure the power as accurately as possible, the −30-dB-down points will be used instead of the normal −3-dB-down points normally adopted to indicate a signal's bandwidth (Fig. 2.42). This will allow for most of the digital signal's power to be measured within its entire communication channel; any power below the −30-dB points can be discounted, and will generally be quite near the noise floor.

After we have obtained the signal's true bandwidth, we can then go about accurately measuring its average power level. Most quality spectrum analyzers have the capability to make digital power measurements. The procedure is

1. Find the *power menu* on your model spectrum analyzer.
2. Select *digital*.
3. Input the digital channel's *center frequency*.
4. Input the digital channel's 30 dB bandwidth.
5. The digital signal's true power will now be indicated on the spectrum analyzer's screen.

Another, slightly less accurate, technique is utilized by spectrum analyzers that do not possess the above automatic power measurement capabilities:

1. Measure the digital signal's bandwidth at the −30-dB-down points.
2. Adjust the analyzer's RBW setting to 1/20 of the signal's −30 dB bandwidth.

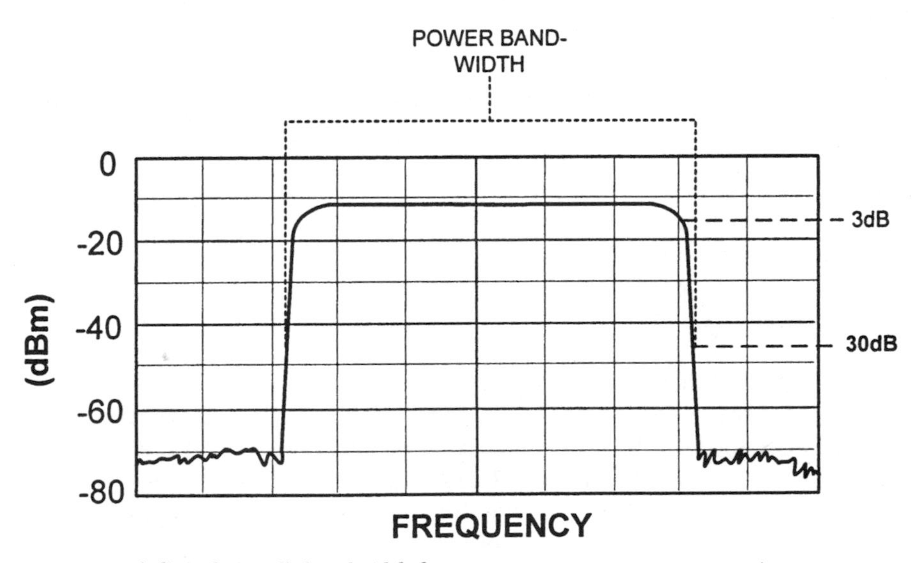

Figure 2.42 A digital signal's bandwidth for an average power measurement.

3. Set the analyzer's SPAN to 1.25 times the signal's −30 dB bandwidth.
4. Decrease the analyzer's video bandwidth (VBW) setting to reduce the signal's displayed noise.
5. Take the digital signal's power measurement with the spectrum analyzer's frequency/amplitude MARKER placed at the *center* of the signal.
6. Now, find the true total average power of the digital signal by taking the power in dBm (as measured in step 5), and adding a bandwidth correction factor (BWCF):

$$\text{BWCF} = 10 \log \left(\frac{\text{signal BW}_{30\text{ dB}}}{\text{RBW}} \right)$$

   To find the true digital output power of the signal, calculate *total digital signal power (dBm) = measured power in dBm (from step 5) + BWCF*

7. For a more accurate digital signal power measurement you can add another correction factor that takes into account the internal RBW and log detection stage losses, inherent in any spectrum analyzer, of approximately 2 dB. The formula for digital power measurement now becomes:

True digital power (dBm) = measured power (dBm) + BWCF (dB) + 2 dB

Constellation and eye diagrams. To view the degradation created by noise and frequency instabilities in a digital signal, as well as other impairments, we can employ constellation and eye diagrams.

To measure or view constellation or eye diagrams requires the ability to tap into the digital receiver demodulator's I and Q outputs, as well as the demodulator's timing clock (Fig. 2.43). The outputs of the demodulator's I and Q may be fed into an oscilloscope with an X-Y display that has the capability to turn on a persistence function for a view of the I/Q outputs over time. Such a setup will allow the operator to confirm the phase and amplitude differences of the output signal—in the form of a constellation diagram—thus allowing the viewing of the signal's quality (lack of distortion, phase noise, or amplitude instabilities). A perfect constellation diagram with no impairments is shown in Fig. 2.44*a*.

Constellation diagrams display the digital modulation's symbol patterns, while eye diagrams (Fig. 2.44*b*) permit the transition of the symbols to be viewed over time. Both measure the baseband signal's modulation condition, and whether impairments are degrading this expected pattern. In eye diagrams the eye itself is rounded, instead of square, because of the necessary limiting of the baseband bandwidth by filters. The eye comprises two lines, one at digital 1 and the other at digital 0, and is only a series of pulses displayed on the phosphor of the test oscilloscope, with each pulse being sent out of the receiver's demodulator with noise and jitter added by the transmitter, the signal path, and the receiver. This makes each pulse slightly different from the

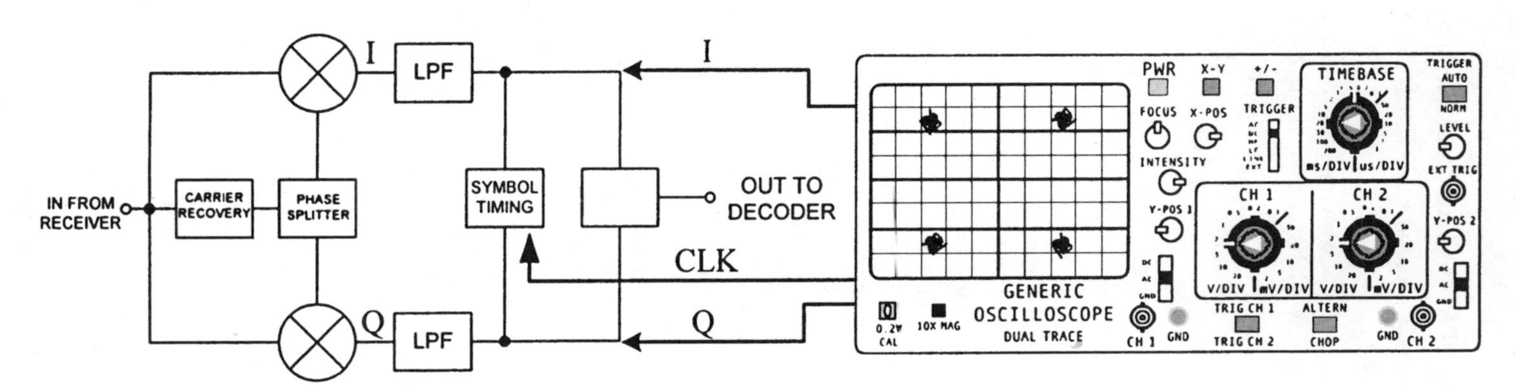

Figure 2.43 Test setup for constellation and *eye* diagrams.

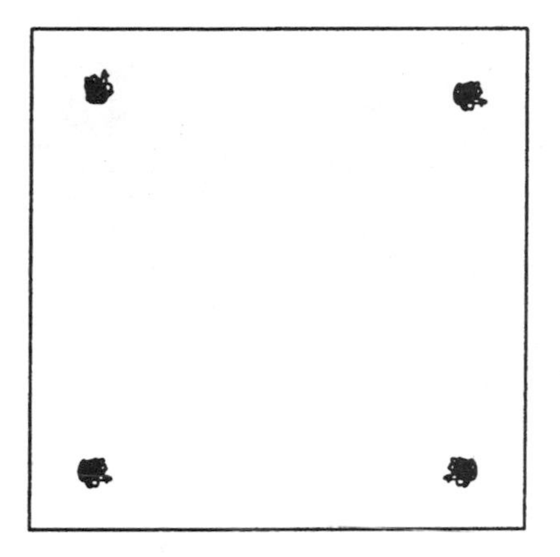

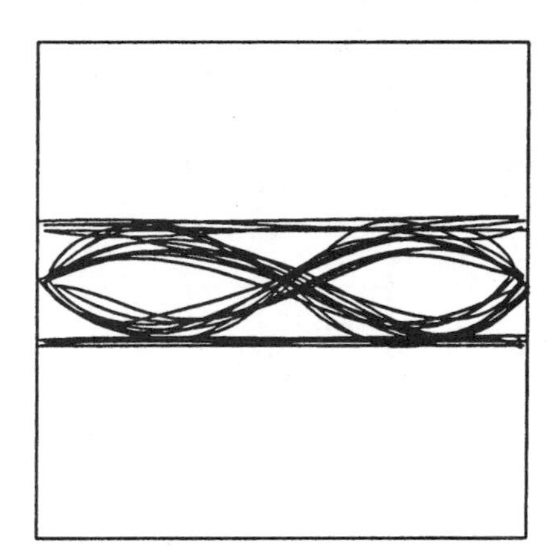

a b

Figure 2.44 (*a*) Constellation diagram; (*b*) eye diagram.

last and, as the received signal is degraded, the eye will begin to close—and the BER will increase.

To view a constellation diagram, attach the *I* to the *X* input of the scope and the *Q* to the *Y* input (which will normally be channel 1 and channel 2, respectively), and turn on the *X-Y* mode and the display's persistence function. Now attach the symbol clock to the external trigger (turn on EXT TRIGGER). The constellation should now be visible, along with faint lines joining the various points. The lines are the actual symbol transitions between the constellation points, called the *symbol trajectories.* To obtain just the constellation points (without these symbol transition lines), a BNC connector on the back plate of some oscilloscopes can be connected to the demodulator's symbol clock so that the oscilloscope's electron beam will be turned on only at the moment of sampling. This will blank the transitions between constellation dots by triggering the beam only at the constellation point instants.

Common impairments, and how they look in a constellation diagram, are shown in Fig. 2.45.

To view an eye diagram, keep the same setup as above, but turn off the scope's *X-Y* function. Set its HORIZONTAL TIMEBASE to obtain 3 symbol times per 10 divisions, and then view the eye diagram. The eye should be open; in fact, the height of the eye can be considered the noise margin of the receiver's output, while the left and right corner of each eye indicate the amount of frequency jitter present. The wider the eye, the less jitter, while the taller the eye, the less noise.

BER tests. The bit error rate test measures the ratio of bad bits to all the transmitted bits, frequently over a complete range of input/output powers. When

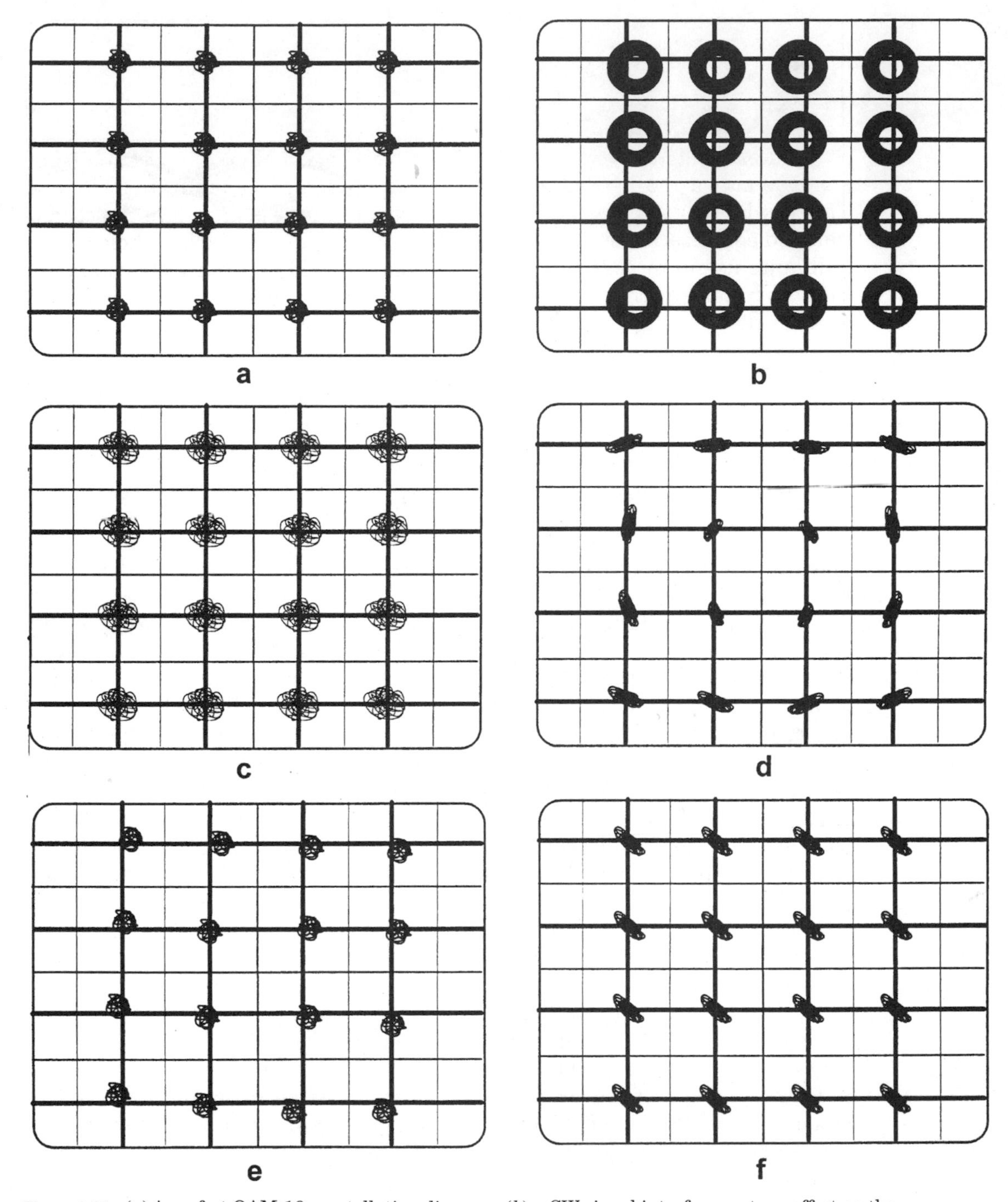

Figure 2.45 (*a*) A perfect QAM-16 constellation diagram; (*b*) a CW signal interference tone effect on the constellation diagram; (*c*) signal degradation caused by a poor SNR; (*d*) signal degradation caused by the digital radio's local oscillator instability; (*e*) slight overdriving of the transmitter's power amplifier; (*f*) multipath causing an uneven amplitude across the passband of the digital signal.

these powers are low, BER rate increases because of degraded SNR; when they are high, amplifier compression is the major contributor to poor BER.

BER testing is an excellent way to check the signal quality of a digital radio and its link. However, any in-line error correction and/or adaptive equalization will make the BER appear far higher than it actually would be without these two processes. Thus, your view of how close to a complete outage the digital signal is may actually be hidden by error correction and equalization circuits. This is why many receiver BER system tests are best done before the correction and equalization stages in order to obtain the raw BER. Nonetheless, it is still quite useful to perform BER tests from one end of a complete communications system to the other—with correction and equalization engaged—to confirm that the entire system meets BER specs over a set time period, and the link is functioning as designed on a system level.

To prevent a digital wireless link from failing, it must be completely tested to confirm that there is enough link budget to overcome any impairments between the transmitter and receiver. Nevertheless, a generally poor BER in a complete end-to-end system measurement will not indicate exactly where the trouble lies. Finding the location of the weak stage in a communications link is done by *error distribution* analysis. The fault could be caused by poor antenna alignment, overdriven amplifiers, low signal output, path obstructions, multipath, cable losses, frequency drift, component malfunction, etc. BER problems can be tracked down by observing the signal through a *vector network analyzer* (VNA), by viewing the receiver's constellation or eye diagrams with an oscilloscope, or by watching the signal in the frequency domain on a spectrum analyzer.

To perform a BER test, a pseudo-random bit sequence is injected into the transmitter's baseband or IF input. Then, the signal to be BER tested can be observed at the transmitter's antenna output into a digital test receiver and on to the bit error rate tester (BERT) or at the receiver's antenna input into a digital test receiver and on to the BERT, or after the receiver's demodulator directly into a BERT. In some high-accuracy, low-BER systems, performing a complete BER test can take anywhere from a few minutes to several hours because of the low amount of bit errors actually generated.

In the design phase, we should confirm that the BER will meet expectations, since it may be found that in order to maintain a proper *link budget* (see Chap. 9, "Communication Systems Design") we must increase transmit power, strengthen the receiver and transmitter antenna gain, and lower receiver NF to preserve the desired path length and quality of service at our preferred BER.

Two-tone test and measurement. The two-tone test to measure IMD has been a vital part of analog radio for years, and is still important in digital radio for preliminary testing.

To measure the two-tone third-order products at the output of a receiver—at a set RF input level—hook up the test gear and receiver as shown in Fig. 2.46. Feed two signals, equal in amplitude and closely spaced in frequency, from the two signal generators into the combiner. The combiner, which will drop approximately 3 dB from each input signal, will send the remainder into the step attenuator. The step attenuator permits the operator to vary both of the two signal generators' amplitudes by exactly the same amount at the same time, speeding up testing. From the attenuator the two-tone test signal is placed into the input of the receiver at the desired amplitude, while the level of the third-order IMD products are measured as dB below the carrier, or dBc. This IMD level will then be checked to confirm that it meets the receiver's design specifications.

P1dB compression point test. To reduce spectral splatter and the BER, amplifiers must be backed off a set amount from their P1dB point, the amount depending on the modulation in use. We can find the P1dB by the following method: Set up test equipment as shown in Fig. 2.47. Set the signal generator frequency to the center of the bandpass of the DUT (the amplifier). Employ attenuator pads if the amplifier's output will overdrive or damage the spectrum analyzer. Increase the signal generator's power output 1 dB at a time until the spectrum analyzer's measurement does not track the input dB for dB. The amplifier's $P1dB_{INPUT}$ will be the signal generator's output level, while the signal amplitude as read on the spectrum analyzer, plus any attenuation dialed into the attenuator, is the $P1dB_{OUT}$ in dBm. The amplifier's third-order intercept point (TOIP) can be *approximated* by adding 10 dB to the $P1dB_{OUT}$ value.

Phase noise tests. Now, more than ever, decreasing phase noise to decrease the BER is not an option: The lower the LO-generated phase noise, the better for the digital radio system. Highly accurate phase noise testing is possible, but takes special, and very costly, equipment and test setups. To perform limited-accuracy phase noise tests on any oscillator, follow the procedure below, which is useful only if the spectrum analyzer employed in the test has a lower phase noise than the DUT:

1. Attach the LO output to the spectrum analyzer's input.
2. Set the spectrum analyzer to the same frequency as the LO.
3. Switch on video averaging on the spectrum analyzer.
4. The difference between the amplitude of the carrier and the noise amplitude, minus *10 log RBW*, is approximately equal to the phase noise in dBc/Hz. (RBW is the resolution bandwidth as set on the spectrum analyzer.)

Reference spur measurement. To check for high reference spurs in the output of a phase-locked loop (PLL), which will damage the BER of the wireless device:

1. Attach the PLL's VCO output to the spectrum analyzer's input.
2. Set the PLL and the spectrum analyzer to the same center frequency. Open the spectrum analyzer's SPAN to allow viewing of all reference spurs (reference spurs are located at f_{comp} above and below the PLL's output frequency, as well as at their harmonics).

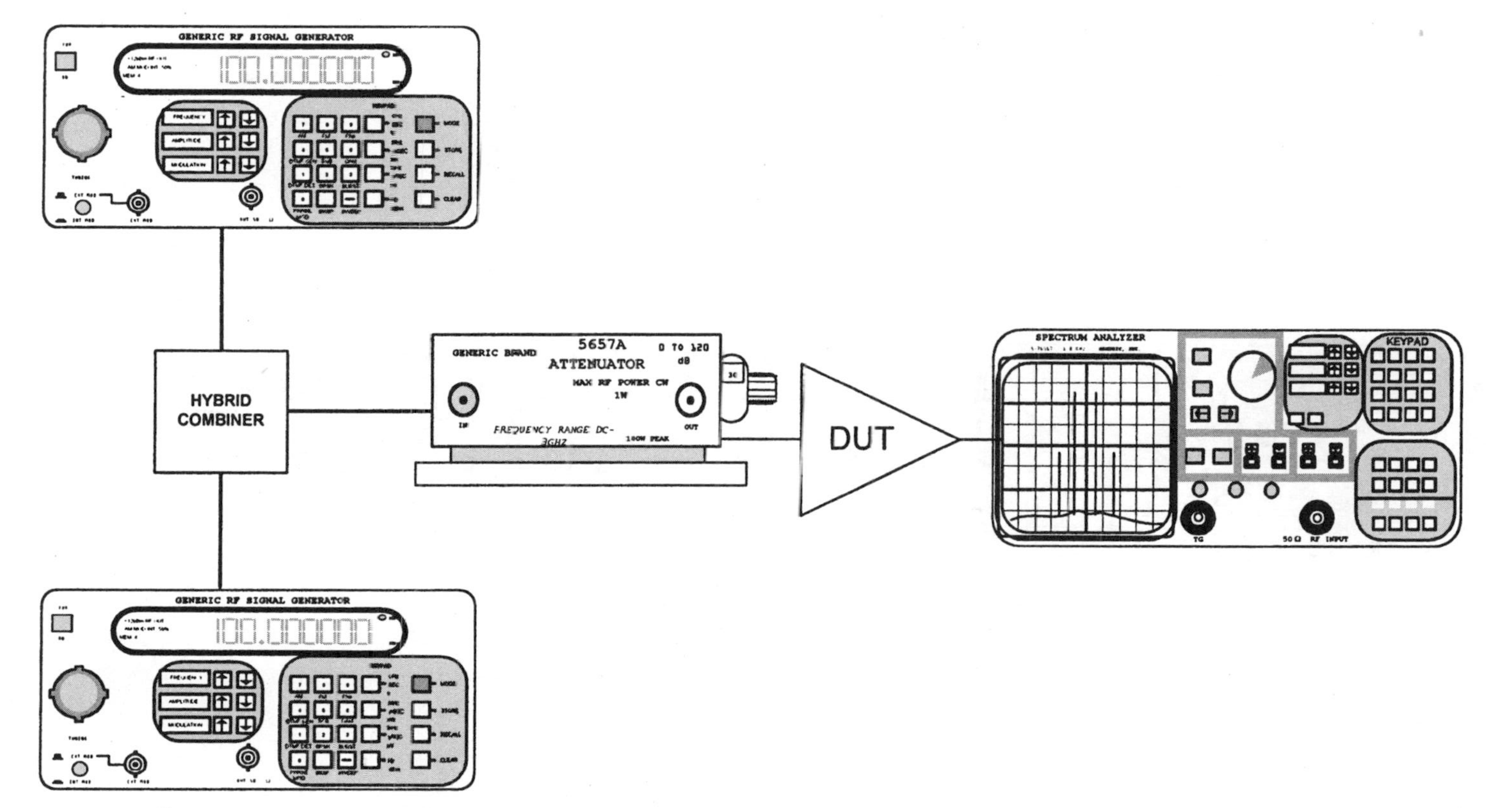

Figure 2.46 Setup to test two-tone third-order products of an amplifier or receiver.

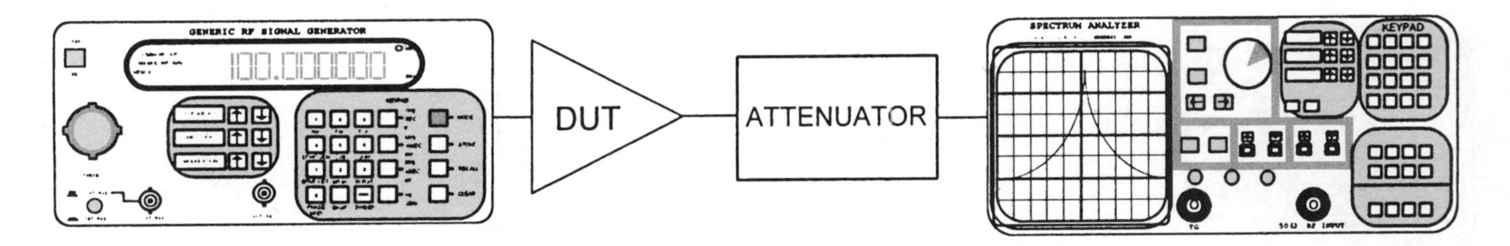

Figure 2.47 Setup to check 1-dB compression point of amplifier.

3. Take the amplitude of the center frequency f_{out} of the VCO, and the output level of the spurs, and subtract the two. This will be the level of the actual reference spurs in dBc.

Desensing test. A receiver must be tested for out-of-band signal rejection, since strong off-frequency interferers can desense the receiver's low noise amplifier (LNA) if its front-end filter is not sufficiently selective, causing BER problems. These powerful out-of-band signals can also cause IMD or mixing products to form in-band because of the overdriving of the LNA. The test is performed by combining two RF signal generators, making one generator the *desired* signal source by setting it to a center in-band frequency at −80 dBm while setting the other signal generator to a frequency at either the lower or upper band edge to function as an *undesired* out-of-band interferer at −20 dBm. Now, confirm that the gain for the desired signal does not decrease below specifications because of amplifier desensing by subtracting the input signal level in dB from the desired signal level output of the receiver.

Digital system test and measurement. To confirm proper system operation, the following are the *minimum* tests that should be performed on a digitally modulated receiver. Most of these tests require that a CW test signal be injected at the receiver's front end, at the center of one of its channels, and at below P1dB power levels, with the output taken from the receiver's last IF output stage. Some of these tests will require a vector network analyzer.

Receiver tests:

1. Gain, measured in dB.
2. Gain flatness across one channel, measured in dB (sweep entire channel with frequency generator).
3. Frequency accuracy after a certain warm-up period, measured in Hz.
4. Frequency stability over a certain temperature range after a specified period of warm-up, measured in Hz.
5. Frequency drift over a set time at 25°C, measured in Hz, from turn-on to full warm-up.
6. $P1dB_{OUT}$ at output, measured in dBm.
7. Phase noise of local oscillator at 10 kHz offset from carrier in dBc, measured in dBc/Hz/10 kHz.
8. Two-tone IMD at output, measured in dBc.
9. In-band internally generated spurs with no input signal, measured in dBm and Hz.
10. 3-dB bandwidth of a channel, measured in Hz.
11. Minimum discernible signal (MDS) with zero SNR, measured in dBm.

12. Carrier-to-noise ratio (CNR) at $P1dB_{INPUT}$, measured 250 kHz from the carrier with a spectrum analyzer (RBW, VBW, VID AVG, SPAN settings must be specified) at the receiver's output, measured in dBc.
13. CNR at *n* miles (using an adjustable attenuator between the CW signal generator and the receiver's front end for simulated free-space path loss) at the receiver's output, measured in dBc.
14. Signal level at *n* miles (using an adjustable attenuator between the CW signal generator and the receiver's front end for simulated free space path loss) at the receiver's output, measured in dBm.
15. VSWR at band center, a dimensionless ratio.
16. Group delay variations (GDV) across entire channel, measured in ns.
17. DC drawn from power supply, measured in mA.
18. Attenuation of potential interferers injected in certain out-of-band channels, and measured at the output in dB of attenuation.
19. BER testing with a BERT.

The following are the minimum tests that should be performed on a digitally modulated transmitter. Most of these tests require that a CW test signal be injected into the transmitter's input IF, at the center of one of its channels, and at below P1dB power levels, with the output taken from the transmitter's RF output.

Transmitter tests:

1. Gain from input IF to output RF, measured in dB.
2. Gain flatness across channel, measured in dB (sweep entire channel with frequency generator).
3. Frequency accuracy after warm-up, measured in Hz.
4. Frequency stability over a certain temperature range after a specified period of warm-up, measured in Hz.
5. Frequency drift over a set time at 25°C, measured in Hz, from turn-on to full warm-up.
6. P1dB output power, measured in dBm.
7. Phase noise of local oscillator at 10 kHz offset from carrier in dBc, measured in dBc/Hz/10 kHz.
8. Two-tone IMD at RF output, measured in dBc.
9. In-band internally generated spurs with no input signal, measured in dBm and Hz.
10. 3-dB bandwidth of channel, measured in Hz.
11. VSWR at band center, a dimensionless ratio.
12. Group delay variations (GDV) across entire channel, measured in ns.
13. LO feedthrough at transmitter output, measured in dBm.
14. DC drawn from power supply, measured in mA.
15. BER testing with a BERT.

Chapter

3

Amplifier Design

An amplifier is an active device that has the ability to amplify voltage, current, or both, at zero frequency (a DC amplifier), low frequencies (an audio amplifier), or high frequencies (an RF amplifier). Since power is $P = VI$, then power amplification is only a normal outcome of this capability, since raising the current and/or the voltage will create power amplification.

AC amplifiers, whether for high or low frequencies, operate by allowing a small fluctuating external input signal to control a much greater DC output bias current. This small input signal changes the amplitude of the larger bias current, with the varying bias current then sent through a high-value output impedance or resistance component, which creates an AC output voltage due to $V = IR$. Depending on the amplifier's designated purpose, these output components may be composed of either a resistor, inductor, or tuned circuit.

There are assorted circuit configurations to allow an amplifier to achieve different frequency responses, input and output impedances, gains, and phase shifts. Various bias circuits can be adopted to produce amplification at different efficiency and thermal stability levels, while special coupling methods can be applied to match impedances and filter out undesired frequencies with other stages or loads.

Amplifier circuit configurations. Amplifiers come in three different basic flavors, each with its own distinct application and capability. They are referred to as *common-base, common-collector,* and *common-emitter* amplifiers, depending on whether the base, collector, or emitter is common to both the input and output of the amplifier circuit.

With an input signal inserted at the emitter and the output taken from the collector circuit, we have our first configuration, the common-base amplifier (Fig. 3.1). The common-base configuration can be found operating as a voltage amplifier for low input impedance circuits. It also possesses a high output impedance and a power amplification due to $P = V^2/R$, but current gain will always be a little less than 1. However, even though the common-base amplifier has superior temperature stability and linearity, and can easily operate at very high frequencies, it is not nearly as common as the other two configurations, the common-emitter and the common-collector amplifier. This is due partly to the common-base amplifier's low input impedance (50 to 75 ohms), but these amplifiers can occasionally be found at the 50-ohm antenna input of a receiver, or sometimes as Class C high-frequency amplifiers.

The JFET version of the BJT common-base amplifier, a common-gate amplifier, can be seen in the IF section of receivers; one such circuit is shown in Fig. 3.2. C_2, C_3, R_2, and the RFC are for decoupling; C_4 and C_6 are for RF coupling; C_5 can be tweaked to obtain a flatter frequency response throughout its passband; T_1 is for matching of the low input impedance, as required.

The most popular amplifier circuit arrangement in all of electronics is the *common-emitter* type of Fig. 3.3. It has the greatest current and voltage gain

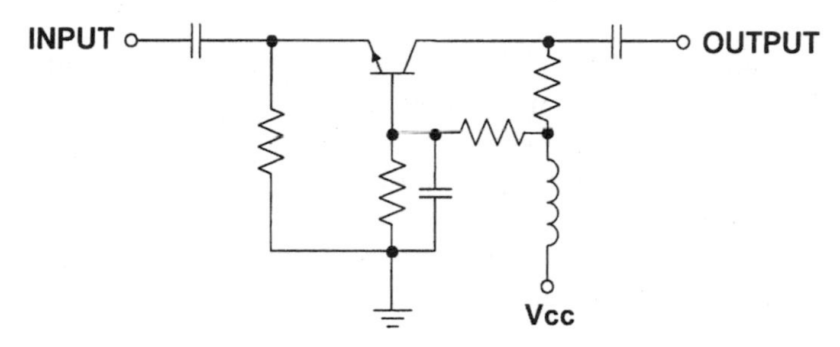

Figure 3.1 A basic common-base amplifier circuit.

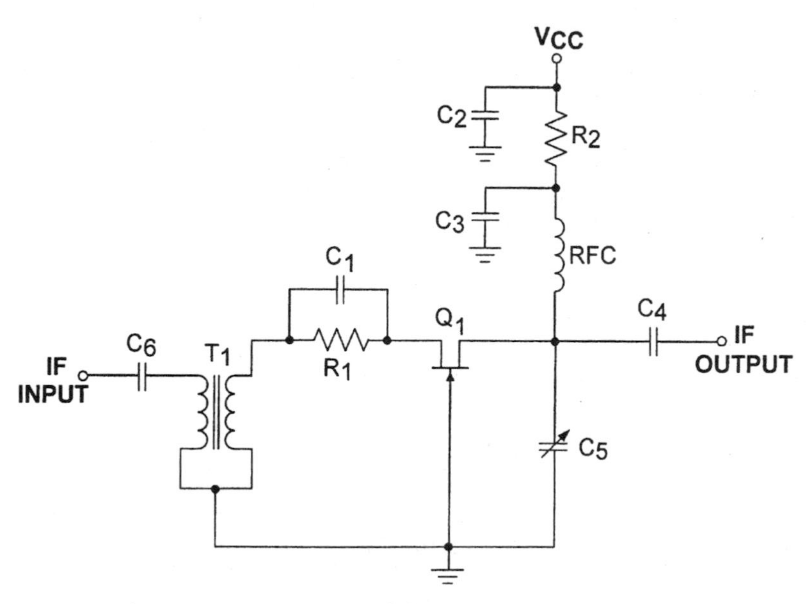

Figure 3.2 A common-gate JFET amplifier.

combination of any amplifier. In fact, common-emitter amplifier configurations are capable of increasing not only voltage and current, but will also make excellent power amplifiers, and have a medium-frequency response. The bias circuit displayed in the figure is only one of the many ways to bias common-emitter amplifiers (see Sec. 3.3, "Amplifier biasing").

The common-emitter amplifier functions as follows: When a signal is placed at the base of the active device (the transistor), an amplified output is extracted from the collector output circuit. The output voltage will have been shifted by 180 degrees in phase compared to the signal present at the amplifier's own input. This is due to the following action. As the signal at the transistor's base turns more positive, an increased current will flow through the transistor. This decreases the transistor's resistance, and thus the voltage that is dropped across its collector-emitter junction, or from the collector to ground. Because the output signal is taken from the voltage that is dropped across the transistor's collector—and the load resistor (R_C) is now dropping the voltage that was formerly available to the collector—a shift in the phase at the amplifier's output is created that is precisely the reverse to that of the input signal.

At RF, a large difficulty in CE amplifiers is an effect called *positive feedback,* which creates amplifier instability and oscillations due to the internal feedback capacitance between the transistor's collector and its base. The collector-to-base capacitance can be as high as 25 pF, or more, in certain types of bipolar transistors. At a certain frequency, this capacitance will send an in-phase signal back into the base input from the collector's output, which will create, for all intents and purposes, an oscillator. To give birth to these oscillations, however, something has to produce a shift in phase, since the CE amplifier already possesses a phase shift from its input to its output of 180 degrees, which would only cause a decrease in the input signal strength (or *degeneration*) if fed back to the BJT's input port. In fact, the internal capacitance and resistance of the transistor, along with other phase delays, can yield a powerful phase shift to

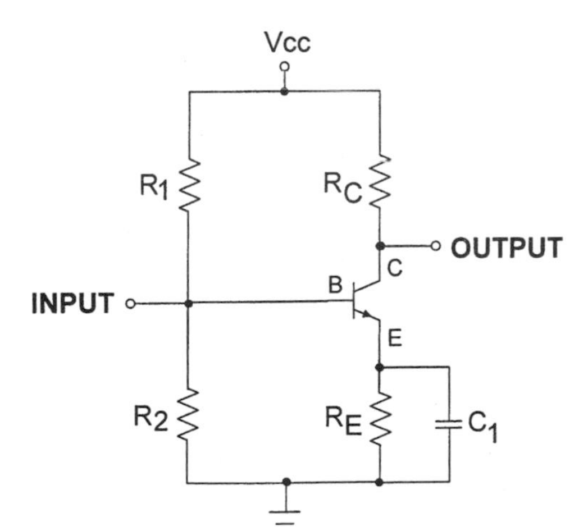

Figure 3.3 A low-frequency type of common-emitter amplifier.

this normally out-of-phase feedback signal. But only those phase delays that are near 180 degrees—furnishing the common-emitter with 0 degrees positive feedback at its base—will bring about the undesired amplifier instability and oscillations. Figure 3.4 illustrates a typical phase-versus-frequency response of a certain single-stage common-emitter amplifier.

Figure 3.5 demonstrates the basic bias circuit configuration of a *common-collector* (CC) *amplifier* (also called an *emitter-follower*). The CC amplifier has the input signal inserted into its base, and the output signal removed from its emitter; which gives a current and power gain, but has a voltage gain of less than 1. This amplifier is used because of its high input impedance and low output impedance, making it beneficial as a buffer amplifier or as an active impedance-matching circuit.

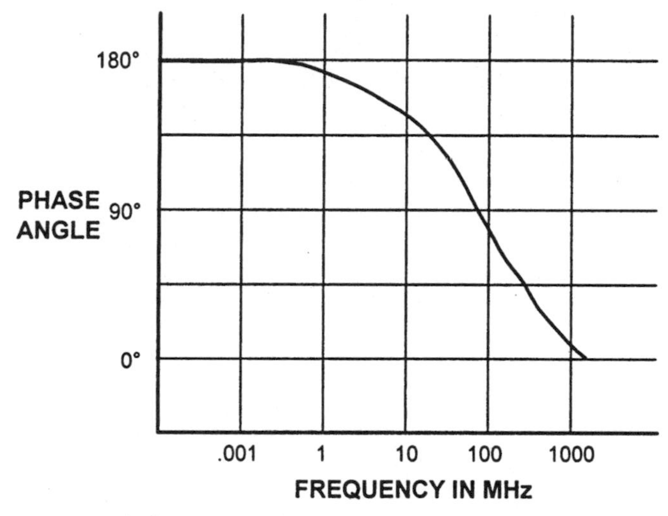

Figure 3.4 A phase-versus-frequency graph for a common-emitter amplifier circuit.

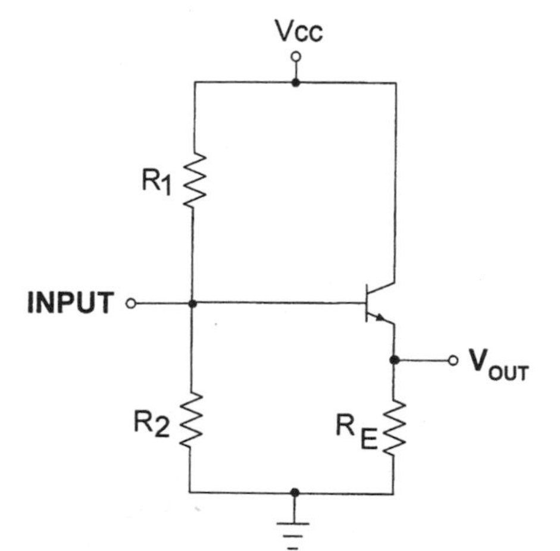

Figure 3.5 A typical common-collector amplifier circuit.

Unlike the common-emitter amplifier, the CC amplifier has no phase inversion between its input and output, since the current through the active device will increase as the input signal to the transistor's base rises in amplitude. This action forces a rise in the current through the emitter resistor, which increases the voltage drop across R_E, resulting in a 0 degree phase shift.

Most common-collector amplifiers do not possess any voltage-robbing collector resistor, nor do they use an R_E bypass capacitor, which would also lower the output voltage at V_{OUT}.

Matching networks. There are numerous matching networks that can be employed to facilitate impedance matching and coupling, and supply some filtering (normally of the low-pass variety) between RF stages as well. Matching allows the maximum power transfer and the attenuation of harmonics to be achieved between stages. Using one of the various topologies of *LC* circuits within a matching network is far less expensive, and can reach far higher frequencies, than the lumped transformer matching that was so popular in the past.

One of the most common *LC* matching topologies, especially for narrowband impedance matching, is the simple *L network,* which can also furnish low-pass (as well as high-pass) filtering to decrease any harmonic output. The low-pass *L* network in Fig. 3.6 is capable of matching a high output impedance source to a low input impedance load. The low-pass *L* network of Fig. 3.7 matches a low output impedance source to a high input impedance load.

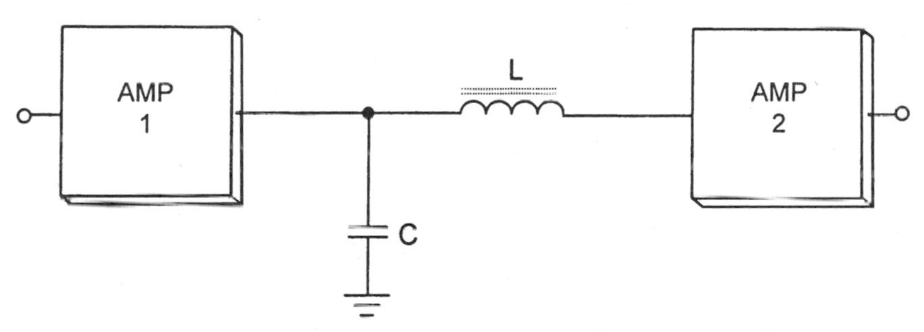

Figure 3.6 A high-to-low impedance-matching *L* network between two amplifiers.

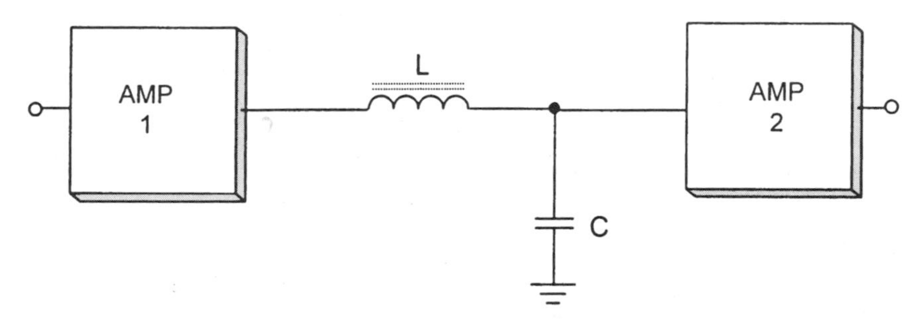

Figure 3.7 A low-to-high impedance-matching *L* network between two amplifiers.

The Q of the L network, which must be low for broadband circuits to obtain a wide bandwidth, and for power amplifiers to minimize high circulating currents and thus high losses, is not selectable, so its usage is limited in such devices.

The *T network* (Fig. 3.8) is another quite popular impedance-matching network; it can be designed to furnish almost any impedance-matching level between stages and a selectable and low Q for wideband and power amplifier use.

Pi networks are chosen for the same reason as the T network, and are found extensively in matching applications of all types. The pi network and its equivalent circuit are displayed in Fig. 3.9. By altering the ratio between capacitors C_1 and C_2 the output impedance of the load can be matched to the source, as well as decreasing the harmonic output. And while the pi network is a low-pass filter, it can have a small resonant peak and a high return loss at this point, at a certain frequency where the loss of the circuit decreases below that of its bandpass value (if the Q is high enough; Fig. 3.10).

Distortion and noise. Two unavoidable, but distinctly undesirable, parts of any circuit are distortion and noise. Distortion can deform the carrier and its sidebands at the transmitter or receiver, causing spectral regrowth and adjacent

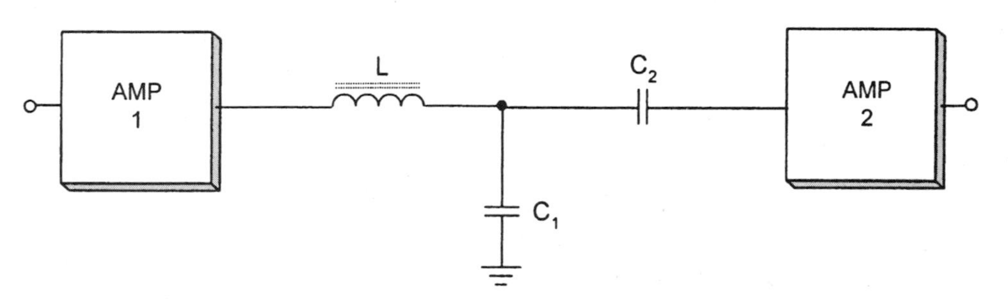

Figure 3.8 The impedance-matching *T* network.

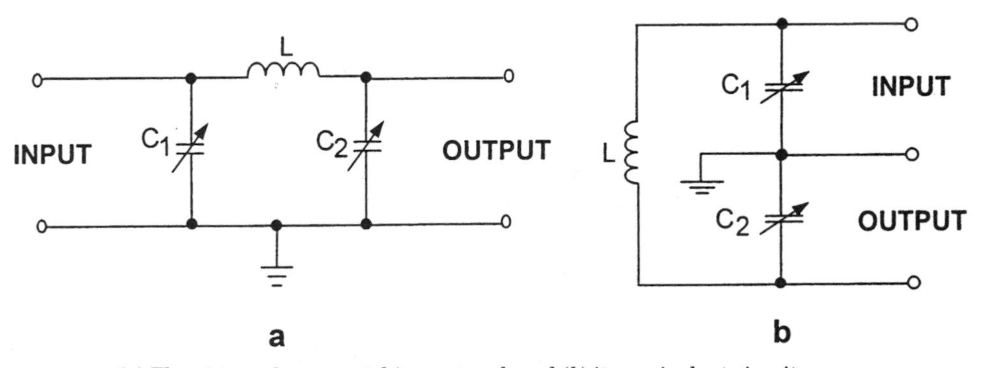

Figure 3.9 (*a*) The pi impedance-matching network and (*b*) its equivalent circuit.

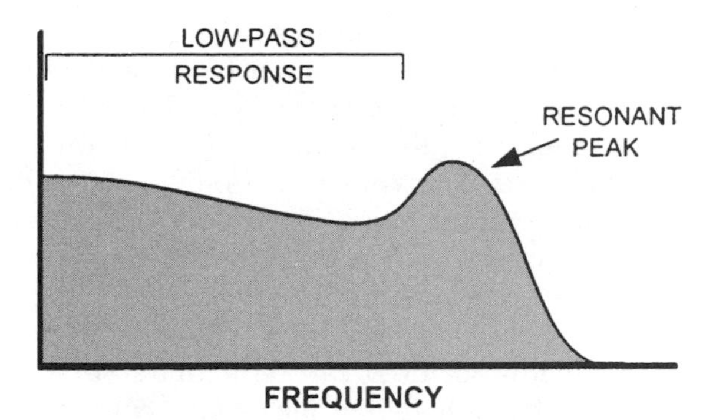

Figure 3.10 The frequency response of the pi network with resonant peak.

channel interference, as well as a faulty, distorted replica of the original baseband signal. This increases the BER, as will any noise, which can be contributed to the system from almost any internal or external source.

Distortion. Distortion can form frequency intermodulation products by the internal nonlinear mixing of any single signal with one or more other frequencies, or create harmonic distortion products when only one frequency is present. Distortion may have different causes, but comparable results: a modulated or unmodulated waveform that is altered in shape or amplitude from the original signal due to improper circuit response.

Frequency distortion will result from circuits that increase or decrease the amplitude of different frequencies better than others. This is normally only a serious problem in IF or RF amplifiers if they are pushed to their extreme frequency limits, since the active device in such a circuit imposes its upper—and many times its lower—frequency limits. Thus, by the inability of a transistor to function properly at higher frequencies (its gain decreases), induced by *transit time* problems and the negative effects of junction capacitance, the frequency distortion will be exacerbated. But since RF transistor circuits in general are matched, filtered, coupled, and decoupled by reactive components—which are frequency dependent—such a reactive circuit will act as a bandpass, bandstop, high-pass, or low-pass filter, thus altering the expected frequency response of the system if improperly designed or tuned.

Amplitude distortion, a form of *nonlinear distortion,* can be produced by unsuitable biasing of an amplifier, causing either saturation or cutoff of the transistor. This is extremely nonlinear behavior, and generates harmonics and intermodulation distortion (IMD) products. Overdriving the input of any amplifier (*overload distortion*) will create the same effect, called *flattopping,* whether the bias is correct or not; creating both saturation and cutoff conditions. The harmonics and IMD so generated will create interference to other services and/or to adjacent channels and increase the system BER in a digital data radio, while a voiceband device will have an output signal with a harsh, coarse output.

Intermodulation distortion (Fig. 3.11), quite similar to *amplitude distortion* above, is produced when frequencies that are not harmonically related to the fundamental are created through nonlinearities in a linear Class A or a nonlinear Class C amplifier, or in a nonlinear mixer's output. IMD is formed by this mixing together of the carrier, any harmonics, the sidebands, IMD from other stages, etc., to produce various spurious responses—both in and out of band. Since these IMD products can fall in band, or cause other signals to fall in band, they can possibly swamp out the desired baseband signal, creating interference, which also causes additional noise which will degrade system performance and BER. In addition, IMD can be manufactured in the power output amplifier of a transmitter when another neighboring transmitter's signal (and/or its harmonics) arrives at its output stage and mixes. This can be particularly problematic in dense urban environments, as there are many signals present that will modulate each other within the nonlinearities of a normal power amplifier, producing a multitude of sum-and-difference frequencies. In these transmitter-to-transmitter cases, the IMD can be attenuated by employing a wavetrap that is tuned to the interfering transmitter's frequency, and/or by shielding and proper grounding to prevent mixing within the other internal stages of the transmitter. However, within a receiver this effect can be much worse: The desired signal and a close transmitter's undesired signal, and/or its harmonics, can be allowed into the receiver's front end, creating reception of unwanted signals and the obliteration of the desired frequency by the IMD products generated by the nonlinear mixing of the two signals. This can be somewhat mitigated by using, at the receiver, an input notch filter, tighter bandpass filtering, amplifiers that are biased for maximum linearity, and confirming that the RF amplifiers are not functioning in a nonlinear region as a result of being overdriven by an input signal.

A more in-depth explanation of "intermod" is warranted because of its vital importance in the design of any linear amplifier. Since intermodulation distortion is produced when two or more frequencies mix in any nonlinear device, this causes not only numerous sum and difference combinations of the original fundamental frequencies (*second-order products:* $f_1 + f_2$ and $f_1 - f_2$), but also intermodulation products of $mf_1 + nf_2$ and $mf_1 - nf_2$, in which m and n are whole numbers. In fact, *third-order intermodulation distortion products,* which would be $2f_1 + f_2$, $2f_1 - f_2$, $2f_2 + f_1$, and $2f_2 - f_1$, can be the most damaging intermodulation products of any of the higher or lower IMD. This is because the *second-order IMD* products would usually be too far from the receiver's or transmitter's pass band to create many problems, and would be strongly attenuated by an amplifier's tuned circuits, the system's filters, and the selectivity of the antenna. As an example: Two desired input signals to a receiver, one at 10.7 MHz and the other at 10.9 MHz, would produce sum and difference second-order frequencies at both 21.6 MHz and 0.2 MHz. These frequencies would be far from the actual passband of the receiver, and will be rejected by the receiver's selectivity. But the third-order IMD formed from these same two signals would be at 10.5 MHz, 11.1 MHz, 32.3 MHz, and 32.5

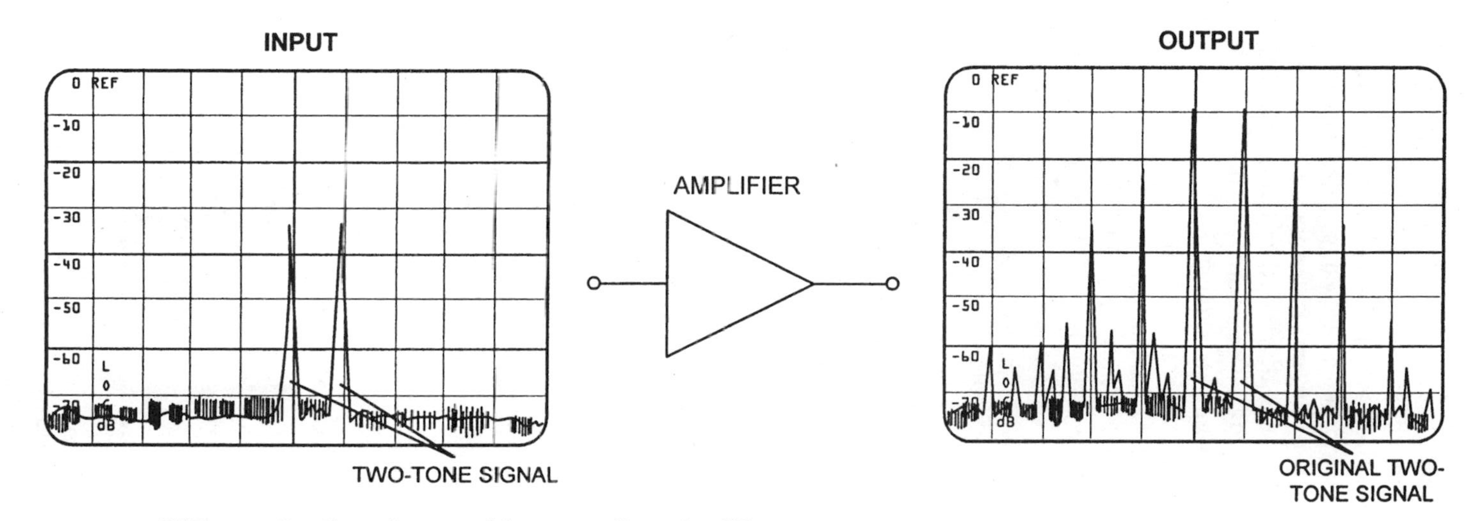

Figure 3.11 IMD generation through an overdriven or nonlinear amplifier.

MHz, with the most destructive frequencies being, of course, at 10.5 MHz and 11.1 MHz. This is well within the passband of this particular receiver. Much higher order IMD is created in receivers and amplifiers, so all IMD up to the seventh order should be accounted for and, if it does fall within band, must be at such a low amplitude that it cannot cause problems.

A low return loss (a high VSWR) can also create IMD in an amplifier or mixer stage because of the reflected waves from the next stage returning and mixing with the output and its sidebands.

Harmonic distortion occurs when an RF fundamental sine wave (f_r) is distorted by nonlinearities within a circuit, generating undesired harmonically related frequencies ($2 \times f_r$, $3 \times f_r$, etc.). Interference to receivers tuned to many megahertz, or even gigahertz, away from the transmitter's output frequency is possible when these harmonics are broadcast into space (Fig. 3.12). The dominant cause of transmitted harmonics is overdriving a poorly filtered power amplifier, with an extreme case of distortion resulting in the sine wave carrier changing into a rough square wave. These *nonperfect* square waves contain not only the fundamental frequency, but numerous odd harmonics, as well as a certain amount of even harmonics.

No amplifier can be completely linear, so a number of harmonics are inevitably produced within all amplifiers, and they must be attenuated as much as possible—especially in a transmitter.

Noise. There are two principal classifications of noise: *circuit generated* and *externally generated.* Both limit the possible sensitivity and gain of a receiver, and are unavoidable—but can be minimized.

Circuit noise creates a randomly changing and wide-frequency-ranging voltage. There are two main causes: *white noise,* created by a component's electrons randomly moving around by thermal energy (heat); and *shot noise,* caused by electrons randomly moving across a semiconductor junction and into the collector or drain of a transistor.

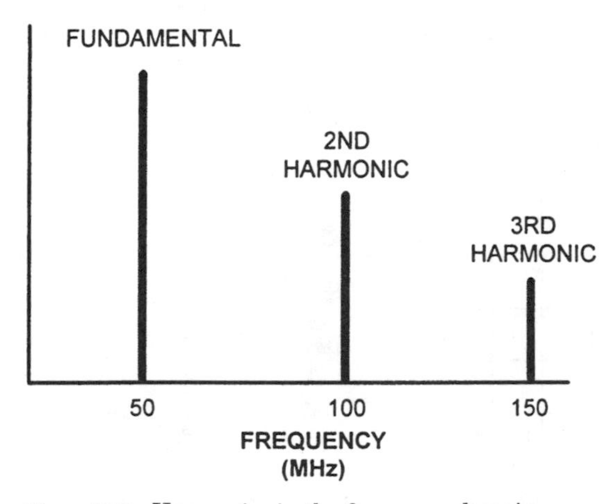

Figure 3.12 Harmonics in the frequency domain.

External noise, produced by atmospheric upheavals like lightning, as well as *space noise* caused by sunspots and solar flares and *cosmic noise* created by interfering signals from the stars, is exacerbated by man-made electromagnetic noise sources such as dimmer switches, neon lights, car ignitions, and electric motors.

Amplifier design considerations. Since the data sheet is the dominant source of information circuit designers have for selecting an active device for their own specialized applications, it is especially important to understand data sheet parameters as they apply to RF transistors.

Check out the device's data sheet with your design to confirm that current, voltage, and power limitations will not be exceeded in your wireless application; whether in a nonsignal DC condition or under a maximum signal situation. Obviously, the transistor's f_T, P1dB, and $G_{A(MAX)}$ (maximum available gain) and, for low-noise amplifier (LNA) applications, the NF, are all vital specifications.

The near maximum output power possible in an amplifier is the 1-dB compression point (P1dB). This is the area where a linear amplifier begins to run out of room for its maximum output voltage swing. Any amplifier will have what is generally considered as a linear P_{OUT} until it reaches this P1dB point, which occurs when a high enough input signal is injected into the amplifier's input. At P1dB, the gain of the amplifier will depart from the gain displayed at lower input powers (Fig. 3.13), and for every decibel placed at the amplifier's input, no longer will there be a linear amplification of the signal. The output gain slope flattens, and soon no significant increase in output power is possible.

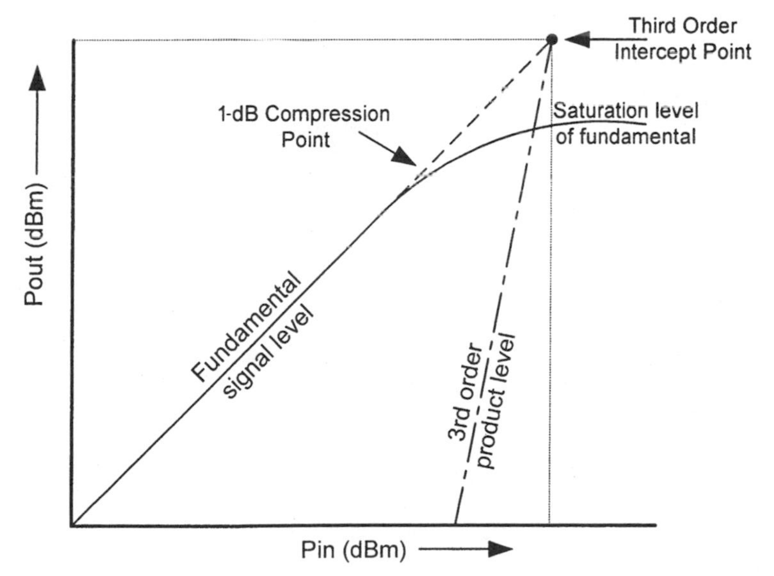

Figure 3.13 The third-order intercept and 1 dB compression points.

When an amplifier is below its P1dB, then for every 1-dB increase in fundamental power into the amplifier, the output second-order products will increase by 2 dB, while the output third-order products will increase by 3 dB. The reverse is also true: For every 1 dB *decrease* in the fundamental input power, the second and third orders decrease in power by 2 and 3 dB, respectively. However, by increasing the desired input signals, there will reach some point where the third-order products must be (theoretically) equal to the fundamental outputs. This is the *third-order intercept point* (TOIP).

The third-order intercept point is approximately 10 to 15 dB *above* the P1dB compression point. The TOIP is the point where, when two different (but closely spaced in frequency) input signals are placed at the amplifier's input port, the *undesired* output third-order products will be at the same amplitude as the *desired* two-tone fundamental input signals. However, the output TOIP itself can never actually be reached. This is because the amplifier will go into saturation before this amplitude is ever truly attained. Even though Fig. 3.13 does not show it, the third-order product's output power will gain-limit, just as the fundamental signal must, when the amplifier goes into saturation.

Another significant amplifier design consideration, especially important in VHF and above in any gain block, is excessive source or emitter inductance. This can create instability (possible oscillatory behavior), as well as gain peaking (Fig. 3.14), and is produced by using an emitter resistor in the amplifier design. It is made worse by the addition of the emitter resistor's own bypass capacitor, long emitter leads, and even long vias to ground (even SMD chip capacitors can have 1 nH of inductance, which can fatally disrupt some amplifiers).

The importance of a good impedance match from amplifier stage to amplifier stage can readily be seen by inspecting the formula below. Any impedance mismatches will end in a loss of power, referred to as *mismatch loss* (ML), and can readily be calculated by:

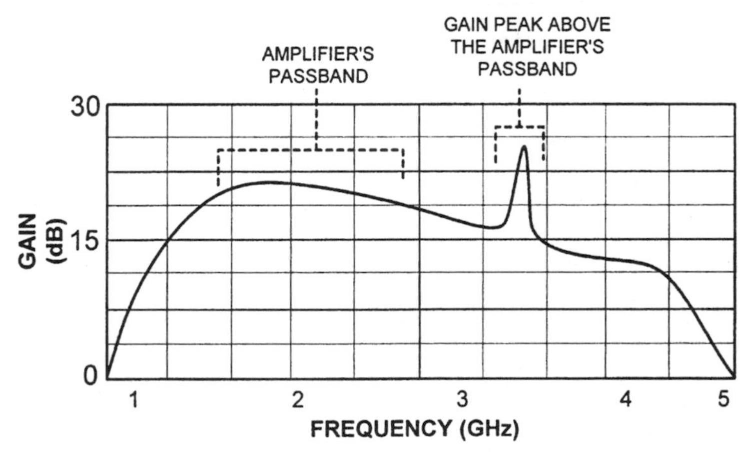

Figure 3.14 Gain peaking in an amplifier's response (*above* its passband), causing general instability.

$$\text{ML} = 10 \log_{10} \left[1 - \left(\frac{\text{VSWR} - 1}{\text{VSWR} + 1} \right) \right]^2$$

where ML = mismatch loss, dB, and VSWR = voltage standing wave ratio, dimensionless units.

Amplifier efficiency is another meaningful specification in many applications. The efficiency of an amplifier is the percentage of the *RF output* power compared to the *RF and DC input* power, and can easily be calculated by:

$$\text{Eff} = \frac{P_{out}}{P_{IN} + P_{DC}} \times 100$$

where Eff = efficiency of the amplifier, %
P_{OUT} = RF output power, W
P_{IN} = RF input power, W
P_{DC} = power supplied to the amplifier by the DC bias, W

As a useful aside: In amplifier design, the desirable specifications, such as a high P1dB, low noise, high efficiency, good gain flatness, proper wideband operation, high gain, and high return loss can frequently be in opposition with each other because of real-life internal transistor design limitations.

3.1 Small-Signal Amplifiers

3.1.1 Introduction

Small-signal amplifiers are needed to increase the tiny signal levels found at the input of a receiver into usable levels for the receiver's detector, or into the proper levels required of the final power amplifier of a transmitter. These amplifiers are Class A or AB for linear operation, high sensitivity, and low distortion in digital, AM, and SSB systems.

A receiver's first RF amplifier will be of the small-signal, high-gain type and must not produce excessive noise, since any noise generated within this first stage will be highly amplified by later stages, decreasing the SNR.

Because of the high operating frequencies, RF amplifiers may sometimes be *neutralized* in order to counteract any possible positive feedback and its resultant self-oscillations. However, designing with a transistor that has unconditional stability at the frequency and impedance of operation has now become much more prevalent.

The voltage gain of the small signal amplifier can be calculated as V_{OUT}/V_{IN}, and when two or more are cascaded, their voltage gain is multiplied. However, the decibel is more frequently used, with these values simply added, or dB + dB, when stages are cascaded.

There are four vital considerations in any discrete RF amplifier design: the choice of the active device, the input and output impedance-matching network, the bias circuit, and the physical layout. Each of these will be discussed in detail.

3.1.2 Amplifier design with *S* parameters

S parameters characterize any RF device's behavior at different frequencies and bias points. With the information that *S* parameters supply, the designer can calculate a device's gain, return loss, stability, reverse isolation, and its input and output impedances. Knowing the transistor's port impedances is required so that the necessary matching network can easily be designed for a proper impedance match from stage to stage. This matching is vital so that maximum power is delivered to the load at a high return loss (minimum power reflected back toward the source from the load).

Impedance matching of active devices is essential because not only will the typical transistor not have a 50-ohm resistive Z_{IN} and Z_{OUT}, but its reactances will also vary over frequency. This means that for maximum power transfer into the system's impedance, which is normally 50 ohms, a matching network must be used to match the active device to the system's impedance, and sometimes over a wide band of frequencies. However, utilizing *LC* components is the dominant matching technique, so the match will be perfect only over a very narrow band of frequencies. There are, nevertheless, techniques for impedance matching that work quite well over a very wide band of frequencies, and these will be discussed. For further information, consult the influential work on practical amplifier design *RF Circuit Design* by Chris Bowick.

As mentioned above, a device that must be matched will normally *not* be at 50 ohms resistive and furthermore will be either inductive or capacitive. This combination resistive and reactive elements in the active device's Z_{IN} and Z_{OUT} is referred to as a *complex impedance.* So the matching network's job is not only to match the active device to the system's impedance, but also to cancel the reactive element to allow for a $50 + j0$ match (or 50 ohms resistive, with no capacitive or inductive reactances). This is called *conjugate matching,* and supplies a perfect impedance match. Nevertheless, in order to decrease the gain of a transistor at various desired frequencies for gain flattening, or to purposely design an amplifier with less gain within its bandpass, as well as for optimal noise figure (NF), a perfect match may not be desired for certain amplifier applications.

It is usually advisable that all circuits, even discrete circuits in the middle of an IF chain, have a Z_{IN} and Z_{OUT} of 50 ohms. Although we could match at any sensible impedance from discrete stage to discrete stage, it would make it quite difficult to perform accurate interstage tests with the 50-ohm test gear commonly available. Thus, after each stage is tested, a 50-ohm circuit can then be confidently placed within the system for reliable cascaded operation.

When designing matching networks, we will take the *S*-parameter 2-port representation of the transistor, ignoring any effect the DC biasing network may have on these parameters in the final design. This is quite valid if only small amounts of RF feedback are produced by high values of R_f (Fig. 3.15; the feedback resistor) in an amplifier's bias network. In this way, the *S* parameters will be satisfactory for computing not only the matching networks, but also for the software simulations of the circuit's responses. However, if an amplifier utilizes a low value resistor for R_f in order to employ heavy feedback,

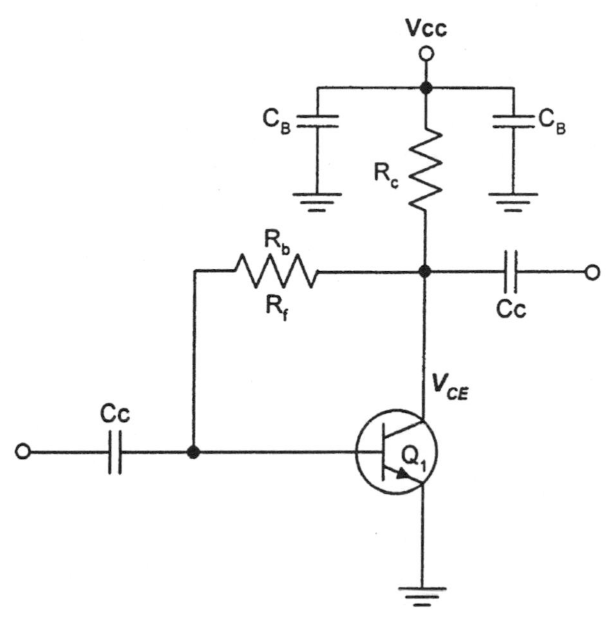

Figure 3.15 Collector feedback bias for a BJT.

then the S parameters may no longer be completely valid. In other words, the amplifier circuit's attributes, as defined by the device's S parameters, are accurate only when the bias network employs the normal, high value of bias resistors within its bias network.

S-parameter files (or *.S2P; Fig. 3.16) contain only the parameters for a few frequencies (usually not more than 20), so when reading S parameters on data sheets, or in the *.S2P text files themselves, we may find that our frequency of interest falls between two values. For accuracy, we will take the *mean* value between the two closest frequencies. As an example: S parameters are given in a certain *.S2P file for 3 GHz and 4 GHz, but our design requires a center frequency of 3.5 GHz. Take the mean value of each S parameter at 3 GHz and 4 GHz. To compute S_{12} at 3.5 GHz:

$$\frac{S_{12}\text{MAG (@ 3 GHz)} + S_{12}\text{MAG (@ 4 GHz)}}{2} = S_{12}\text{MAG (@ 3.5 GHz)}$$

and

$$\frac{S_{12}\theta\text{ (@ 3 GHz)} + S_{12}\theta\text{ (@ 4 GHz)}}{2} = S_{12}\theta\text{ (@ 3.5 GHz)}$$

The simple analysis of S parameters will furnish the designer with a lot of data on the active device of interest, such as three different ranges of possible gains: the *maximum available gain* (MAG) that the transistor can attain when perfectly

```
! MANUFACTURER'S NAME (Small Signal Semiconductors)
! MODEL OF TRANSISTOR
! TYPE OF TRANSISTOR AND PACKAGE
! S-PARAMETERS at Vce=3.0 V  Ic=50 mA.   LAST UPDATED 08-05-00
! Common Emitter S-Parameters:
# GHz  S  MA  R  50
!  f          S11              S21              S12              S22
! GHz      MAG    ANG      MAG    ANG      MAG    ANG      MAG    ANG
 0.010 0.1968    0.2    8.019  178.9  0.0777    0.6  0.3276   -0.5
 0.020 0.1967    0.2    8.070  178.3  0.0777    1.0  0.3266   -1.2
 0.030 0.1975   -1.4    8.032  177.5  0.0777    1.4  0.3270   -2.4
 0.050 0.1926   -2.3    8.027  175.8  0.0777    2.5  0.3251   -4.3
 0.100 0.1938   -4.8    8.002  171.6  0.0781    4.9  0.3245   -8.6
 0.150 0.1953   -6.2    7.960  167.6  0.0789    7.3  0.3231  -12.7
 0.200 0.1920   -9.6    7.924  163.4  0.0799    9.5  0.3188  -16.8
 0.250 0.1923  -12.0    7.824  159.4  0.0809   11.9  0.3177  -21.2
 0.300 0.1947  -14.4    7.753  155.4  0.0823   13.8  0.3126  -25.0
 0.400 0.1931  -19.5    7.528  147.5  0.0859   17.8  0.3047  -33.3
 0.500 0.1912  -25.9    7.284  139.9  0.0898   21.1  0.2948  -41.2
 0.600 0.1895  -31.9    7.019  132.7  0.0945   23.9  0.2828  -48.6
 0.700 0.1885  -37.7    6.707  125.8  0.0999   26.0  0.2719  -55.9
 0.800 0.1850  -45.0    6.437  119.1  0.1054   27.9  0.2603  -62.9
 0.900 0.1787  -52.4    6.124  112.8  0.1114   29.2  0.2491  -69.8
 1.000 0.1753  -60.3    5.836  106.8  0.1175   30.2  0.2389  -76.2
```

Figure 3.16 *.S2P *S*-parameter file for set bias conditions and set frequencies. Lines preceded by ! are comments for the user, and are ignored by simulation programs.

matched (MAG is considered a figure of merit only); *transducer gain,* which is the true gain of an amplifier stage, together with the effects of impedance matching and device gain—but not including power lost within real-world components; and *transducer unilateral gain,* which is the dB measurement of an amplifier's power gain into an *unmatched* 50-ohm load—a worst-case gain evaluation.

Another very meaningful piece of information that *S* parameters are easily able to reveal is whether the active device will remain stable under *any* impedance presented at its input or output port; or whether the device may begin to oscillate at some impedance combination. Stability calculations using *S* parameters yield the *Rollet stability factor,* or simply *K*. Any transistor with a *K* of over 1 will be *unconditionally stable* at the particular frequency and DC bias point chosen for the transistor—with *any* input and output impedance it may be presented with. In other words, it will never begin to oscillate under (almost) any circumstance. However, if the value of *K* is under 1, then there will be some value of input and output impedance that will cause the amplifier to become *potentially unstable.* In other words, the amplifier may begin to oscillate. What values of impedance will cause this instability will not be disclosed by the formula. This will be discussed in more detail in the pages that follow.

Matching and gain. In order to begin the design of any amplifier, we should first discover whether the active device we have chosen will remain stable at our frequency and bias of interest over all impedance variations. This stability should also be maintained over a very wide region of frequencies, both low and high, for wide-ranging stability.

The K stability formula is

$$K = \frac{1 + (|D_S|^2 - |S_{11}|^2 - |S_{22}|^2)}{2|S_{21}||S_{12}|}$$

where

$$D_S = S_{11}S_{22} - S_{12}S_{21}$$

As explained below, vector algebra is used to calculate D_S, while scalar quantities (magnitudes only) are used to calculate K.

Thus, if $K > 1$, then the active device will be unconditionally stable for all Z_{IN} and Z_{OUT} presented at its ports. This is by far the easiest transistor to design an amplifier with. But if $K < 1$, then the device is potentially unstable. If this is so, Z_{IN} and Z_{OUT} must be very cautiously selected. Alternatively, you can pick a different active device with a $K > 1$, or opt for another transistor bias point that will give a $K > 1$, or use a neutralizing circuit, or place a low-value resistor at the amplifier's input (to decrease gain).

The following is an example of how to rapidly calculate whether a chosen transistor will be stable at 1.5 GHz, with a $V_{CE} = 10$ V and an $I_C = 6$ mA.

1. The S parameters at that particular frequency and bias point are found to be (by looking at its S parameter file):

$$S_{11} = 0.195\angle 167.6°$$

$$S_{22} = 0.508\angle -32°$$

$$S_{12} = 0.139\angle 61.2°$$

$$S_{21} = 2.5\angle 62.4°$$

2. First calculate D_S*:

$$D_S = (0.195\angle 167.6° \times 0.508\angle -32°) - (0.139\angle 61.2° \times 2.5\angle 62.4°) = 0.25\angle -61.4°$$

3. Then calculate K†:

$$K = \frac{1 + |0.25|^2 - |0.195|^2 - |0.508|^2}{2|2.5|0.139|} = 1.1$$

---

*Use *full* vector algebra ($Z\angle \pm 0°$) in S-parameter calculations (for example, $S_{11} = 0.35\angle -45°$). How to multiply, subtract, divide, and add vectors is explained below.

†Do *not* use full vector algebra; employ only the S-parameters' magnitudes (for example, $S_{11} = 0.35$). $|S_{11}|$ means to ignore the *sign* of the magnitude, and always make it positive.

Since K is greater than 1, we see that we have a stable transistor at 1.5 GHz with the transistor's bias conditions as stated in the S-parameter file. For ease of design, it is always recommended that, if at all possible, we exploit only unconditionally stable transistors in our amplifier circuits.

For the calculation of maximum available gain to be valid, K must be greater than 1, or unconditionally stable. Thus, if K is over 1 for a transistor, we can proceed to the MAG calculation to see if the transistor will give us the gain value we desire. MAG is, of course, never attained in practice, so only when the MAG is 20 percent or more above our required gain would we want to work with that particular transistor.

To calculate the MAG of the transistor, or the maximum gain that the transistor can attain when *perfectly* matched:

1. Calculate

$$B_1 = 1 + |S_{11}|^2 - |S_{22}|^2 - |D_S|^2$$

2. B_1 determines whether + or − will be adopted in the MAG equation in step 3. If B_1 returns a negative answer, use the + sign after K; if B_1 is positive, utilize the − (negative) sign after K.

3. $\text{MAG} = 10 \log \dfrac{|S_{21}|}{|S_{12}|} + 10 \log (|K \pm \sqrt{K^2 - 1}|)$

In the equations for B_1 and MAG, do *not* use full vector algebra; employ only the S-parameters' magnitudes (for example, $S_{11} = 0.35$). $|S_{11}|$ means to ignore the *sign* of the magnitude, and always make it positive.

As an example of a MAG calculation:

1. $B_1 = 1 + |0.195|^2 - |0.508|^2 - |0.25|^2 = +0.717$
2. Since B_1 has returned a positive number, the sign after K (1.1) in the equation below will be negative.
3. Complete for MAG:

$$\text{MAG} = 10 \log \frac{|2.5|}{|0.139|} + 10 \log (|1.1 - \sqrt{1.1^2 - 1}|)$$

$$= 12.56 + (-1.92) = 10.63\text{dB}$$

Thus, the amplifier will supply a maximum available gain of 10.63 dB.

After finding that the transistor has a K greater than 1 at our desired frequency, and with a MAG greater than 20 percent of that required for our application, the actual Z_{IN} and Z_{OUT} of the transistor can then be calculated. These impedance calculations will take into account the *reflected impedances* caused by S_{12}, the transistor's value of isolation in the reverse direction, since only if S_{12} has a value of zero will it have no effect on the transistor's individual Z_{IN}

and Z_{OUT}; however, this is never the case. So we will want to perform a *simultaneous conjugate match* to prevent the matching of the input port from changing the matching of the output port, and vice versa.

To begin the calculation of the transistor's input and output impedances:

1. Determine the value of C_2, which is used in one of the following equations, by:

$$C_2 = S_{22} - (D_S S_{11}^{\Psi}) \qquad D_S = S_{11} S_{22} - S_{12} S_{21}$$

(S_{11}^{Ψ} equals the complex conjugate of S_{11}. In other words, just change the sign of the *angle,* but not the *magnitude's* sign, of the S_{11} value (for example, $S_{11} = +12 \angle +18°$, so $S_{11}^{\Psi} = +12\angle -18°$). C_2 and D_S are calculated as vectors.

2. Calculate B_2, which is also used in one of the following equations:

$$B_2 = 1 + |S_{22}|^2 - |S_{11}|^2 - |D_S|^2$$

3. Then calculate the *magnitude* of the load reflection coefficient (Γ_L), which is the value of the impedance that the transistor must see at its output to be perfectly matched:

$$|\Gamma_L| = \frac{B_2 \pm \sqrt{|B_2|^2 - 4|C_2|^2}}{2|C_2|}$$

Note that the sign used for $B_2 \pm$ is the opposite of that obtained in the B_2 calculation of step 2. This formula supplies magnitude.

4. The angle is the same as that calculated in step 1 for the C_2 angle, but simply reverse that answer's sign.

Now, to calculate the output impedance of the transistor (Z_{OUT}). Use the following formula [all signs (±) must be strictly maintained for all calculated numbers. $1 - \Gamma_L$ will subtract 1 from the *real* term of Γ_L, and will simply change the sign of the imaginary term, while $1 + \Gamma_L$ will add 1 to the real term of Γ_L and ignores the imaginary number completely]:

$$Z_{OUT} = Z_{LOAD} \left(\frac{1 + \Gamma_L^{\Psi}}{(1 - \Gamma_L^{\Psi}} \right)$$

where Z_{LOAD} = transistor's load placed at its output (typically 50 ohms; written as $50 + j0$).

To calculate the value of the transistor's *input impedance* (Z_{IN}) for the transistor's *output impedance* as calculated above:

1. $$\Gamma_S = \left[S_{11} + \frac{S_{12} S_{21} \Gamma_L}{1 - (\Gamma_L \cdot S_{22})} \right]^{\Psi}$$

2. $Z_{IN} = Z_{SOURCE}\left(\frac{1 + \Gamma_S^{\Psi}}{1 - \Gamma_S^{\Psi}}\right)$

where Z_{SOURCE} = transistor's source (the prior stage) placed at its input (typically 50 ohms; written as $50 + j0$ for this formula). Vector algebra is used to calculate Γ_S, Z_{IN}, Z_{OUT}, and Z_{LOAD}.

Note that all signs (±) must be strictly maintained for all calculated numbers. $1 - \Gamma_L$ will subtract 1 from the *real* term of Γ_L, and will simply change the sign of the imaginary term. Γ_L^{Ψ} or Γ_S^{Ψ} equals the complex conjugate of Γ_L or Γ_S respectively. In other words, just change the sign of the *angle,* but not the *magnitude's* sign (for example, $+12 \angle +18° = +12\angle -18°$). If Ψ is outside a bracket, then the *answer* to everything within the bracket must be converted into this complex conjugate.

Now that we have discovered the input and output impedances of our chosen device, we can begin to impedance-match its ports to obtain a simultaneous conjugate match for the transistor's required source and load. So the next step is to design the matching networks for our circuit.

As an example, the active device, a transistor, has these S parameters at 1.5 GHz and a V_{CE} = 10 V and an I_C = 6 mA:

$$S_{11} = 0.195\angle 167.6°$$

$$S_{22} = 0.508\angle -32°$$

$$S_{12} = 0.139\angle 61.2°$$

$$S_{21} = 2.5\angle 62.4°$$

We will want the amplifier to run between two 50-ohm terminations. To design an input and output matching network to maintain maximum amplifier gain:

1. Calculate K. Confirm unconditional stability ($K > 1$) by calculation or lookup in a table, if supplied. (All negative real number results *must* be used in all calculations as negative real numbers.):

   a. $D_S = S_{11} S_{22} - S_{12} S_{21} = 0.25 \angle -61.4°$

   b. $K = \dfrac{1 + |D_S|^2 - |S_{11}|^2 - |S_{22}|^2}{2|S_{21}||S_{12}|}$

   c. $K = \dfrac{1 + (0.25)^2 - (0.195)^2 - (0.508)^2}{2(2.5)(0.139)} = +1.1$

   Use vector algebra for D_S. Do not use vector algebra for K, use magnitudes of the S parameters.

2. Calculate MAG:

a. $\text{MAG} = 10 \log_{10} \left| \frac{S_{21}}{S_{12}} \right| + 10 \log_{10} | K \pm \sqrt{K^2 - 1} |$

b. $B_1 = 1 + |S_{11}|^2 - |S_{22}|^2 - |D_S|^2 = +0.717$

c. $\text{MAG} = 10 \log \left(\frac{2.5}{0.139} \right) + 10 \log (1.1 - \sqrt{(1.1)^2 - 1} = 10.63 \text{ dB}$

Do not use vector quantities to calculate MAG and B_1, use magnitudes only. Since $B_1 = +0.717$, then the sign between 1.1 and the square root in step 2c is negative. 10.63 dB MAG is all right for our needs, so we continue.

3. Calculate $\angle_L$ (load reflection coefficient) required for a conjugate match for the transistor. As stated above, the $\angle_S$ and $\angle_L$ are the values that the transistor must have at its input and its output for a perfect match:
   a. $C_2 = S_{22} - (D_S S_{11}{}^{\Psi}) = (0.508 \angle -32°) - [(0.25 \angle -61.4°)(0.195 \angle 167.6°)]$
   b. $C_2 = 0.555 \angle -33.5°$
   c. $B_2 = 1 + |S_{22}|^2 - |S_{11}|^2 - |D_S|^2 = 1 + (0.508)^2 - (0.195)^2 - (0.25)^2$
   d. $B_2 = +1.157$
   e. Therefore, since B_2 equals +, the sign equals − in the $\angle_L$ equation below:

$$\angle_L = \frac{B_2 \pm \sqrt{(B_2)^2 - 4|C_2|^2}}{2|C_2|} = \frac{1.157 - \sqrt{(1.157)^2 - 4\,(0.555)^2}}{2\,(0.555)} = 0.748$$

So the answer for the magnitude of $\angle_L$ is:

$$\angle_L = 0.748$$

   f. Now find the angle of Γ_L: The angle equals the same value as in C_2 = $(0.555 \angle -33.5°)$, but is opposite in sign. Therefore, the angle of Γ_L = $\angle +33.5°$.
   g. Our complete answer is $\Gamma_L = 0.748 \angle +33.5°$

   Use magnitudes, not vectors, to calculate B_2 and Γ_L.

4. Calculate the source reflection coefficient (Γ_S):

   a. $\Gamma_S = \left[S_{11} + \frac{S_{12} S_{21} \Gamma_L}{1 - (\Gamma_L \cdot S_{22})} \right]^{\Psi}$

   b. $\Gamma_S = \left[0.195\angle 167.6° + \frac{(0.139\angle 61.2°)(2.5\angle 62.4°)(0.748\angle 33.5°)}{1 - (0.748\angle 33.5°)(0.508\angle -32°)} \right]^{\Psi}$

   $= (0.61\angle 160.8°)^{\Psi}$

   c. $\Gamma_S = 0.61\angle -160.8°$

Use vector quantities to calculate Γ_S.

5. Calculate the input impedance of the transistor (Fig. 3.17):

$$Z_{IN} = Z_{SOURCE} \frac{1 + \Gamma_S^{\Psi}}{1 - \Gamma_S^{\Psi}} = 50 + j0 \left[\frac{1 + (-0.576 + j0.2)}{1 - (-0.576 + j0.2)} \right]$$

$$= 50 + j0 \left(\frac{0.424 + j0.2}{1.57 - j0.2} \right) = 12.4 + j7.9$$

where Z_{SOURCE} = impedance placed at the transistor's input by the prior stage (typically 50 ohms, written as 50 + j0 for this formula). Use vector quantities in the equation for Z_{IN}.

6. Now match Z_{SOURCE} to Z_{IN} with the matching procedures presented further in this chapter.

7. Calculate the transistor's output impedance (Fig. 3.18):

$$Z_{OUT} = Z_{LOAD} \left(\frac{1 + \Gamma_L^{\Psi}}{1 - \Gamma_L^{\Psi}} \right) = 50 + j0 \left[\frac{1 + (0.624 - j0.413)}{1 \quad (0.624 - j0.413)} \right] = 70.5 - j132$$

where Z_{LOAD} = impedance of the transistor's load.

8. Now match Z_{OUT} to Z_{LOAD} with the procedures presented later in this chapter.

9. At this time it is possible to calculate the *transducer gain* (G_T, the actual gain of an amplifier stage, which includes the effects of impedance matching and device gain, but does not include power losses in real-world components). G_T will be quite close to the MAG value):

$$G_T = 10 \log_{10} \left[\frac{|S_{21}|^2 (1 - |\Gamma_S|^2)(1 - \Gamma_L|^2)}{|(1 - S_{11}\Gamma_S)(1 - S_{22}\Gamma_L) - S_{12}^* \times S_{21}^* \times \Gamma_L^* \times \Gamma_S^* \times|^2} \right]$$

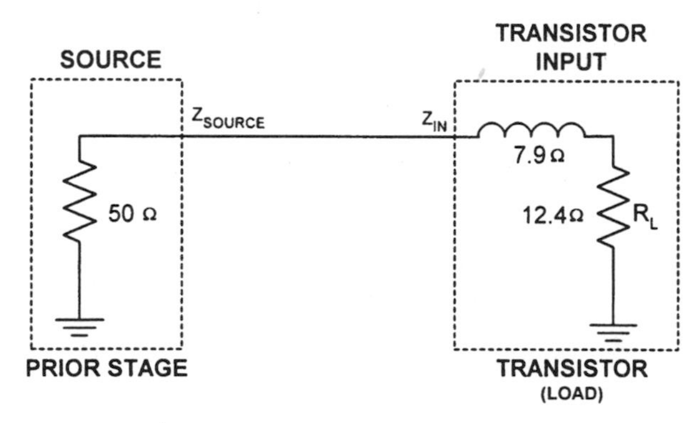

Figure 3.17 A transistor's input at RF.

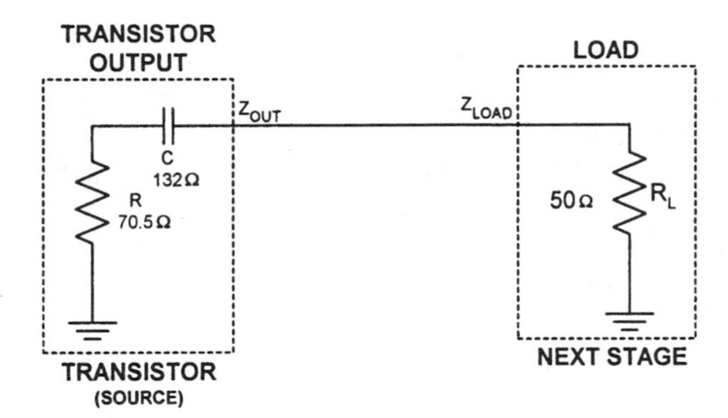

Figure 3.18 A transistor's output at RF.

3.1.3 Vector algebra

When it is necessary to utilize full complex numbers ($Z \angle \pm\theta^\circ$) in our calculations, we can perform the required mathematical functions by the following methods.

To multiply polar quantities: First, multiply the magnitudes; then add the phase angles.

To divide polar quantities: First divide the magnitudes; then subtract the phase angles.

To subtract polar quantities: First, convert to rectangular notation ($R + jX$; see "conversions" in Sec. 3.1.4), then subtract $R_1 - R_2 = R_T$, and $jX_1 - jX_2 = jX_T$; then convert the rectangular answer back to polar.

To add polar quantities: Perform as in subtraction; but *add* the rectangular values ($R_T = R_1 + R_2$; $jX_T = jX_1 + jX_2$).

Stability. All active devices are quite stable when presented with a 50-ohm source and load over the entire frequency range in which the device exhibits gain. Most problems with stability occur when the circuit designer does not take into account the elevated *low-frequency* gain of a normal amplifier; and its inherent instability when presented with anything other than 50-ohm terminations. This lack of 50-ohm termination as the frequency is decreased can be due to two main reasons: (1) the amplifier's impedance matching circuits are good only for a narrow band of frequencies, so they will present 50 ohms to the transistor over a relatively restricted range. (2) The inductor adopted for decoupling of the power supply (which is a very low impedance source) from the amplifier becomes closer to a short circuit as the frequency is *decreased.* This can create instability at low frequencies because a distributed choke, or even a low-value RF lumped choke employed for RF decoupling will give a true choke response only over a higher band of frequencies. This means that as the frequency of operation is decreased, the "open" circuit of the RF choke will begin to look more like a piece of straight wire than as a choke, causing the amplifier to have a load that no longer appears as 50 ohms, which can create oscillations in a conditionally stable amplifier. One way to lessen this effect is to add a lumped RF choke *above* the distributed RF choke to choke out lower frequency RF, since the amplifier has more gain here than at the higher frequencies (Fig. 3.19). Another technique to ensure that no low-frequency oscillations are created by amplifier stability problems is to use a 50-ohm resistor at the DC end of the bias circuit (Fig. 3.20), as this allows the amplifier to operate into a 50-ohm termination at frequencies so low that the distributed RF choke would have no effect. The termination resistor works to stabilize the conditionally stable amplifier, and is required only if there is no other series voltage dropping resistor between the collector and V_{CC}. Since this resistor will have the full collector current running through it, subtract the voltage dropped across it from the V_{CC} value to obtain the voltage at the amplifier. Looking at Fig. 3.20 again, we see that capacitor C_B also helps to shunt low-frequency RF to ground to decrease the disruptive low-frequency RF gain. Figure 3.21 is another configuration that is able to maintain decoupling from the power supply at low frequencies. The circuit accomplishes this by using both a low- and a high-frequency choke to sustain a high impedance into the power supply.

To lower the chances of any instability in an amplifier, a potentially unstable transistor will require a degenerative feedback network, a source and load impedance that assures stability, or another bias point for the transistor. None of this is required with an unconditionally stable transistor.

However, to completely maintain stability within an amplifier circuit, it should be remembered that the circuit elements themselves can add a larger feedback path for oscillations than even the transistor itself. Since Barkhausen's criterion for oscillations is a loop gain of unity or higher, and an in-phase (regenerative) feedback from output to input, then we can see that at certain frequen-

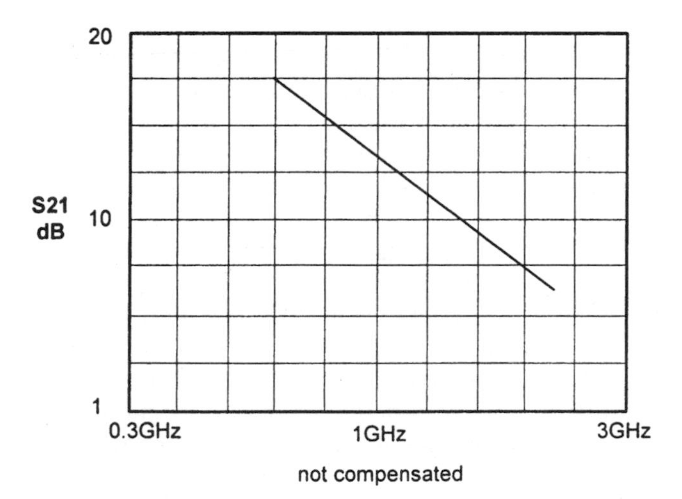

Figure 3.19 No gain compensation in an amplifier.

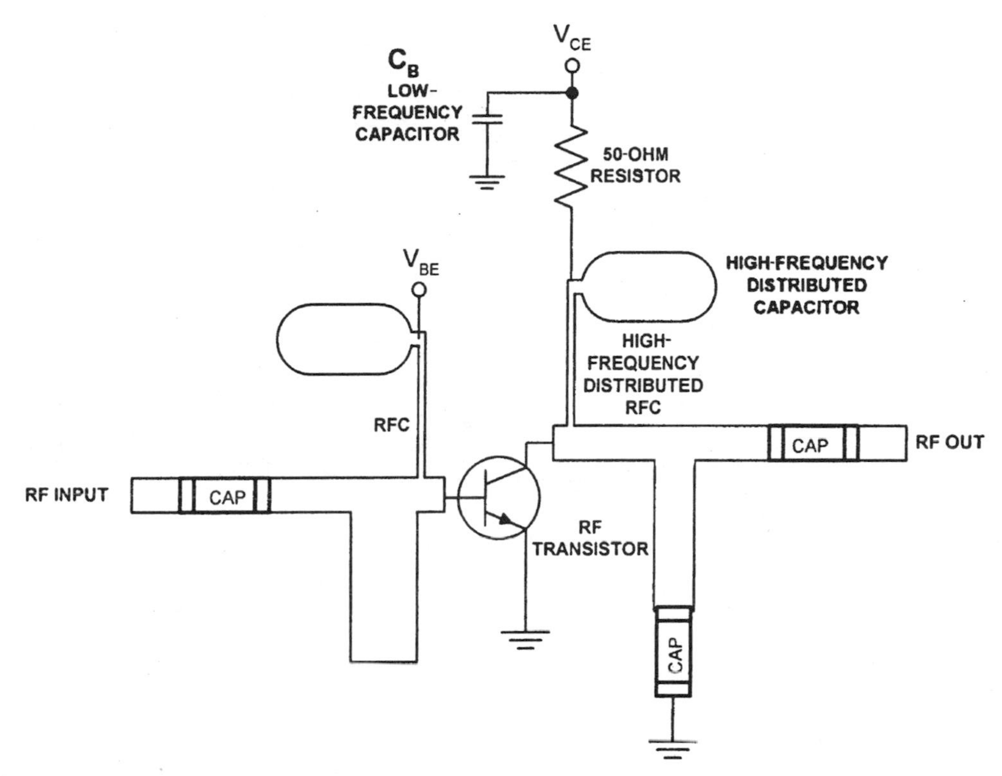

Figure 3.20 Terminating low frequencies into 50 ohms to prevent instability.

cies, and with a high enough feedback path somewhere on the board, oscillation can become a problem if the layout is poor. In addition, the higher the gain of an amplifier stage, the more likely oscillations will begin to break out; 25 dB of gain is considered the maximum for stability from a single stage.

Neutralization (degenerative feedback), as mentioned above, is sometimes used to stabilize a potentially unstable amplifier. Nonetheless, the amplifier neutralization procedure will be successful only if the positive feedback path that created instability and oscillations is internal to the transistor, and not if poor layout and/or lack of input/output shielding creates the return path. Neutralization is also problematic with wideband transistor amplifiers because of the variations in input and output capacitance of a bipolar transistor with changes in frequency and bias currents, as well as the neutralization retuning requirements for transistors in different production lots.

Another viable technique for creating a stable amplifier is to simply reduce the gain of the stage. This works because an amplifier, as mentioned above, must reach the Barkhausen criterion to oscillate (just as an oscillator must). This means that reducing the feedback and/or the gain will stabilize an amplifier. Unfortunately, reducing stage gain appreciably is often an unacceptable solution, both from an economic and an efficiency standpoint.

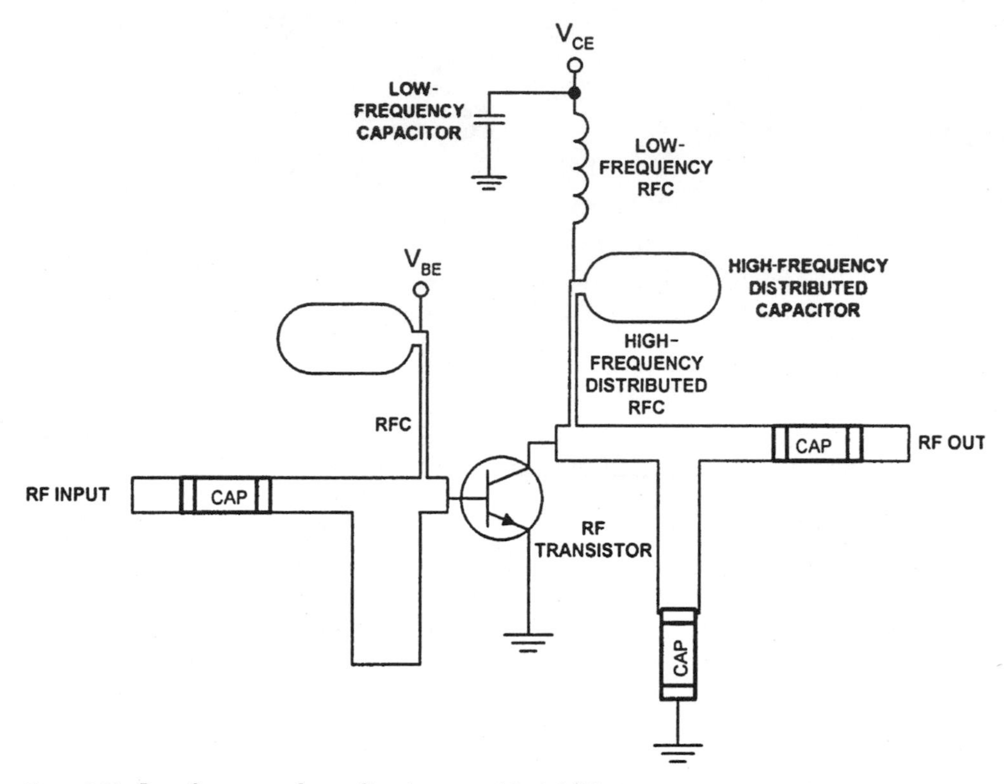

Figure 3.21 Low-frequency decoupling to prevent instability.

Approximations. When searching for a small-signal transistor with a specific gain and stability, we may not want to be as formal with our calculations as we have been up to now, because of time constraints. Ballpark figures will sometimes suffice. There is a much faster method to obtain gain and stability figures, called *S*-parameter *scalar approximations,* that can be utilized for amplifiers to obtain approximate design values. For the following formulas, only the magnitudes of the *S* parameters are employed, and not the phase angles.

1. G_{tu} (transducer unilateral gain) is the dB measurement of an amplifier's power gain into an unmatched 50-ohm load—a worst-case gain value—and can be roughly calculated by:

$$G_{tu} = 10 \log_{10} |S_{21}|^2$$

2. Mismatch losses (αp) at the transistor's input or output, in dB, are calculated by:

$$\alpha p_{\mathrm{IN}} = -10 \log_{10} (1 - S_{11}{}^2)$$

$$\alpha p_{\mathrm{OUT}} = -10 \log_{10} (1 - S_{22}{}^2)$$

The total mismatch loss for the entire unmatched transistor is

$$\alpha p_{\text{TOTAL}} = \alpha p_{\text{IN}} + \alpha p_{\text{OUT}}$$

3. MAG (maximum available gain) is calculated by:

$$\text{MAG} = G_{tu} + \alpha p_{\text{TOTAL}}$$

4. Since designing for MAG is never recommended because an amplifier can be unstable at this high gain value, we would like to be able to compute the MSG, or the maximum stable gain:

$$\text{MSG} = 10 \log_{10} (|S_{21}| \div |S_{12}|)$$

Thus, if the MAG is smaller than the MSG, then the amplifier will be unconditionally stable, unless poor circuit layout produces an external feedback path.

Scalar approximation is a more rapid technique than the methods presented in the prior pages, since we are employing only the magnitude of the S parameters, and not their phase angle. As an example, we are given a transistor with the following S parameters:

$S_{11} = 0.195\angle 167.6°$
$S_{22} = 0.508\angle -32°$
$S_{12} = 0.139\angle 61.2°$
$S_{21} = 2.5\angle 62.4°$

Therefore $S_{11} = 0.195$, $S_{22} = 0.508$, $S_{21} = 2.5$, $S_{12} = 0.139$, and

$G_{tu} = 10 \log_{10} |2.5|^2 = 7.96$ dB
$\alpha p_{\text{IN}} = -10 \log_{10} (1 - 0.195^2) = 0.168$ dB
$\alpha p_{\text{OUT}} = -10 \log_{10} (1 - 0.508^2) = 1.29$ dB
$\alpha p_{\text{TOTAL}} = 0.168 \text{ dB} + 1.29 \text{ dB} = 1.46$ dB

which demonstrates that about 1.46 dB will be gained by proper impedance matching.

$$\text{MAG} = 7.96 \text{ dB} + 1.46 \text{ dB} = 9.42 \text{ dB}$$

(10.63 dB was calculated for this same transistor with the full MAG method described earlier.)

$$\text{MSG} = 10 \log_{10} (|2.5| \div |0.139|) = 12.55 \text{ dB}$$

With MAG < MSG by over 3 dB—even with our approximation methods—this transistor will be very stable.

3.1.4 Matching networks

The most common amplifier matching networks are lumped and distributed *LC*-type L, T, and pi circuits (Fig. 3.22).

Once all of the information on the parameters of the device to be matched is assembled, we will need to design the matching network. This is so the device's impedances will match the impedances of the circuit it will be inserted into, so that we may obtain the maximum power transfer from one stage to another, with no power reflections: $Z_{SOURCE} = R + jX$ must equal $Z_{LOAD} = R - jX$ (a conjugate match; Fig. 3.23).

However, there is only one frequency that will be *perfectly* matched from source to load, since X_C and X_L are frequency dependent, or:

$$L = \frac{X_L}{2\pi f} \qquad and \qquad C = \frac{1}{2\pi f X_C}$$

Nonetheless, we may normally obtain quite a decent return loss over a very wide band of frequencies by proper matching techniques with the correct matching network.

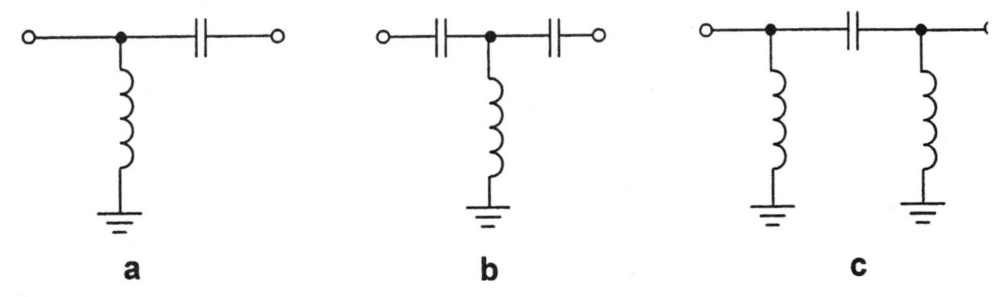

Figure 3.22 Popular matching networks: (*a*) L; (*b*) T; (*c*) pi.

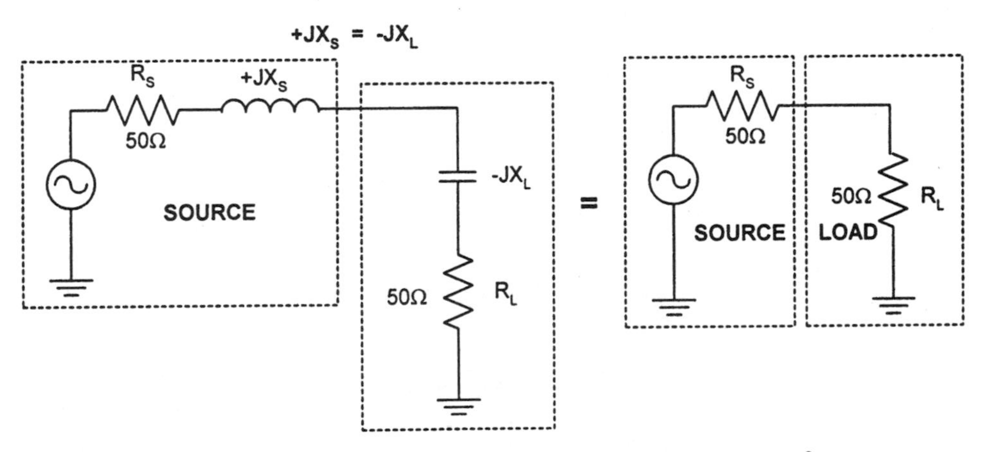

Figure 3.23 Canceling reactances and equal resistances maximizes power transfer.

In interstage matching, it is possible to choose between two methodologies: match the output impedance of the first stage (usually a "high" impedance) to the input impedance of the second stage (usually a "low" impedance). This uses the fewest components. Or, match everything to 50 ohms for standardization. This permits the testing of the final physical design from stage to stage with normal 50-ohm test gear.

Lumped L matching. The simple, but very popular, L matching network has a disadvantage in that the Q of the circuit cannot be selected as it can in the more complex networks shown below. A low Q is desired to increase the bandwidth of the amplifier, as well as to decrease lossy circulating currents in power amplifiers. Still, the value of Q is usually naturally low in an L network, and thus will suffice for most semiwideband matching needs.

First, to design a basic *resistance-matching-only* L network for matching the two-different-value resistances of R_S and R_P (Fig. 3.24), the network topology must initially be chosen. For a high-to-low impedance transformation choose Fig. 3.6; for a low-to-high impedance transformation, choose Fig. 3.7:

1. Find the natural Q of the circuit by the following formula, in which Q_S and Q_P must be a positive number:

   a. $Q_S = Q_P = \sqrt{\dfrac{R_P}{R_S} - 1}$

   b. $Q_S = Q_P = \sqrt{\dfrac{58}{12} - 1}$

   c. $Q_S = Q_P = 1.96$

2. Find the reactance of element X_P of the L network (Fig. 3.25):

   a. $X_P = \dfrac{R_P}{Q_P}$

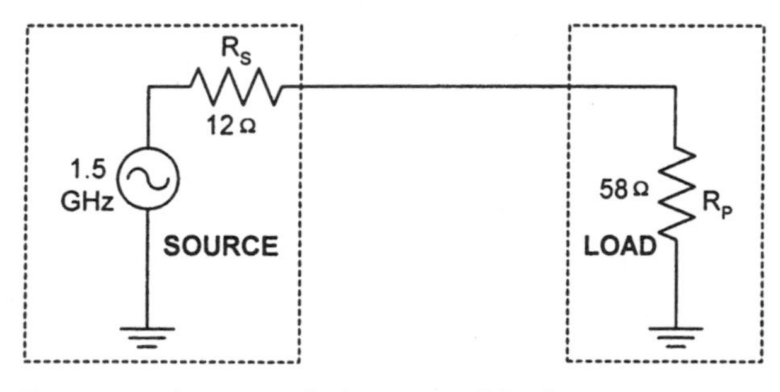

Figure 3.24 An unmatched source and load.

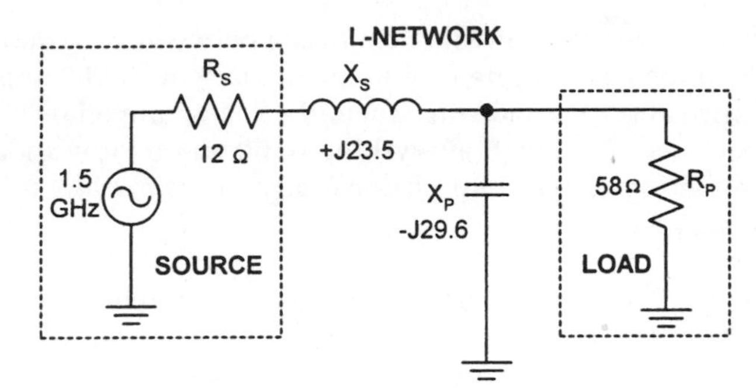

Figure 3.25 Matching two different source and load resistances with an L network.

b. $X_P = \dfrac{58}{1.96}$

c. $X_P = 29.6$ ohms

3. Find the reactance of element X_S of the L network:
   a. $X_S = Q_S R_S$
   b. $X_S = 1.96 \times 12$
   c. $X_S = 23.5$ ohms
4. To convert the calculated X_S reactance into an inductor value:

   a. $L = \dfrac{X_S}{2\pi f}$

   b. $L = \dfrac{23.5}{2\pi\ 1.5\ \text{GHz}}$

   c. $L = 2.5$ nH
5. To convert the calculated X_P reactance into a capacitor value:

   a. $C = \dfrac{1}{2\pi f X_P}$

   b. $C = \dfrac{1}{2\pi(1.5\ \text{GHz})\ 29.6}$

   c. $C = 3.58$ pf

The completed matching network is shown in Fig. 3.26.

When Fig. 3.26 must be chosen in a high-to-low impedance matching situation, simply change the R_P designation to R_S, and R_S to R_P, and then use the same calculations as above (this switch is needed because the L network's matching capacitor X_P is now in parallel with the *source's* resistance, instead of the *load's* resistance).

When two different, but pure, resistances must be matched, the above technique is easily and rapidly applied to perform this task. However, if reactances must also be canceled within one or both of these circuits—as well as the resistances matched—then one (or both) of the two following methods may be employed.

Absorption uses the reactances of the impedance matching network itself to absorb the undesired load and/or source reactances (Fig. 3.27). This is accomplished by positioning the matching inductor in series with any load or source inductive reactance. In this way, the load or source's X_L actually becomes *part*

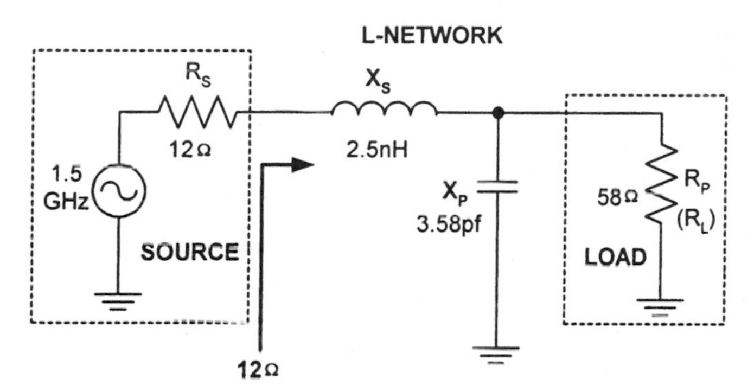

Figure 3.26 The final L network component values for a matched source and load.

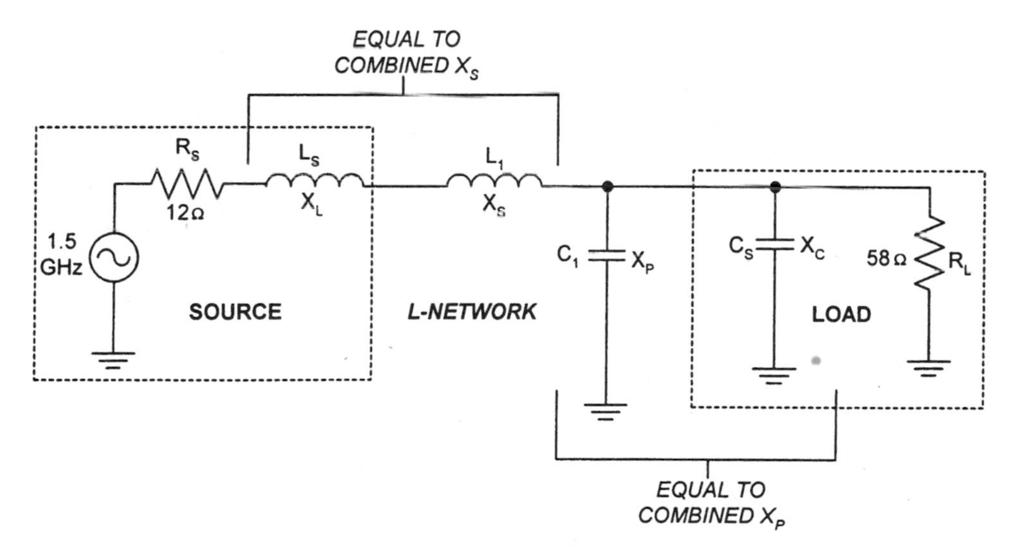

Figure 3.27 Circuit that requires the addition of components to absorb reactances.

of the matching inductor. The same outcome can be attained by positioning a matching capacitor in *parallel* with any load or source X_C, thus combining the two values into one larger value. This allows the internal stray reactances of both devices to actually contribute to the matching network, with these internal reactances now being subtracted from the calculated values of the *LC* matching components. In other words, the transistor's own stray reactances are now becoming an additive part of the matching network. This method is useful only if the stray internal reactances of the device are *less* than the calculated reactances required for a proper match, which is normally the case.

The other technique is use *resonance* to resonate out the stray reactances of the device or circuit to be matched (at our desired frequency), with a reactance that is equal in value but opposite in sign, and then continuing as if the matching problem were a completely resistive one ($R + j0$). This will make the internal stray reactances of the two devices or circuits disappear, thus allowing only the pure resistances to be easily dealt with.

The first approach, *absorption*, is demonstrated with the practical example of Fig. 3.28*a:*

1. Disregard *all* source and/or load internal reactances.
2. Place an L network in series with the internal stray X_L of the source, and the capacitance in parallel with the internal stray X_C of the load (Fig. 3.28*b*).
3. While still neglecting all of the stray reactances, use the formulas and methods of resistive lumped L matching as outlined above to calculate and match R_S to R_L.
4. Subtract the internal stray reactance values from the L network's calculated values of L_1 (2.5 nH) and C_1 (3.58 pF), which in this case will be 2.5 nH − 1 nH = 1.5 nH = L_1 and 3.58 pF − 1.5 pF = 2.08 pF = C_1.
5. The new L network component values are now the actual values required to obtain the proper $12 - j9.4$ conjugate match for the $12 + j9.4$ source (or $Z_L = 12 + j0$).

To design a matching network employing the second method, the *resonance* approach, view the example circuit of Fig. 3.29:

1. Resonate out the 1.5 pF of stray capacitance within the load by employing a shunt inductor with a value of

$$L = \frac{1}{|2\pi f|^2 C_{\mathrm{STRAY}}}$$

   or L = 7.5 nH (Fig. 3.30). The internal stray capacitance can now be considered as no longer existing within the load.
2. Since the source is purely resistive ($Z_S = R_S + j0$), and the load is now as well ($Z_L = R_L + j0$), we can utilize the formulas for basic resistive lumped matching to design an L network to match the source to the load.

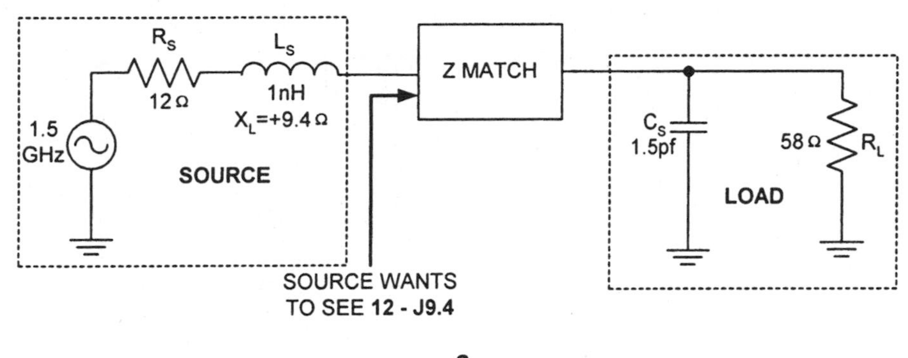

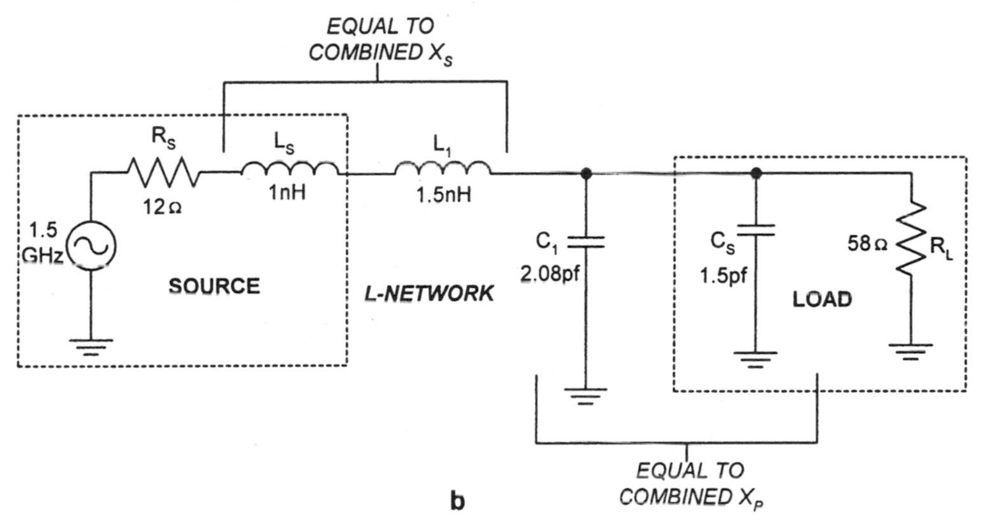

Figure 3.28 (*a*) Circuit that requires matching; (*b*) the addition of components to absorb reactances.

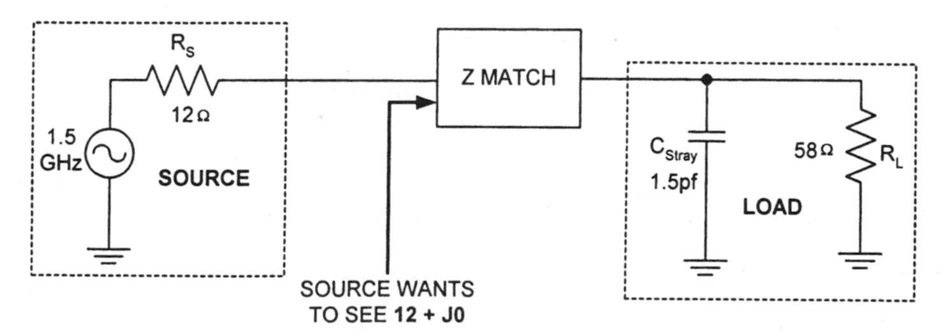

Figure 3.29 Example circuit for the resonance impedance-matching design approach.

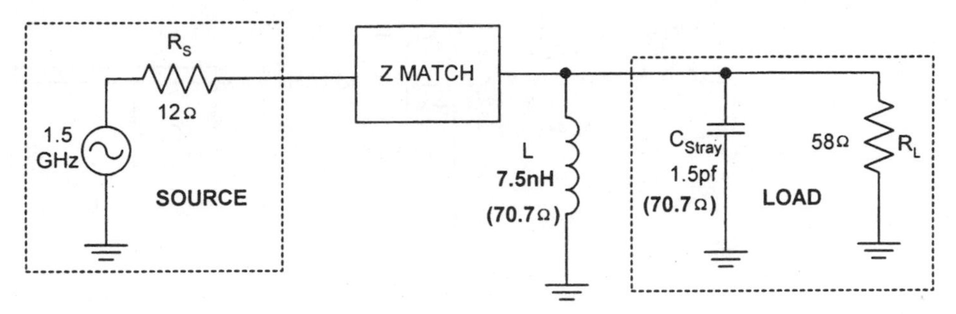

Figure 3.30 Canceling the load's stray reactance.

3. Simplify by combining both of the inductors we now have (Fig. 3.31) with a single inductor by (Fig. 3.32):

$$L_{NEW} = \frac{L_1 L_2}{L_1 + L_2}$$

or 2.22 nH $= L_{\text{NEW}}$
The 50-ohm source is now perfectly matched to the complex load.

Pi and T network matching. Three-element impedance matching (pi or T) networks are popular in narrowband applications. The narrowband popularity is due to the higher Q over that which L networks possess, yet pi and T networks also permit almost any Q to be selected. Nonetheless, T and pi circuits can never be lower in Q than an L network. The Q desired for a particular application may be calculated with the following formula, assuming the utilization of high-Q inductors:

$$Q = \frac{f_c}{(f_2 - f_1)}$$

where Q = loaded quality factor of the circuit
f_C = center frequency of the circuit
f_2 = upper frequency that we will require to pass with little loss
f_1 = lower frequency that we will require to pass with little loss

Employ the guidelines below to design a pi network capable of matching two different pure *resistances* (Fig. 3.33). With the following design methodology, consider the pi network as two L networks attached back to back, with a *virtual* resistor in the center; which is used only as an aid in designing these networks, and will not be in the final design:

1. Find "*R*", the virtual resistance, as shown in Fig. 3.34. (*Note:* In this example, the Q of the pi network is chosen to be 10):

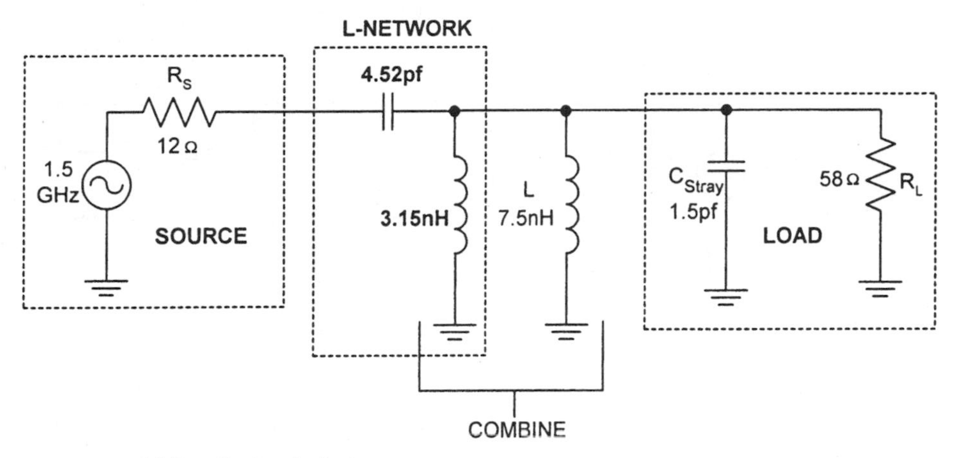

Figure 3.31 Adding the two inductors.

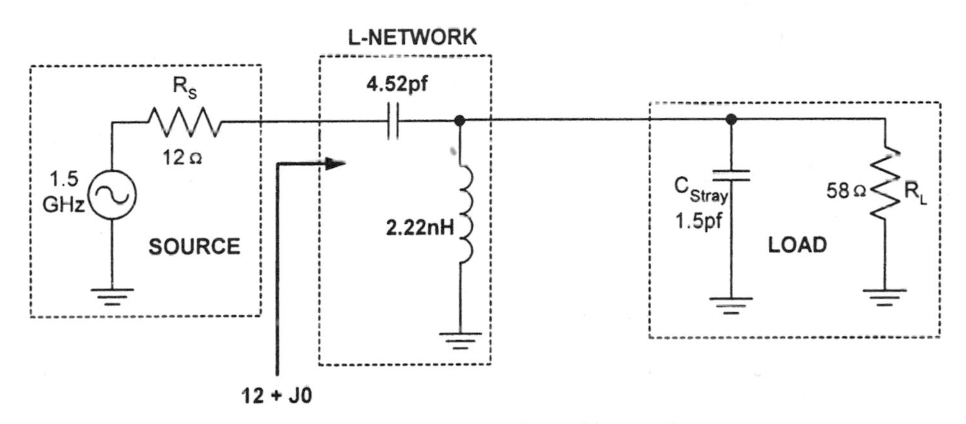

Figure 3.32 Combining the two inductors into one for an L network.

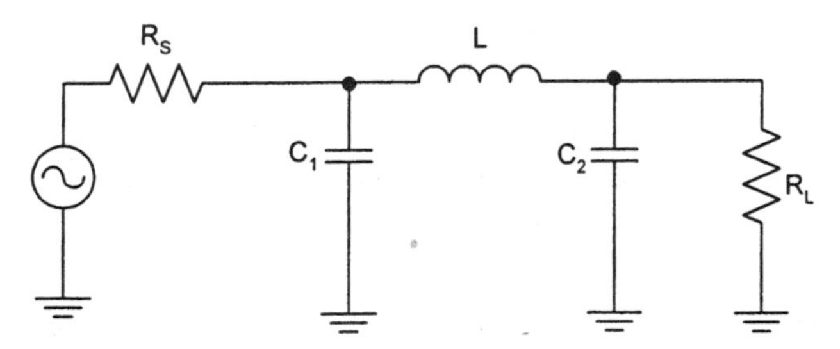

Figure 3.33 A pi matching network between a source and its load.

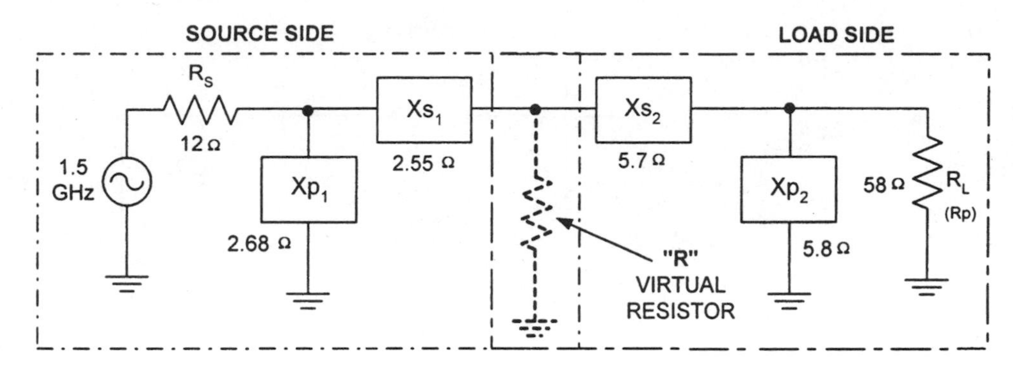

Figure 3.34 Using a virtual resistor to design a pi network.

$$"R" = \frac{R_H}{Q^2 + 1} = \frac{58}{10^2 + 1} = 0.57 \text{ ohms}$$

R_H is equal to whichever source or load resistance is *larger*, R_L or R_S.

2. Find X_{P2} and X_{S2} by:

$$X_{P2} = \frac{R_L}{Q} = \frac{58}{10} = 5.8 \text{ ohms} \qquad \text{and} \qquad X_{S2} = Q"R" = 5.7 \text{ ohms}$$

for the *load*-side values.

3. Find the value of X_{P1} and X_{S1}:

$$X_{P1} = \frac{R_S}{Q_1} \mid Q_1 = \sqrt{\frac{R_S}{"R"} - 1} = \frac{12}{4.48} \mid 4.48 = \sqrt{\frac{12}{0.57} - 1} = 2.68 \text{ ohms}$$

$$X_{S1} = Q_1"R" = 4.48 \times 0.57 = 2.55 \text{ ohms}$$

4. Combine X_{S1} and X_{S2} ($X_{S1} + X_{S2}$) (Fig. 3.35).
5. One of four different pi matching configurations can be chosen, depending on the following requirements: Must we get rid of stray reactances, pass or block DC, or filter excess harmonics (Fig. 3.36)?
6. Convert the reactances calculated to L and C values by:

$$L = \frac{X_S}{2\pi f} \text{ and } C = \frac{1}{2\pi f X_P}$$

To match two stages with a pi network, while canceling *reactances* and *matching resistances* (Fig. 3.37), observe the following procedures. Convert the load/source to/from parallel or series equivalences as required to make it

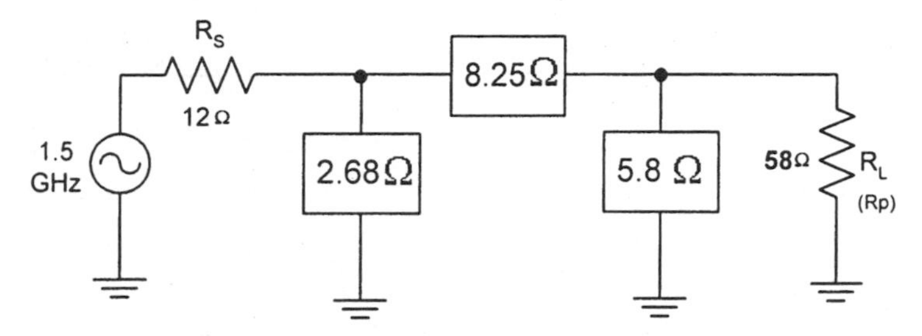

Figure 3.35 Reactance values as calculated for pi network.

easier to absorb any reactances. These conversion equations are found later in this chapter.

1. Select a proper network topology that will *absorb* both stages' reactances. In this case, we would choose a pi network with two parallel capacitors (see discussion above on absorption).
2. Choose a desired Q and frequency of operation.
3. Find "R," the virtual resistance:

$$"R" = \frac{R_H}{Q^2 + 1}$$

Note: R_H is equal to whichever source or load resistance is larger, R_L or R_S.

4. Find X_{C2} and X_{L2} by:

$$X_{C2} = \frac{R_L}{Q} \quad \text{and} \quad X_{L2} = Q"R"$$

for the *load*-side values.

5. Find X_{C1} and X_{L1} by:

$$X_{C1} = \frac{R_S}{Q_1} \mid Q_1 = \sqrt{\frac{R_S}{"R"} - 1} \quad \text{and} \quad X_{L1} = Q_1"R"$$

6. As shown in Fig. 3.38, add X_{L1} and X_{L2} to form X_{LNEW}; combine $X_{CSTRAY1}$ and X_{C1}; then combine $X_{CSTRAY2}$ and X_{C2} (X_{C1} and X_{C2} must be smaller than $X_{CSTRAY1}$ and $X_{CSTRAY2}$ respectively, since adding two capacitors' reactances in parallel involves:

$$\frac{X_C \times C_{STRAY}}{X_C + C_{STRAY}} = X_C$$

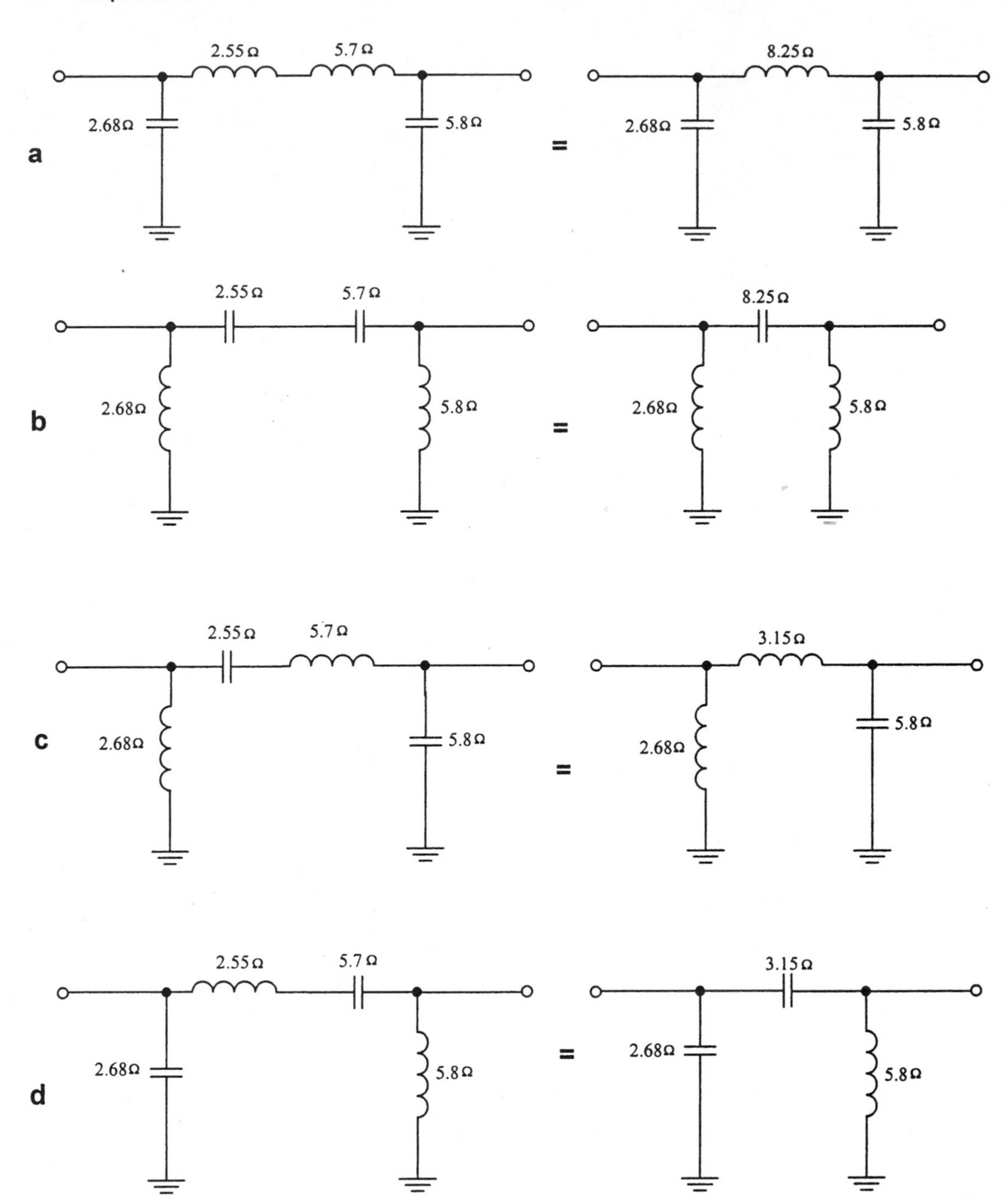

Figure 3.36 Various legitimate pi networks before and after combining components.

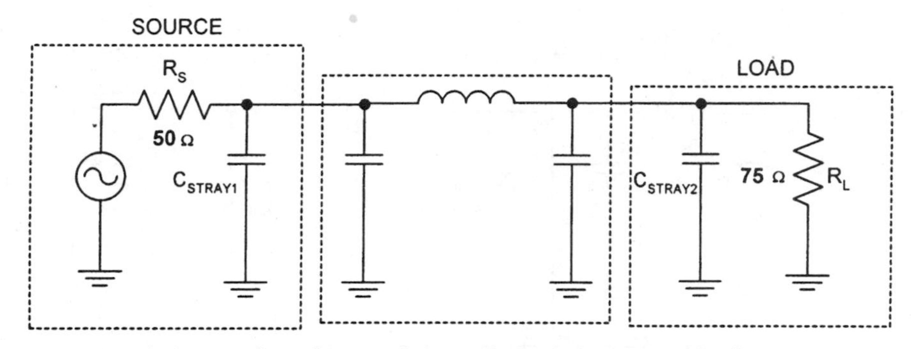

Figure 3.37 A pi network used in a resistive and reactive source and load.

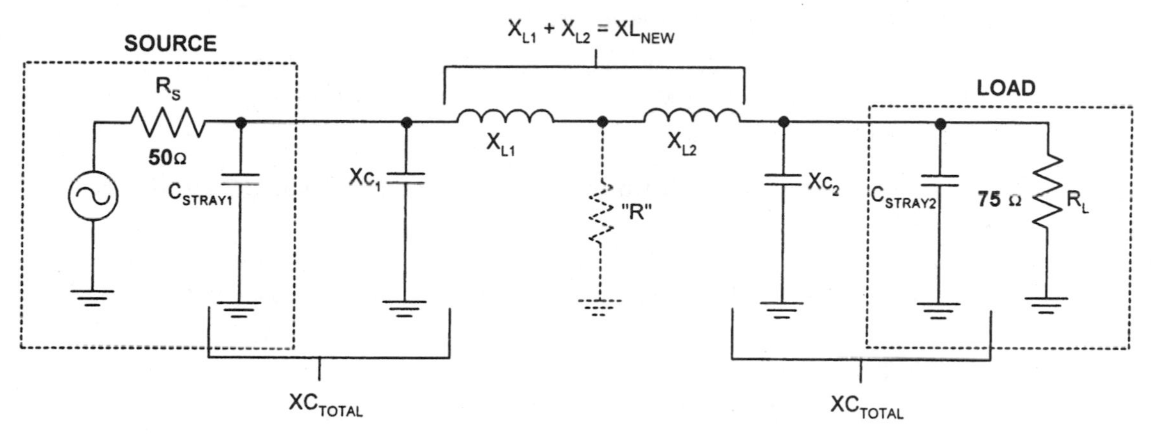

Figure 3.38 Pi network before combining calculated components.

Thus, if $X_{CSTRAY} < X_{C1}$, then X_{CTOTAL} will not be able to reach the proper X_C value. Also, increase X_{C1} until:

$$\frac{X_{C1} \cdot X_{CSTRAY1}}{X_{C1} + X_{CSTRAY1}} = X_{CTOTAL} = X_{C1} \quad \text{or} \quad \frac{X_{C1} \cdot X_{CSTRAY1}}{X_{C1} - X_{CSTRAY1}} = X_{CNEW}$$

This is so X_{C1} and $X_{CSTRAY1}$, in *parallel,* will still equal the computed value of X_{C1} (X_{CTOTAL}).

7. Convert the reactances calculated to L and C values by:

$$L = \frac{X}{2\pi f} \quad \text{and} \quad C = \frac{1}{2\pi f X}$$

The completed network is as shown as Fig. 3.39.

T networks are required when two low impedances need to be matched with a high Q, and must be of a higher Q than that available with the L network type.

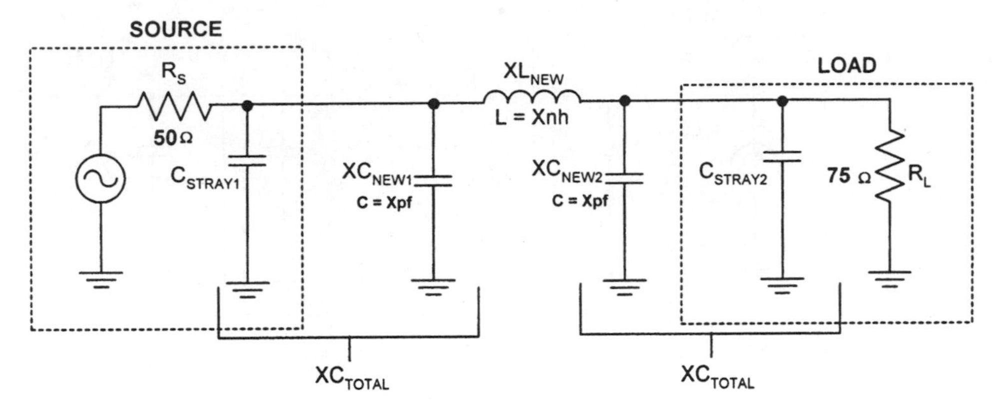

Figure 3.39 Pi matching network after combining components.

Follow this procedure to match two unequal and pure *resistances,* as shown in the example of Fig. 3.40.

1. Decide on the loaded Q (in this case 15), and the frequency (in this case 1.5 GHz).
2. Find the "R" value by "R" = R_{SMALL} $(Q^2 + 1)$; "R" = $12\,(15^2 + 1)$; "R" = 2712 ohms. R_{SMALL} is the smaller value of the two resistances, whether it is R_S or R_L.
3. Find $X_{S1} = QR_S = 15 \cdot 12 = 180$ ohms.
4. Find X_{P1} = "R"/Q = 2712/15 = 181 ohms.
5. Find:

$$Q_2 = \sqrt{\frac{"R"}{R_L} - 1} = \sqrt{\frac{2712}{58} - 1} = 6.76$$

6. Find:

$$X_{P2} = \frac{"R"}{Q_2} = \frac{2712}{6.76} = 401 \text{ ohms}$$

7. Find:

$$X_{S2} = Q_2 R_L = 6.76 \cdot 58 = 392 \text{ ohms}$$

8. X_{P1} and X_{P2} are combined by:

$$X_{\text{TOTAL}} = \frac{X_{P1} X_{P2}}{X_{P1} + X_{P2}} = \frac{181 \cdot 401}{181 + 401} = 125 \text{ ohms}$$

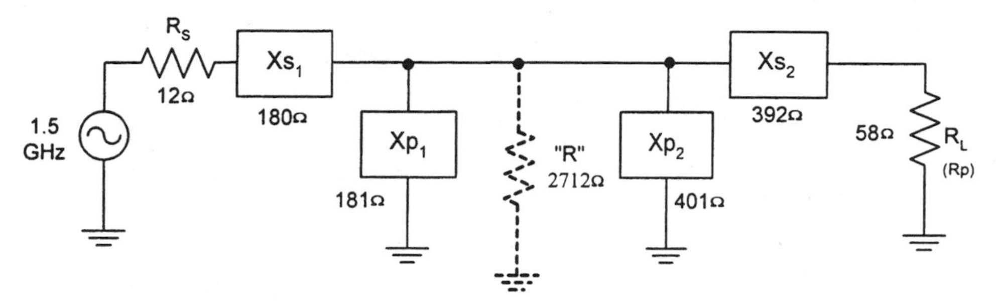

Figure 3.40 Designing a T network for use between a resistive source and load.

9. The circuit is shown completed in Fig. 3.41. Other possible circuit configurations can be used as required (Fig. 3.42). Figure 3.42*a, b,* and *c* are combined as in step 8 above, but the signs must be maintained for *b* and *c* because of the opposite reactance employed (+ for inductors and − for capacitors).

Wideband matching. Sometimes it may be necessary to design a low-*Q*, very wideband matching network. This can be done as follows, by using Fig. 3.43*a* for a pure resistive load that is smaller than the pure resistive source or by employing Fig. 3.43*b* for a pure resistive load that is larger than the pure resistive source. X_{S1} and X_{P1} can be considered as a separate L network from X_{S2} and X_{P2}, so each L may be oriented any way that is convenient. For instance, X_{S1} may be an inductor, so X_{P1} must then be a capacitor; however, X_{S2} may be the capacitor, with X_{P2} being the inductor:

1. Solve for "*R*":

$$"R" = \sqrt{R_S R_L} - 8.7 \text{ ohms}$$

2. Solve for loaded *Q*:

$$Q = \sqrt{\frac{"R"}{R_{\text{SMALLER}}} - 1} = 2.2$$

3. Complete for Fig. 3.43*a*:

$$X_{P2} = \frac{"R"}{Q_2} \mid Q_2 = \sqrt{\frac{"R"}{R_L} - 1} = 3.97 \text{ ohms } (Q_2 = 2.19) \quad \text{and}$$

$$X_{S2} = Q_2 R_L = 3.28 \text{ ohms}$$

$$X_{P1} = \frac{R_S}{Q_1} \mid Q_1 = \sqrt{\frac{R_S}{"R"} - 1} = 22.9 \text{ ohms } (Q_1 = 2.18) \quad \text{and}$$

$$X_{S1} = Q_1 "R" = 18.96 \text{ ohms}$$

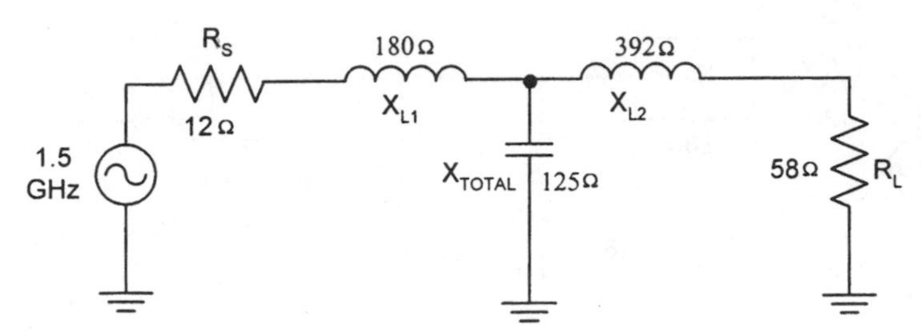

Figure 3.41 Values for a completed T network.

4. Or complete for Fig. 3.43*b:*

$$X_{P2} = \frac{R_P}{Q} = 22.7 \text{ ohms} \qquad \text{and} \qquad X_{S2} = Q''R'' = 19 \text{ ohms}$$

$$X_{P1} = \frac{''R''}{Q} = 3.95 \text{ ohms} \qquad \text{and} \qquad X_{S1} = QR_S = 3.3 \text{ ohms}$$

It is possible to match for increasingly wider bandwidths by adding sections as shown in Fig. 3.44:

1. Maximum bandwidth is always achieved if the ratios of each of the two ensuing resistances are equal, or:

$$\frac{''R''_1}{R_{\text{SMALLER}}} = \frac{''R''_2}{''R''_1} = \frac{''R''_3}{''R''_2} = \cdots = \frac{R_{\text{LARGER}}}{''R''_n}$$

2. Design as in Fig. 3.43*b* for this circuit if $R_L > R_S$, or adopt the Fig. 3.43*a* design procedure and circuit elements if $R_L < R_S$.

Impedance matching with distributed circuits. Even though it is possible to design low-value distributed series capacitors into a microwave circuit, it is ordinarily too difficult and inaccurate a procedure. This means that, wherever possible, we will want to employ shunt distributed capacitors when matching impedances in our microwave designs. But what would we do if, for instance, we find that the series input impedance of a device is inductive, and we would like to tune this inductance out? This would generally require a conjugate series capacitance to cancel the device's series input inductance. However, since we would like to get away from using a lumped series capacitor, we can convert the *series input impedance* (Fig. 3.45) of the device to an equivalent *parallel input impedance* (Fig. 3.46), which will now permit us to exploit a *shunt* distributed element to resonate out the input reactance of the device. The formulas to accomplish this conversion are:

$$R_P = R_S + \frac{X_S^{\ 2}}{R_S} \qquad \text{and} \qquad X_P = \frac{R_P R_S}{X_S}$$

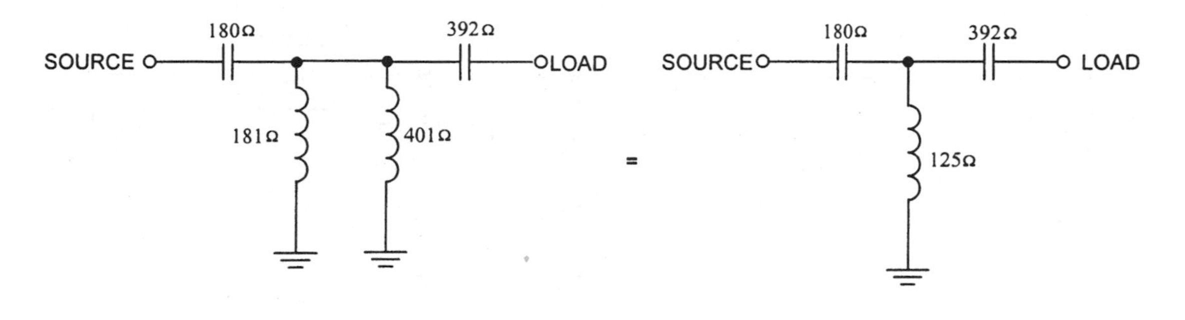

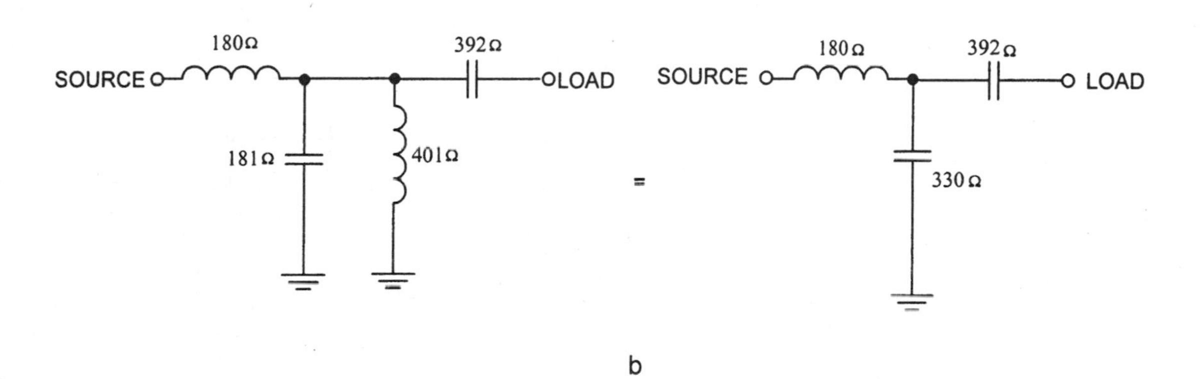

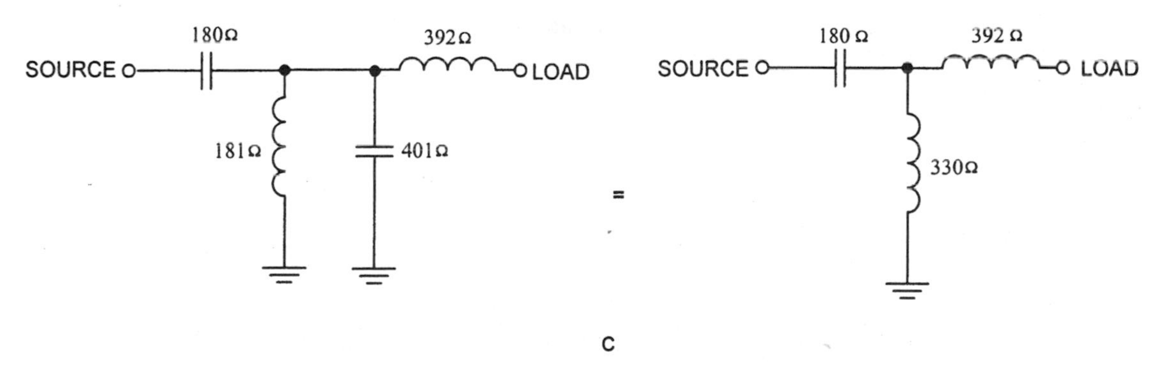

Figure 3.42 The T network and its various legal configurations before and after combination.

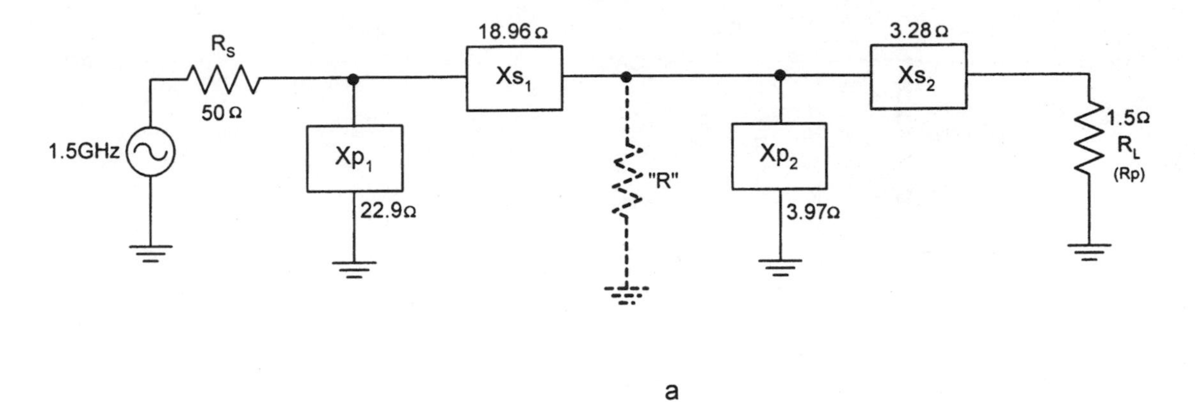

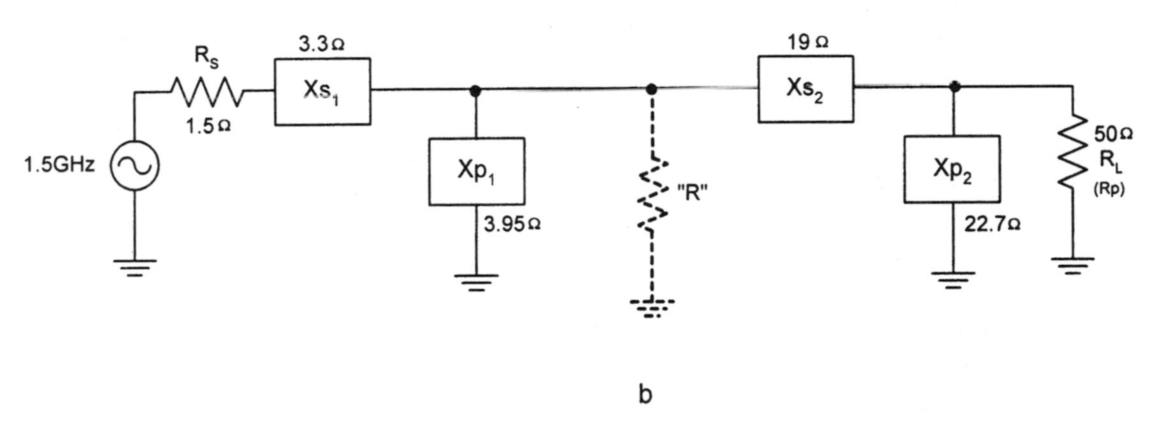

Figure 3.43 Two different configurations for a wideband matching network: (*a*) high to low; (*b*) low to high.

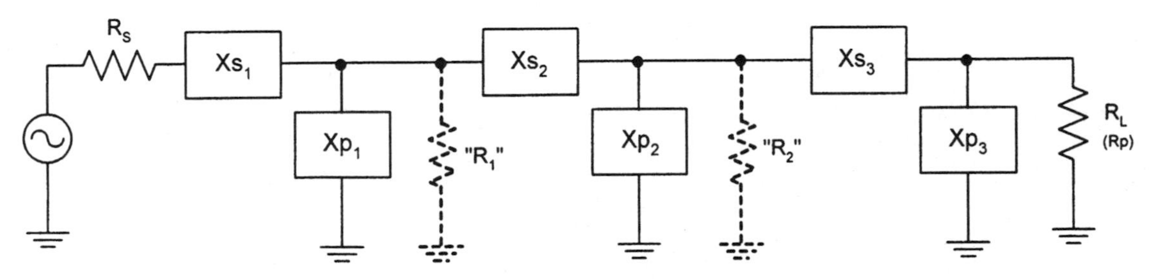

Figure 3.44 Matching for very wide bandwidths.

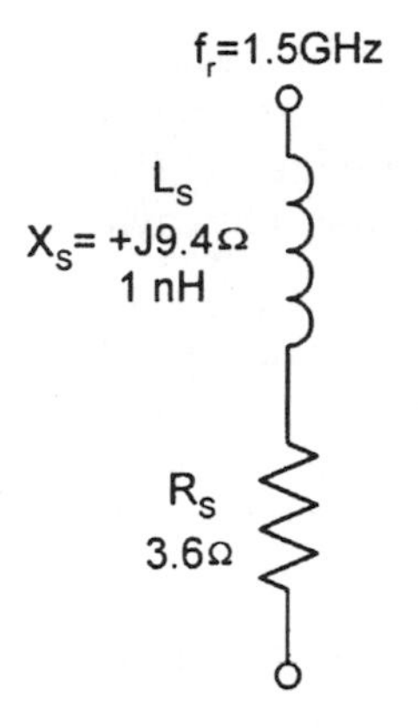

Figure 3.45 Series input impedance.

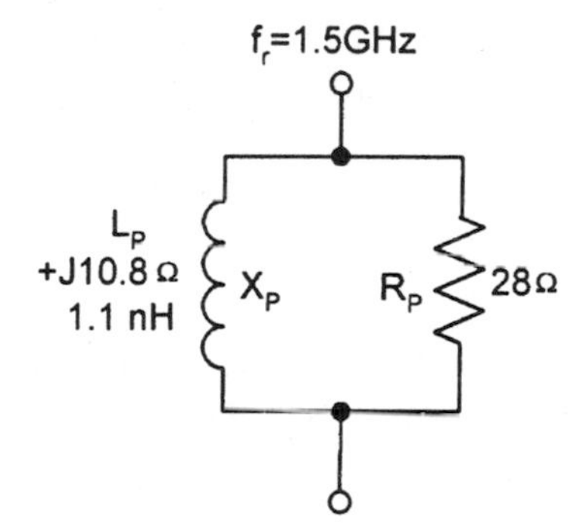

Figure 3.46 Parallel input impedance.

where R_P = equivalent parallel resistance, ohms
R_S = series resistance, ohms
X_S = series reactance, ohms
X_P = equivalent parallel reactance, ohms

Indeed, we could design the distributed matching circuit as we would the lumped type, and simply substitute the equivalent lumped distributed components as explained in Sec. 1.3.2, "Microstrip as equivalent components." However, we may find that the calculated L and C values may be beyond the normally maximum 30 degree per wavelength length limit imposed on accurate equivalent distributed components, and will be either unrealizable or inferior to a lumped part. Thus, it may be far easier to utilize microwave *quarter-wave line matching* as outlined below.

Microwave quarter-wave line matching. For small- and large-signal device impedances, matching can be accomplished as follows:

1. Calculate the input/output impedances of the device to be matched (they will be series impedances, or $R \pm jX$), or obtain these values from the data sheet.

2. Convert series $R \pm jX$ to parallel as required. Whether to employ parallel or series will depend on whether it would be easier, with microstrip, to resonate out the reactance in series or in a parallel equivalence. If a distributed part must be used for this purpose, a shunt capacitor is always desired.
3. Calculate the required microstrip width and length, at the frequency of interest, to simulate a lumped value that will cancel out the reactive component of the device being matched, making the input or output $R + j0$. Lumped microwave capacitors and inductors can also be utilized if the microstrip part is unrealizable because it would be inordinately over 30 degrees in length.
4. Then, match the now real (resistive) part of the transistor's input or output by employing a microstrip transformer (Fig. 3.47). The transformer microstrip section is placed between the two mismatched impedances (in this case, 50 ohms for the system's transmission line impedance, and 20 ohms for the transistor's input resistance). The transformer segment will be $(\lambda/4)V_P$ long (V_P = propagation velocity; Sec. 1.3.2, see "Microstrip as Transmission Line"), and as wide as a microstrip transmission line would be with an impedance of $Z = \sqrt{R_1 R_2}$, which in this case is 31.6 ohms).

Reflection coefficients. The magnitude of the reflection coefficient (signified by ρ or Γ) of a circuit or transmission line is simply the ratio between the reflected wave and the forward wave of a signal, or:

$$\rho = \frac{V_{\mathrm{REFL}}}{V_{\mathrm{FWD}}} \quad \text{and} \quad \rho = \frac{\mathrm{VSWR} - 1}{\mathrm{VSWR} + 1}$$

The reflection coefficient will always be some value between 0 and 1, since the reflected wave's amplitude will never be higher in amplitude than the

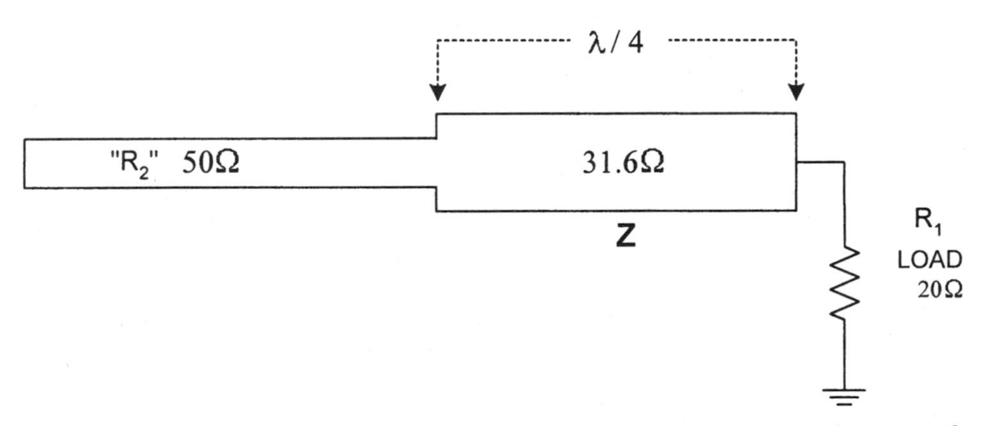

Figure 3.47 Using a distributed transformer to match a 50-ohm resistive source and an unequal resistive load.

height of the forward wave. Most values of ρ, however, will contain both magnitude and phase, instead of simply magnitude as above. These reflection coefficients are an indicator of the quality of the match between one impedance and another, or V_{REFL}/V_{FWD}, with a perfect match equaling zero and the worst match equaling 1. They can be expressed in rectangular ($\Gamma = R \pm jX$) or polar ($\Gamma = P \angle \pm 0$) forms.

Nevertheless, calculating just the magnitude ratio will allow the computation of the return loss and mismatch for any circuit.

$$\text{Return loss: RL (in dB)} = 10 \log_{10} \rho^2$$

$$\text{Mismatch loss: ML (in dB)} = 10 \log_{10} (1 - \rho^2)$$

Conversions. There may be occasions when we will need to convert from the old *Y* parameters (another way to characterize a transistor) into the newer *S* parameters, even though this is rarely required today.

$$Y_{11} = \left[\frac{(1 + S_{22})(1 - S_{11}) + S_{12} S_{21}}{(1 + S_{11})(1 + S_{22}) - (S_{12} S_{21})} \right] \times \frac{1}{50}$$

$$Y_{12} = \left[\frac{-2S_{12}}{(1 + S_{11})(1 + S_{22}) - (S_{12} S_{21})} \right] \times \frac{1}{50}$$

$$Y_{21} = \left[\frac{-2S_{21}}{(1 + S_{11})(1 + S_{22}) - (S_{12} S_{21})} \right] \times \frac{1}{50}$$

$$Y_{21} - \left[\frac{(1 + S_{11})(1 - S_{22}) + S_{12} S_{21}}{(1 + S_{22})(1 + S_{11}) - (S_{12} S_{21})} \right] \times \frac{1}{50}$$

Use *full* vector values of *S* ($Z \angle \pm 0°$) in calculations (for example, $S_{11} = 0.35 \angle -45°$).

We will have many instances when we have to convert from rectangular ($Z = R \pm jX$) to polar ($Z = R \angle \theta$) notation, and back, when designing amplifiers. The manual technique below—good only for *positive* real numbers—is one method. However, a simple scientific calculator performs the job much faster and more accurately.

1. To convert rectangular into polar form ($R \pm jX$ to $Z \angle \theta$):
   a. $Z = \sqrt{R^2 + X^2}$
   b. $\theta = \tan^{-1} X/R$ (*Note:* $\tan^{-1}$ = arc tangent)
2. To convert polar into rectangular form ($Z \angle \theta$ to $R \pm jX$):
   a. $R = Z\,(\text{COS}\ \theta)$
   b. $X = Z\,(\text{SIN}\ \theta)$

There are also many times when we must convert a *series* resistance and reactance into a *parallel* resistance and reactance. This is in order to make a certain impedance-matching problem easier to solve. Below is another technique for doing this.

To convert from series to parallel, using the example of Figs. 3.45 and 3.46 ($f_C = 1.5$ GHz):

1. Find the Q of the series circuit:

$$Q = \frac{X_S}{R_S} \qquad \text{or} \qquad Q = 2.62$$

2. If $Q < 10$, use $R_P = (Q^2 + 1)R_S = 28$ ohms
3. If $Q > 10$, use $R_P = Q^2 R_S$
4. $X_P = R_P \div Q_P = 10.8$ ohms (*Note:* $Q = Q_S = Q_P$)
5. $C_P = 1 \div 2\pi f X_P$
6. $L_P = 2\pi f X_P$

To convert from parallel to series using Figs. 3.45 and 3.46:

1. Find $Q = R_P \div X_P$
2. $R_S = R_P \div (Q^2 + 1)$ if $Q < 10$
3. $R_S = R_P \div Q^2$ if $Q > 10$
4. $X_S = R_P\,(X_P R_P / X_P^{\,2} + R_P^{\,2})$
5. $C_S = 1 \div 2\pi f X_S$
6. $L_S = 2\pi f X_S$

Selective mismatching. Designing an amplifier for a specific gain can be accomplished by selective mismatching at either its input or output port. This is an important technique, since we do not always require all of the gain that can be supplied by a particular transistor. Thus, a stage can be designed for a certain gain (or NF) by actually *not* matching the load to the source by some predetermined amount. This technique is a powerful and legitimate one, but it is wise to attempt it only when we are using an unconditionally stable transistor. However, if the extra parts can be afforded in the amplifier design, fixed attenuators can also be adopted for this purpose where noise figure is not a concern.

To carry out selective output mismatching of a transistor amplifier in order to lower its gain by mismatch losses, follow this procedure (Fig. 3.48):

1. Choose gain desired (G_{DESIRED}) for the amplifier.
2. Calculate $M_L = G_{\text{MAX}}\,(\text{dB}) - G_{\text{DESIRED}}\,(\text{dB})$

   where M_L = mismatch loss, dB

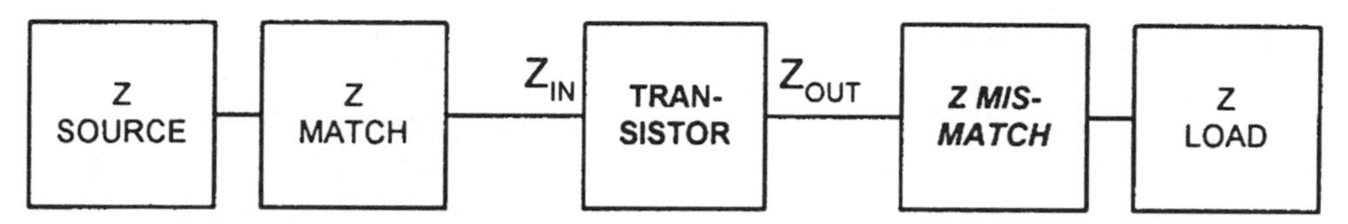

Figure 3.48 Selective output mismatching of an active device to lower stage gain.

G_{MAX} = dB gain from the transistor's data sheet, or use MAG
G_{DESIRED} = dB gain desired from the amplifier

3. Calculate RATIO, which is the ratio between the transistor's *real* output impedance $R_{Q(\text{OUT})}$ and the matching network's input, $R_{\text{IN(MATCH)}}$ (to be calculated in the next step):

$$\text{RATIO} = \frac{1 + \sqrt{1 - (10^{-M_L/10})}}{1 - \sqrt{1 - (10^{-M_L/10})}}$$

4. Find the $R_{\text{IN(MATCH)}}$ (or R_v) of the matching network:

$$R_v = \frac{R_{Q\,(\text{OUT})}}{\text{RATIO}}$$

where $R_{Q(\text{OUT})}$ = *real* part of the transistor's Z_{OUT}
R_v or $R_{\text{IN(MATCH)}}$ = *virtual* resistance at the matching network's input
RATIO = ratio of the transistor's real output impedance $R_{Q(\text{OUT})}$ to the matching network's input, $R_{\text{IN(MATCH)}}$

(R_v is used in calculations only, and is not a real circuit element.)

5. Cancel the reactance of the transistor's output by placing a reactance of the opposite value in series (Fig. 3.49; X_L). Now design the transistor's T matching network of Fig. 3.50 (L_1, L_2, C_1) to cancel all reactances in the load, but designed as if the transistor's true output impedance was now the new value of R_v.
6. Remove R_v from the design (it is only used for the initial calculations). Combine all series reactances.
7. An impedance mismatch is now formed, creating a drop in amplifier gain. This is due to mismatch losses caused by designing the transistor's output T matching network as if the transistor had an output impedance of R_v, instead of its true value. The completed mismatched output impedance amplifier is as shown in Fig. 3.51.
8. Design the input matching network for the transistor normally.

As an example, follow Figs. 3.48 to 3.51 above. Design a transistor amplifier with a gain of 6 dB at 1.5 GHz with the following device *S* parameters:

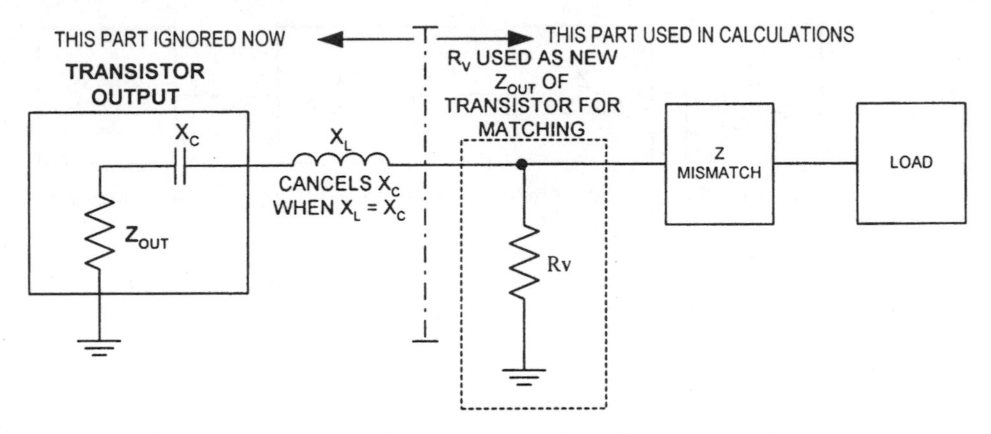

Figure 3.49 Using a virtual resistor (R_v) as the active device's temporary output resistance.

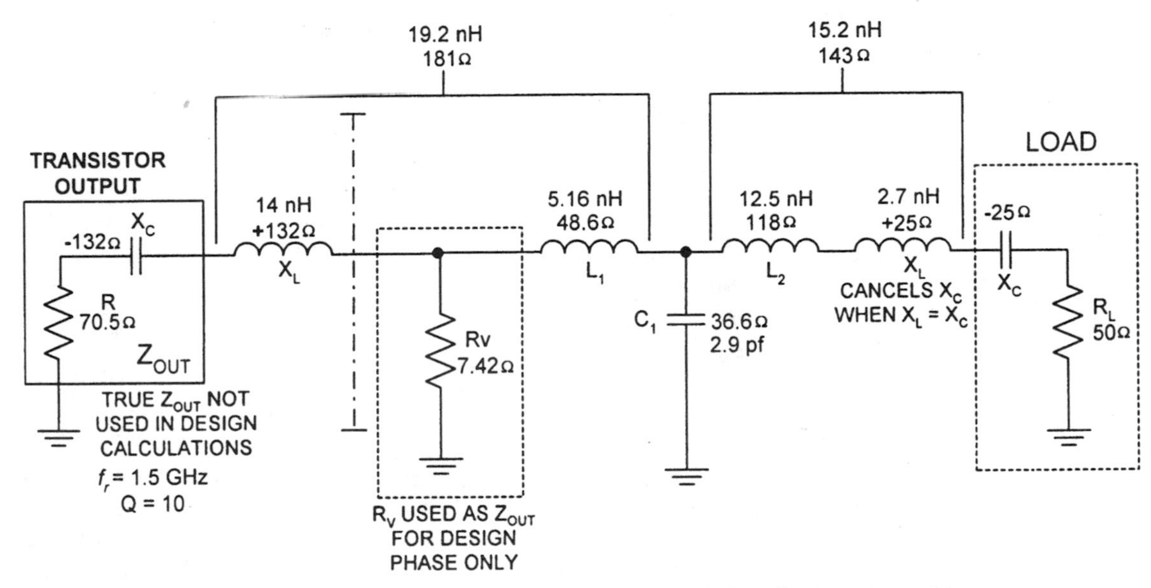

Figure 3.50 Combining the calculated values for a T matching network for selective mismatching.

$V_{CE} = 10$ V; $I_C = 6$ mA; $S_{11} = 0.195 \angle 167.6°$; $S_{22} = 0.508 \angle -32°$; $S_{12} = 0.139 \angle 61.2°$; $S_{21} = 2.5 \angle 62.4°$; MAG = 10.63 dB; MSG = 12.55 dB.

1. Choose $G_{\text{DESIRED}} = 6$ dB (or whatever value you would like the gain to be).
2. Calculate $M_L = G_{\text{MAX}}$ (dB) $- G_{\text{DESIRED}}$ (dB) $= 10.63 - 6 \approx 4.63$ dB
3. Calculate

$$\text{RATIO} = \frac{1 + \sqrt{1 - (10^{-4.63/10})}}{1 - \sqrt{1 - (10^{-4.63/10})}} = 9.5$$

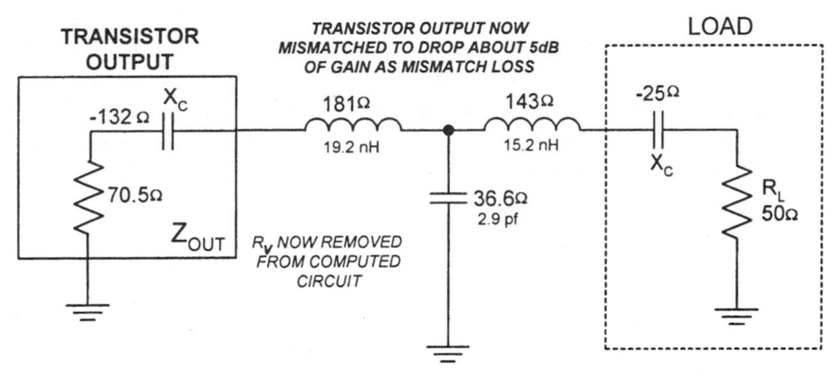

Figure 3.51 The completed mismatched amplifier with our desired gain.

4. Calculate the virtual resistance R_v required at the input to the impedance matching network:

$$R_v = \frac{R_{Q(\mathrm{OUT})}}{\mathrm{RATIO}} = 70.5/9.5 = 7.42 \text{ ohms}$$

where $R_{Q(\mathrm{OUT})}$ = real part (resistance) of the transistor's output.

5. Design matching network to cancel all reactances at the *transistor's* output and the *load's* input, while employing R_v as the new Z_{OUT} of the transistor for the impedance-matching network. Remove R_v, and combine all series reactances.

6. Design *input* matching network for a conjugate match to the transistor's input.

LNA low-noise design. Designing for the lowest noise figure for a small-signal amplifier is required when planning a receiver system for VHF and above. And since an exact 50-ohm match will rarely be used in a low-noise amplifier (LNA), only transistors with a K of 1 or more should be adopted in this application.

A minimum noise figure can be obtained from any transistor by carefully choosing its source load R_S and its bias point. This optimum source load and bias point can be found either by source resistance (R_S) versus collector current (I_C) charts or by I_C versus NF charts, both available at only a limited number of frequencies on the active device's data sheet. The optimal combination of R_S versus I_C for certain frequencies may also be available on a small Smith chart printed on the data sheet. $\Gamma_{S(\mathrm{opt})}$, which is the optimum source reflection coefficient for the lowest NF, can also be found on many low-noise transistor data sheets.

To design an LNA, first locate a transistor with a low NF at the desired frequency, then find—in the data sheets—the I_C and R_S (or $\Gamma_{S(\mathrm{opt})}$) that will give the best NF. Now design the bias network to supply the chosen I_C, while the optimal source impedance of R_S is matched at the transistor's input instead of the true impedance of the source. In other words, if we match the low-noise active device to $\Gamma_{S(\text{opt})}$ by making the LNA "see," for instance, the antenna's 50-ohm output impedance as this $\Gamma_{S(\text{opt})}$ value, we can make the LNA transistor supply a low NF (but at an increased VSWR over a perfect power match). We then terminate the transistor's *output* by:

$$\Gamma_L = \left[\frac{S_{22} + S_{12} S_{21} \Gamma_{S(\text{opt})}}{1 - (S_{11} \Gamma_{S(\text{opt})})} \right]^{\Psi}$$

where Γ_L = load reflection coefficient

$\Gamma_{S(\text{opt})}$ = optimum source reflection coefficient for the lowest NF (as found on the transistor's data sheet).

Ψ = the complex conjugate of the result

Vector quantities should be used for S.

$\Gamma_{S(\text{opt})}$ will be close to the complex conjugate of the device's input impedance, and will, at times, be so close to S_{11}^{Ψ} as to be almost identical. This would be the optimal situation, since NF and VSWR will then be optimized. This situation is more common with GaAs FETs that are operated above 2 GHz (GaAs FETs must be employed in the microwave regions above 2 GHz if BJTs cannot furnish the necessary gain and noise performance).

When the S_{11}^{Ψ} (perfect input impedance match for power) and $\Gamma_{S(\text{opt})}$ (perfect NF "match") diverge excessively from each other, system performance may suffer, with an associated noise figure and increased mismatch loss. A way to compromise between VSWR and NF is to employ a small amount of source lead inductance for the low-noise FET. If source lead inductance is used under 2 GHz, where it is most needed in both FETs and BJTs, S_{11}^{Ψ} and $\Gamma_{S(\text{opt})}$ will come quite close together, thus hopefully decreasing both NF and VSWR, while slightly decreasing gain—with minimal impact on device stability. This increased source inductance need only be added in the form of slightly longer source leads (before they contact the PCB's ground plane). The length must be found empirically, or approximated with certain RF software packages. As stated above, stability under 2 GHz will not be dramatically affected by this increase in source inductance, but at higher frequencies, this will not be the case, and instability of the LNA can result. Too high a source inductance value will, nevertheless, create amplifier instability at high *or* low frequencies, so the best value should be chosen carefully. Therefore, with both the BJT and the FET, a compromise between stability, NF, and gain must be reached, since any inductance at the emitter or source will lower gain and increase stability at low frequencies, but will actually increase gain and decrease stability at high frequencies in a wideband LNA.

If stability is still a concern, adding a very low value resistor (13 ohms) between the transistor's collector and its output matching network will decrease gain, but force stability over a very wide band of frequencies. In fact, safety from oscillation in high-gain amplifiers from VHF upward will normally demand some small value of resistive collector loading. The higher the value of the resistor, the more stability, but the less amplifier gain. Values of between 5 to 20 ohms should suffice.

As an example of LNA design, the ideal bias point for the lowest NF for a certain low-noise transistor was found to be $V_{CE} = 10$ V and $I_C = 6$ mA on the device's data sheet. The optimal Γ_S for low-noise is $\Gamma_{S(opt)}$ and equals $0.65 \angle 138°$ for this transistor, also as printed on the data sheet, for the frequency and bias of interest ($\Gamma_{S(opt)}$ may likewise be referred to as Γ_0 and Γ_{opt}). Along with $\Gamma_{S(opt)}$, we will also find the G_a expected of the transistor at our operating frequency; this is the *associated gain* at the minimum noise figure, in dB. The S parameters of the transistor at the above bias conditions are $S_{11} = 0.35 \angle 160°$; $S_{22} = 0.37 \angle -36°$; $S_{12} = 0.05 \angle 61°$; $S_{21} = 3.4 \angle 62°$. Design an LNA that has a Z_{IN} of 50 ohms and a Z_{OUT} of 50 ohms at 500 MHz. Calculate impedance matching networks and gain.

1. Is the transistor unconditionally stable ($K > 1$)?
2. The $\Gamma_{S(opt)}$ for optimal NF, as stated on the data sheet, is $0.65 \angle 138°$, which equals $-0.48 + j0.43$.
3. Find the input matching network's optimum NF matching from source to the transistor's input using $\Gamma_{S(opt)}$ in step 2; make the transistor's input think that it is seeing $\Gamma_{S(opt)}$ as its source.
4. Find

$$\Gamma_L = \left(S_{22} + \frac{S_{12}\, S_{21}\, \Gamma_{S(opt)}}{1 - S_{11}\, \Gamma_{S(opt)}}\right)^{\Psi}$$

where Γ_L = load reflection coefficient
$\Gamma_{S(opt)}$ = value as shown in step 2 above
Ψ = complex conjugate of the result

Use vector quantities for S.

5. Match the transistor's output to the next stage.

3.2 Large-Signal Amplifiers

3.2.1 Introduction

Linear Class A power and small-signal amplifiers can be designed by S parameters. Nonlinear, Class B and C power amplifiers cannot reliably exploit these parameters, but instead must depend mainly on *large-signal input/output parameter* design. These values can be found in the power transistor's data sheet in rectangular notation (such as $1.1 - j3.2$) for its series input and output impedances at a number of frequencies, and at a specific V_{CC} and P_{OUT}. The series input and output impedances can be made available as a Smith chart representation, or in a much easier to read tabular format on the device's data sheet. If desired, the series impedance values can be converted to a shunt impedance (as presented earlier in this chapter), if this information is not so provided. When provided, they may be in the form of a separate graph of input and output parallel equivalent *resistance versus frequency,* and a graph of input and output parallel equivalent *capacitance versus frequency,* at a set V_{CC} and P_{OUT}.

Designing Class C power amplifiers with small-signal S parameters will result in a circuit that is not optimized and will not function as intended. This is because any transistor's input resistance, capacitance, gain, and output resistance will be significantly different when the device is run as a large-signal power amplifier as opposed to a small-signal, Class A amplifier. However, sometimes S parameters may be placed on power MOSFET data sheets, but are only to be utilized to *approximate* a beginning design, which must then be tweaked in software, and then hardware.

Impedance matching, especially in RF power amplifiers, is required so that the transfer of energy to the next stage is accomplished with as little wasted power as possible. Matching that increases the return loss (decreasing the VSWR) of a system or amplifier prevents ripples in the passband of filters from forming and permits the active device to perform as designed with a flat gain, proper NF, low distortion, and high stability.

Since the input and output impedances of a power transistor are a complex impedance (Fig. 3.52), and this input impedance can be at very low values (which can necessitate an impedance transformation ratio of up to 20 times for a BJT), it can be seen that proper matching of a power amplifier to its source and load is anything but trivial. And the higher the desired output power of the large-signal transistor, the lower will be its output impedance, which can make matching difficult, especially at these high power levels. In fact, the design of power amplifiers just a few short years ago involved much trial and error in the tweaking of matching and bias networks in order to obtain an efficient and workable amplifier that did not self-destruct and had viable component values. To a lesser extent, this is still the case, but usually only in order to fine-tune the amplifier for low VSWR, high gain, maximum efficiency, and maximum output power because of real-world component tolerances as well as the effects of stray reactances on the completed circuit.

The transistor's input and output impedances will also decrease with an increase in frequency, which further complicates the design of matching networks for high power amplifiers—especially since these impedances can go down to $^1/_2$ ohm or less.

However, by choosing a transistor with a high collector voltage requirement, we can increase its output impedance over a transistor that operates at a lower value of collector voltage, or:

$$Z_{OUT} = \frac{V_C^{\,2}}{2P_{OUT}}$$

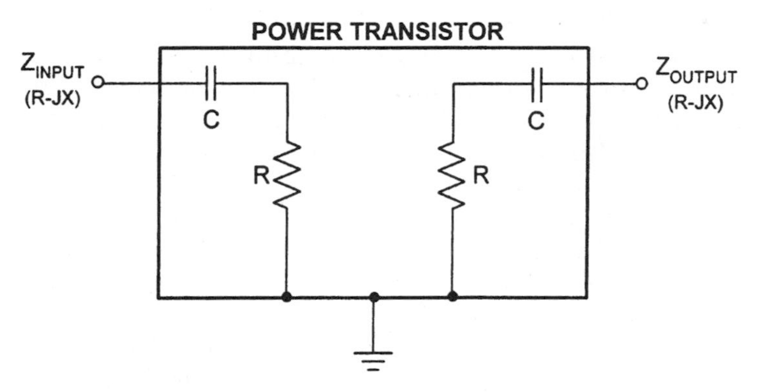

Figure 3.52 The complex series impedance of a power transistor.

where Z_{OUT} = output impedance of the transistor, ohms
V_C = DC voltage at the collector
P_{OUT} = output power, watts

There is, of course, a practical limit to this high collector voltage concept due to the available on-board (PCB) voltages, as well as internal transistor design issues.

In selecting a power transistor for our design, certain factors and specifications must be taken into consideration. The most important are: power output (P_{OUT}), V_{CC}, packaging, cost, gain, frequency of operation, power input ($P_{IN(MAX)}$), class (AB, B, C, or A), ruggedness, and built-in matching networks. The gain at the frequency of operation for the transistor must, of course, fit the requirements as specified, but choosing a power transistor with an excessive f_T will result in a more delicate device. This is because one way for the transistor designer to increase the frequency of its operation is by making the device physically smaller—and a smaller device lowers its safe power dissipation levels. Most power transistors will also be specifically characterized for different Q points—normally Class C or Class AB. If the transistor is used at another bias Q point, its parameters, such as gain, impedance, and even device lifetime, will change. In addition power gain is ordinarily at its peak with Class A amplifiers, and begins dropping as the forward bias is decreased; with Class C having the lowest power gain of any amplifier type. However, this change in bias will also affect the transistor's tolerance to impedance mismatches, which will be the greatest for Class C–biased amplifiers, decreasing as the device gets closer to Class A.

As most power transistors run at low supply voltages, current draw can be quite high, which demands chokes and inductors capable of handling these currents. Another problem with power amplifiers over small signal types is that any high-Q circuits at the amplifier's output tank will result in high circulating currents within the tank, causing very high dissipative losses and low amplifier efficiency. Unfortunately, this is in direct conflict with any requirements to attenuate output harmonics with high-Q matching in many applications. Choosing the Q of the matching networks to be either high or low will depend on whether the amplifier will be operated in a broadband application. If it is, then the Q should be as low as possible in order to pass as wide a band of frequencies as possible, while also enhancing the amplifier's stability. This stability should not be compromised if we do not allow the matching network Q to exceed 5, even in designs for narrow bandwidths.

The physical PCB layout of power amplifiers must be carefully watched. Excessively long emitter leads in a common-emitter amplifier can cause degeneration—and instability in higher frequency applications—with the effect of lower gain due to the added lead inductance. In Class C common-base power amplifiers, the effects can be even more pronounced, and will rapidly lead to complete instability.

Indeed, power amplifier stability can become an almost impossible task if the transistor is operated significantly below its own power or frequency rating. This is due to the increased gain over a safe, stable value when the transistor is not operated closer to its design specifications.

Many power transistors today are protected against instant destruction caused by brief intervals of mismatch and instability by modern fabrication techniques. Protection is important, since instability oscillations will create high peak voltages and collector currents, causing damage to an unprotected device.

A typical single-ended Class C power amplifier, with matching networks, collector bias, and decoupling circuits, is shown in Fig. 3.53.

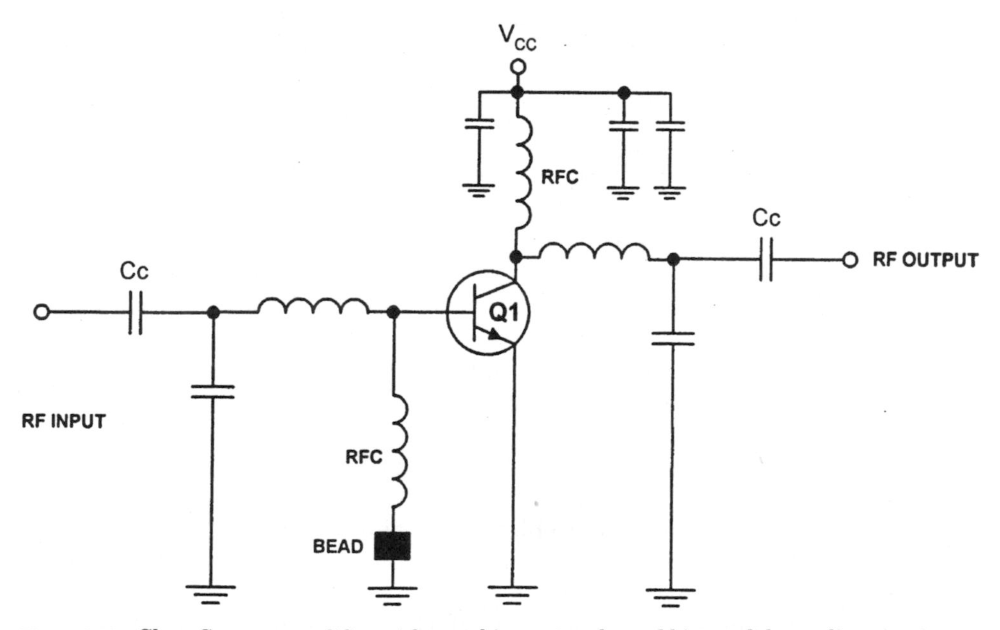

Figure 3.53 Class C power amplifier with matching networks and bias and decoupling circuits.

3.2.2 Amplifier design with large-signal series equivalent impedances

The dominant procedure for design of high-power, nonlinear amplifiers is the *large-signal series equivalent impedance* method, which characterizes a Class C, common-emitter power transistor's equivalent input and output impedances (Fig. 3.52). The large-signal series equivalent impedance is found in the data sheet of the power device, and merely represents the complex conjugate, at a specific V_{CC}, frequency, power output, power input, and bias where the transistor will supply maximum gain. This does not, however, guarantee that maximum efficiency will result from such a match, since in wideband amplifier design the lower frequencies—where gain is naturally at its highest level—may be purposefully mismatched; while the higher frequencies will be conjugately matched to peak their gain.

In designing power amplifiers, the concept of *load resistance* is sometimes employed (Fig. 3.54). This simply means that the output of the source (the driver) stage must see a certain impedance at the input of its load stage (the PA) in order to be capable of supplying the requisite input power. This is because:

$$R_L = \frac{(V_{CC} - V_{SAT})^2}{2P}$$

or, with less accuracy,

$$R_L = \frac{V_{CC}^2}{2P}$$

where R_L = load resistance (the input of the power amplifier)
V_{CC} = supply voltage of the driver
V_{SAT} = driver transistor's saturation voltage
P = P_{OUT} level required of the driver.

This will then allow the load stage, the power amplifier (PA), to output the proper power into the antenna because it has received the necessary power level from the driver. However, we must still consider the input and output

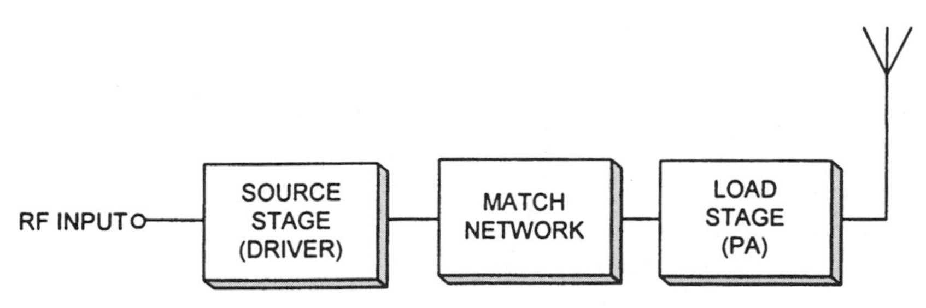

Figure 3.54 A power amplifier and driver with matching network, with the power amplifier as the driver's load resistance.

impedances of both of the transistors as a complex $R \pm jX$ value, and we must form a matching network that forces the driver transistor's output impedance to match the load resistance (R_L) value as demanded for maximum power output from the driver at its V_{CC}. In other words, the PA's input now appears to the driver stage to be the R_L that it must be in order to be able to output the required power as needed by the PA stage. This must, as well, match the resistive input to the PA and absorb or resonate out any $\pm jX$ part for a conjugate match.

Matching networks for power amplifiers should normally consist of the T type, rather than the pi type. Pi-type matching networks for high-powered amplifiers sometimes result in unrealistic component values at the higher operating frequencies encountered today into a 50-ohm load; T networks are capable of much higher frequency operation before this becomes a major problem. Both T and pi networks can be used, however, if the output impedance of the transistor is higher than its load, or the power output of the amplifier is under 15 W.

To begin the design of a power amplifier, follow these steps for power transistor impedance matching:

1. Look through the transistor's data sheet for the *output power versus input power* graph (an *output power versus frequency* graph is similar) to find out how much input power is needed to drive the amplifier for a specific output power, and at the desired frequency of operation. As necessary, apply the common formula to see the gain in dB:

$$\text{dB} = 10 \log \left(\frac{P_{\text{OUT}}}{P_{\text{IN}}} \right)$$

2. Search the data sheet for the *series equivalent impedance* on a Smith chart or in a tabular chart to obtain the transistor's series Z_{IN} and Z_{OUT} (Z_{OL}) at the frequency, power, and V_{CC} of interest (Fig. 3.52).
3. Now follow the same matching network design procedures as for small-signal amplifiers to obtain a conjugate match.
4. If a wideband power amplifier is required, then frequency-flatten as discussed at the end of this chapter.
5. Complete the bias network design for a Class C amplifier as presented in Sec. 3.3, "Amplifier Biasing."

Stability, tests, and cures. Class C power amplifiers *must* remain stable under any load or V_{CC}, since instability can "smoke" the transistor because of increased collector currents and high voltages. One way to test stability is to decrease V_{CC} to a quarter of its typical value while, with the output placed into its characteristic impedance, varying the input drive level. If the amplifier remains stable, then there is a very good likelihood that it will not oscillate under almost any adverse load condition (any power amplifier must, especially in today's competitive market, not self-destruct in a short- or open-circuited state). In addition, if the power transistor has appropriate heat sinking, the amplifier will have a much stronger chance of withstanding very poor return losses caused by a missing or short-circuited load.

In order to minimize the chances of instability, the base of a power BJT Class C amplifier should be grounded through a low-Q choke (Fig. 3.53), with a ferrite bead that is operational at these frequencies attached to the grounded end of the base lead. Decreasing low-frequency gain, which is naturally at an increased level, will also assist in stability. This is discussed under "Gain flattening," below. And the proper RF grounding of the transistor's emitter leads will help in maintaining gain and avoiding oscillations, since the smallest amount of inductance in this path to ground can prove disastrous to a power transistor. In fact, even the naturally occurring parasitic inductances and capacitances in the passive elements used for biasing, coupling, and decoupling should be modeled in software to prevent unnecessary and expensive tweaking of the completed power amplifier.

Gain flattening. All wideband power amplifiers should incorporate some type of compensation to maintain a flat gain across their entire bandwidth to within 2 dB or better. This is needed because of an amplifier's inclination to possess a higher gain at its lower frequencies than at its higher frequencies; gain decreases at 6 dB per octave as frequency increases. The high gain, as mentioned above, can cause low-frequency instabilities and subsequent transistor damage. By far the simplest method is to add a *losser network* (Fig. 3.55) between the driver and the power amplifier. This will send "excess" low-frequency power to R at an almost perfect amplitude compensation value of 6 dB per octave, thus flattening the gain response of the power amplifier.

Since this circuit is merely a high-pass network with a load, design it to pass—without attenuation—the highest frequency of interest. The natural

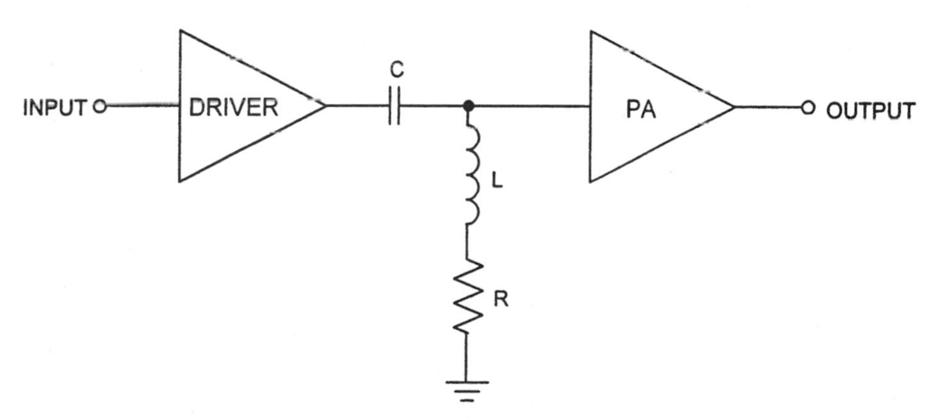

Figure 3.55 Gain flattening with an LR losser network.

rolloff of this circuit will then continue to flatten the gain as the undesired frequencies decrease in value. However, the low frequencies may begin to display an increasingly degraded return loss, so some empirical tweaking may be required, both in software and on the lab bench, for optimized values of *L, C,* and *R*.

3.3 Amplifier Biasing

3.3.1 Introduction

Classes of operation. Special classes of amplifier bias levels are utilized to achieve different objectives, each with its own distinct advantages and disadvantages. The most prevalent classes of bias operation are Classes A, AB, B, and C. All of these classes use circuit components to bias the transistor at a different DC operating, or *Q,* point (Fig. 3.56).

As shown in Fig. 3.57, Class A bias permits a signal's amplified current to flow for the entire cycle, or 360 degrees, of the input signal. This allows the amplified output signal to never reach saturation or cutoff, and thus stay within linear operating parameters. The output will be a relatively accurate amplified representation of the input signal.

Because of their low efficiency, Class A single-ended amplifiers are ordinarily used only in small-signal, nonpower applications, especially as low-distortion linear RF and IF amplifiers. This lack of efficiency is caused by the large amount of continuous DC supply power required at all times to produce the constant current that is always flowing through the amplifier—with or without any input signal present.

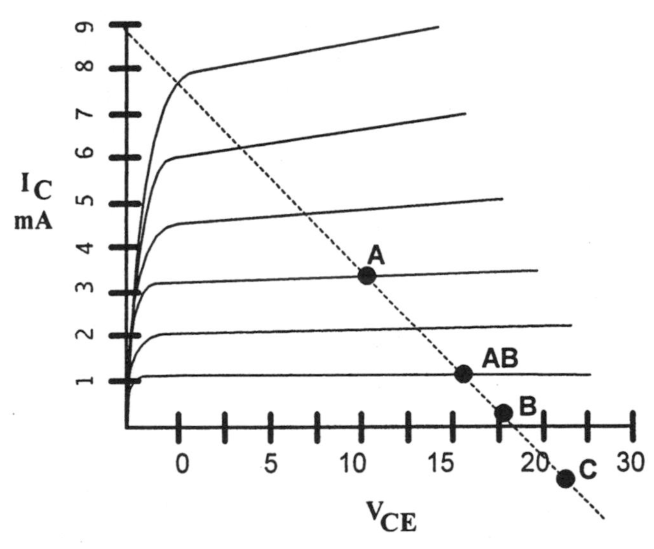

Figure 3.56 The locations of various bias *Q* points for different amplifier classes.

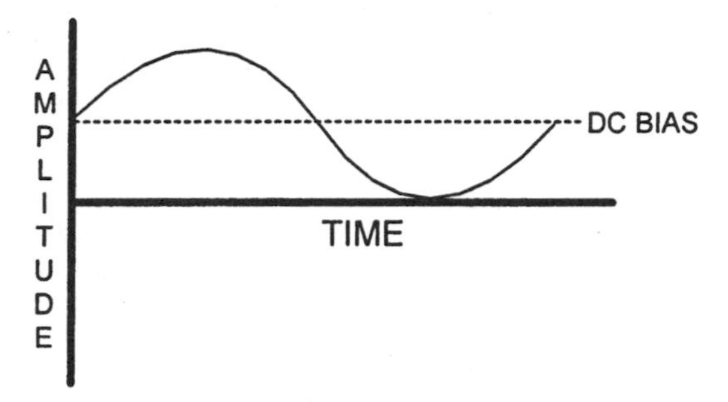

Figure 3.57 A Class A amplifier's output waveform.

Simply by decreasing the Q point of the amplifier a small amount, Class AB operation is reached (Fig. 3.58). This class of operation has a little higher efficiency than Class A since the static output current (I_C) through the amplifier will be smaller, and will also flow for something less than a complete cycle when a signal is present, normally around 300 degrees in power amplifier applications. This type of bias can also be used in small-signal linear amplifiers because the modest input signal amplitude is unable to push the amplifier into cutoff. But any Class AB single-ended power amplifier will display more output distortion than a Class A type because of the output clipping of the signal's waveform. However, Class AB is a common bias for push-pull audio power amplifiers, as well as very linear RF *push-pull* power amplifiers.

Class B bias efficiency is quite high: with no input signal, nearly zero power dissipation occurs within the amplifier. This is a result of the almost complete absence of collector current flow, since the bias is just barely decreased to overcome the 0.6 V of the base-emitter junction. When a signal is placed at the input, the output current will flow for approximately 180° of a full cycle (Fig. 3.59). This conduction will only occur when a half cycle of the signal forward biases the base, while the other half cycle will reverse-bias the emitter-base, creating a lack of output. However, considering that the Class B amplifier acts as a half-wave rectifier—amplifying only half of the incoming signal—it is normally found only in push-pull power amplifier arrangements.

Class C amplifiers are even more efficient than Class B bias, since they consume only a small leakage current when no input signal is present. When an input signal is inserted, a Class C will amplify for less than half of the input signal's cycle, and will really supply only a *pulse* at its output port. The conduction angle will be 120 degrees or less (Fig. 3.60), because the emitter-base junction is, in fact, slightly *reverse* biased. Many Class C schemes, however, may not use any bias at all, since silicon transistors, because of their 0.6-V emitter-base barrier voltage, will not conduct until this voltage is overcome by the input signal. As a pulsed output is unusable for most wireless purposes, this pulse must be changed back into a sine wave by a tuned circuit (see "Flywheel effect" in the Glossary) or filter, which will also decrease the harmonic output level. With the flywheel effect reconstructing the missing alternation, the output of a Class C

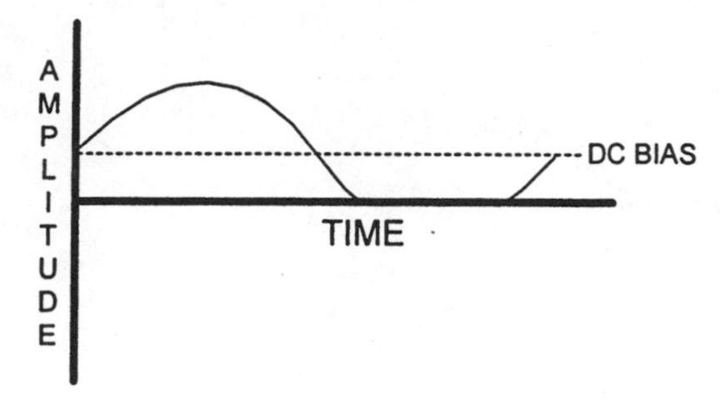

Figure 3.58 A Class AB amplifier's output waveform.

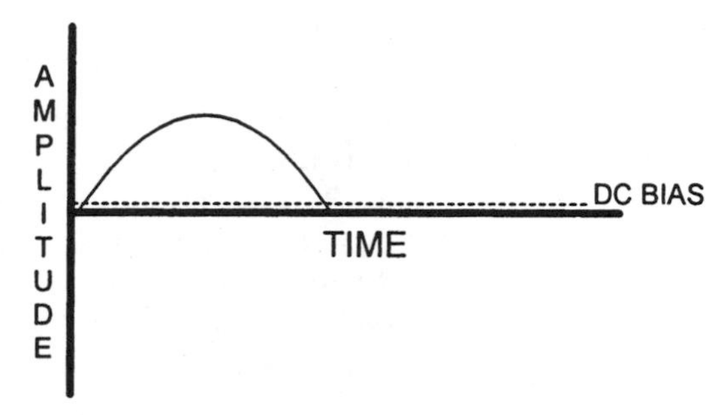

Figure 3.59 A Class B amplifier's output waveform.

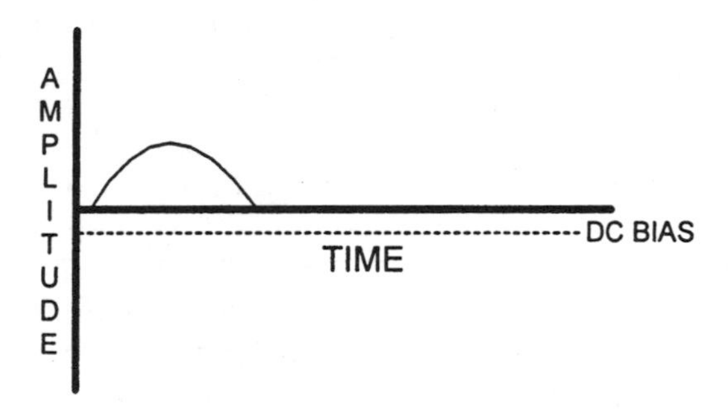

Figure 3.60 A Class C amplifier's output waveform.

amplifier will also have a peak-to-peak voltage that is double the V_{CC} of the power supply.

Class C amplifiers are found in FM driver stages, single-ended nonlinear RF power output stages, mixers, and active frequency multipliers.

Amplifier biasing circuits. A transistor amplifier must be biased with resistors and a power supply for a couple of reasons. Foremost, we would need two separate voltage supplies to furnish the desired class of bias for both the emitter-collector and the emitter-base voltages. This is still done in certain applications, but biasing was invented so that these separate voltages could be obtained from a single supply. Second, transistors are remarkably temperature sensitive, inviting a condition called *thermal runaway.* Thermal runaway will rapidly destroy a bipolar transistor, since collector current quickly and uncontrollably increases to damaging levels as the temperature rises; unless the amplifier is temperature stabilized to nullify this effect.

The dominant biasing schemes to obtain both temperature stabilization and single-supply operation are *base-biased emitter feedback, voltage-divider emitter feedback, collector feedback, diode feedback,* and *active bias.* All five are found in Class A and AB operation, while Class B and C amplifiers can implement other methods. Which bias circuit to adopt depends on the desired circuit costs, complexity, stability, and other considerations.

Base-biased emitter feedback (Fig. 3.61) works in the following way: The base resistor R_B, the 0.7-V base-to-emitter voltage drop V_{BE}, and the emitter resistor R_E are all in series, in addition to being in parallel with the power supply (V_{CC}), as shown in Fig. 3.62. As the collector current I_C increases because of a rise in the transistor's temperature, the emitter current through the emitter resistor will also increase, which increases the voltage dropped across R_E. This action lowers the voltage that would normally be dropped across the base resistor, and, since the voltage drops around a closed loop must always equal the voltage rises, the reduction in voltage across R_B decreases the base current, which then lowers the collector current. The capacitor C_E located across R_E bypasses the RF signal around the emitter resistor to stop excessive RF

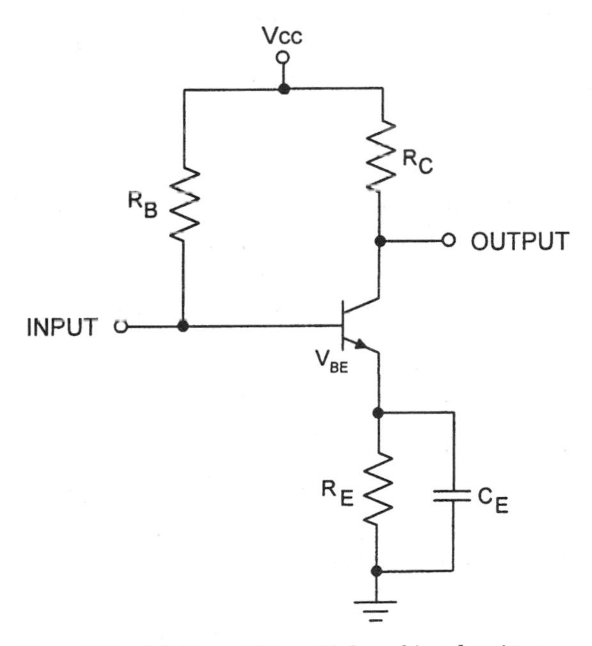

Figure 3.61 A C-E amplifier with base-biased emitter feedback biasing.

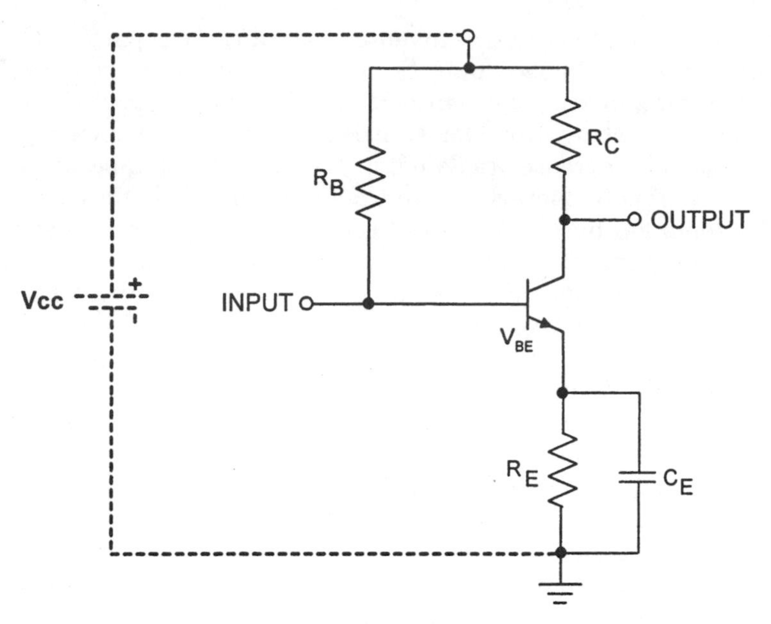

Figure 3.62 A C-E amplifier displaying its V_{CC} connection.

gain degeneration in this circuit. The higher the voltage across R_E, the more temperature stable the amplifier, but the more power will be wasted in R_E because of V_E^2/R_E, as well as the decreased AC signal gain if R_E is not bypassed by a low-reactance capacitor. Standard values of V_E for most HF (amateur band) designs are between 2 to 4 V to stabilize ΔV_{BE}. However, UHF amplifiers and above will try to completely avoid emitter resistors.

One voltage source is also supplying all of the biasing required for the base-biased emitter feedback circuit for the proper operation of the NPN transistor, since R_B and R_C are accurately allocating the suitable voltages to both the collector and the base—with the appropriate polarity—through a single power supply. This is due to the following: The collector resistor, the collector-emitter junction, and the emitter resistor are all in series with each other, and share V_{CC}'s voltage. Thus, the collector-to-emitter voltage is equal to V_{CC}, minus the voltage drop across the collector and emitter resistors of R_C and R_E, forcing the collector to be correctly reverse biased. The base circuit is also properly forward biased by the following action: The base resistor, the emitter-base junction, and the emitter resistor are in all series and share the V_{CC} power supply's voltage. So, the voltage drop across R_B will be equal to V_{CC} minus the normal emitter-base voltage drop of 0.7 V and the voltage drop across the emitter resistor. And since the voltage drop across the emitter-base and the emitter resistor are kept relatively low, most of the power supply's voltage is dropped across R_B, properly forward biasing the transistor's base. In fact, the base current, and thus the collector current, can be increased by decreasing the value of the base resistor. However, because of the inclusion of the emitter resistor

R_E and the emitter resistor's bypass capacitor C_E and their small—but unavoidable—values of stray inductance, the base-biased emitter feedback circuit is not normally employed in microwave amplifiers; gain reduction and possibly instability problems are caused by these reactances.

One of the more common of the low-cost bias schemes, with a higher temperature stability than the above method, is the *voltage divider emitter feedback biasing* circuit of Fig. 3.63. This circuit is temperature stable because the current through the voltage divider of R_1 and R_2 is significantly higher than the base current, and any rise in the device's temperature, which will increase the base current, will not substantially vary the voltage across R_2, which is equal to the voltage at the base in respect to ground; thus maintaining a constant voltage from base to ground. In addition, just as in the *base-biased emitter feedback* discussed above, when the emitter current rises with an increase in the transistor's junction temperature, the top of the emitter resistor will turn more positive. But as the base is always around 0.7 V *more* positive than the emitter itself, the base-emitter junction will now have an actual decrease in the voltage dropped across it when referenced to the common emitter lead, thus reducing I_C to its desired amplitude.

For sensitive applications, we can go even further to increase temperature stabilization. The most common method is *diode temperature compensation,* shown in Fig. 3.64. Two diodes, D_1 and D_2, which are attached to the transistor's heat sink or to the device itself, will carefully track the transistor's

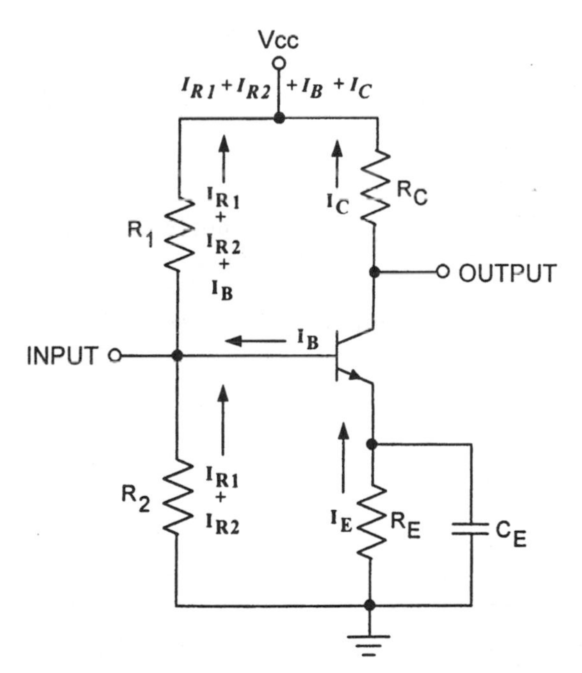

Figure 3.63 A voltage divider emitter feedback–biased C-E amplifier with current flow.

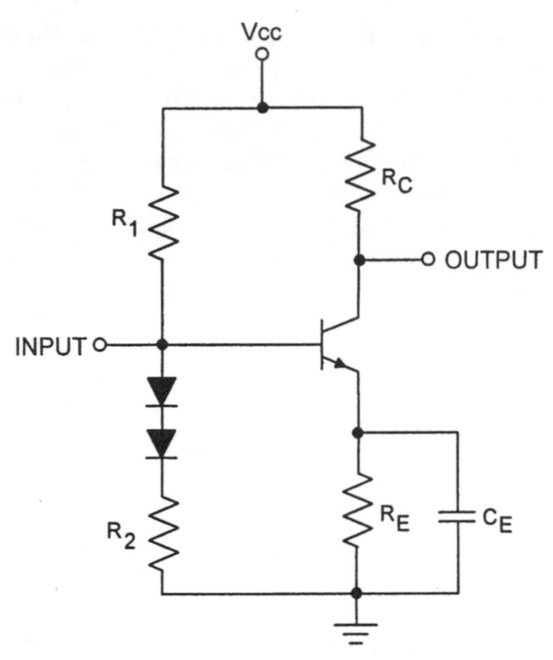

Figure 3.64 A diode temperature-compensated C-E amplifier with voltage divider.

temperature changes. This is accomplished by the diode's own decrease in its internal resistance with any increase in heat, which reduces the diode's forward voltage drop, thus lowering the transistor's base-emitter voltage, and diminishing any temperature-induced current increase in the BJT. Only one diode, or transistors or thermistors, may also be found in temperature compensation circuits for amplifiers.

A prevalent and very low cost biasing scheme for RF and microwave circuits, but with less thermal stability than above, is *collector feedback bias.* The circuit, as shown in Fig. 3.65, employs only two resistors and a transistor, and has very little lead inductance because of the emitter's direct connection to ground. Its temperature bias stabilization functions thus: As the temperature increases, the transistor will start to conduct more current from the emitter to the collector. But the base resistor is directly connected to the transistor's collector, and not to the top of the collector resistor as in the above biasing techniques, so any rise in I_C permits more voltage to be dropped across the collector resistor. This forces less voltage to be dropped across the base resistor, which decreases the base current and, consequently, I_C.

The discussion on active bias can be found in "Class A active bias for microwave amplifiers" in Sec. 3.3.2.

FETs can utilize a common Class A biasing technique called *source bias,* a form of self-bias (Fig. 3.66). With field-effect transistors, unlike bipolar junction transistors, no gate current will flow with an input signal present; so the drain current will always be equal to the source current. However, source current does flow through the source resistor R_S, creating a positive voltage at the top of this resistor. Now, since the common-source FET's source is shared by both the drain and the gate circuits, and the gate will always be at zero volts with respect to ground—since no gate current equals no voltage drop across R_G—the gate is now *negative* with respect to the common source. This allows the FET to be biased at its Class A, AB, or B Q points, depending on the value chosen for R_S, while a capacitor can be inserted across R_S to restrain the bias voltage to a steady DC value.

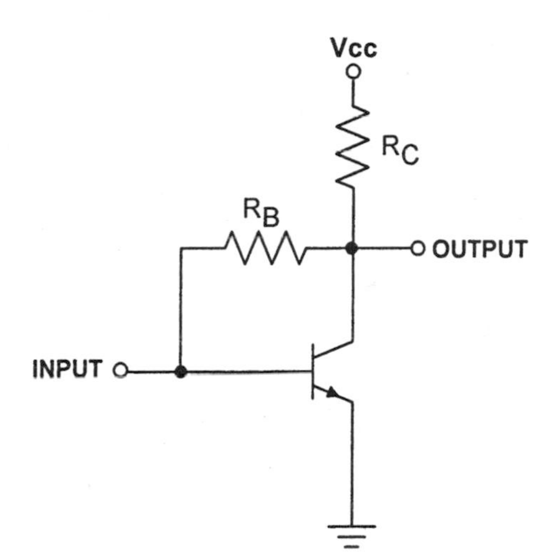

Figure 3.65 A common-emitter amplifier with collector feedback bias.

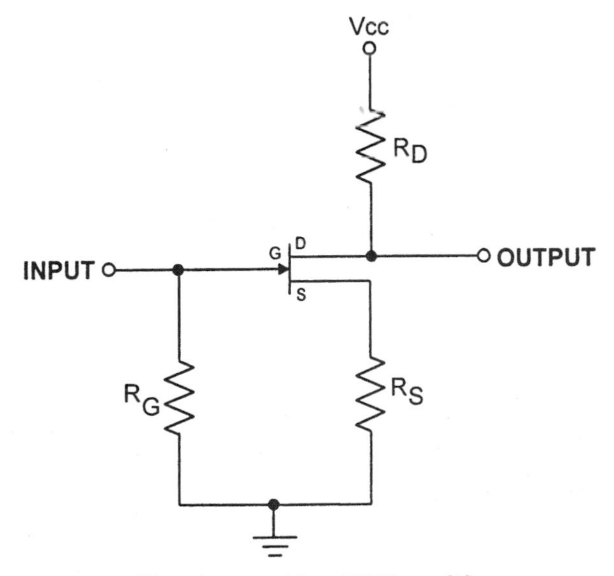

Figure 3.66 Class A source-biased FET amplifier.

For bipolar transistors, *Class C* amplifiers permit the use of three biasing techniques: *signal, external,* and *self-bias.* Nonetheless, the average Class C transistor amplifier is normally not given any bias whatsoever (Fig. 3.67), but in order to lower the chances of any BJT power device instability, the base should also be grounded through a low-Q choke, with a ferrite bead on the base lead's grounded end (Fig. 3.68). All of these biasing techniques will still require

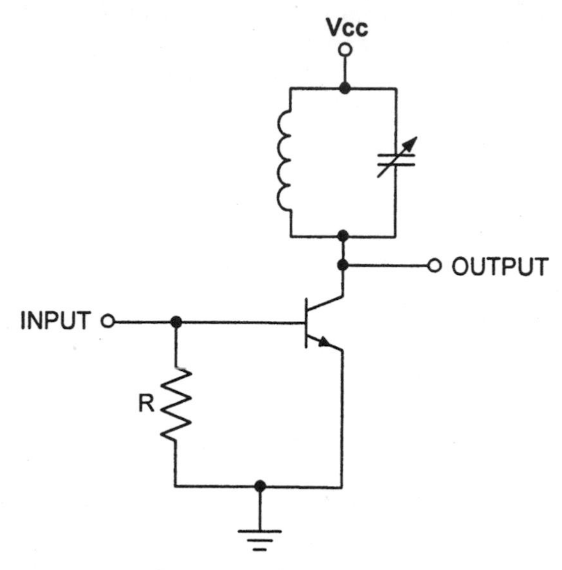

Figure 3.67 A Class C amplifier showing lack of bias.

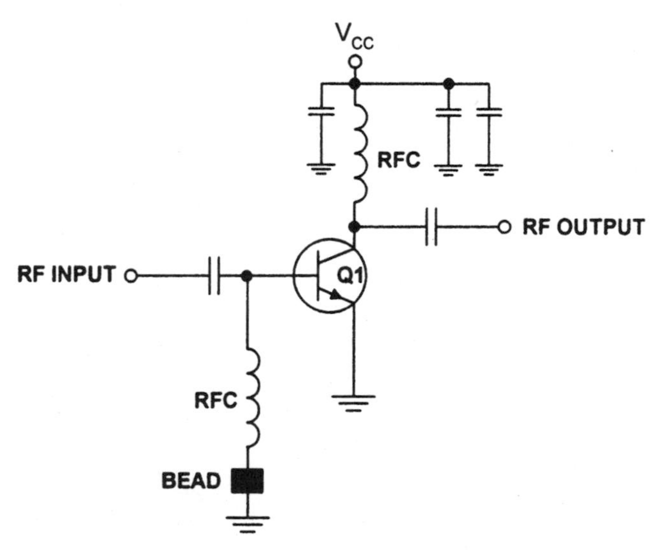

Figure 3.68 Class C power amplifier with ferrite bead on ground end of base lead.

a signal with a high amplitude to overcome the reverse or complete lack of bias at the Class C amplifier's input.

Signal-bias (Fig. 3.69) actually makes use of the signal itself to obtain the negative bias required for Class C operation. When a strong signal reaches the input of the transistor it begins to conduct, charging up the series capacitor, *C*. However, when the signal voltage does not possess the amplitude to turn on the transistor, or when the signal creates a reverse bias, *C* will then discharge through the shunt resistor, *R*. When this discharge occurs, a negative potential will form at the top of *R*, which produces the negative bias necessary for Class C operation of the amplifier. By manipulating the *RC* time constant of *R* and *C*, we can increase the negative bias so much that only the highest peaks of the input signal will turn on the transistor.

A less common method is external-bias, shown in Fig. 3.70. This circuit uses a negative bias supply to bias the base and the standard positive supply for the collector circuit. The radio-frequency choke (RFC) acts as a high impedance for the RF signal so that it does not enter the bias supply.

Self-bias (Fig. 3.71) uses the emitter current to form a voltage drop across the emitter resistor and, because of the direction of the current flow from emitter to collector, makes the top of the emitter resistor positive. With the emitter positive, which is the common element, the base—being at DC ground through RFC—is now negative in respect to the emitter. This action creates Class C operation. The capacitor C_E, placed across the emitter resistor, also has the same voltage across its terminals as R_E, and stops the bias voltage from being affected by the signal's amplitude swing.

Class B biasing is normally utilized only with push-pull amplifiers, such as that shown in Fig. 3.72, to obtain linear amplification characteristics. Any

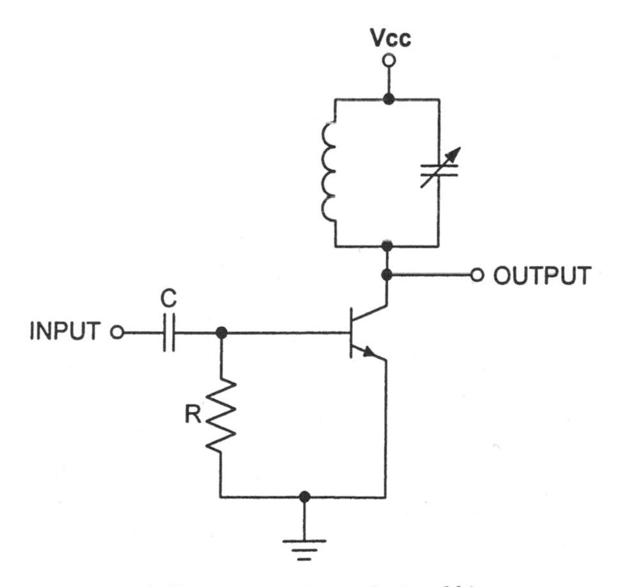

Figure 3.69 A Class C amplifier with signal bias.

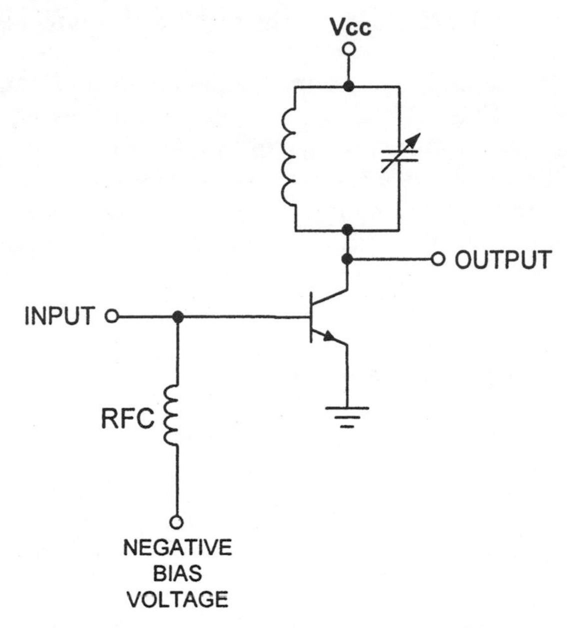

Figure 3.70 A Class C amplifier with external bias.

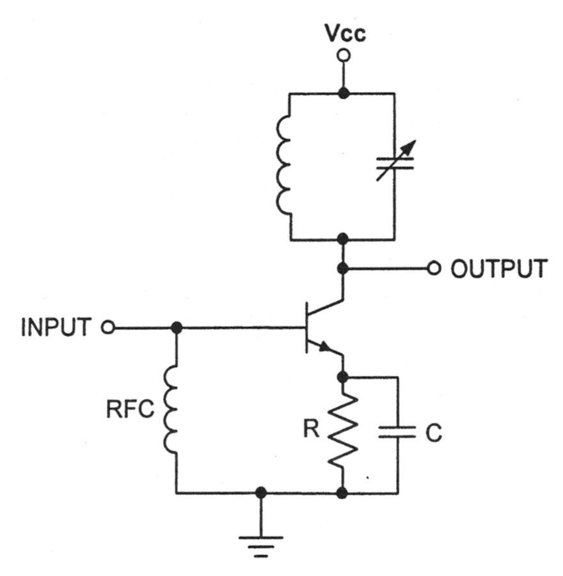

Figure 3.71 A Class C amplifier with self-bias.

scheme that biases the amplifiers at just above cutoff can be used for Class B operation, such as that shown in the push-pull amplifier circuit of Fig. 3.73 utilizing biasing *stabistors*. Stabistors are just two series diodes that maintain an even 0.7 V on each of the transistor's emitter-base junctions, while also helping to protect against the destructive effect of thermal runaway. But for

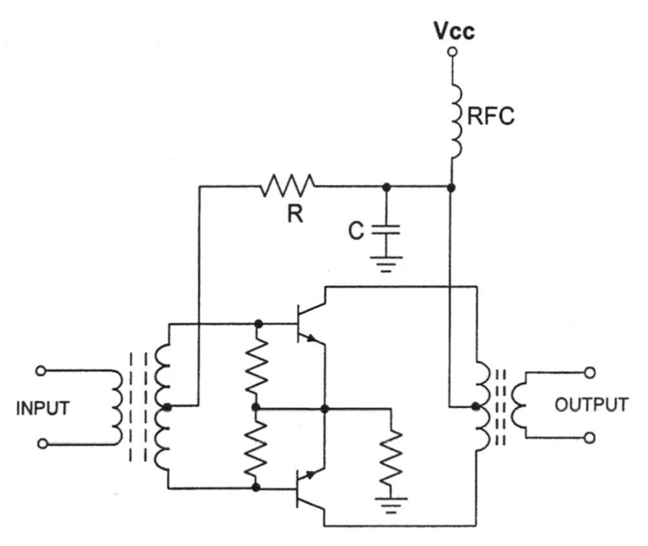

Figure 3.72 A Class B push-pull amplifier.

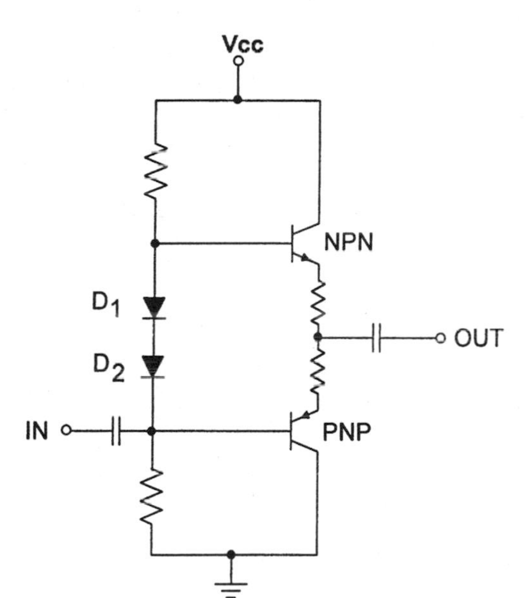

Figure 3.73 A Class B push-pull amplifier stabilized with stabistors.

nonlinear operation, Class B can employ single-ended amplifiers fed into high-Q tuned RF circuits, with any biasing arrangement that sets the transistor at approximately 0.7 V above cutoff.

Bias circuit considerations. To reiterate some of the more important aspects of high-frequency bias design: Any emitter bias resistor and emitter capacitor can create low-frequency instability and bias oscillations, in addition to increasing the NF and decreasing the gain of the amplifier. This demands that high-frequency transistors have a directly grounded emitter lead, with no emitter feedback caused by lead wire inductance. A transistor bias circuit must not only supply bias voltages to the collector and to the base, but it must also control the effects of the amplifier's temperature variations, since DC current gain, h_{FE} (ß, or I_C/I_B) of a transistor will increase by about 0.5 percent for every Celsius degree in an un-temperature-biased circuit, as demonstrated in Fig. 3.74. This graph shows the typical h_{FE} versus temperature behavior of a standard silicon transistor. Moreover, RF transistors can change their S_{21} (RF gain), stability, and NF quite dramatically as bias varies because of this temperature sensitivity. In fact, bias has a very large effect on all S parameters, as evidenced by all *.S2P S-parameter files being taken at a certain collector-to-emitter voltage and collector current. This places special demands on LNAs, which must have a very stable bias arrangement so that NF is not degraded along with temperature.However, if the transistor is expected to operate only in slightly elevated room temperature environments, then relatively primitive temperature stabilization bias schemes are all that may be required for most LNAs and general RF amplifiers.

Looking further into amplifier temperature effects, and since the two transistor characteristics that have such a large consequence on its DC operating point over temperature are ΔV_{BE} and Δß, then any good temperature-stable bias design will obviously tend to decrease these variations, as discussed above. With normal transistors, changes in beta with temperature can be drastic, and will vary the I_C by as much as ±25 percent for a temperature variation of ±50°C. In addition, part-to-part variations in ß

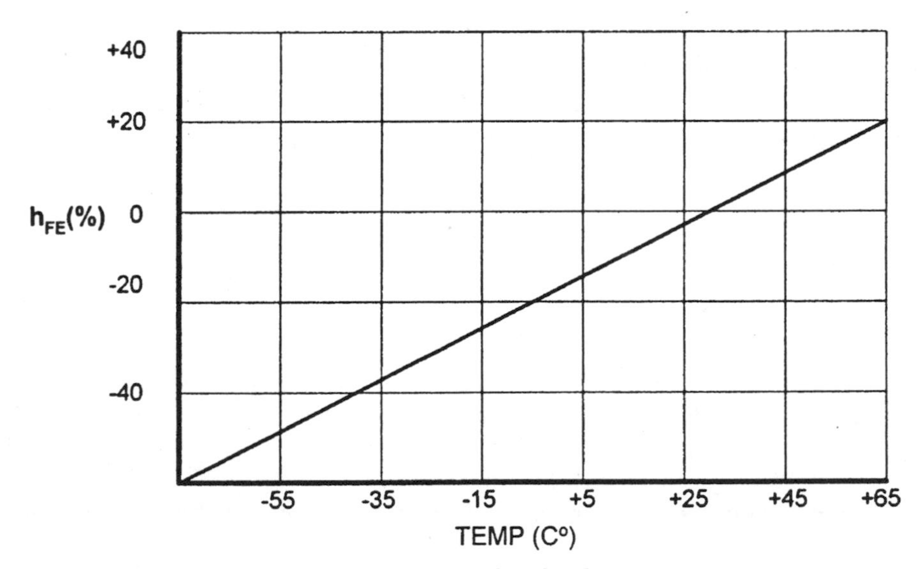

Figure 3.74 Change in h_{FE} versus temperature for a bipolar transistor.

for even a single transistor model can be of a 10-to-1 range, with ß variations of 40 to 400. Thus, in a manufacturing atmosphere, a way must be found to design an amplifier that ignores beta variations from transistor to transistor, as well as over temperature. The following formula can be adopted to calculate the change that can be expected in the I_C of a transistor, and to give us an idea of a proper bias design that will stabilize these ß variations (refer to Fig. 3.75):

$$\Delta I_C = I_{C1}\left(\frac{\Delta \beta}{\beta_1 \beta_2}\right)\left(1 + \frac{R_B}{R_E}\right)$$

where $I_{C1} = I_C$ (at $\beta = \beta_1$)
$\beta_1 = \beta_{LOWEST}$ expected of the transistor
$\beta_2 = \beta_{MAX}$ expected of the transistor
$R_B = R_1$ and R_2 in parallel
R_E = transistor's emitter resistor value
$\Delta\beta = \beta_2 - \beta_1$

We can see that the entire domination over these beta variations, which affects I_C, is only in the ratio of R_B/R_E:

$$\frac{\left(\dfrac{R_1 R_2}{R_1 + R_2}\right)}{R_E}$$

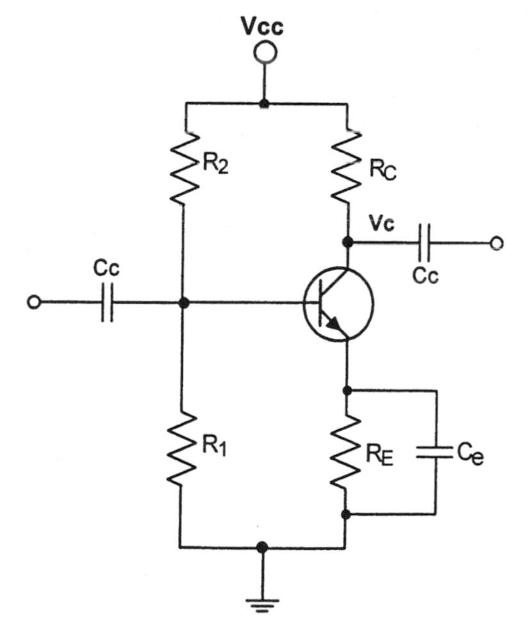

Figure 3.75 Basic C-E circuit for bias stabilization calculation.

As this ratio decreases, the ß variations stabilize—but the gain of the amplifier will decrease. An R_B/R_E ratio of 10 or less will usually give a very stable ß design.

Class A amplifiers with either inductor or *LC* resonant tank collector loads (Fig. 3.76) are able to have a lower V_{CC} and less power losses than circuits employing a resistive load at the collector. This is because the DC voltage drop across the collector load (the inductor) is at a very low value—equal to its DC resistance. Since the inductor or inductor/capacitor combination "forces" the average voltage to be approximately V_{CC} at the transistor's collector—instead of half the V_{CC} when a collector resistor is used—the RF will swing $1V_{CC}$ *above* this average V_{CC} value present at the collector down to approximately 0 V. This effectively doubles the voltage at the output of the transistor.

In designing small signal amplifiers, the collector current does not necessarily have to be at the middle of the transistor's $I_{C(MAX)}$, as it will be amplifying only low signal levels. The I_C can be chosen to be in the most linear part of its characteristic curve, and at a low enough amplitude that DC power dissipation is at a minimum; but not so low that any RF signal will be too near cut-off, or at excess distortion levels, or where the stage gain will suffer. However, most I_C values, as well as V_{CE} values, will be chosen to conform to the *S*-parameter files available for ease of design and simulation.

It must be kept in mind that after calculating the matching network for an amplifier with the existing *S* parameters, we must also calculate the bias components with the very same V_{CE} and I_C that were used to originally measure these *S* parameters, and as are shown in the *.S2P file, or the active device's impedances will not be correct, since Z_{IN}/Z_{OUT} varies with changes in I_C and V_{CE}.

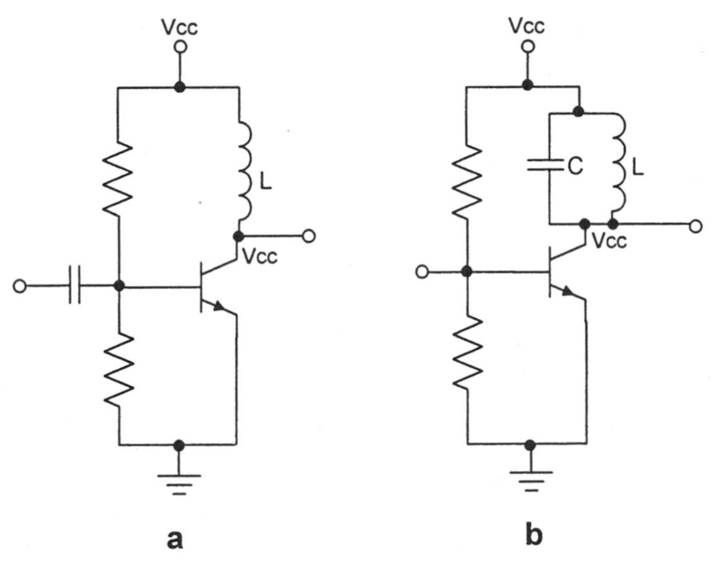

Figure 3.76 Class A transistor amplifiers with (*a*) inductor load and (*b*) tank load.

Some lower-frequency RF amplifiers will split the single emitter feedback resistor into two emitter resistors (Fig. 3.77), with only one of these resistors having an AC capacitor bypass, while the other one is providing constant degenerative feedback to enhance amplifier stability, reducing the chance of oscillations. This also allows the designer to solidly set the gain, irrespective of the transistor's varying batch-to-batch gain tolerances, to:

$$20 \log \left(\frac{R_C}{R_4} \right) = \text{gain in dB}$$

Buffer amplifiers. A buffer amplifier is designed to isolate the load from the source, which makes a high S_{12} (isolation) and a high S_{11} and S_{22} (return loss) important for a good, nonreflective match. Typically the buffer will be placed between an LO and its mixer (Fig. 3.78), preventing the LO frequency from being affected by a poor match at the mixer's port, as well as supplying some additional gain (many buffers, however, may have little or no gain). Some buffer amplifiers will have a high Z_{IN}, and are adopted mainly to block the loading of the output of an oscillator. An ordinary buffer may have an S_{12} of –20 to –50 dB, and an S_{11} of –10 to –20 dB. High-isolation MMICs, instead of discrete components, are sometimes appropriate in this isolation buffer role.

3.3.2 Bias designs

There are many different ways to bias an amplifier, depending on the required temperature stability, efficiency, costs, active device, power output, linearity, etc.

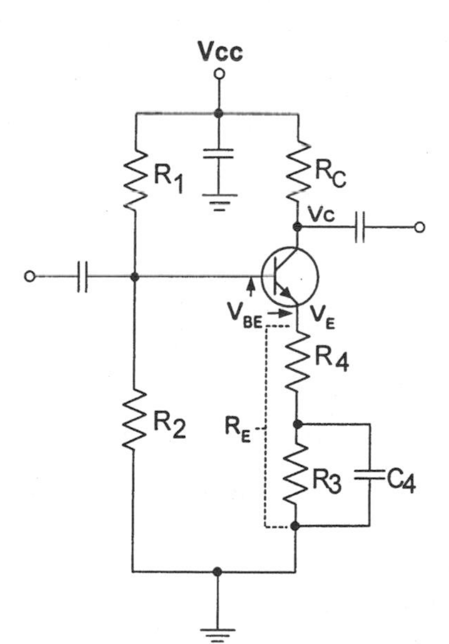

Figure 3.77 Split emitter feedback for bias and gain stabilization.

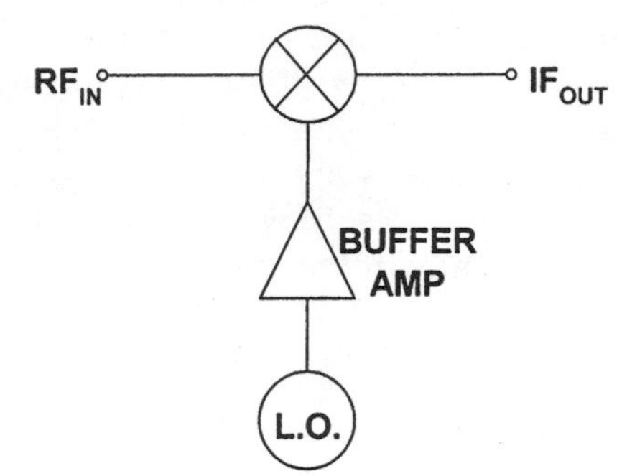

Figure 3.78 A buffer amp in an LO/mixer stage.

The following are the most popular bias circuits and design methods. In order to form a complete amplifier, input and output matching, including the decoupling/bypass capacitors and inductors, as well as the coupling capacitors, is all that is required. These concepts are all discussed in the appropriate sections.

Class A highly temperature stable diode BJT amplifier for HF and below (Fig. 3.79):

1. Choose the transistor's operating point. Example: $V_{CC} = 12$ V; $I_C = 10$ mA; $V_C = 6$ V; ß = 50. (I_C and V_C should be the same as the available *S*-parameter files for the active device, as found in *.S2P format.)
2. $R_1 = R_2 = 10$ kilohms
3. $V_B = \left(\dfrac{R_2}{R_1 + R_2}\right)(V_{CC} - 2V_F) + 2V_F \qquad (V_F \approx V_{BE} \approx 0.65\text{ V})$
4. $R_E = \dfrac{I_E}{V_B - V_{BE}} = \dfrac{I_E}{\left(\dfrac{R_2}{R_1 + R_2}\right)V_{CC}} \qquad (I_E \approx I_C)$
5. $V_C = \dfrac{V_{CC}}{2}$
6. $R_C = \dfrac{V_{CC} - V_C}{I_C}$
7. $V_F = 0.7$ V

Class A HF, VHF, UHF temperature-stable BJT amplifier design (Fig. 3.80):

1. Choose the transistor's operating point. Example: $V_{CC} = 12$ V; $I_C = 10$ mA; $V_C = 6$ V; ß = 50. (I_C and V_C should be the same as the available *S*-parameter files for the active device, as found in *.S2P format.)
2. Use a value for V_{BB} and I_{BB} to supply a constant stabilizing current (I_B): $V_{BB} = 2$ V; $I_{BB} = 1$ mA

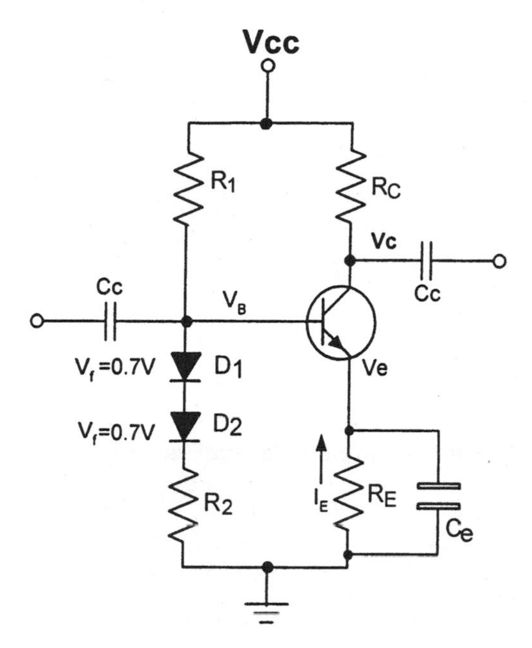

Figure 3.79 Class A diode-temperature-stabilized amplifier for bias calculations.

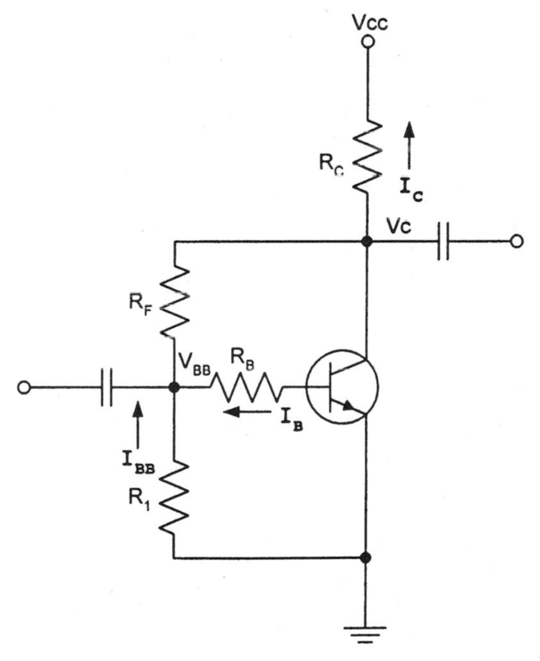

Figure 3.80 Class A HF, VHF, and UHF moderately stable amplifier design with voltage feedback.

3. Calculate $I_B = I_C/ß$

4. Calculate $R_B = \dfrac{V_{BB} - V_{BE}}{I_B}$

5. Calculate $R_1 = V_{BB}/I_{BB}$

6. Calculate $R_F = \dfrac{V_C - V_{BB}}{I_{BB} + I_B}$

7. Calculate $R_C = \dfrac{V_{CC} - V_C}{I_C + I_B + I_{BB}}$

To design a Class A or AB HF temperature-stable amplifier (Fig. 3.81):

1. Select the operating point of the transistor: V_{CC} = 12 V; ß = 50; V_C = 6 V; I_C = 10 mA. (I_C and V_C should be the same as the available *S*-parameter files for the active device, as found in *.S2P format.)
2. Give a V_E value of 2 V for bias temperature stability.
3. Give $I_E \approx I_C$ for typical or higher ß transistors.
4. Calculate $R_E = V_E/I_E$
5. Calculate $R_C = \dfrac{V_{CC} - V_C}{I_C}$

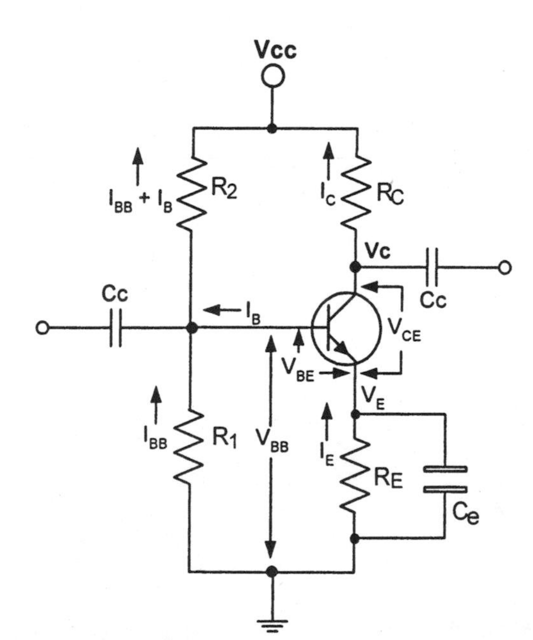

Figure 3.81 A low-frequency Class A amplifier for bias design example.

6. Calculate $I_B = I_C/ß$
7. Calculate $V_{BB} = V_E + V_{BE}$
8. Give a value for I_{BB} of 1.5 mA. (The larger in value I_{BB} is, the more improved will be the stability, but power dissipation increases.)
9. Calculate $R_1 = V_{BB}/I_{BB}$
10. Calculate $R_2 = \dfrac{V_{CC} - V_{BB}}{I_{BB} + I_B}$
11. $X_{CE} < 1$ ohm

To design a moderately temperature stable collector feedback Class A or AB amplifier for HF, VHF, UHF and above (Fig. 3.82):

1. Select a Q point for the transistor, such as: $I_C = 10$ mA; $V_{CC} = 12$ V; $ß = 50$; $V_C = 6$ V. (I_C and V_C should be the same as the available S-parameter files for the active device, as found in *.S2P format.)
2. Calculate $I_B = I_C/ß$
3. Calculate $R_F = ß\left(\dfrac{V_C - 0.7}{I_C}\right)$
4. Calculate $R_C = \dfrac{V_{CC} - V_C}{I_B + I_C}$

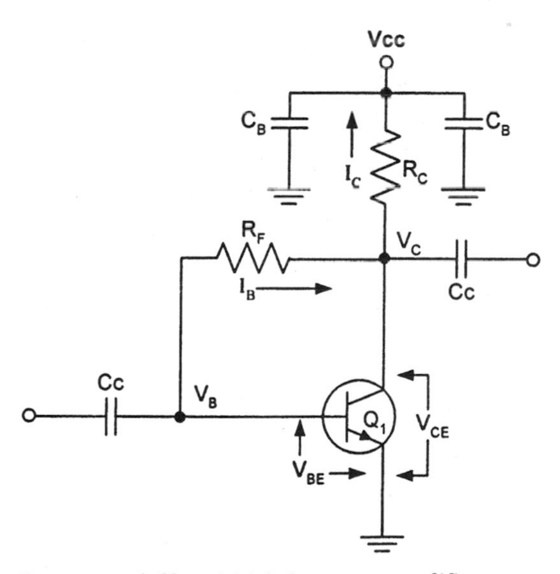

Figure 3.82 A Class A high-frequency amplifier.

5. $X_{CB} = X_{CC} < 1$ ohm

Alternatively:

1. Select a V_{CE} of $V_{CC}/2$ and an I_C that is shown by the data sheets to be at a maximum gain, NF, or P1dB. Or choose a V_{CE} and an I_C that are the same as in the *S*-parameter file available for the active device. (Do not confuse V_{CE} in *S*-parameter calculations with V_{CC}. V_{CC} can be as high as desired, but V_{CE} must be as stated in the *S*-parameter file available for the chosen device for accurate design or simulations.)

2. Calculate $R_C = \dfrac{V_{CC} - V_{CE}}{I_C}$

3. Calculate $R_F = ß \left(\dfrac{V_{CE} - 0.6}{I_C} \right)$

4. $X_{CB} = X_{CC} < 1$ ohm

Class A common-collector buffer amplifier (Figs. 3.83 and 3.84):

1. Do not overdrive buffer: 0.3 V RMS maximum input.
2. $I_e \approx 5$ mA (10 mA for more-linear output)
3. $V_e = V_{ce} = 0.5V_{CC}$

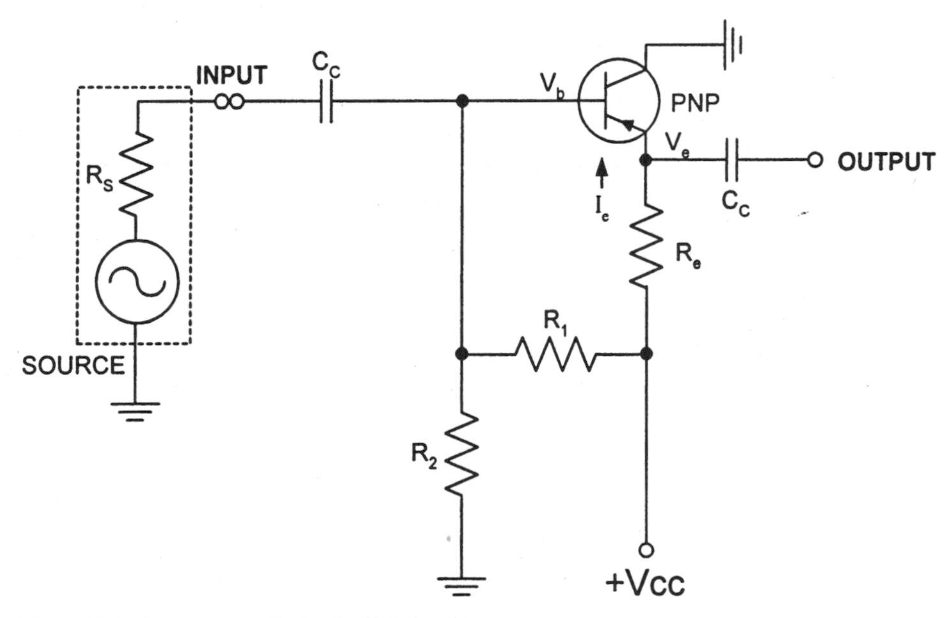

Figure 3.83 A common-collector buffer circuit.

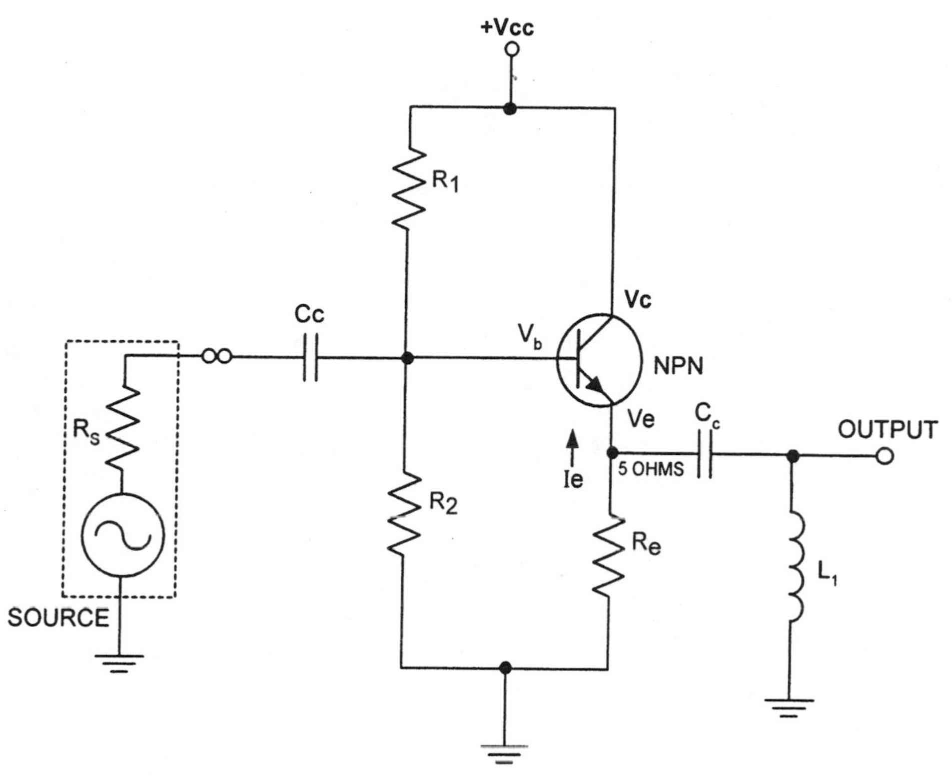

Figure 3.84 Another common-collector buffer circuit.

4. $R_e = \dfrac{V_{CC} - V_e}{I_e}$

5. $V_b = V_e - 0.6$

6. $R_2 = \dfrac{\beta V_b}{10\, I_e}$

7. $R_1 = \dfrac{\beta(V_{CC} - V_b)}{11\, I_e}$

8. $C_c = 1/2\,\pi f$

9. $Z_{IN} \approx R_e \times h_{fe}$

10. $Z_{OUT} \approx 5$ ohms or $R_{OUT} \approx \dfrac{R_S}{\beta + 1}$ or $Z_{OUT} \approx \left(h_{ib} + \dfrac{R_b \,||\, R_S}{h_{fe}}\right) || R_e$

where $R_b = \dfrac{R_1 R_2}{R_1 + R_2}$

h_{ib} = $0.025/I_e$
R_S = output impedance of the *source*
L_1, C_1 = match to 50 ohms

Class A active bias for microwave amplifiers. Both the lumped amplifier of Fig. 3.85 and the distributed amplifier of Fig. 3.86 can function as linear Class A amplifiers. They can perform with high temperature stability without the assistance of the gain-reducing and stability-robbing emitter resistor. (The emitter resistor possesses a small value of inductance, which is a big problem in amplifier applications at high VHF and above.) No bias resistors are required because of the inclusion of the DC active bias circuit of Fig. 3.87, which includes a PNP biasing transistor and its associated diode. Figures 3.88 and 3.89 show the completed and biased amplifiers, both lumped and distributed.

To design the active biasing network of Fig. 3.87 for a high-frequency Class A lumped or distributed amplifier:

1. Select an I_D through the diode of 2 mA.
2. Select an appropriate I_C for Class A bias of the RF transistor amplifier of Fig. 3.85 or 3.86.
3. Select a V_{CC} for the active bias network that is approximately 2 or 3 V greater than the V_{CE} required for the RF transistor of Fig. 3.85 or 3.86.
4. Select an RFC for the active bias circuit with an appropriate self-resonant frequency (SRF) that is greater than the frequency of operation.
5. Select both a silicon PNP transistor with a ß of at least 30 (a PNP is used so that the V_{CC} may be a positive voltage) and a low-frequency silicon diode.

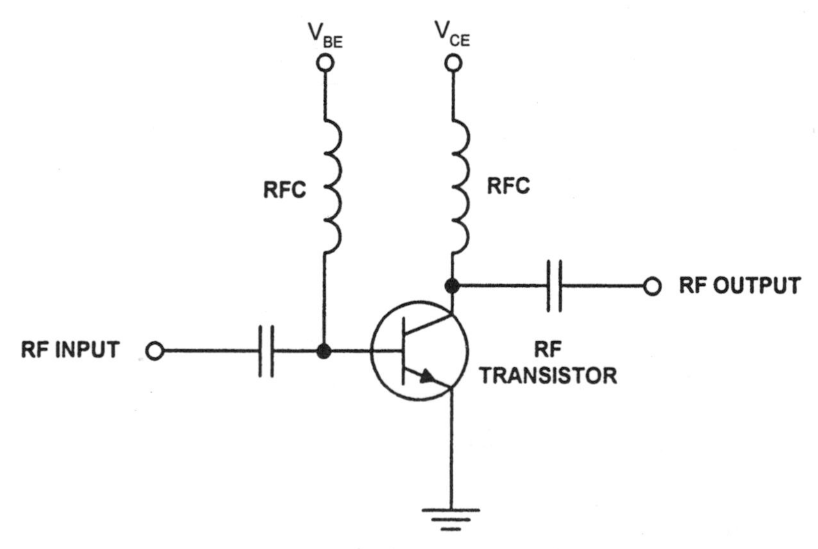

Figure 3.85 Class A lumped linear amplifier without bias circuit.

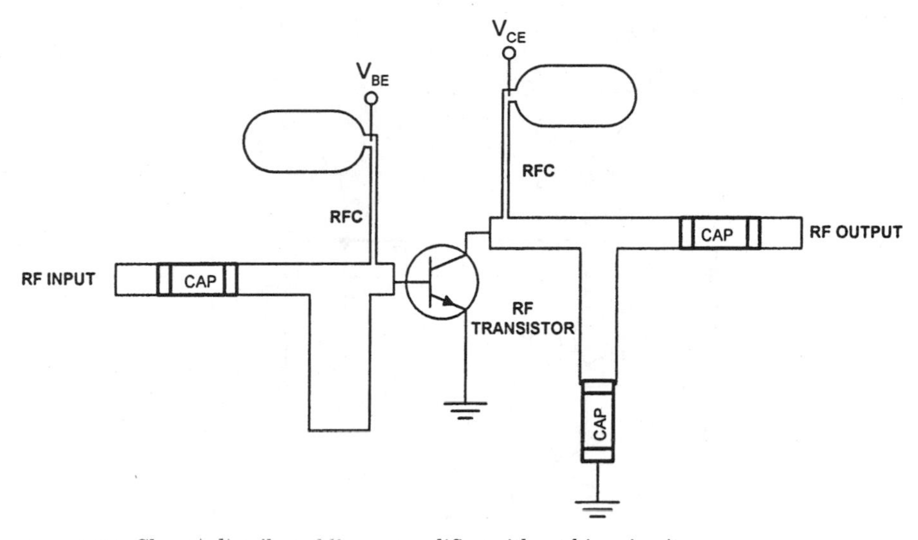

Figure 3.86 Class A distributed linear amplifier without bias circuit.

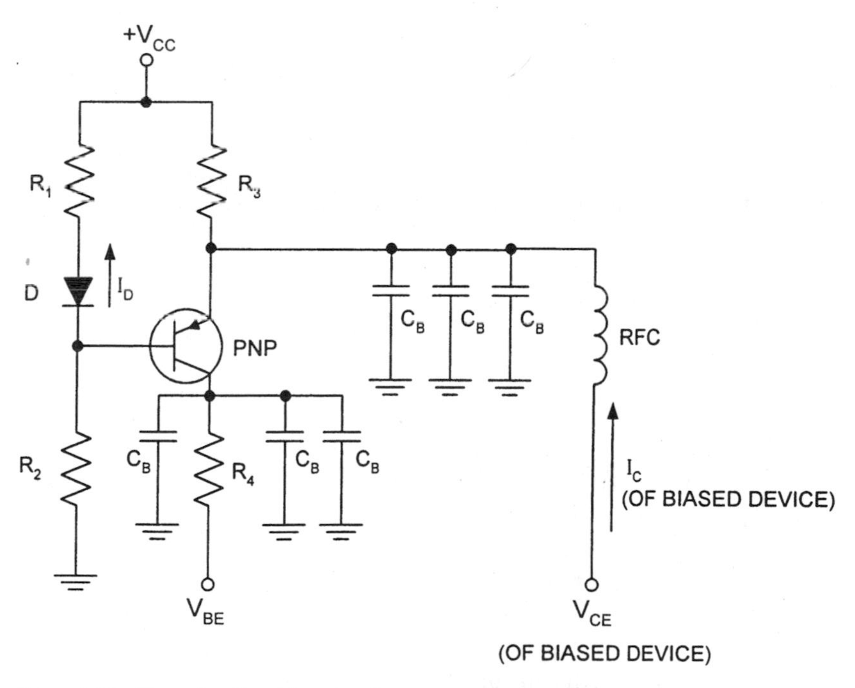

Figure 3.87 Class A active bias circuit.

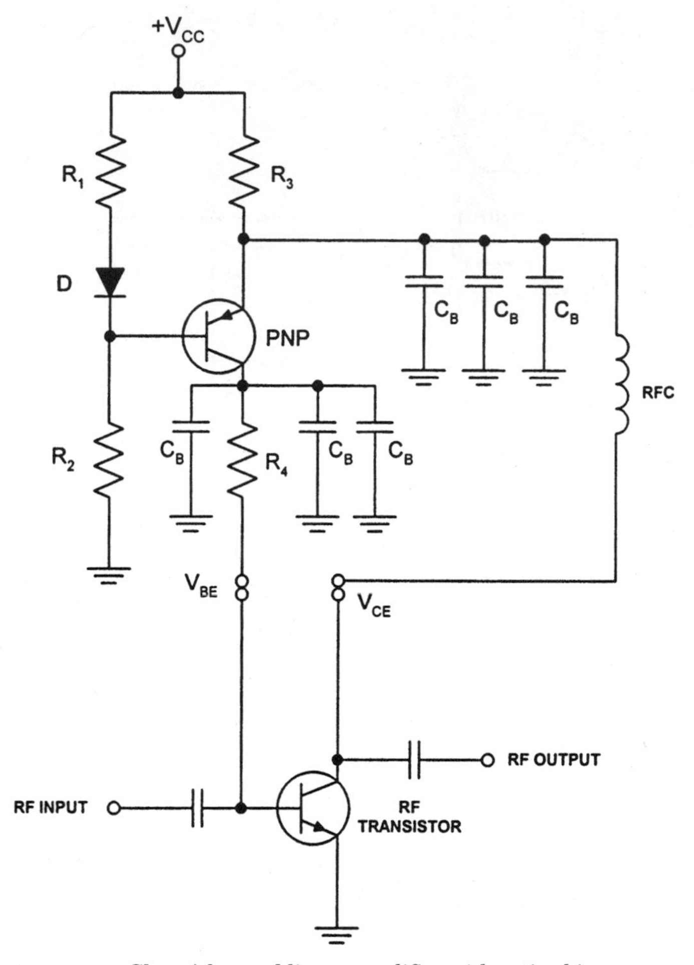

Figure 3.88 Class A lumped linear amplifier with active bias.

6. $R_1 = \dfrac{+V_{CC} - V_{CE}}{I_D}$

7. $R_3 = \dfrac{+V_{CC} - V_{CE}}{I_C}$

8. $R_2 = \dfrac{V_{CE} - 0.7}{I_D}$

9. $R_4 = \dfrac{\beta_{MIN} V_{CE} - 1}{I_C}$

10. The collector current of the biased device will be

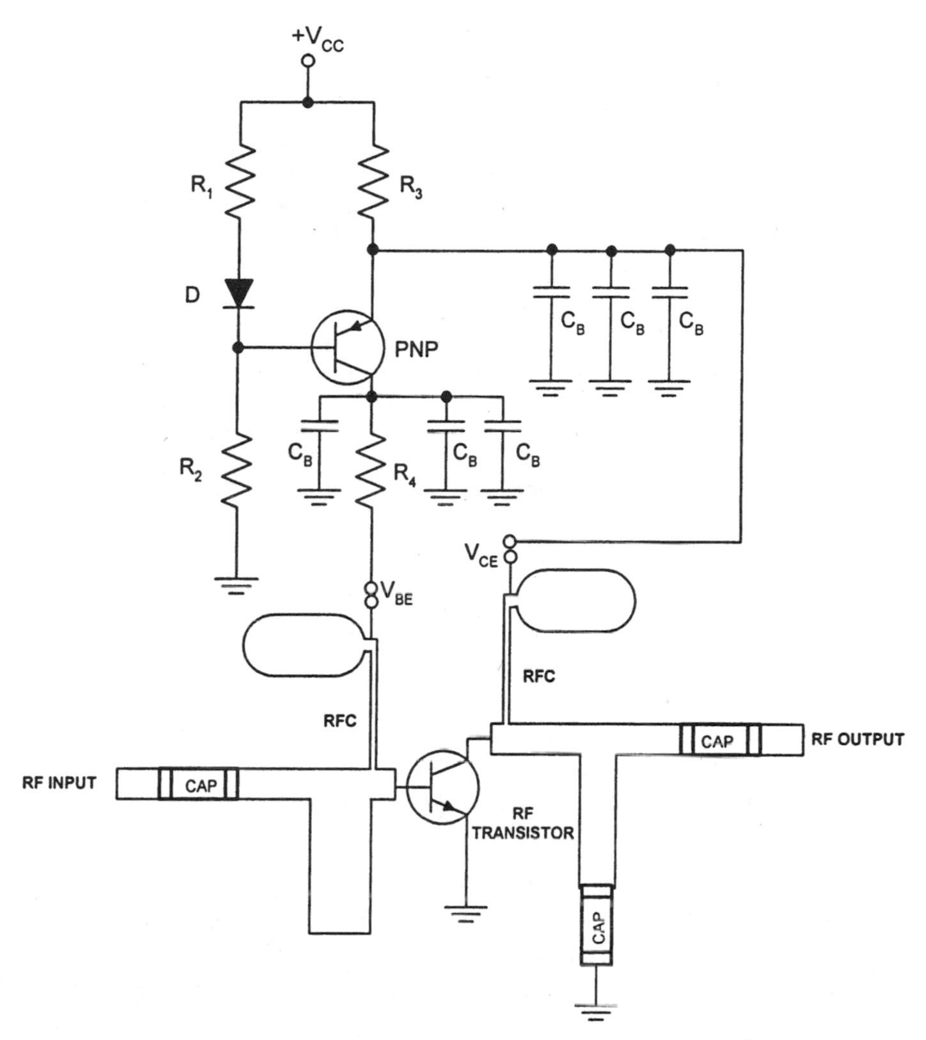

Figure 3.89 Class A distributed linear amplifier with active bias and matching circuits.

$$I_C = \frac{R_1 (V_{CC} - 0.7)}{R_3 (R_1 + R_2)}$$

11. The collector-to-emitter voltage of the biased device will be

$$V_{CE} = V_{CC} - I_C R_3$$

12. C_B = <1 ohm

JFET self-bias common-source Class A amplifier for VHF and below (Fig. 3.90)

1. Select a V_{dd} and an appropriate V_{gs} for Class A operation from the data sheet for the JFET selected, and note the I_d for this chosen V_{gs}.

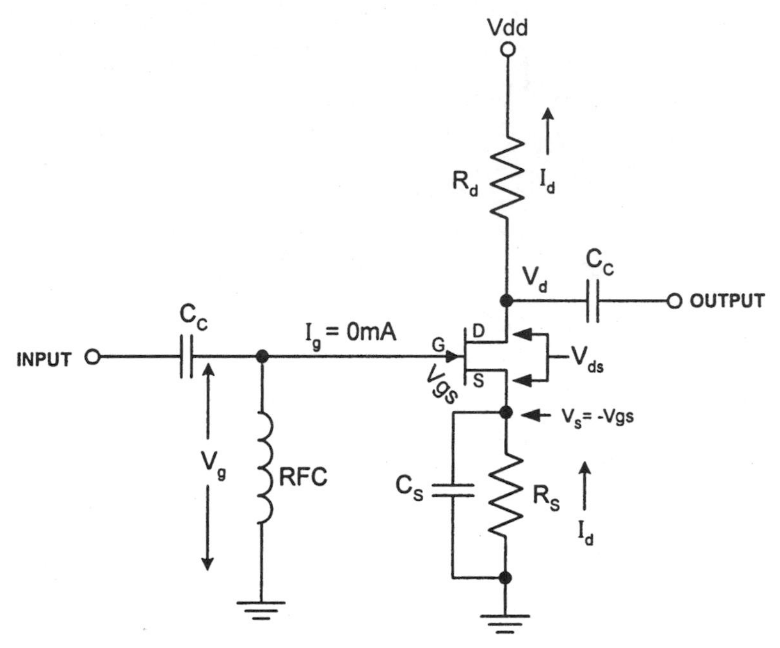

Figure 3.90 JFET biasing up to medium frequencies.

2. Compute $R_S = V_{gs}/I_d$.
3. Select a V_{ds} of $(V_{dd} - 2V_{gs})/2$ for a V_d of $V_{dd}/2$.
4. Calculate $R_d = (V_{dd} - V_{ds} - V_{gs})/I_d$. (If R_d computes to lower than 1 kilohm, an RFC must be used between the top of R_d and the V_{dd} in order to sustain a minimum RF impedance into the power supply.)
5. Place a high-impedance RFC (with an appropriate SRF) or a high-value resistor (1 megohm) from the FET's gate to ground.
6. $C_C < 1$ ohm.
7. Add bypass capacitor C_S to increase the gain of the amplifier.

Note: Always confirm that the FET will safely dissipate the power required, which is $P = I_d V_d$.

Since V_{gs} for a specific I_d is not always available, use the following equation to find V_{gs} when I_{dss} and V_p are known (look in the JFET data sheet for I_{dss} and V_p):

$$V_{gs} = V_p\left(1 - \sqrt{\frac{I_d}{I_{dss}}}\right) \qquad (V_p \approx V_{gs(off)} \text{ and } V_s = -V_{gs})$$

The I_d and V_{ds} will normally be chosen as a duplicate of the values used in any available *S*-parameter file for the device to be modeled. In fact, many manufacturers of FETs will have *S* parameters that are taken at different values of V_{ds} and I_d. The I_d can be quoted as a percent of I_{dss}—the maximum I_d—such as "50 percent of I_{dss}," which would work well for Class A bias.

JFET HF Class A stable amplifier (Fig. 3.91)

1. Select a Q point (V_{dd}, I_d, $V_d = V_{dd}/2$).
2. Calculate

$$R_d = \frac{V_{dd} - V_d}{I_d}$$

   where

$$I_d = I_{dss}\left(1 - \frac{V_{gs}}{V_p}\right)^2$$

3. Find V_p and I_{dss} from data sheet.
4. Calculate

$$V_{gs} = V_p\left(1 - \sqrt{\frac{I_d}{I_{dss}}}\right)$$

5. Select V_s to be 2 or 3 V.
6. Calculate $R_S = V_S/I_d$.

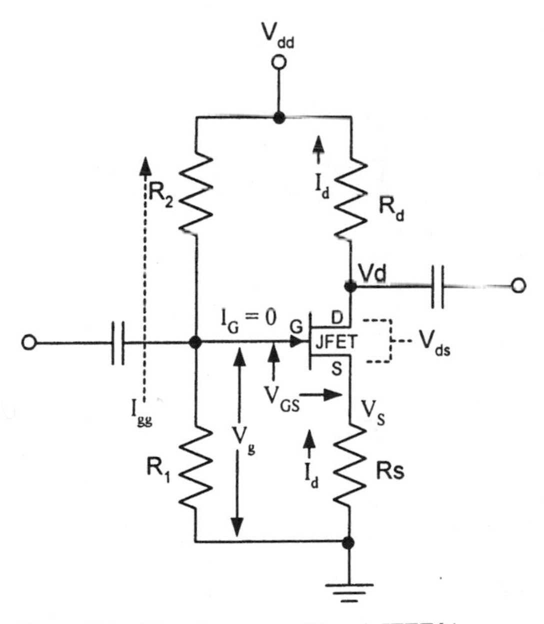

Figure 3.91 A low-frequency Class A JFET bias circuit.

7. Calculate $V_g = V_{gs} + V_s$.
8. Use an R_1 value of 220 kilohms (this effects the DC input resistance).
9. Calculate

$$R_2 = \frac{R_1 (V_{dd} - V_g)}{V_g}$$

Note: The I_d and V_{ds} will usually be chosen as a duplicate of the values used in any available *S*-parameter file for the device to be modeled. In fact, many manufacturers of FETs will have available *S* parameters taken at different values of V_{ds} and I_d. (I_d is usually quoted as a percent of I_{dss}—the maximum I_d—such as "50 percent of I_{dss}," which would work well for Class A bias.)

Bias design of a BJT Class C power amplifier (Fig. 3.92). Since the average silicon transistor will naturally run in Class C mode if no base bias at all is supplied, the bias network for such an amplifier is quite simple, as shown in Fig. 3.92. In order to minimize the chances of instability, the base of this BJT Class C amplifier should be grounded through a low-*Q* choke. A ferrite bead that is operational at the frequencies of interest should be attached to the ground end of the base lead.

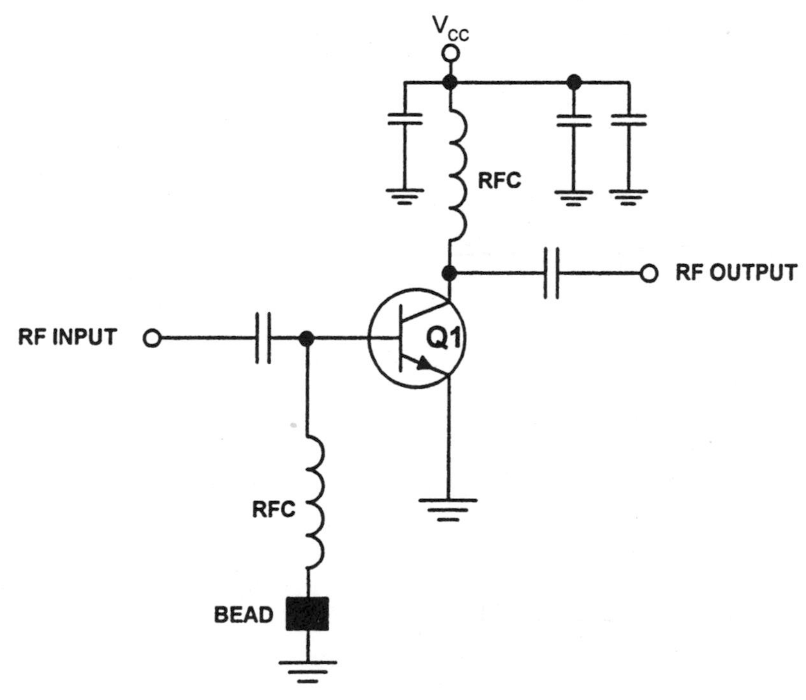

Figure 3.92 A high-frequency Class C power amplifier.

General Bias Notes

1. Utilization of an emitter resistor is avoided at VHF and above because its small inductance would create instability and decrease gain. A stripline opposed emitter (SOE) transistor package helps minimize this inductive effect in the transistor's leads themselves. However, some series lead inductance will improve stability at lower frequencies. For instance, at 2 GHz an inductance of up to 2 nH is good, but this value is usually present on the bare emitter leads and the plated via hole to ground anyway.
2. The collector-to-base breakdown voltage of a transistor should be chosen to be about 3 times its V_{CC}.
3. S_{21} will fall at 6 dB/octave in any active device, which translates into high gain at low frequencies. This can mean low-frequency instability, necessitating a gain flattening network in the transistor's base circuit (See "Gain Flattening" in Sec. 3.2.2).

3.4 MMICs

3.4.1 Introduction

MMICs are *monolithic microwave integrated circuits,* typically containing a 50-ohm small-signal amplifier that requires very few support components for biasing, and none for impedance matching.

Figure 3.93 illustrates a very common MMIC package, with integral microstrip leads, for high-frequency operation. Some MMICs may have a separate DC power input pin on the package itself, which may be of the eight-pin dual in-line package (DIP) variety.

Even though a majority of amplifier MMICs are unconditionally stable, it is wise not to assume that *all* MMICs are. However, the manufacturer will usually warn you if the amplifier is potentially unstable, even if the warning is in small print at the bottom of the data sheet.

Taking the example of a high-quality and stable MMIC in the Agilent INA series of RFIC gain blocks, the internal structure is as shown in Fig. 3.94. This Agilent design employs a single transistor driving a Darlington pair, with a

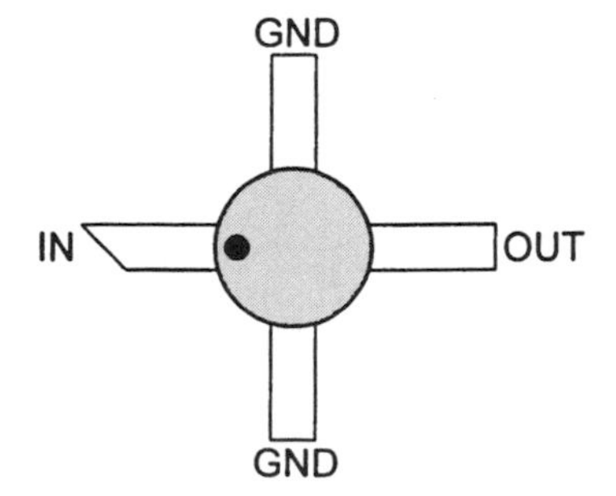

Figure 3.93 Standard MMIC amplifier package.

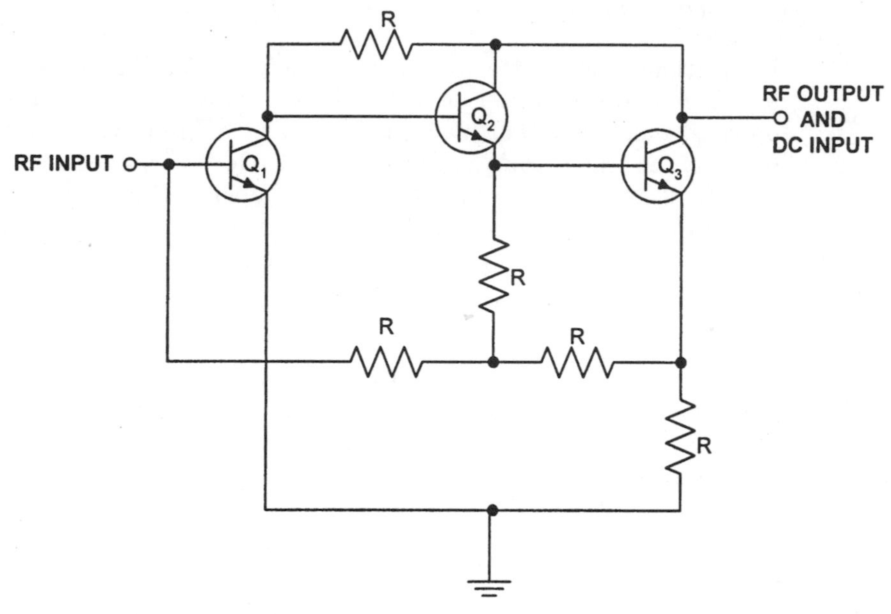

Figure 3.94 Internal circuit arrangement of an average MMIC.

small resistive feedback to set the RF parameters. The single transistor has only a small amount of negative feedback for low-noise performance, with high gain into the DC coupled input of the Darlington pair. The Darlington pair has strong degenerative feedback, and sets the gain and matching of the RFIC, as well as the gain flatness.

3.4.2 MMIC biasing

The *current-biased* MMIC (by far the most common type), will attempt to draw more current as the temperature rises. To preserve a MMIC's drain current (I_d) so that its Q-point bias does not vary with temperature or large input signals, the DC bias circuitry consisting of R_{BIAS}, Fig. 3.95) must maintain a constant I_d to the MMIC under all conditions. This bias stabilization will lower the MMIC's DC voltage (V_d) with higher device temperatures, and increase the voltage with lower device temperatures, thus preventing the bias current from decreasing or from increasing to a level that can actually destroy the MMIC. Either way, changes in I_d will also vary both the gain and the P1dB of the device. However, the R_{BIAS} in the DC bias line will drop more of the MMIC's V_d voltage from the V_{CC} supply (since $V = IR$), which decreases V_d and, consequently, the MMIC's bias current to the desired levels. Since the effectiveness of this temperature bias control is dependent on the voltage drop across R_{BIAS}, a value of up to 4 V may be required for proper stabilization over a temperature range of $-25°\text{C}$ to $+100°\text{C}$. By using a higher value resistance of R_{BIAS} along with a higher supply voltage, stabilization is improved. Nevertheless, R_{BIAS} should not have such

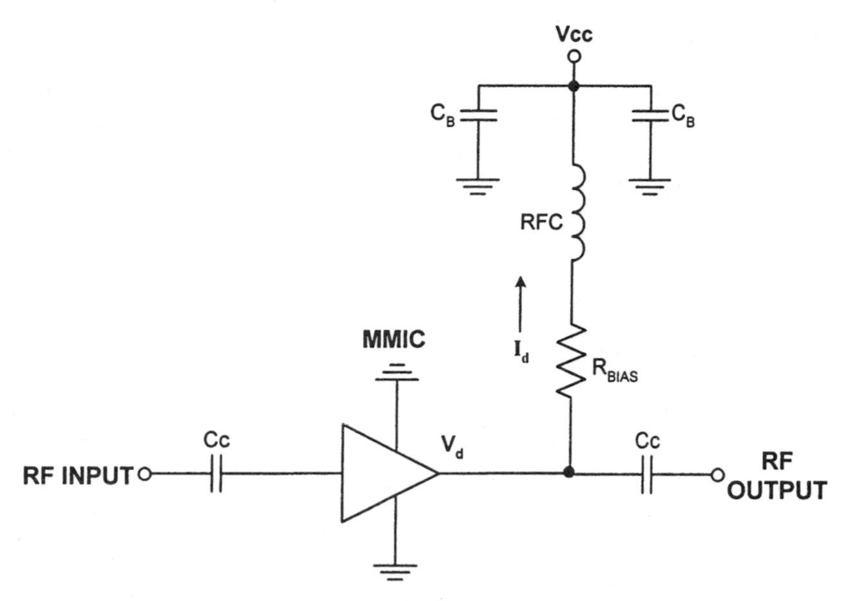

Figure 3.95 Standard MMIC gain block with biasing.

poor temperature stability that it varies its resistance dramatically over temperature. Carbon composite resistors are recommended because of their small resistance variation over wide temperature extremes.

If the R_{BIAS} does not add up to 500 ohms or more, then the gain of the MMIC stage will suffer. This is because all power supplies are virtually a short circuit to RF, and R_{BIAS} decreases this gain "shorting" effect on the output of the MMIC by being at a high value. However, if R_{BIAS} does not compute to be at or over 500 ohms (and it rarely is), then an RFC should be added to increase the output to this value, or $R_{\text{BIAS}} + X_L \geq 500$ ohms.

R_{BIAS}, since it drops the excess voltage from V_{CC}, also smoothes out any voltage fluctuations to the MMIC—which would cause an unstable bias point—by acting almost as a constant current source. In addition, the added RFC blocks most of the RF from entering the bias V_{CC} line by behaving as a high impedance to the RF, while the two C_B's bypass any extraneous RF to ground.

The manufacturer's approved DC bias current for the MMIC should be followed closely because of problems with decreased gain and improper matching at lower I_d levels, and device damage at higher I_d levels:

$$I_d = \left(\frac{V_{CC} - V_d}{R_{\text{BIAS}}}\right)$$

Agilent, through empirical studies, recommends placing R_{BIAS} at the output to the MMIC, *followed* by the RFC for improved performance. The bypass capacitors, of course, should always be placed *after* the RFC, and not before, or the RF gain will decrease severely.

MMIC biasing procedure (Fig. 3.95)

1. Choose a V_{CC} that will allow at least 2 V, and preferably 4 V, to be dropped across R_{BIAS} for stability, while also supplying the MMIC with the proper V_d level. If R_{BIAS} does not reach 500 ohms, use an RFC for a combined impedance of 500 ohms for both R_{BIAS} and the RFC:

$$R_{BIAS} = \frac{V_{cc} - V_d}{I_d}$$

where V_d = DC voltage at the MMIC's power pin
I_d = DC current into the MMIC's power pin
V_{cc} = power supply voltage.

2. Check the power dissipation within the bias resistor R_{BIAS} to allow for the appropriate safety headroom of at least *double* the calculated R_{BIAS} wattage, or $P = 2\,(I^2R)$
3. Use coupling capacitors at the MMIC's input and output as described in Sec. 3.4.3, "MMIC Coupling and Decoupling."

As mentioned above, most MMIC amplifiers' gain is moderately affected by a change in I_d. By looking at the I_d versus S_{21} (dB) curves for a particular device, this susceptibility can readily be seen. This also offers a way, with these particular amplifiers, to operate them as variable-gain amplifiers (VGAs)—as long as stability is not adversely affected. Gain variations of 5 to 15 dB are possible, depending on the MMIC, by varying I_d through an AGC circuit. A MMIC should be used as a VGA only for low-level signals, since the P1dB will also decrease along with the I_d and gain of the MMIC. The exact value of the gain variations obtained will differ slightly with the input frequency.

The above describes biasing and operation of the most prevalent MMIC, the current-biased MMIC. However, some MMICs, such as Agilent's MGA-85563 LNA MMIC (Fig. 3.96), are *voltage-biased.* This type of MMIC operates quite well when only low values of V_{CC} are available (since no R_{BIAS} is required) at low current draw levels. This makes it perfect for portable battery-powered applications.

Some MMICs can be adopted to limit output signal amplitudes for modulations that employ a constant modulation envelope, like common FM. A MMIC with a hard saturating characteristic, as well as high gain, is required for this application—such as the INA series of MMICs. Since almost all MMICs will vary in both gain and saturation level, depending on bias current draw, the bias point of these MMIC limiters must not be allowed to vary with large RF drive transitions, and the factory-recommended bias current levels should be maintained to limit harmonic output. Maintaining this constant bias point in limiter applications can best be accomplished by using the biasing circuit as shown in Agilent's Application Note AN-S003.

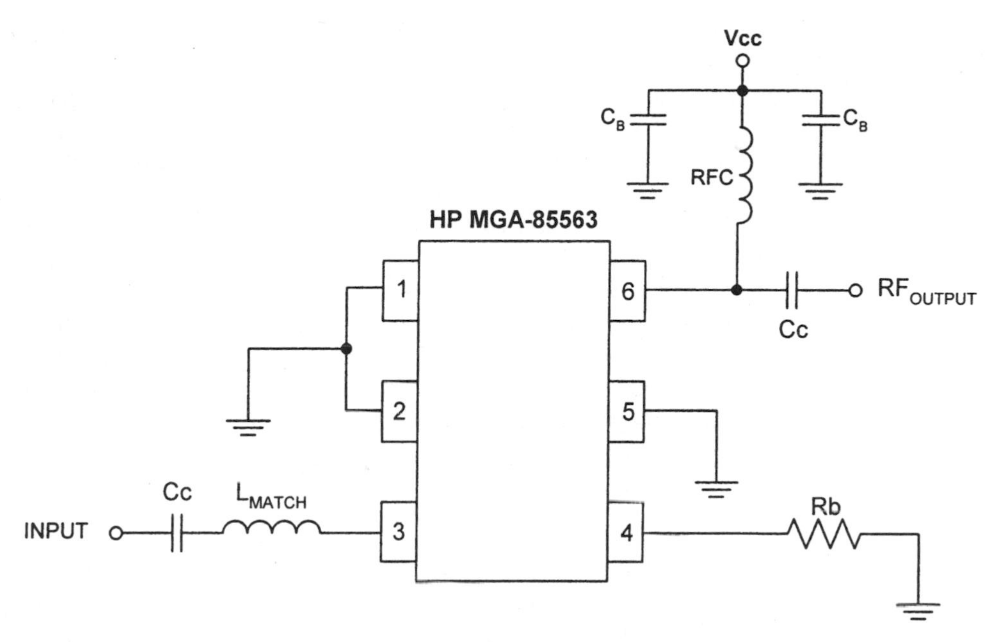

Figure 3.96 Agilent (HP) voltage-biased LNA MMIC.

3.4.3 MMIC coupling and decoupling

Coupling and decoupling is just as important in MMICs as in discrete amplifier circuit design. As shown in Fig. 3.95, two C_C's are utilized at both the input and output of the MMIC to block DC from reaching any other devices—which would disrupt the biasing of the next stage, or just be shorted to ground—while coupling RF with no voltage drop. The capacitors are typically chosen to supply 1 ohm or less X_c at the lowest frequency to be passed, while the highest frequency should not be close to the capacitor's parallel resonant frequency. In fact, for narrow frequency use, the capacitor's own series-resonant frequency is sometimes chosen to be the same as the amplifier's signal frequency, thus allowing lower value capacitors to be selected for microwave coupling, while also minimizing undesired lower frequencies from passing to the next stage.

Any RF allowed to enter the bias power supply can cause various circuit instabilities throughout a system. To decouple, or stop, AC from entering the supply (while permitting DC an unimpeded flow) we can use the RFC and the C_B of Fig. 3.95. Normally more than one value of C_B will be selected, as shown, so that a wide band of frequencies will be blocked (shunted to ground), while also filtering any power supply ripple or electromagnetic interference (EMI) from entering the MMIC stage itself.

As stated, it is quite important that decoupling and coupling capacitors not be near their *parallel*-resonant mode, or they will act as a high impedance to RF instead of as an RF short, while the decoupling inductors should not be close to any *series*-resonant modes, or they will begin to function as short circuits, and not as high impedances.

When a MMIC (or discrete) amplifier must be capable of operating properly across a very wide bandwidth, then two RFCs may be required in its decoupling circuit (Fig. 3.97). A low-impedance coil (L_1) that functions suitably at very high frequencies—without hitting any series resonances—and a high-impedance coil (L_2) used to block the lower frequencies. The high-impedance coil will begin to lose its ability to block the upper frequencies of the passband because of its inherently elevated turn-to-turn capacitance in this large, low-frequency coil. In fact, at high frequencies, this high-impedance coil begins to look more like a short. The smaller-value, but much higher frequency, coil (L_2) now takes over. An added bypass capacitor (C) to ground may also be placed between the large coil and ground to further decouple any RF into the supply. These precautions will allow a very wide passband to maintain a relatively flat gain over frequency.

3.4.4 A MMIC amplifier circuit

The Agilent (HP) voltage-biased MGA-85563, shown in Fig. 3.96, is capable of operation from 800 MHz to 5.8 GHz, with a V_{CC} of 3 V at 15 mA and an NF of approximately 1.6 dB. It has 18 dB of gain with unconditional stability.

Looking at the MGA-85563 circuit, we find at the RF input a DC blocking capacitor, C_C—required only if DC is present at the input from the prior stage. The inductor L_{match} is chosen to cancel the natural capacitive reactance of the device's input to supply a $50 + j0$ input. However, the match should actually be chosen to give optimum source impedance for the lowest NF if the MMIC

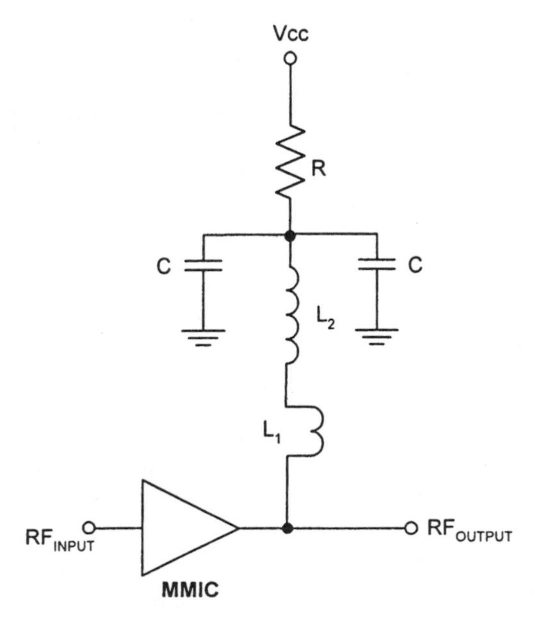

Figure 3.97 Decoupling of a MMIC at both high and low frequencies.

will be used in receiver front-end applications. For optimum NF, some simulation runs and tweaking on the prototype board are usually required for any matching network. Agilent's recommended L_{match} value for 800 MHz is 22 nH, 900 MHz is 18 nH, 1.5 GHz is 8.2 nH, 1.9 GHz is 5.6 nH, and 2.4 GHz is 2.7 nH. Above 3 GHz no inductor is required. The RF output of the MGA-85563 is 50 ohms, so no matching network is required at this port. Pin 6 outputs the RF through the DC blocking capacitor C_C, while the DC (voltage-based) bias is also injected into pin 6. The RFC blocks the RF from entering V_{CC}, and the two capacitors C_B bypass any RF that makes it through the RFC, while also filtering power supply EMI from entering the MMIC. Pin 4 of the MGA-85563 can be utilized to increase the IP3 at the RF output by increasing the MMIC's bias current from its normal 15 mA up to 35 mA. Since this mode obviously consumes more current, it is employed only when higher output powers are required. With pin 4 left floating, the device will have an IP3 of +12 dBm, while an R_b of 15 ohms will cause an I_d of 30 mA and raise the IP3 to +17 dBm.

3.4.5 MMIC layout

In most MMIC and RFIC layouts, it is normally undesirable to tie all the ground pins together to a single through-hole via to the ground plane, particularly if the ground pins are separated by any distance. This is because the feedback from one internal stage to another creates instabilities within the MMIC itself. Since most MMICs have at least two internal amplification stages, direct grounding by the shortest route possible is vitally important to prevent not only MMIC instability but also undesirable frequency *gain peaking,* as well as decreased input return loss caused by high impedance ground loops between these various internal amplifier stages, causing regenerative feedback. In addition, at microwave frequencies, it is always advisable to use more than one ground via in order to lower the inductance to ground (Fig. 3.98); and with a single via as close to each ground pin as possible.

When laying out a printed circuit board (PCB) for a MMIC, lead lengths should be kept as short as possible to minimize lead inductance, especially when operating above 1 GHz. Also, depending on the PCB's dielectric constant and its thickness, the 50-ohm microstrip that will interface with the MMIC

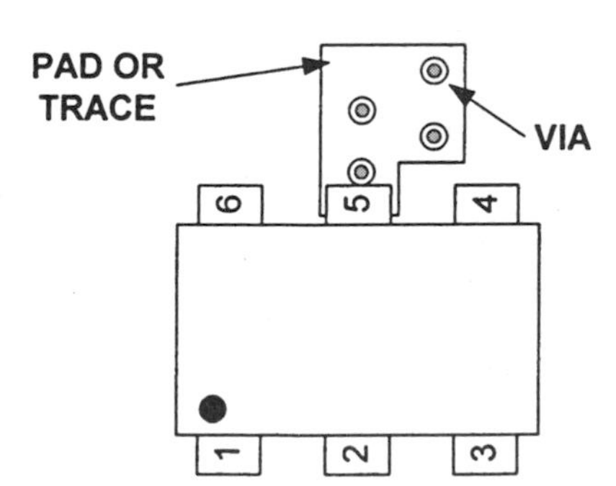

Figure 3.98 Proper board grounding of an RFIC.

may not be the same width as the MMIC lead itself. This will form *step discontinuities* at the MMIC's input and output ports, creating 0.01 to 0.2 nH of added series inductance. These step discontinuities can be minimized by tapering the wider microstrip rapidly down to the narrower width of the MMIC lead (Fig. 3.99). To minimize undesired inductance to the ground plane, the MMIC's ground leads are placed directly over the PCB's through-hole vias, as long ground leads lower gain and result in a poor P1dB. As a matter of proper microwave and RF design, all grounds on the top side of the board should lead directly to the bottom ground plane by the shortest possible route—which is almost always through a direct through-hole via. In fact, some high-frequency MMICs are so sensitive to this effect that substrates of not thicker than 32 mils are recommended in order to lessen the inductive properties of even a simple via. This is because a via passing through a 32-mil board can have an inductance of approximately 0.15 nH, while vias through a 62-mil board may have up to 0.5 nH of damaging inductance.

3.5 Wideband Amplifiers

3.5.1 Introduction

A wideband MMIC or discrete amplifier is designed to have an extremely broad band of frequencies that it can pass with flat gain and (preferably) a decent return loss response—along with perfect stability.

In order to properly design a discrete wideband RF amplifier we must suppress the lower frequencies, where the gain is the highest (Fig. 3.100*a*). One way to do this is by giving these frequencies a poor impedance match, while with the higher frequencies—where the gain is much less—we can give a perfect match. This will flatten the gain of the amplifier (Fig. 3.100*b*), but will not

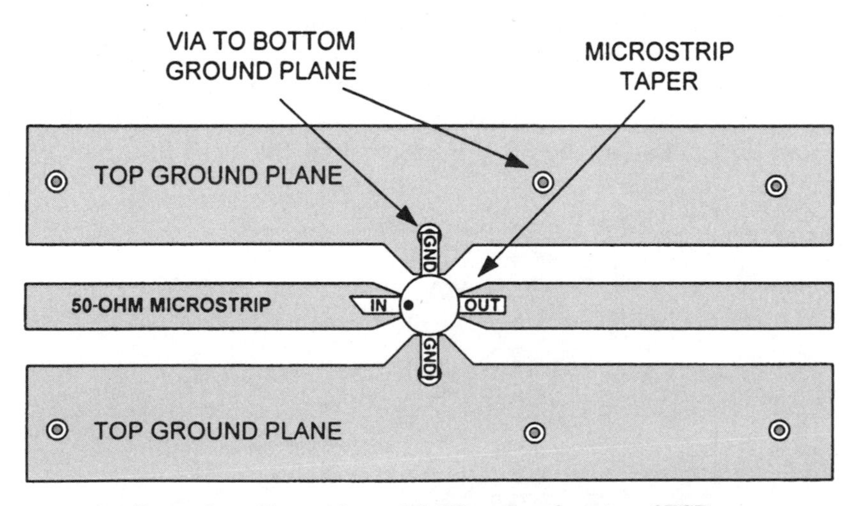

Figure 3.99 Proper board layout for an MMIC as seen from top of PCB.

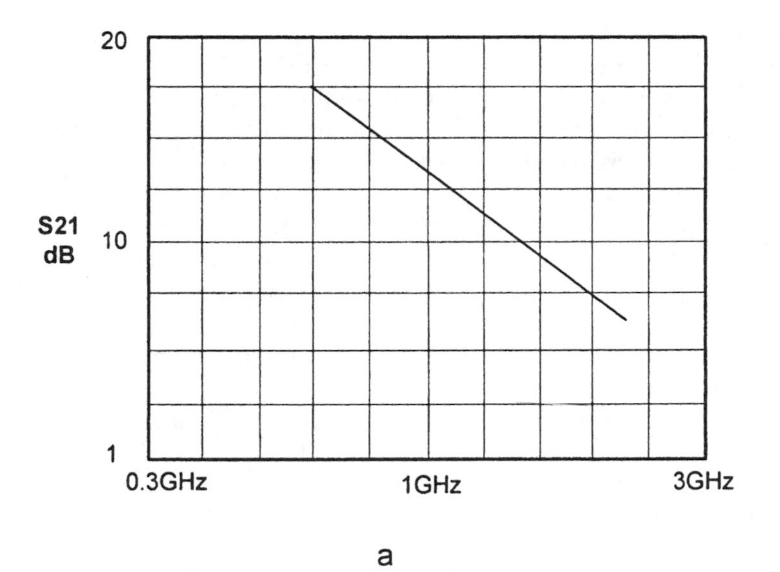

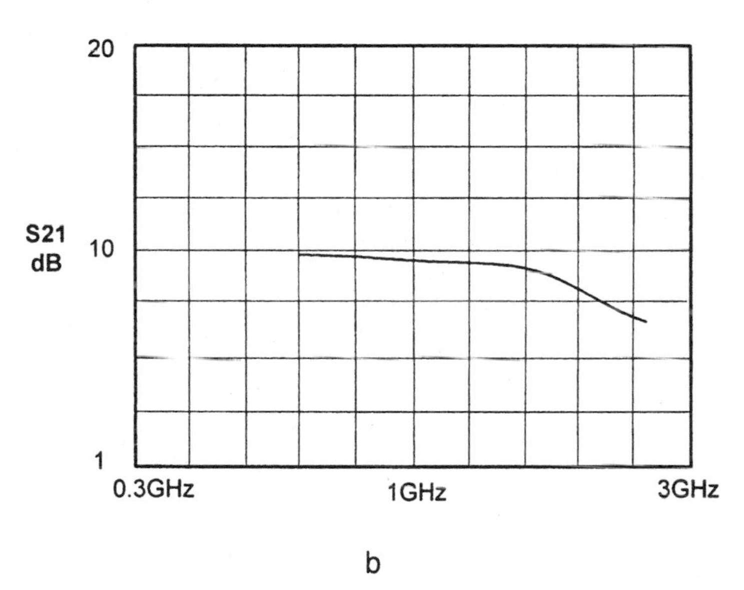

Figure 3.100 Gain flattening of an RF amplifier: (*a*) not compensated; (*b*) compensated.

provide a high return loss at all frequencies. In other words, matching the transistor at high frequencies will supply more gain at these frequencies, while mismatching at the lower end of the spectrum will decrease the gain at these frequencies because of mismatch losses. However, stability is always of prime importance; since we now have various impedance mismatches across a wide bandpass—and also at frequencies both below and above the bandwidth of the amplifier—we want to be assured of stability at *all* frequencies. We can do this by checking with our S parameters that the small-signal active device will be unconditionally stable at all frequencies that are plotted within that particular S-parameter file. This, however, will not include the very lowest of frequencies, which are far below the measurements of most S-parameter files.

Huge instability problems internal to the amplifier can occur at frequencies between 1 and 20 MHz, where the transistor's gain can be as high as 40 dB. This is mainly a problem with power amplifiers, but this high gain—combined with even the slightest of internal or external in-phase feedback to the base—will be an obvious recipe for oscillations. These oscillations are viewable on a spectrum analyzer as a single carrier surrounded by sidebands; with the injected carrier modulated by the low-frequency oscillations. So, we must find a way to lessen either the gain or the feedback—or both—of the power amplifier at these lower frequencies. Two helpful methods of accomplishing this are to choose a transistor with a low h_{FE} and to use the lowest value of collector choke that will still supply a virtual short circuit for low-frequency AC, but a virtual open circuit to RF (Fig. 3.101). The inductor itself should be paralleled by a low-value resistor of 330 to 560 ohms for de-Qing purposes and preventing parasitic oscillations. A high-value capacitor must be attached to the top of the choke to send any of these low frequencies to ground through the very low capacitive reactance.

Another extremely potent method of decreasing low-frequency gain that is especially suitable for wideband power amplifiers is to use negative feedback (Fig. 3.102). The capacitor (C) in this circuit is adopted to block the DC bias, while easily allowing the dangerous low-frequency AC to pass back to the base. The resistor (R) element controls the amount of feedback to the base, and can

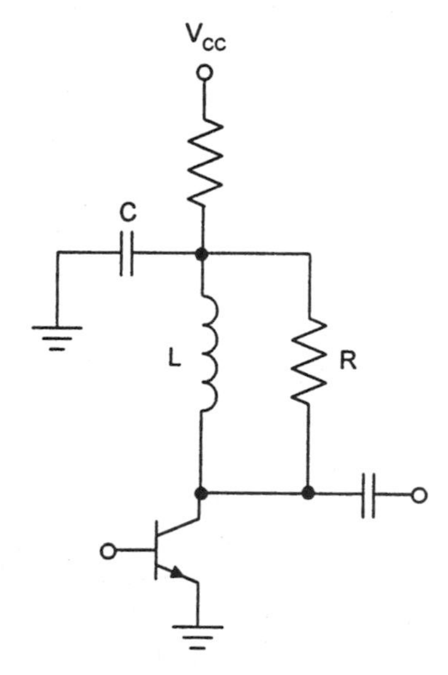

Figure 3.101 A collector inductor load for decreasing low-frequency oscillations.

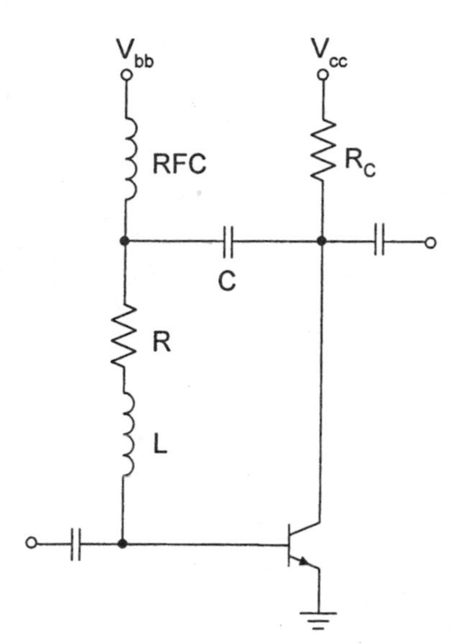

Figure 3.102 A negative feedback circuit.

be found empirically (50 to 500 ohms will be the normally employed value). The inductor (L) is the important component that actually controls the feedback. It functions by having a very high reactance to the amplifier's desired frequencies, so it will not permit degenerative feedback that would lower the gain levels at these higher frequencies. However, as the frequency is *decreased,* the reactance of the inductor will, of course, decrease. This increases the level of degenerative feedback to the base, decreasing the gain of any low-frequency signals. Feedback of any kind may sound dangerous, but a common emitter amplifier will have a perfect 180 degree phase shift at its lower frequencies of operation ($\ll f_T$), thus assuring that *regenerative,* or oscillatory, feedback will not occur. The only caution would be to confirm that the inductor is capable of blocking the RF at its highest frequency, and not allow the RF to pass through because of parasitic capacitances within this choke.

But if we keep in mind that almost any amplifier will be completely stable *if* all of the frequencies it is passing can see a perfect 50-ohm resistive impedance at both its input and output (a difficult objective with power amplifiers), then we can use this knowledge to our advantage. Excellent stability—even at frequencies above or below our interests—can be assured by supplying the transistor with a good 50-ohm source and load. This can be in the form of a 51-ohm load resistor that will be seen only by the unnecessary lower frequencies through a one-pole low-pass filter, or by the use of a diplexer, or by employing a 1- or 2-dB 50-ohm pad at the amplifier's output.

Another viable technique to design stable wideband amplifiers is by using resistive components to match an amplifier over its entire bandwidth (Fig. 3.103). However, this technique has a large disadvantage of producing much lower gain compared to the typical LC matching method, with less reverse isolation and a higher noise figure. Its greatest advantage is that it has a very

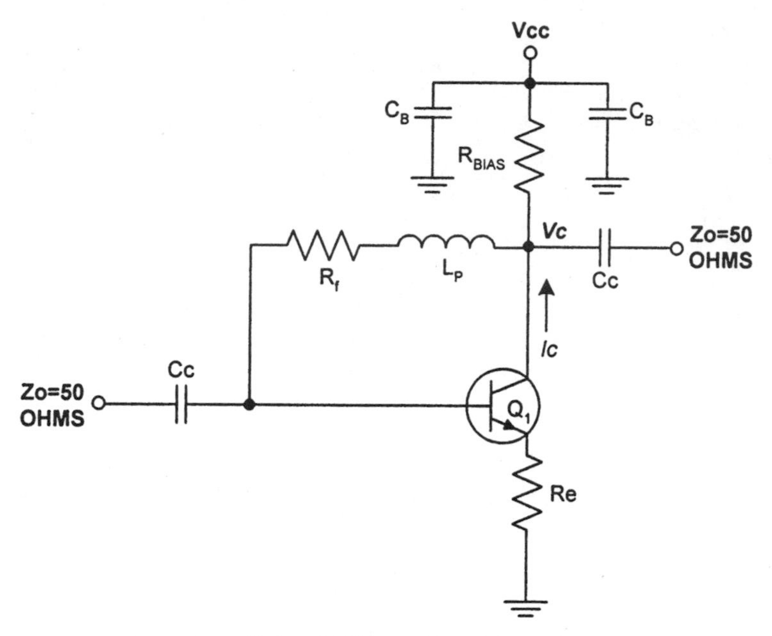

Figure 3.103 Resistive wideband RF amplifier with 50-ohm input and output.

wide bandwidth, with a decent return loss, and much increased stability. As shown, L_p is a peaking coil that *decreases* the negative feedback at higher frequencies, holding up the gain of the amplifier just as it begins to fall. But the insertion loss will not be quite as good at these peaked frequencies. This design is referred to as a *resistive negative feedback amplifier,* and can be utilized at almost any frequency, but is more common at 600 MHz and below, where gain is inexpensive.

Another technique to increase high-frequency gain while maintaining a flat, wide bandwidth is to use a low value of bypass capacitor in parallel with the amplifier's emitter resistor. This will increase the emitter resistor's degenerative feedback as the frequency is decreased, leveling out the gain. The application of emitter components of any kind is viable only at 2 GHz and below because of the stability-robbing presence of their added lead inductance.

3.5.2 Design of wideband amplifiers

Design of wideband *LC* matching networks for amplifiers is covered in detail in Sec. 3.1.4, "Matching networks."

The type of resistive ultrawideband amplifier discussed above (Fig. 3.103) can initially be designed by the following formulas. It will need to be extensively optimized, however, within a CAD program, and physically tweaked on the PCB, to obtain an almost perfect 50-ohm match across a wide bandwidth.

1. $R_{\text{BIAS}} = \dfrac{V_{CC} - V_C}{I_C}$

2. Choose R_e to be between 5 and 10 ohms (the higher the R_e, the less gain).

3. $R_f = \dfrac{Z_0^{\,2}}{R_e}$

4. $Z_0 = \sqrt{R_f R_e}$

5. $X_{LP} \approx \dfrac{R_F}{8}$ at f_{MAX}

6. $L_P = \dfrac{X_{LP}}{2\pi f}$

7. $V_{\text{GAIN}} = 20 \log_{10}\left(\dfrac{R_f - R_e}{50 + R_e}\right)$ (in a 50-ohm system)

3.6 Parallel Amplifier

3.6.1 Introduction

Single-ended amplifier configurations cannot always supply us with all of the power we may need for certain applications, for we may require up to several hundred watts of output power. This can be accomplished with RF *parallel amplifiers* (Fig. 3.104).

With parallel amplifiers, each transistor is on or off at the *same* time, unlike push-pull, which sequentially distributes the power back and forth for equal, but alternating, time periods.

Since the output current of the parallel circuit is shared evenly between the transistors (when perfectly matched), this will double the power handling

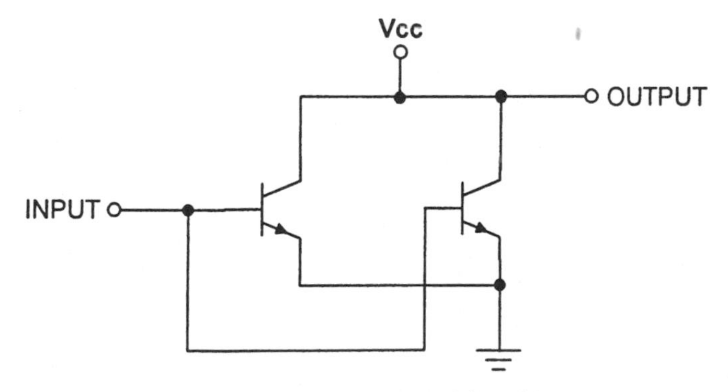

Figure 3.104 A parallel amplifier circuit without bias components.

capabilities ($P = IE$) compared to a single-ended amplifier configuration. Parallel amplifiers allow the entire parallel circuit to function as if it were a single high-power transistor, at any bias desired (Class A, AB, B, and C). Figure 3.105 demonstrates a complete parallel circuit, with impedance matching, biasing, and filtering. Parallel amplifier circuits must have excellently matched active devices, and their input and output capacitances will also be double that of a single device—which can be problematic with high-frequency operation.

Gain will stay the same whether a single power amplifier or a parallel power amplifier configuration is used. The advantage of paralleling amplifiers is that the output power *capability* (P1dB) will increase by 3 dB for two amplifiers (Fig. 3.106), and by 6 dB for four amplifiers. However, the input drive power into these paralleled stages must also increase to take advantage of this attribute.

The input and output impedances of the combined amplifiers will decrease as more devices are paralleled. With 50-ohm MMICs, the Z_{IN} and Z_{OUT} of the total paralleled stages can be computed by $50/N$, with N being the number of amplifiers in parallel. A matching network for the MMICs would therefore be necessary if we must match them into a 50-ohm system, or gain would suffer.

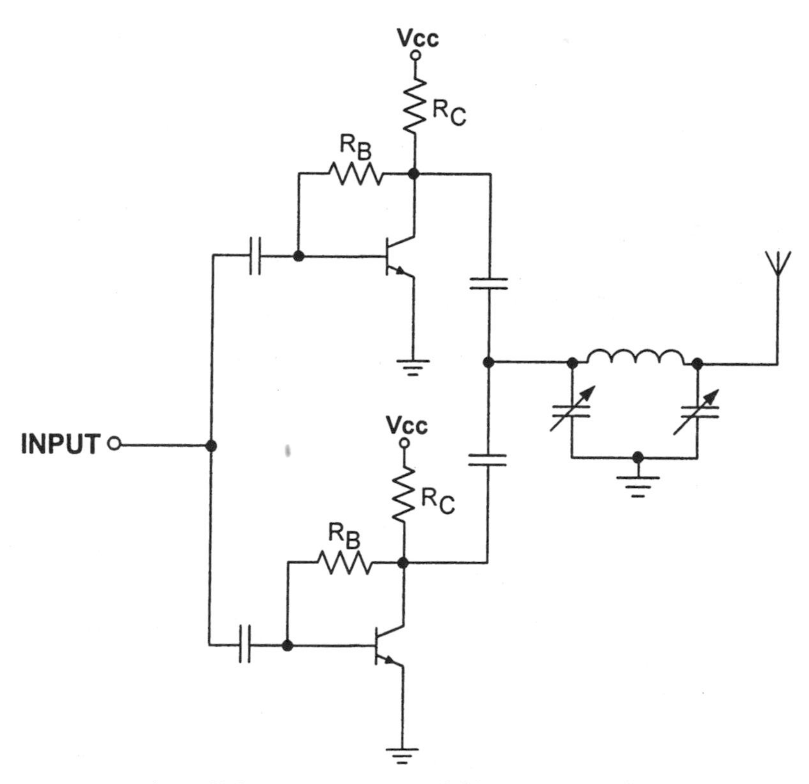

Figure 3.105 A parallel power amplifier with bias components.

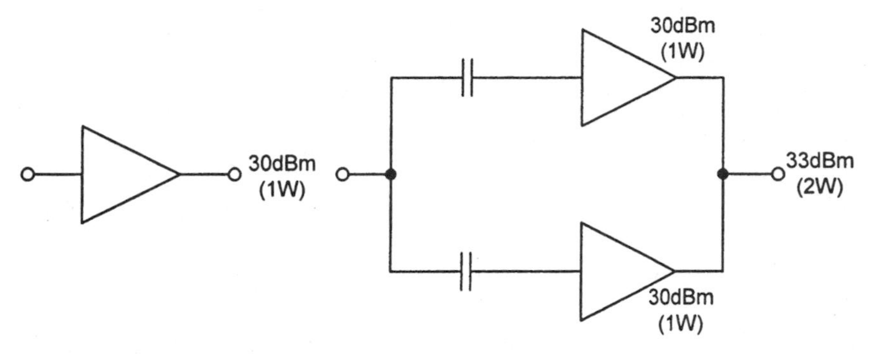

Figure 3.106 The output power of a single MMIC compared to two paralleled MMICs.

As stated, paralleling two MMICs will increase their P1dB by 3 dB as compared to a single device, and Fig. 3.107 shows a recommended circuit for such a two-MMIC setup. This design maintains the MMICs' input and output impedances by using a lumped *Wilkinson power divider/combiner network,* along with a single DC bias line. The Wilkinson network, unfortunately, will also make the amplifiers much more narrowband. And, since MMICs are not always completely resistive, some tuning out of the small MMIC reactances may be needed by the tweaking of C_1 and C_6 to peak the gain at the desired frequency of operation.

Since directly paralleling discrete high-frequency power transistors makes their already very low input/output impedances even lower, effective low-loss matching can become a problem. With these types of discrete designs, it is better to first match each active device first, preferably to 100 ohms in a 2× parallel amplifier, and then directly combine, as shown in Fig. 3.105. Or, match each active device to 50 ohms, and then utilize the combiner of Fig. 3.107 to blend the single amplifiers into a parallel amplifier stage (see "Splitters and Combiners").

Formerly, indiscriminant direct paralleling of active devices would lead to uneven current distribution among the transistors, and the subsequent thermal destruction of the affected transistor with the most current draw. This is not a problem with the vast majority of modern power transistors because of emitter ballast resistors that are placed internally within the semiconductor die itself.

3.6.2 Design of parallel MMIC amplifiers

As mentioned above, paralleling two amplifiers will double the available output power, increasing it by 3 dB. To design the 50-ohm MMIC parallel power amplifier of Fig. 3.107, bias as instructed in Sec. 3.4.2 under "MMIC Biasing Procedure," couple as in Sec. 3.4.3, "MMIC Coupling and Decoupling," and use the following power splitter/combiner formulas to design the input/output network:

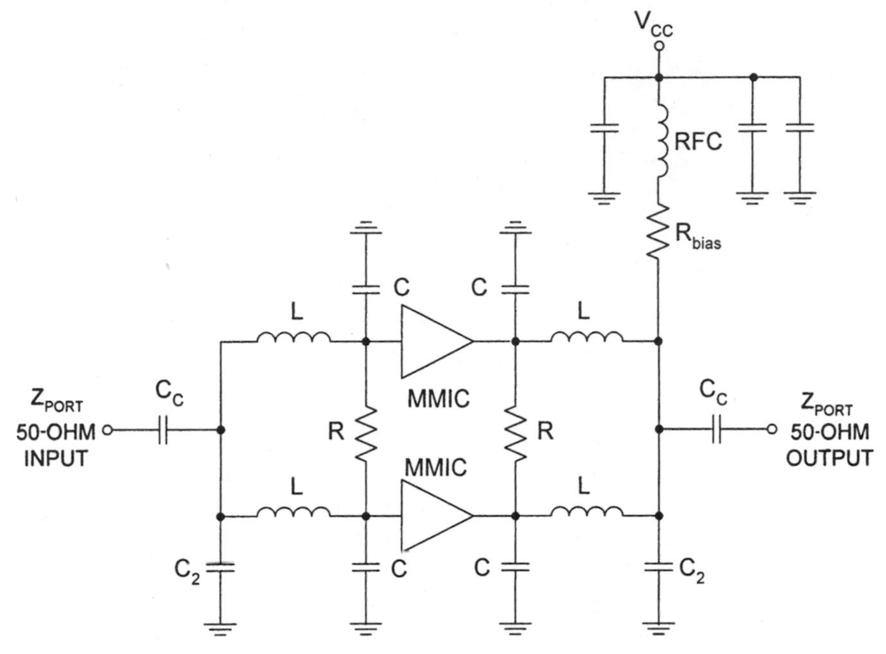

Figure 3.107 Paralleled MMIC amplifiers with a splitter and a combiner.

1. $L = \dfrac{50}{1.4\,\pi f}$

2. $C = \dfrac{1}{2.83\,\pi f 50}$

3. $C_2 = 2C$
4. $R = 2Z_{PORT} = 100$ ohms
5. Make impedance X_C of C_C less than or equal to 1 ohm.

3.7 Audio Amplifiers

3.7.1 Introduction

Many nonpower audio amplifier stages need not be matched to their source nor to their load. Since matching is used to maximize power transfer between stages and to reduce standing waves, audio amplifiers are much more concerned with reducing distortions and isolating each stage from the effects of the next. However, matching with low-frequency transformers is quite common with discrete audio power stages in order to obtain high efficiencies, while *RC* matching is also employed in audio voltage amplifiers. An acceptable single-stage audio voltage amplifier can be designed by using the low frequency bias design formulas as presented in Sec. 3.3, "Amplifier Biasing."

Operational amplifiers are far more common for voice-frequency amplification for both low-level voltage signals and high-level power signals. They can be acquired from many manufacturers, and are obtainable in an optimized single-voltage supply package for ease of biasing.

3.7.2 Design of an IC audio amplifier

The National LM386 is a low-voltage audio amplifier that is perfect for low-frequency amplification. In voiceband radios this IC can amplify the audio signal all the way from the detector stage to the 8-ohm speaker or headphones. The National device has very low quiescent current drain (4 mA), accepts a wide range of V_{cc} (4 to 12 V), has adjustable voltage gain (20 to 200), decent distortion levels [<10 percent total harmonic distortion (THD)], and has an output driving power of 700 mW with a 9-V supply into 8 ohms.

To design an audio IC amplifier, use Fig. 3.108 as a model. C_B will bypass any RF that escaped the detector, while it can also be chosen to limit the audio-frequency response of the amplifier. Many demanding cases will require either an active op-amp or a passive *RC* low-pass or bandpass filter placed before the amplifier in order to limit the frequency response and noise even further; voice should be band-limited to between 300 and 2500 Hz, since this will reduce the higher-frequency noise and heterodyne outputs, as well as any low-frequency 60- and 120-Hz hum. The gain for the amplifier as shown will be 200, but can be adjusted downward by the addition of a resistor in series with the capacitor between pins 1 and 8 (for instance, a 1.2 kilohm resistor will set the gain to 50). Removing the capacitor, and leaving pins 1 and 8 open entirely, will decrease the gain to 20. Adjusting the pot at the input to the LM386 will alter the amplitude of the output signal into the speaker.

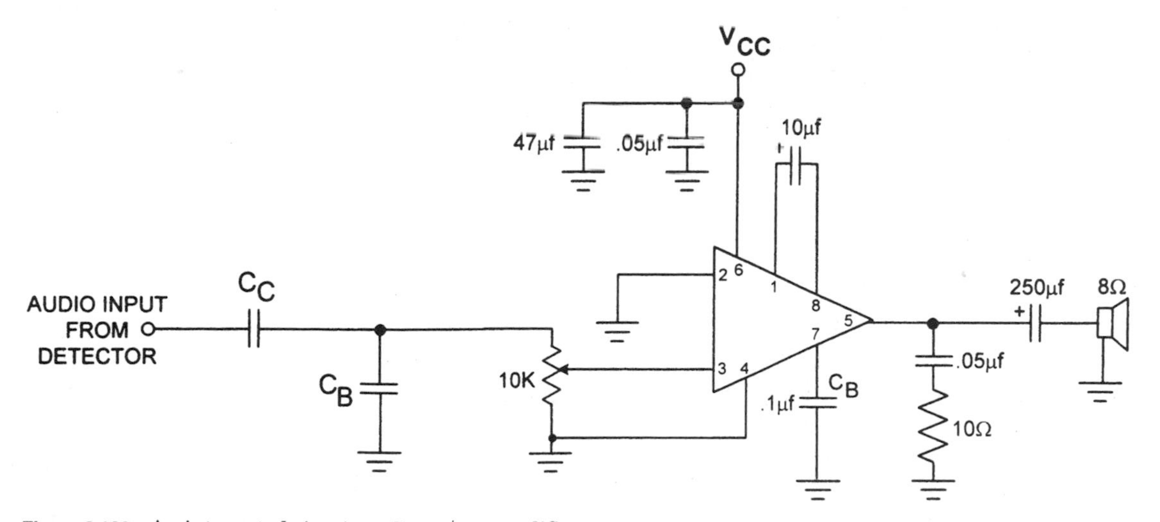

Figure 3.108 An integrated circuit audio power amplifier.

3.8 VGA Amplifiers

3.8.1 Introduction

Variable gain amplifiers (VGAs) can be designed in one of two ways: either by varying the active device's bias voltage to its base, which controls its collector current, and thus the gain of the transistor (see "Automatic gain control"), or by placing a voltage- or current-controlled variable attenuator at the input to a fixed-gain amplifier. Since the latter design usually results in a more linear amplifier response over gain, especially with large input signals, it is preferred over the variable-bias design in many applications. An added disadvantage to the variable-bias type is that any modification to the bias of a transistor will also alter its S parameters. This means that not only will the gain be varied, but so will the return loss of the amplifier—which can prove catastrophic if the VGA is attached to a filter circuit (a filter's response is dependent on its source and load impedance).

In using PIN diode attenuators, a few cautions are in order. As with most devices using PINs, the *minimum* usable frequency is normally *above* 10 MHz. Some special PIN diodes may attain lower frequencies, while some will not operate properly until much higher frequencies are reached. However, as the frequency is decreased in any PIN, its IMD and insertion losses will increase. As well, many PIN attenuator designs used for AGCs should be tested for IMD performance, considering that a PIN has better intermodulation specs at higher bias currents and, as high received signal levels result in increased attenuator IMD, decreasing the bias current to attenuate the signal will add to this problem.

3.8.2 Design of VGA amplifiers

A VGA amplifier with low distortion for 10 MHz and above (Fig. 3.109). A low-cost attenuator type of VGA that does not vary the bias of the amplifier is shown. With an AGC voltage of zero the gain will be high for this stage; as the AGC voltage increases, the gain will decrease because of the shunting effect of the decreasing resistance of the PIN diode. However, the return loss will also decrease, so an unconditionally stable transistor is the safest for this circuit. Match and bias the transistor as normal, and calculate R_P as:

$$R_P = \frac{V_{\text{AGC(HIGH)}} - 0.8}{25 \text{ mA}}$$

where $V_{\text{AGC(HIGH)}}$ = maximum AGC voltage expected.

Another method is the best, and the most nonreflective, type of VGA: Employ an absorptive attenuator design in front of any fixed gain discrete or MMIC amplifier. An AGC voltage of zero at the attenuator's DC control input will result in low gain (even a negative gain), while a control voltage of greater than zero results in a steadily increasing gain. The return loss will remain quite usable up to very high attenuation levels. However, a rise in the noise figure of a VGA circuit—whether bias or attenuator controlled—is unavoidable

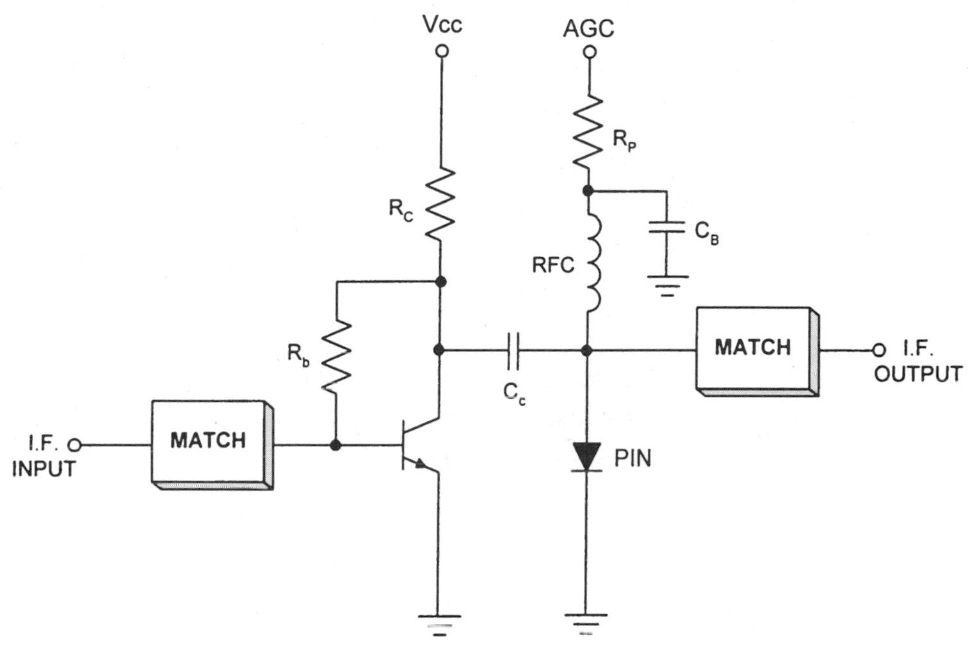

Figure 3.109 A simple but low-cost VGA circuit.

as the gain is decreased. Design the amplifier and the attenuator circuits as described in the various attenuator and amplifier sections.

A low-cost variable-bias VGA amplifier (reverse gain control, Fig. 3.110)

1. Choose R_C to drop half the V_{CC} when the transistor is at its desired gain:

$$R_C = \frac{\left(\frac{V_{CC}}{2}\right)}{I_C}$$

2. Choose an R_B of 10 kilohms.
3. Find the voltage required at the AGC (AVC) port of the amplifier that causes the base current (I_B) to create a collector current for the desired full gain. The transistor's *characteristic curves* in its data sheet will contain the information on I_B versus I_C, while the *current gain* graphs will show the I_C versus h_{FE}:

$$\text{AVC} = (I_B \cdot 10 \text{ kilohms}) + 0.7$$

4. Choose the limits of AVC voltage that will supply the required range of gain by substituting the desired I_B in the above equation with the minimum and maximum I_B-related gain values.

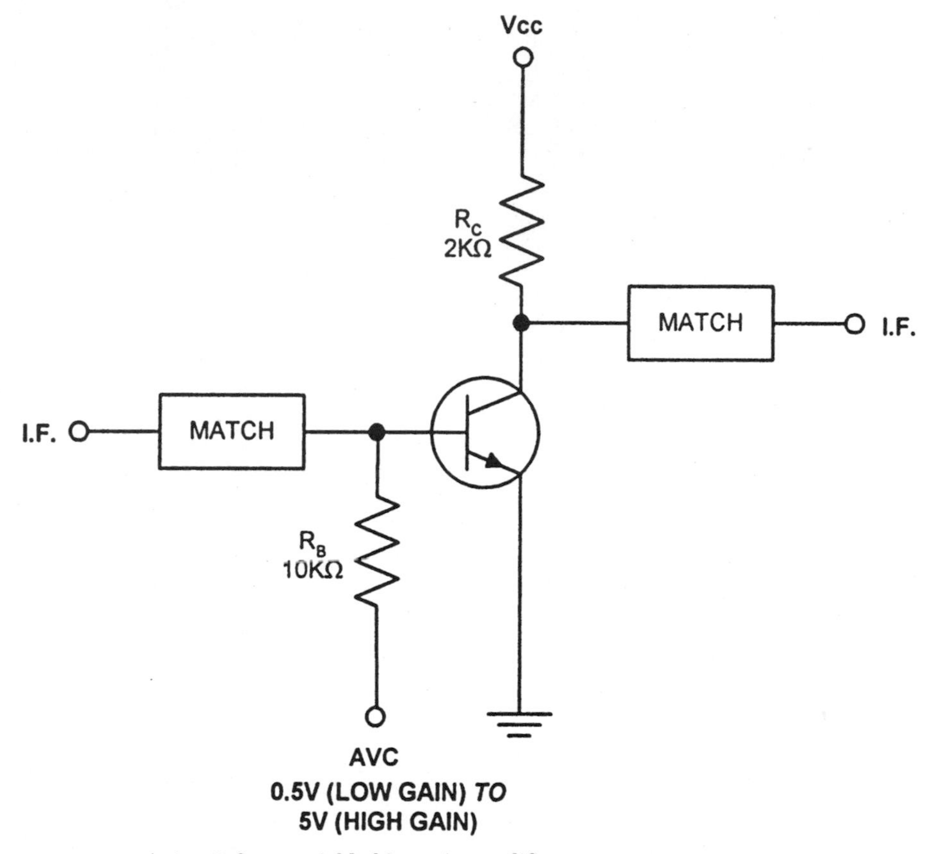

Figure 3.110 A circuit for a variable-bias gain amplifier.

5. Design the transistor's input/output matching networks for the desired bias at maximum gain and for the frequency of operation.

The gain of the circuit is now controlled completely by the AGC voltage at the transistor's base. But since gain is managed by altering the I_C through the AGC at the transistor's base, then the input and output impedances, as well as the stage's stability, will also vary. Distortion and gain compression may also occur with strong input signals.

3.9 Coupling/Decoupling of Amplifiers

3.9.1 Introduction

Amplifier coupling. To prevent DC biasing of consecutive amplifier stages from adversely affecting other stages, a method of coupling an AC signal into or out of another amplifier—or another source and its load—must be found. Unmatched RF amplifiers also demand a good impedance match for maximum power transfer and the reduction of reflections (decreased VSWR), as well as for supplying filtering for harmonic attenuation.

The actual type of coupling in a discrete circuit, sometimes combined with impedance matching, will depend on the sort of signal being amplified. DC, low-frequency AC, high-frequency AC, and wideband amplification, will all have specific requirements.

3.9.2 Design of decoupling/coupling circuits

Decoupling. RF must not enter an amplifier's power supply, and the power supply's voltage to the amplifier must not influence any circuits that are before or after the amplifier. Both occurrences would have a negative effect on system operation, as any alternating current into the amplifier's DC bias supply can cause circuit instabilities and noise throughout a system; while passing DC beyond the immediate amplifier stage area would affect the bias of any following amplifier—or be sent to ground as a short. In performing as coupling and decoupling elements, capacitors and inductors prevent any of the above from occurring.

However, in order to function as desired, coupling and decoupling capacitors must not be near their *parallel* (high-impedance) resonant mode, while decoupling inductors must not be close to any *series* (low-impedance) resonant modes—nor should the inductors be above their maximum frequency of operation, in which case they would start to become capacitive.

Inductors are far from perfect components, and possess parasitic capacitances. So when an amplifier must be able to function properly across a wide band of frequencies, two RFCs (Fig. 3.97) will normally be required: a low-inductance coil that works at very high frequencies without encountering any series resonances, and a high-inductance coil used to block the lower frequencies. This is necessary because the low-frequency, high-impedance inductor will begin to pass the higher frequencies of the passband through the natural turn-to-turn capacitance of any coil. An additional bypass capacitor to ground may be placed between the larger coil and ground to further decouple any RF into the amplifier's supply. This type of decoupling will permit the amplifier's wide passband to sustain a nearly flat gain response over its entire frequency range.

Coupling. There are various coupling techniques that can be used between stages, depending on frequency, cost, performance, and impedance-matching needs.

Capacitor coupling (Fig. 3.111), also referred to as *RC coupling,* is found in AC and RF amplifiers only, and is capable of amplifying over a very wide bandwidth (the amplifier's required impedance matching circuit will limit this bandwidth, however). As shown in the figure, the series coupling capacitor C_C blocks the DC bias to the next stage, but allows the RF signal to pass unattenuated. The coupling capacitor and R_6 form a voltage divider, allowing most of the RF signal to be dropped across the high resistance of R_6 located at the input to the next stage. The voltage divider functions as described because the capacitor has a much lower impedance to the RF than does the resistor. This signal across R_6 will then add to or subtract from the second stage's emitter-base junction, forcing its collector current to vary through R_7, producing an amplified output voltage.

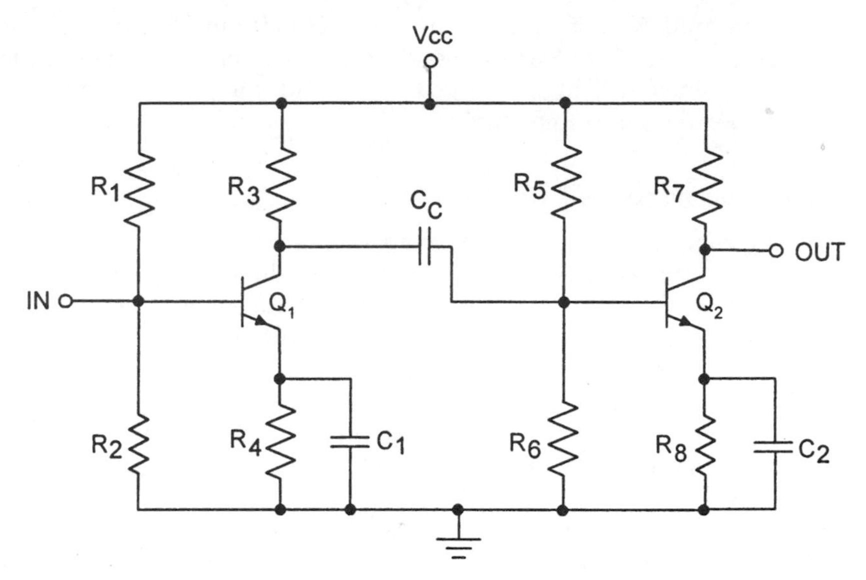

Figure 3.111 Capacitive coupling between two stages.

RC coupling is a simple method to transfer energy from one circuit to another, but has great difficulty matching the stage's impedances. And, unless we employ a low-value, but series-resonant, coupling capacitor at the stage's input and output, this coupling method does not prevent harmonics from being transferred from stage to stage and being further amplified. (This small resonant coupling capacitor will also help to stabilize the amplifier chain. It accomplishes these tasks by attenuating the lower and higher undesired RF frequencies with its low capacitance value and series-resonance operation—but easily passes the frequencies of interest.) An L, T, or pi network is normally added for harmonic attenuation and impedance-matching requirements.

Inductive coupling, also referred to as *impedance coupling* (Fig. 3.112), is found in AC and RF circuits only, and is comparable to RC coupling, but instead of exploiting a resistor in the collector circuit, it has a collector inductor. Inductive coupling has the advantage in that the collector inductor wastes little DC power because of its very small DC series resistance, thus permitting far more efficient amplifier operation. This high-value inductor works as a transistor's collector load because of its high reactance to the alternating collector current, which produces an AC voltage drop. This action will subtract from or add to the voltage from the transistor's emitter-collector. However, inductive coupling is practical only over a relatively narrow band of frequencies, since X_L changes directly with frequency—and stage gain would thus change as well.

Direct coupling (Fig. 3.113), also referred to as *DC coupling,* is valuable for very low frequency and DC amplification. R_3 functions as a collector resistor

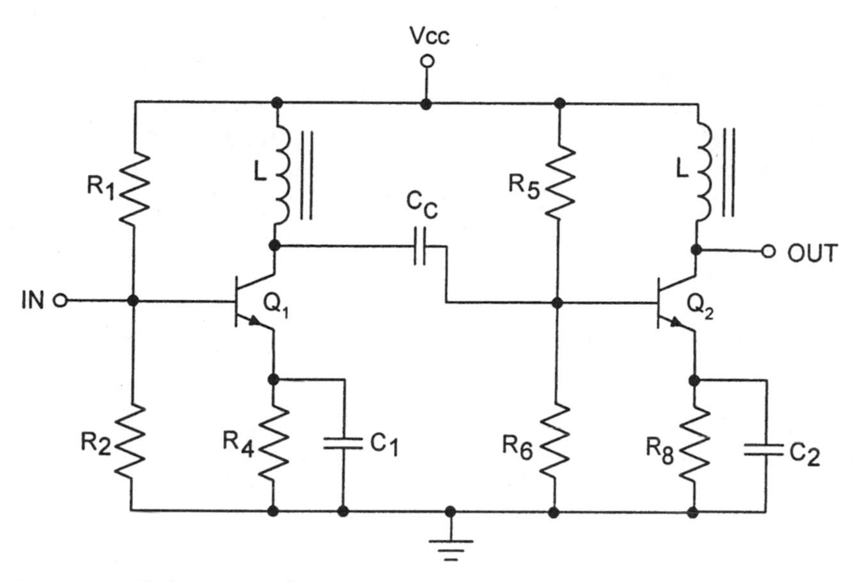

Figure 3.112 Inductive coupling between two stages.

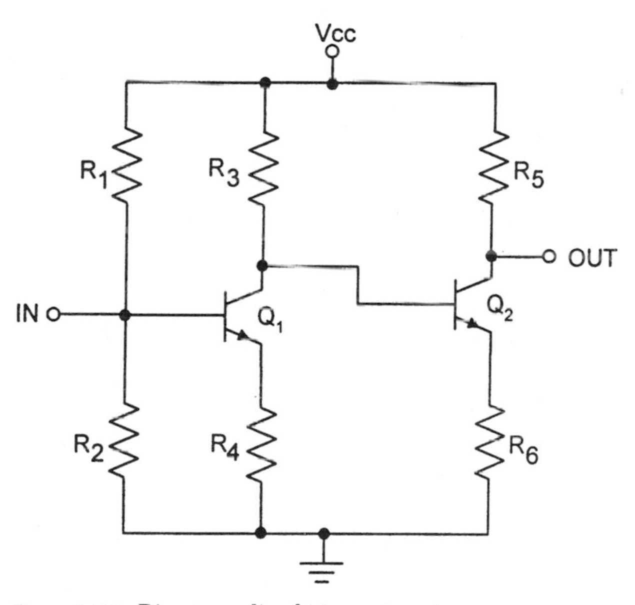

Figure 3.113 Direct coupling between two stages.

for Q_1 and as a base resistor for Q_2, and thus must be carefully chosen to function in both roles, since a small temperature-induced current change in Q_1 is directly amplified by Q_2. Precision components and tight placement of parts to allow each component the same changes in temperature must be used to stabilize this circuit.

Low-frequency transformer coupling (Fig. 3.114) employing laminated iron cores is adopted for low-frequency AC. This method, because of the interstage transformer, will not pass the DC bias, and can also be used to match the relatively high output impedance of Q_1 to the low input impedance of Q_2. One end of the transformer's secondary is connected to the base of Q_2, while the other end is connected to the top of Q_2's bias resistor R_6. The signal being amplified by Q_1, and being output from T_1, will then subtract from or add to the base bias. This results in a varying base current, which causes a much higher amplitude collector current, creating amplification at the output of Q_2. However, since a low-frequency iron-core transformer is both expensive and heavy, this coupling method is rarely found today in most audio coupling applications.

High-frequency transformer coupling (Fig. 3.115) with tuned circuits, employing ferrite, powdered-iron, and air cores, is not as popular as it once was because of the expense and size of the transformer, but can still be found in some RF and IF amplifiers up to a maximum frequency of 150 MHz. Transformers provide the required impedance matching for the efficient and maximum power transfer between amplifier stages, and block the DC bias from stage to stage as well. These transformers function the same as the low-frequency iron core transformers above, except for the frequency-selective narrowband resonant tanks formed by C_2 and the primary of T_1, as well as C_3 and the secondary of T_1.

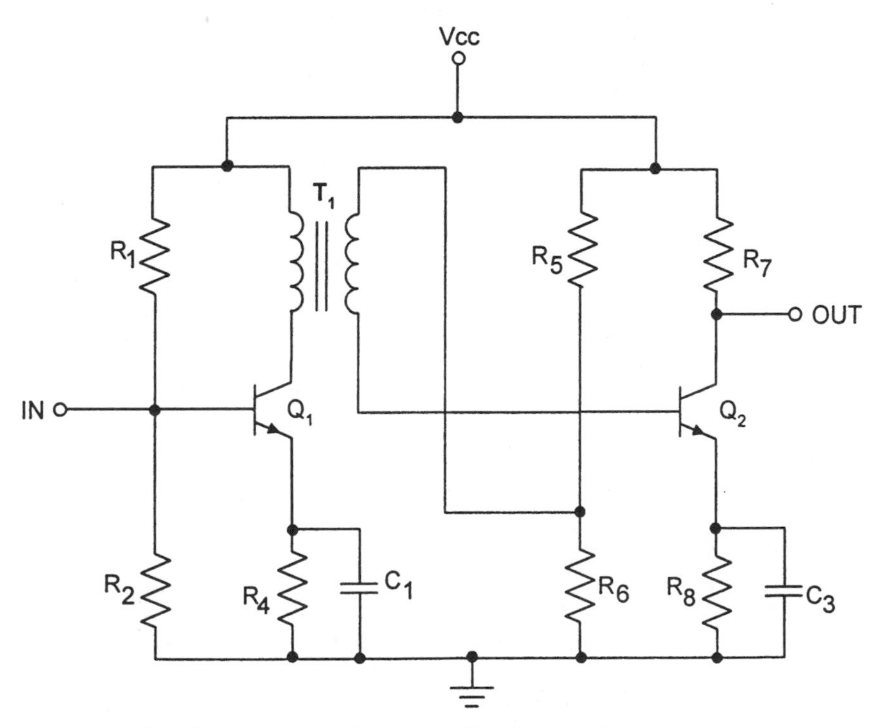

Figure 3.114 Low-frequency transformer coupling between two stages.

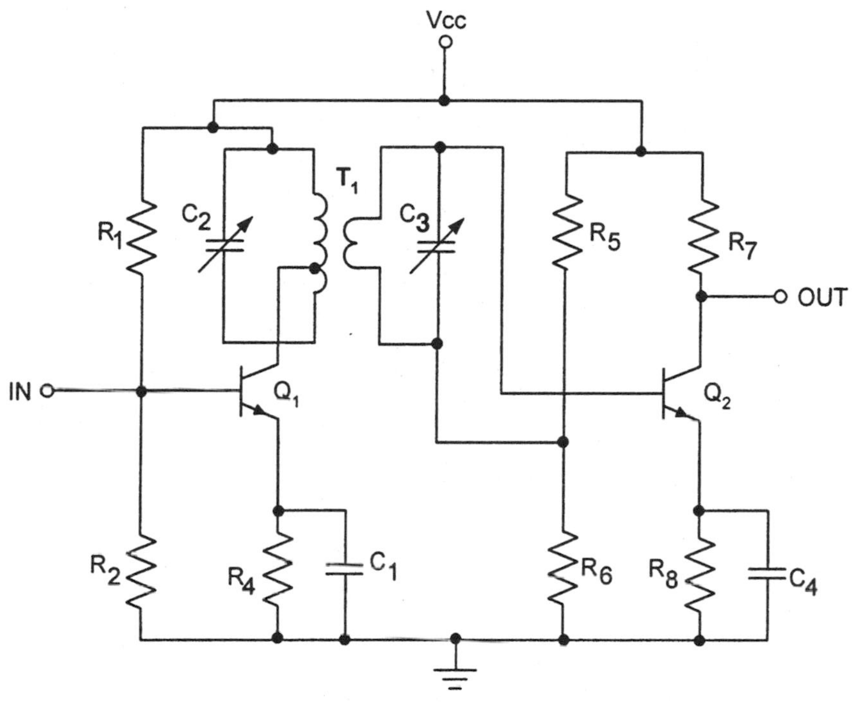

Figure 3.115 High-frequency transformer coupling between two stages.

Degree of coupling. The degree of coupling between a tuned transformer's primary and secondary, which is mainly governed by the distance between the windings, will affect the signal's amplitude and bandwidth as it passes through the transformer. Indeed, as the coefficient of coupling increases (*overcoupling,* Fig. 3.116*a*) or, in other words, as the windings are brought closer together, more flux lines from the primary will cut the secondary. This will produce a higher output voltage and a wider bandwidth over that of the *loose coupling* of Fig. 3.116*c*. The wider bandwidth is caused by the high capacitance now present between the closely spaced primary and secondary, while the high signal amplitude is due to the increased flux lines that cut the secondary. However, as the coefficient of coupling is decreased toward loose coupling, the amplitude and the bandwidth of the signal diminish. Nonetheless, loose coupling can be used to lower the capacitive coupling into the next stage, thus lowering harmonic output, and give the narrower bandwidth that may be required for certain applications. For typical narrowband uses, *optimum coupling* (Fig. 3.116*b*) will be found a good compromise between bandwidth and amplitude.

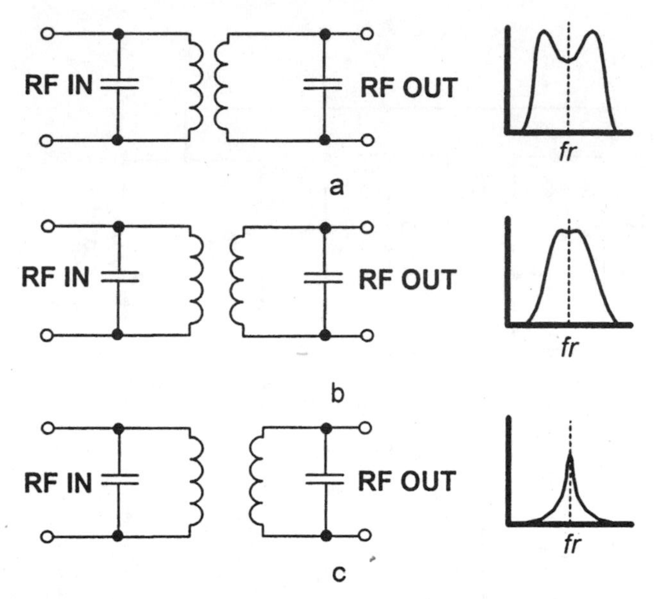

Figure 3.116 The degree of coupling and the effect on the output signal.

Chapter

4

Oscillator Design

Oscillator design is perhaps one of the least understood practices of wireless engineering in general, and is definitely considered to be the most complex. Indeed, until Randall W. Rhea released his groundbreaking book *Oscillator Design and Computer Simulation* in 1990, oscillator design was strictly a hit-or-miss affair for many engineers. As anyone in RF is well aware, it is quite easy to design an oscillator—just design a poor amplifier and turn on the power, and it will probably begin to oscillate. But the problem is to design an oscillator that will oscillate at the desired frequency and amplitude, that will start reliably and not wander, that will not be plagued with spurious responses and harmonics, that will not be excessively affected by normal changes in temperature, and that will be consistent in operation when built over a long production run.

This chapter will concentrate on the design, simulation, and verification of voltage-controlled oscillators (VCOs), *LC* oscillators, and crystal oscillators over a wide range of frequencies. But first, a memory refresher on basic oscillator theory.

Sine-wave oscillators. When a pulse is applied to a tank circuit, it will ring at the tank's resonant frequency, creating a decaying sinusoidal wave (Fig. 4.1). But if amplification from an active device, such as a transistor, is used to amplify and sustain this output, then an oscillator can be formed. The natural resonant frequency of the tank circuit is established by the tank's L and C components, or:

$$f_r = \frac{1}{2\pi \sqrt{LC}}$$

Thus, oscillators will use a small part of their output signal from the active device in order to send a regenerative, or in-phase, feedback into their own

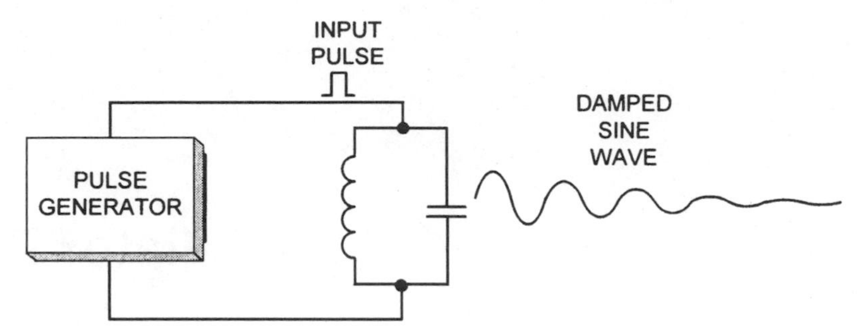

Figure 4.1 A damped sine-wave output of a tuned tank after insertion of a single pulse.

input. This will create a continuous oscillation, with the transistor constantly amplifying its own feedback.

Considering that the typical oscillator functions by feeding a 180 degree out-of-phase signal back to its input, with the phase shift caused by the common-emitter configuration of the oscillator's own amplifier (Fig. 4.2), then we will require a method to shift this out-of-phase output signal back to zero degrees in order to obtain the necessary regenerative feedback (Fig. 4.3). Utilizing the reactance of inductors and capacitors to carry out this phase shifting is the easiest way to construct an RF oscillator.

Most oscillators produce an output power around 5 to 10 dBm, and are biased at Class A or AB in common-emitter configuration (some higher-frequency oscillators are common base, however), though a few may be biased at Class C.

An oscillator is self-starting, and must be quite reliable in this regard. A Class A sine-wave oscillator starts by the following mechanism:

1. Power is applied to the oscillator's active device.
2. Noise and/or transients cause the oscillator to start, beginning the low-power output of sinusoidal waves, after which an oscillator is just translating its DC input power into output sinusoidal oscillations.
3. These sinusoids build to a very high level, which causes saturation of the active device, and surplus loop gain is dissipated. But the surplus loop gain must not be so high that excessive clipping of the output waveform occurs.
4. The oscillator generates sinusoidal waves of stable frequency and amplitude.

There are three main sine-wave oscillator classifications: The *LC* oscillator (and VCO), the crystal oscillator, and the *RC* oscillator. We will concentrate on the first two, since *RC* oscillators function only at audio frequencies.

General oscillator design considerations. Biasing of an oscillator's amplifier section is employed for multiple reasons: to allow the use of a single V_{CC}, to set the bias point for a certain class of operation, to swamp out any device varia-

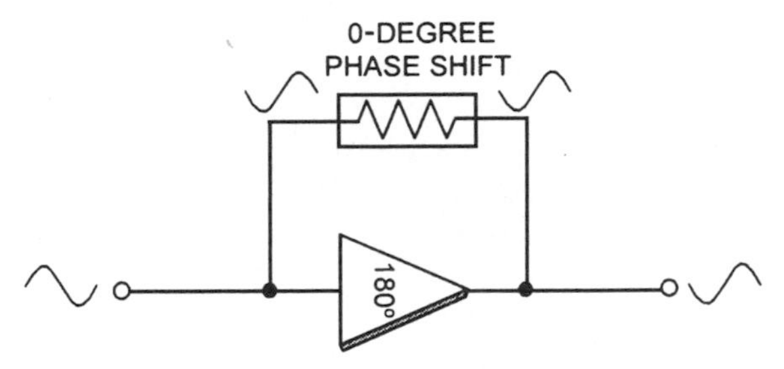

Figure 4.2 An amplifier with degenerative feedback cannot oscillate.

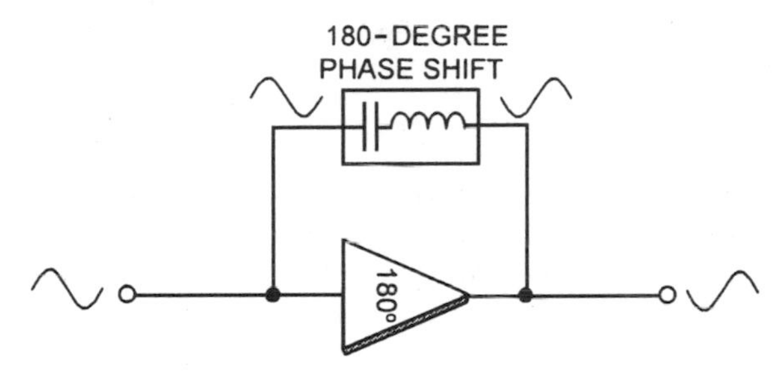

Figure 4.3 An amplifier with regenerative feedback can oscillate.

tions in β, and to stabilize the active device over wide temperature variations. All of these variables must be set by a proper bias network.

A vital parameter of any oscillator circuit will be its Q. A high-Q feedback oscillator [such as the crystal or surface acoustic wave (SAW) types] will have a more frequency-stable output than an LC oscillator. This is because variances in the transistor's reactances caused by changing V_{CC}, temperature, as well as lot variations from transistor to transistor, will cause far less frequency shifts than a low-Q (LC) oscillator.

The proper choice of each of the components for an oscillator is very important, since even the passive components can have a significant impact on oscillator operation. Unless frequency compensation is desired, the oscillator's feedback network capacitors should be NPO for minimum frequency drift under normal temperature variations. The proper choice of the active device is also critical. Transistors with a very high f_t—compared to the oscillation frequency—work much better in oscillator circuits than those with marginal f_t specifications. This is due to the transistor's ability, at a high f_t, not only to maintain its 180 degree phase shift at higher frequencies (an amplifier's phase shift begins to drop from 180 degrees as frequency increases), but also to have a higher feedback gain.

Most oscillators should be extensively decoupled from any noise and intermittent voltage variations of the power supply, and injection of the oscillator's own RF into the mains should be avoided, in all modern wireless applications.

4.1 Oscillator Simulation

4.1.2 Introduction

The following technique of open-loop oscillator design, as well as more LC and crystal oscillator topographies, can be found in further detail in the classic oscillator text by Randall W. Rhea, *Oscillator Design and Computer Simulation.*

These design procedures, as popularized by both Rhea and R. Matthys, have made oscillator design a simple and far more repeatable procedure than in the past. Formally, it was either a horribly mathematically intensive procedure—with an uncertain outcome—or simply copying a particular oscillator design and empirically swapping out the frequency-determining components until the oscillator functioned as close as possible to the desired specifications.

4.1.3 Open-loop design

Open-loop design of an oscillator involves opening the feedback loop of the oscillator from the transistor's output, back to the resonant phase-shifting network (Fig. 4.4), within our linear simulation software. We then insert a software tool called a *Bode plotter* within this open loop. After the software design and simulation of the oscillator is complete, the open loop will be closed, and a fully functioning oscillator will have been completed.

Much of the preliminary design optimization of an oscillator can be assisted by software programs, such as the included PUFF linear simulator, or the high-end Genesys simulator available from Eagleware. Within these two programs we can display the gain and phase of a signal as it passes through the oscillator's open-loop circuits. The tool that allows us to view this gain and phase is the Bode plotter (Fig. 4.5). The Bode plotter inserts a reference signal into the input of a circuit while sweeping through a range of frequencies, and can be found in both a Spice and a linear simulator. The signal that is placed into the circuit from the Bode plotter can be considered to be at zero gain and zero phase shift. Thus, any gain—either positive or negative—or any phase shift that occurs to the Bode's original swept input frequency after it passes through the circuit will be read on the Bode plotter's window and displayed as frequency versus gain and frequency versus phase shift in dB and degrees, respectively. This allows us to view what happens to a signal at the output of a circuit as the input of this same circuit is swept in frequency at a constant

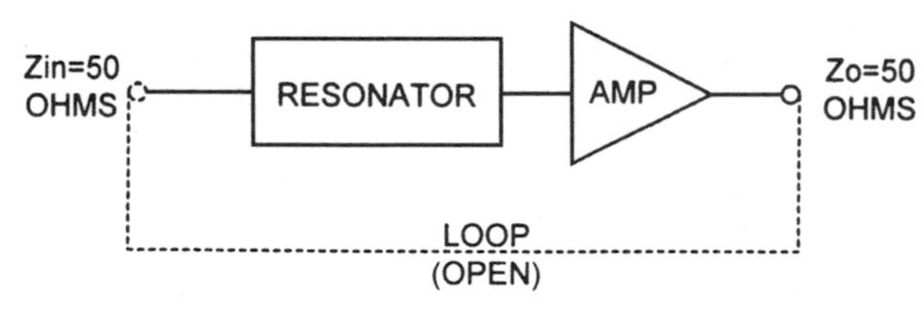

Figure 4.4 Proper input and output of a feedback oscillator.

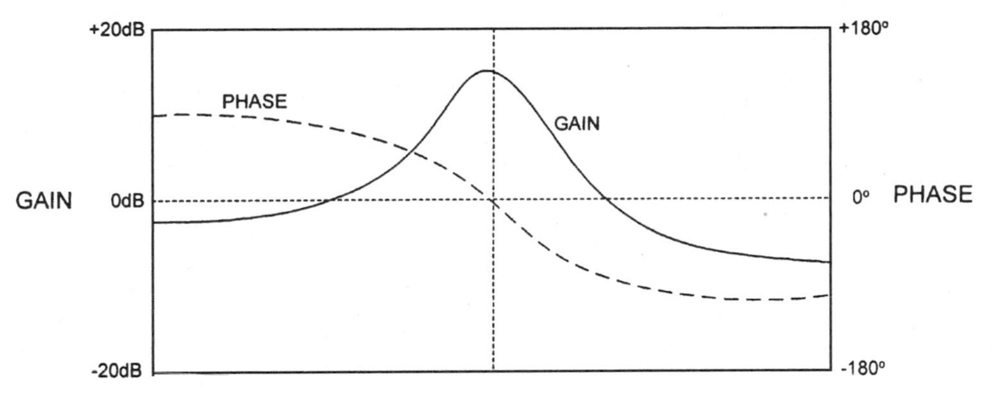

Figure 4.5 Bode plot of oscillator gain and phase.

amplitude and phase. In a nutshell, we can now see the circuit's effect on the gain and phase of a signal *after* it passes through an amplifier, filter, or open-loop oscillator.

With an S-parameter linear software program's Bode plotter we can not only view the effect a circuit has on the gain (S_{21}) and phase [ang (S_{21})] of an inserted signal, but we can also observe the input (S_{11}) and output (S_{22}) return losses, as well as the reverse gain (S_{12}).

For accurate gain and phase responses under simulation, the input and output impedances of the circuit under simulation must be at exactly 50 ohms, since that is the default value of most linear simulators. This is an important concept, because it is not always possible to obtain 50-ohm input/output impedances with an open-loop oscillator, and the linear program's S-parameter port impedances may have to be changed to some higher or lower value to equal the actual input/output impedance of the open-loop oscillator circuit. (The true open-loop input/output impedances are always indicated in the oscillator design procedures of this book). If the terminating impedances of the circuit or simulator were not taken into account, the gain and phase margins displayed on the Bode plotter would be incorrect, and so would the Q and the indicated input/output oscillator's port matching. However, the resonant frequency peak as suggested on the Bode plotter will remain relatively unaffected.

A linear S-parameter program, such as the Puff or Eagleware package, have a further advantage over normal Berkeley Spice programs: They will immediately indicate whether the open loops of the oscillator's input and output are properly matched to each other by displaying a Smith chart tool. The Smith chart will confirm that both the input (S_{11}) and the output (S_{22}) impedances of the open loop are matched at the frequency of interest.

Figure 4.6 demonstrates how to best analyze an oscillator with a linear simulator by using the Bode tool in an open-loop circuit. Most software simulators will not actually have a physical Bode tool as displayed in the figure, but it will be shown as only input and output ports, with the designed oscillator circuit located between these two 50-ohm terminated ports. The Bode plotter display itself will be exhibited as a separate graphical window within the program. In using the Bode tool—by injecting a signal into the oscillator's input and checking the phase and gain at the oscillator's output—we will have a very good indication that our design is valid. This is accomplished, as described above, by breaking the feedback loop of the oscillator and attaching the Bode plotter between the broken input/output points of the oscillator. To obtain a proper reading, set the frequency and phase of the Bode plotter to a *linear* scale, adjust the magnitude to display a gain of −20 to +20 dB (or, if need be, higher values), set the display to show phase values from −180 to +180 degrees (Fig. 4.5), and adjust the frequency sweep to approximately ±25 percent of the expected oscillation frequency (narrow or widen as necessary to obtain the display as shown in the figure). This *open-loop Bode response* test is a good indication that the oscillator will oscillate and function as intended, since the Bode plotter is outputting a 0 degree phase angle signal at the frequencies of interest into the input of the oscillator's resonator, which changes its phase by 180 degrees before it reaches the input of the transistor; the transistor, being in common-emitter configuration, changes it another 180 degrees, making for a phase change of 360, or 0, degrees, for regenerative oscillatory feedback. The proper phase change, at the appropriate amplitude, can be confirmed on the Bode plotter as shown in Fig. 4.7. The Bode plotter is displaying the maximum gain peak at the frequency of the desired oscillation, which should occur at the same frequency as the phase trace when it crosses 0 degrees from the output to the input of the oscillator (in order to sustain oscillatory feedback). At its maximum amplitude the gain trace is called the *gain margin* when it is located at the same frequency as the point where the phase trace crosses the 0 degrees phase point on the Bode plotter screen, and is measured in dB.

The higher the gain margin, the more tolerance the oscillator will have and still start or continue to oscillate when components on the assembly line vary in specifications, or the load varies in impedance. Temperature will also have far less of a deleterious effect with this higher gain margin. A typical, safe value would be 10 dB or more; however, any gain above 0 dB at the 0 degree phase crossing will still allow the oscillator to start because of noise amplification buildup. Nevertheless, temperature, load, and parts variations will make start-up erratic and/or slow when the loop is actually closed for the completed oscillator if this gain margin is too low. In fact, if an oscillator has a sufficiently high gain margin, closing the loop should cause only a minor shift in the design frequency, with the high open-loop gain being reduced to unity when the oscillator reaches its steady state.

In simulating the open-loop oscillator, not only should the gain peak be at the point where the phase is zero, but it should also be as close to the *center* of the phase slope as possible in order to maintain the oscillator's long-term stability and low noise characteristics. The amount of excess phase above or below this center of the phase slope is referred to as the *phase margin* and is as important as the gain margin.

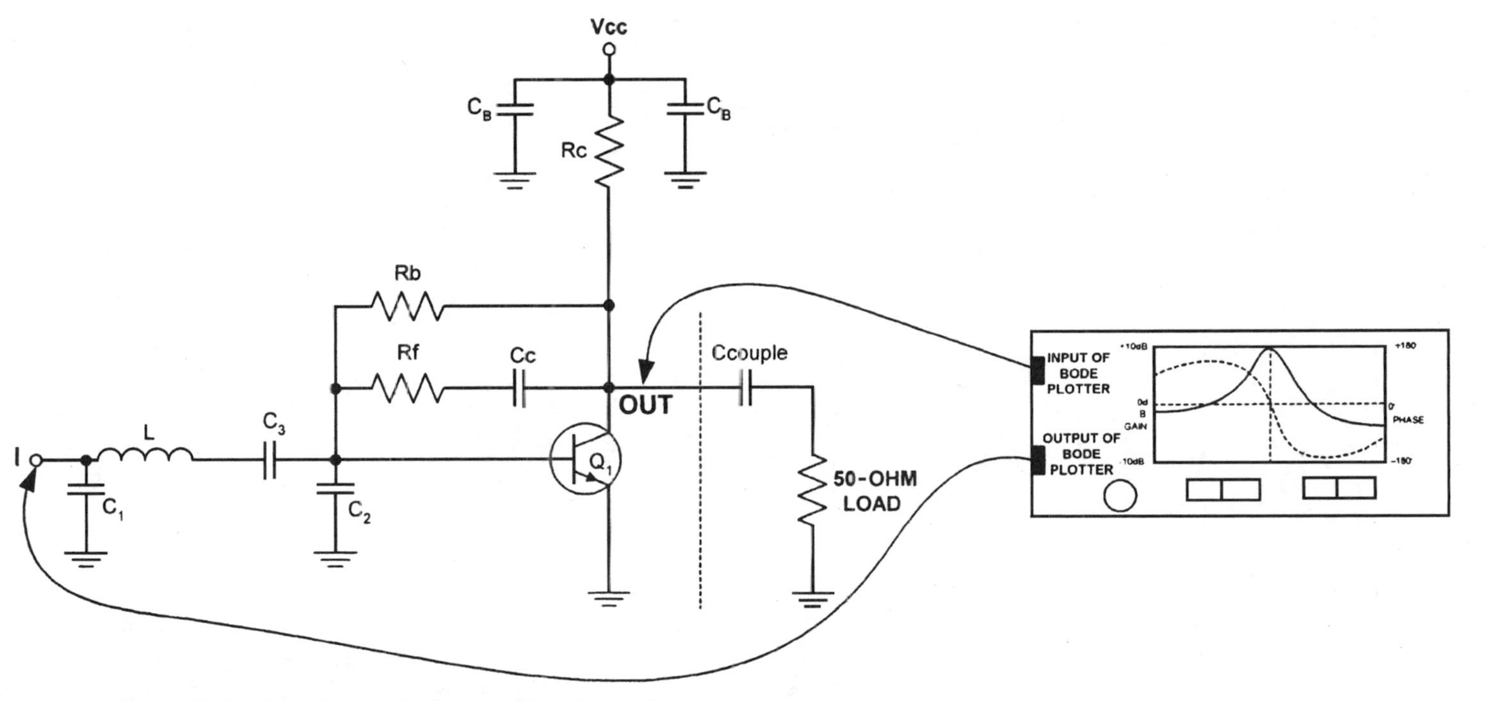

Figure 4.6 Virtual Bode plotter inserted into an oscillator's open loop.

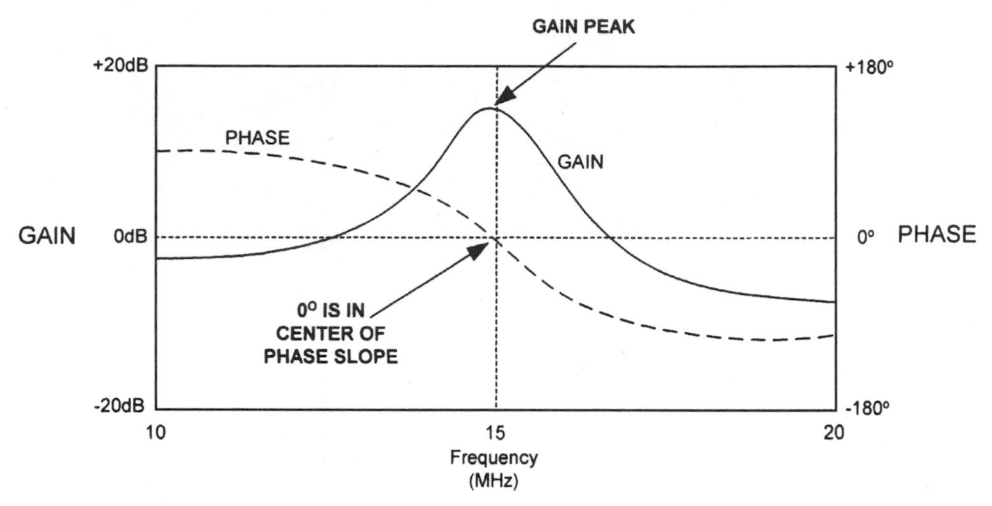

Figure 4.7 Perfect Bode plot of a correctly designed oscillator.

As mentioned above, employing a linear S-parameter-based program will supply the designer with information on the open-loop input and output return loss and impedance, as well as the gain and phase margins. These added functions will permit the engineer to not only confirm gain peaking at zero phase of the open loop, but also whether the input and output of the open loop are both at, preferably, 50 ohms. This will result in maximum gain and more accurate simulations. In addition, 50-ohm test equipment, such as a network analyzer, may be used on the physically completed 50-ohm oscillator for test verification.

After this open-loop simulation procedure is accomplished successfully in the linear S-parameter simulator, then the loop may be closed (Fig. 4.8), and energy may then be tapped from the oscillator and placed into a load. The energy tapped from the oscillator, however, will decrease the available feedback from the loop. A series X_L or X_C of approximately 100 ohms may be placed at the oscillator's output, with a 50-ohm load attached, and the simulation attempted with the Spice oscilloscope and fast Fourier transform (FFT) tool. These software tools are connected to the 50-ohm load to confirm oscillation, output voltage, starting, harmonics, etc. (*Caution:* A Spice frequency source must normally be included somewhere in the Spice simulation, or the oscillator will not function—just attach the Spice frequency source to the input of the oscillator through a series 1 megohm resistor to "fool" the simulator. Spice simulators may take 20 minutes or more of computer time for the oscillator to reach full amplitude; so be patient when employing such simulators within the time domain.) Figure 4.8 is using a series C_{COUPLE} to couple the energy to the load from the oscillator. Coupling out the oscillator's energy, without decreasing the feedback to excessively low levels, will be discussed later in this chapter.

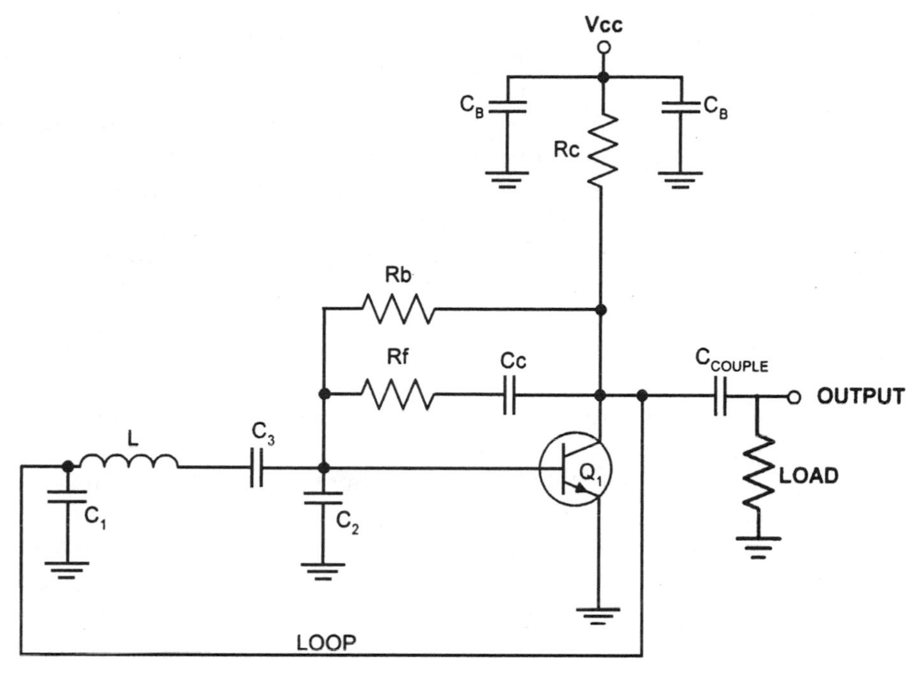

Figure 4.8 Loop closed after Bode analysis for a functionig oscillator.

Since the loaded Q of an oscillator will govern its phase noise and frequency drift, then the higher the loaded Q the more stable the oscillator is over temperature, and the lower the phase noise figures. The loaded Q of an oscillator can be measured by employing the open-loop test method above, along with the formula:

$$Q_L = \frac{f_0}{3 \text{ dB BW}}$$

where Q_L = loaded Q in dimensionless units
f_0 = center frequency of the oscillator
3 dB BW = bandwidth of the oscillator's gain (S_{21}) at its half-power points

This formula is accurate only if the phase slope crosses zero at the center of its fall, which is an optimum, or high-phase-margin, condition. It is evident from the formula that the highest loaded Q of the oscillator depends on the bandwidth being narrow within the gain response plot for a high-quality and stable oscillator design.

The loaded Q of an LC (or VCO) oscillator must never be allowed to degrade below 5 or 10—and should preferably be much higher—to stabilize the oscillator and to lower its phase noise. And by increasing the oscillator's power output with higher transistor bias current we can also decrease the phase noise, since the carrier will now be at a higher relative amplitude above this noise.

As open-loop oscillator design accuracy depends on both ends of the oscillator loop being at the same impedance (as well as both terminating impedances in the linear simulator being equal), then it can be seen that ignoring cascaded input and output impedances will result in a nonoptimized oscillator design. However, a new technique that allows the oscillator designer to not have so much dependence on the oscillator's terminating impedances with the linear simulator for an accurate prediction of the oscillator's gain and phase has just been recently presented. This new technique employs *Harada's equations,* and assumes that S_{12} equals zero, which is never the case. Therefore, it is recommended that for ease of computations, and for very acceptable accuracy, that normal open-loop analysis (as demonstrated by Rhea) should be followed for the design of most oscillators.

In utilizing open-loop oscillator design, it is assumed that the open loop is stable. In other words, the amplifier section (with bias) should not be unstable, since it is only when the loop is closed from input to output that oscillations are meant to occur. Proper frequency stability may become quite erratic with an unstable device, so choose only unconditionally stable transistors.

When simulating a crystal oscillator, we must first select the proper crystal by obtaining certain crystal parameters (Fig. 4.9), such as the crystal's motional capacitance (C_M), motional inductance (L_M), series resistance (R_M), and parallel plate capacitance (C_P or C_0) from the manufacturer for the crystal's desired frequency of operation, its holder type, and quartz cut (typically AT). The manufacturer will also need to be informed if the crystal is to be utilized in a series or parallel resonance oscillator (see "Pierce Crystal Oscillator Design" in Sec. 4.3.3), and whether the crystal is being run on its fundamental or on one of its overtone frequencies. The crystal's required aging specification in parts per million per year (ppm/year), initial frequency accuracy in ppm, and the frequency accuracy over temperature in ppm are all important as well.

Since many linear computer simulation packages may not necessarily have crystal models available, we must model the crystal as shown in Fig. 4.9, and place it where the crystal would be within the oscillator circuit. This equivalent *LCR* model, while simplistic, is more than adequate to realistically repre-

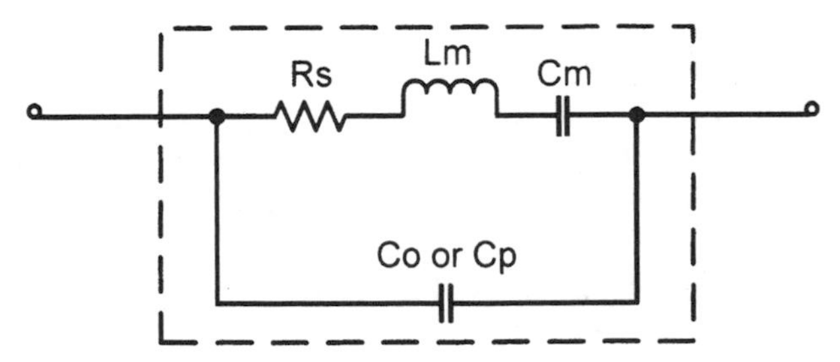

Figure 4.9 Equivalent internal structure of a crystal.

sent a typical crystal for an oscillator using the manufacturer's above motional specifications of L_m, C_m, R_m, and C_p.

Note that if a crystal's holder or its package is changed, as might be required when redesigning for a smaller oscillator, this will have an effect on the above motional properties of the crystal, and a new simulation must be performed with these latest values to assure proper operation.

A word on open-loop simulation using a Spice simulator. We can obtain accurate design results when simulating low-frequency oscillators with Spice, especially when we must employ a low-frequency transistor for which there are no S-parameters available. Bode plotters, as well as the *AC analysis* tool in Spice simulators, may not give accurate loop gain measurements unless we attach a basic Spice frequency source (set to 1 Hz) through a 50-ohm resistor to the oscillator's "input" (the *LC* resonator at the node between C_1 and L_1 of an *LC* oscillator). However, the resonant frequency at maximum gain as indicated in the Bode plotter and the AC analysis tool window will vary slightly from the other, as will the 0 degree phase-crossing frequency.

Oscillator output coupling. Since adding a load to the output of the oscillator can drastically affect the oscillator's frequency, power, and whether it will even start or not, then proper coupling is important even at the simulation phase. Tapping the oscillator's output power at the proper location is crucial, as is a suitable amount of output coupling—not too much, not too little. In other words, we do not want to load down the oscillator by supplying excess RF power to the load or, conversely, by supplying too little power. In performing the software open-loop analysis, it is best to add the desired series output coupling reactance, along with the output load, to the final simulation run to confirm proper gain margin will still exist. Using the appropriate coupling capacitor or inductor in series with the load, with a high enough X_C or X_L, is recommended to maintain oscillations (Fig. 4.10): A reactance of 200 ohms, as an example, should not degrade the oscillator's vital feedback amplitude excessively, while minimally decreasing the effective output power of the oscillator. However, if

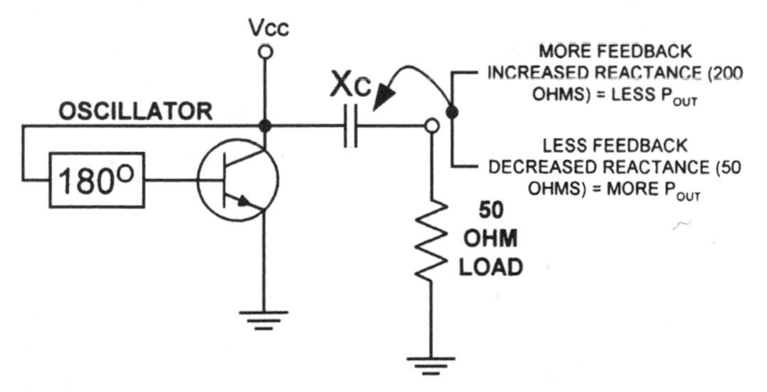

Figure 4.10 Comparisons between a high and a low coupling reactance on the output power of the oscillator.

simulations predict that a reactance of 50 ohms is possible for the output capacitor or inductor into the load, while still maintaining the proper gain margin, then more power can safely be output into the load. Optimize this coupling reactance value in the open-loop Bode simulations by decreasing the reactance of the coupling element until the gain margin is still at a safe level of at least 5 dB or greater. This also means that a true conjugate match directly from the oscillator output to the load input is usually impossible, since it would excessively load down the oscillator, and prevent it from starting or from running properly.

The approximate output power of the oscillator can be found in Spice by employing the oscilloscope tool across the oscillator's load and calculating:

$$P = \frac{V_{RMS}^{2}}{R} \text{ (in watts)} \qquad \text{or} \qquad P = 10 \log \frac{P}{1 \text{ mW}} \text{ (in dBm)}$$

4.2 VCOs and *LC* Oscillators

4.2.1 Introduction

Today, *LC* oscillators are normally *variable frequency oscillators* (VFOs) of the *voltage-controlled oscillator* (VCO) type since they can readily be tuned by adjusting the capacitance of a varactor diode to set the frequency of oscillation. Still, very low cost products that utilize fixed-frequency *LC* oscillators can be found, sometimes up to 2.4 GHz; but they have poor frequency stability over temperature and poor phase-noise specifications because of their very low *Q*.

Indeed, for any *LC* oscillator to be even remotely frequency stable, and have decent phase noise, it will require a high capacitance-to-inductance ratio in its *LC* circuit (for a higher *Q*), a steady and clean power supply, stable temperature conditions, and strong isolation from its load. Still, *LC* oscillators will drift in frequency by up to 1 percent or more because of aging components or (if unshielded) when a conductive surface is in close proximity. For any quality wireless device this is unacceptable—unless some form of frequency regulation is employed.

As mentioned above, *LC* oscillators are found mainly in voltage-controlled oscillator circuits, since VCOs are heavily utilized in frequency synthesis for phase-locked loops, and in any application where a DC control voltage is required to alter the output frequency of an oscillator.

4.2.2 Types of *LC* oscillators

There are numerous kinds of *LC* oscillators. However, both the *Hartley* and the *Colpitts* oscillators are very common, and an understanding of their function will allow a good grasp of other *LC* designs.

The Hartley oscillator, as shown in Fig. 4.11, exploits a tapped coil in its tank circuit, made of L_1 and C_1, to change the phase of the feedback to the transistor's base into a regenerative signal, and to set the frequency of oscillation. C_2

and C_3 block the DC, but couple the AC feedback, while L_2 and C_6 decouple the oscillator output, preventing it from being injected into the power supply. L_2 also functions as the transistor's collector load, and R_1 and R_2 supply the forward bias. R_E and C_4 further increase the temperature stability of this circuit without allowing the AC gain to be decreased, as it would be if just R_E were used.

The *LC* tank of the Hartley furnishes the required 180 degree phase shift for regenerative feedback, thus allowing oscillations. This is because L_1's tapped coil forces the signal between the center tap and the top of the coil (due to the current flow with respect to the grounded tap; Fig. 4.12) to be opposite in polarity to the center tap and the bottom of the coil. The location of the tap on the inductor sets the amplitude of the positive feedback.

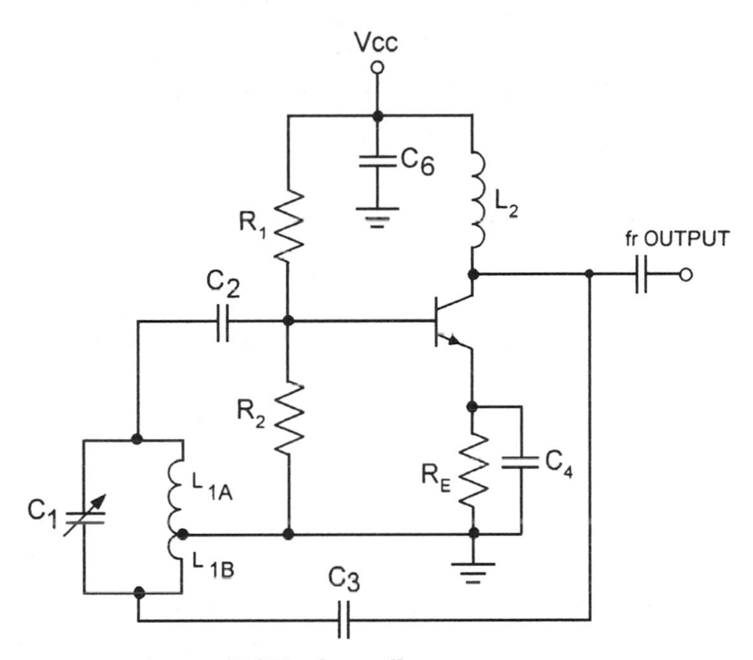

Figure 4.11 A type of *LC* Hartley oscillator.

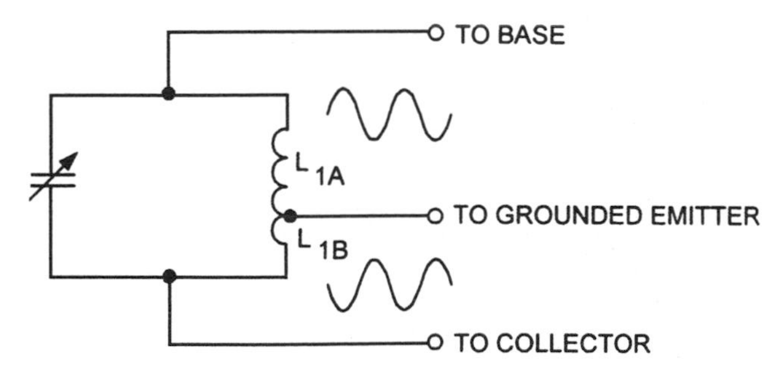

Figure 4.12 A tapped coil showing the phase relationships.

As an alternative to a tapped coil, we can use tapped twin capacitors, as in the Colpitts oscillator of Fig. 4.13. This performs the same function as the tapped coil by creating a 90 degree phase shift across each capacitor, thus furnishing the positive feedback we need into the transistor's base for oscillations. Considering that the capacitance ratio changes the feedback voltage, the two capacitors should be ganged to change the frequency by more than a few percent of its center frequency, or a tunable inductor can be used, with the capacitors at fixed values.

VCOs are simply *LC* oscillators that can vary their output frequency with a DC input control voltage. *Varactor diodes* can be used for this purpose. These diodes, in step with a DC voltage, vary their capacitance up and down. When placed in the tank of an *LC* oscillator, the resonant frequency of the tank, and thus of the oscillator, can be made to change either above or below a rest, or center, frequency. This rest oscillation frequency is established by the DC bias across the varactors, normally setting their initial capacitance at some intermediate value. So, by adding to or subtracting from the rest bias value, the frequency of the oscillator can be altered over a wide range.

As an example of a VCO, consider Fig. 4.14. Q_1 and its associated components are designed as a Hartley oscillator, with Q_2 acting as a buffer at the output to prevent the Hartley from being loaded down by the low input impedance of the next stage. The back-to-back varactors shown are commonly employed in a VCO so that, at low bias levels when one varactor is being affected by the AC, the other is actually being reverse biased, which will

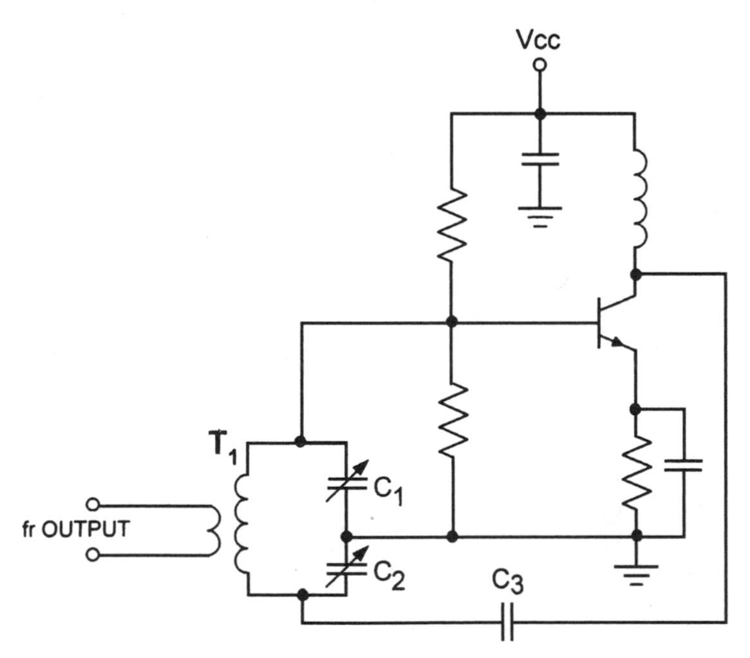

Figure 4.13 A type of *LC* Colpitts oscillator.

decrease formation of distortion products. However, this cuts the varactor capacitance values in half.

There are two different types of varactor diodes utilized in VCO circuits. The *abrupt* form has a very high Q (and thus low phase noise), and can take a wide voltage tuning range (up to 0 to 55 V) to travel through the full range of capacitance values, meaning that abrupt diodes possess low tuning sensitivity. Abrupt diodes also have a low capacitance range, but with low distortion characteristics. The *hyperabrupt* varactor type, on the other hand, has a complete tuning range of 0 to 20 V for increased sensitivity, so it is the varactor of choice for wideband applications. However, it has a lower Q, and thus more phase noise, than the abrupt type.

Both varactor types may have a 0 V capacitance specification, but because of nonlinearity and Q problems, at least 0.1 V should always be across any varactor—and sometimes more.

4.2.3 Designing *LC* oscillators and VCOs

Designing *LC* oscillators and VCOs with the following procedures, while verifying their operation as described in Sec. 4.1, "Oscillator Simulation," will permit the engineer to design and build stable and reliable circuits for a variety of requirements.

***LC* BJT oscillator design (Fig. 4.15).** This oscillator will function reliably up to 500 MHz.

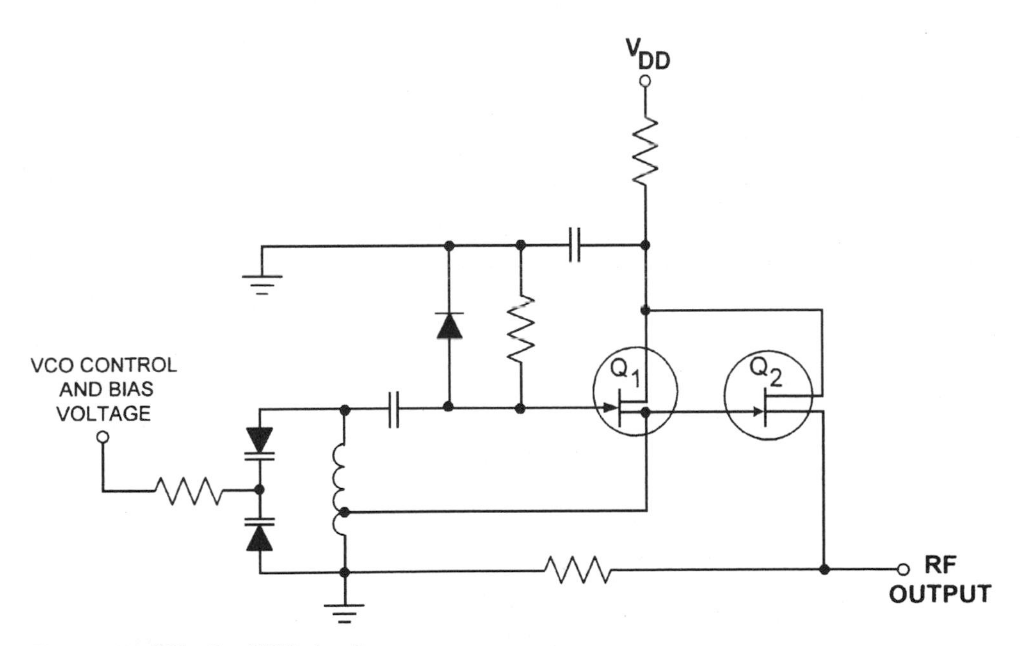

Figure 4.14 A Hartley VCO circuit.

1. Choose a proper high-frequency transistor with an f_T that is much higher than the oscillation frequency (5 times higher is a good choice).
2. Bias the active device Class A by the following procedure:
   a. Choose the supply voltage. Select a Q point for the transistor that is consistent with the available S-parameter file for I_C and V_C. Example: (I_C = 10 mA; V_C = 6 V; V_{CC} = 12 V. Find transistor's typical β, such as β = 50.
   b. Calculate $I_B = I_C/\beta$
   c. Calculate $R_B = \beta \dfrac{V_C - 0.7}{I_C}$
   d. Calculate $R_C = \dfrac{V_{CC} - V_C}{I_B + I_C}$
3. Calculate the values for the LC resonator and other components by:

$$L = \frac{190}{2\pi f} \qquad C_1 = \frac{1}{48\pi f} \qquad C_2 = \frac{1}{48\pi f}$$

$$C_3 = \frac{1}{300\pi f} \qquad C_C < 1\ ohm\ (X_C) \qquad R_f = \frac{2500}{(0.025/I_C)}$$

   (R_f should be tweaked in the preliminary open-loop S-parameter analysis until both the input and output are close to 50 ohms on the Smith chart.)
4. $C_{COUP} \approx$ 50 to 200 ohms (X_C) for a 50-ohm load. Find the necessary value of C_{COUP} by simulating the oscillator into a 50-ohm load, and use the lowest C_{COUP} reactance value that will still allow the oscillator to maintain a decent gain margin (>5). If a high input impedance buffer amplifier follows C_{COUP}, then $C_{COUP} = C_C$ (however, the phase noise will go up).
5. Simulate and optimize as explained in Section 4.1, "Oscillator Simulation."

Notes. Increasing C_1 and C_2, as well as L, while decreasing C_3, will increase the loaded Q of the oscillator (high loaded Q reduces phase noise and frequency drift, and reduces temperature effects).

While this oscillator is capable of operation at up to 500 MHz, it does have a lower frequency limit of about 25 MHz, at which point it is advisable to work with a Colpitts design.

MMIC *LC* oscillator (Fig. 4.16). This is an oscillator that is capable of up to 1 GHz frequency operation. It is much simpler to design, but will cost more, than the LC BJT oscillator above. The MMIC oscillator is only used when higher-frequency operation is required.

1. A V_{cc} should be chosen that will allow at least 2 V (preferably 4 V) to be dropped across R_{BIAS} for stability. (If R_{BIAS} does not reach 500 ohms, employ an RFC for a combined impedance of 500 ohms for both R_{BIAS} and the RFC):

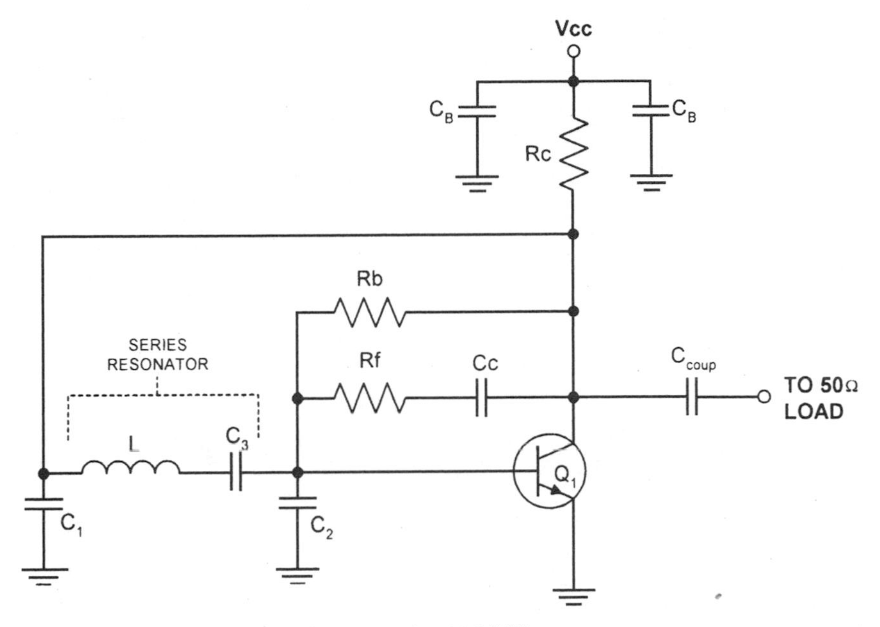

Figure 4.15 An *LC* oscillator design good to 500 MHz.

$$R_{\text{BIAS}} = \frac{V_{CC} - V_{\text{MMIC}}}{I_{\text{MMIC}}}$$

where V_{MMIC} = DC voltage required at the MMIC's power pin and I_{MMIC} = DC current required into the MMIC's power pin. The power dissipation within the R_{BIAS} resistor should be checked so that a proper resistor can be chosen with at least double the dissipation expected:

$$P = I_C^{\,2} R_{\text{BIAS}}$$

2. Calculate the component values for the *LC* resonator of the oscillator:

$$L = \frac{190}{2\pi f} \qquad C_1 = \frac{1}{48\pi f} \qquad C_2 = \frac{1}{48\pi f}$$

$$C_3 = \frac{1}{300\pi f} \qquad C_B < 1\ \Omega\ (X_C)$$

3. $C_{\text{COUP}} \approx$ 50 to 200 ohms (X_C) for a 50-ohm load. Find the necessary value of C_{COUP} by simulating the oscillator into a 50-ohm load, and use the lowest C_{COUP} reactance value that will still allow the oscillator to maintain a decent gain margin (>5). If a high input impedance buffer amplifier follows C_{COUP}, then $C_{\text{COUP}} = C_C$ (however, phase noise will go up).
4. Simulate and optimize as explained in Sec. 4.1, "Oscillator Simulation."

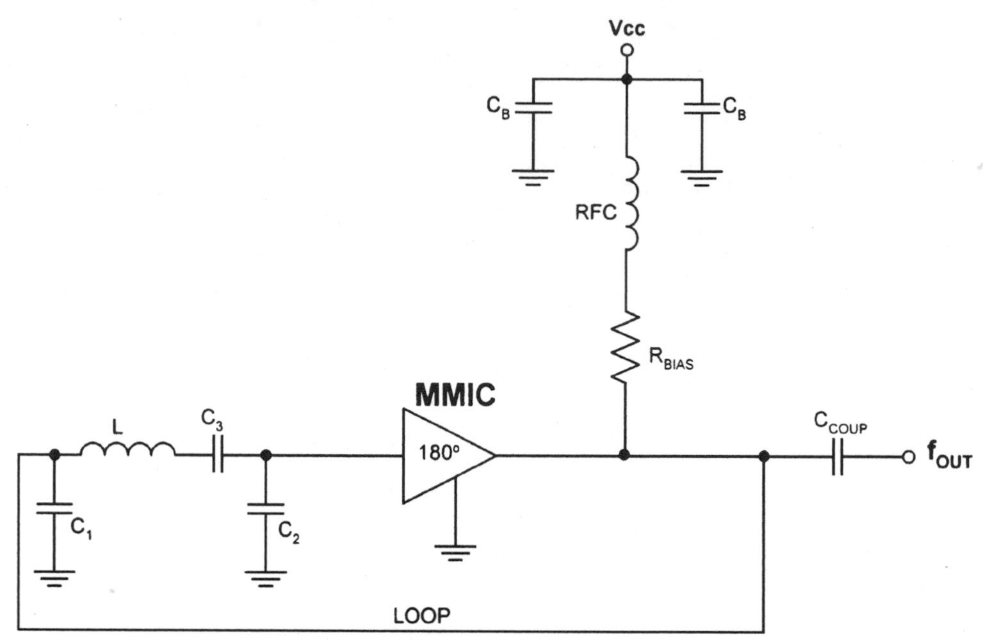

Figure 4.16 A complete MMIC *LC* oscillator.

JFET *LC* and VCO Colpitts oscillator (Fig. 4.17). This oscillator is good to 50 MHz, has a loaded Q of 20 to 25, and can tune to over 100 percent of f_{OUT}. L_1 should give as high a Q as possible to maximize gain and phase margin. This VCO can easily be made into a simple *LC* oscillator by replacing the varactor D_1 with a capacitor of similar value, and removing the varactor's bias network of R, both C_c's, and L_2.

The oscillator's open-loop Z_{OUT} and Z_{IN} will be around 150 ohms, so the terminating impedances of the S-parameter linear simulation program should be changed to this value to obtain the proper open-loop gain and phase. However, when the mismatch is slight, the actual loop gain of the final closed-loop oscillator will be negligible. If the active device does not have S-parameter files available because of its low frequency of operation, then employ Spice models in a Spice simulator as explained in Sec. 4.1, "Oscillator Simulation."

1. Self-bias the oscillator.
   a. Choose an appropriate V_{GS} for Class A operation of the active device, and note the I_D for the selected V_{GS}.
   b. Compute $R_S = \dfrac{V_{GS}}{I_D}$.
   c. Choose a V_{DS} of $(V_{dd} - 2\,V_{GS})/2$ for a V_d of $V_{dd}/2$.
   d. Calculate $R_d = (V_{dd} - V_{ds} - V_{GS}/I_d)$.
   e. Since V_{GS} for a specific I_D is not always available, use the following equation to find V_{GS} when I_D, I_{DSS}, and V_P are known (look in the JFET's data sheet for these values):

$$V_{GS} = V_P\left(1 - \sqrt{\frac{I_D}{I_{DSS}}}\right)$$

f. RFC equals 1000 ohms.
g. $C_C < 1$ ohm (X_C)
h. L_2 = RFC

2. Confirm that the JFET device will safely dissipate the power of:

$$P = I_D V_D$$

3. To design the resonant LC network:

$$L_1 = \frac{258}{2\pi f} \qquad C_{D1} = \frac{1}{480\pi f} \qquad C_2 = \frac{1}{48\pi f}$$

4. Couple the output of the oscillator to its 50-ohm load through a 200- to 600-ohm reactance (C_{COUP}), which can be either inductive or capacitive.
5. Simulate and optimize as explained in Sec. 4.1, "Oscillator Simulation."

BJT 500-MHz VCO (Fig. 4.18). This voltage-controlled oscillator will function reliably up to 500 MHz.

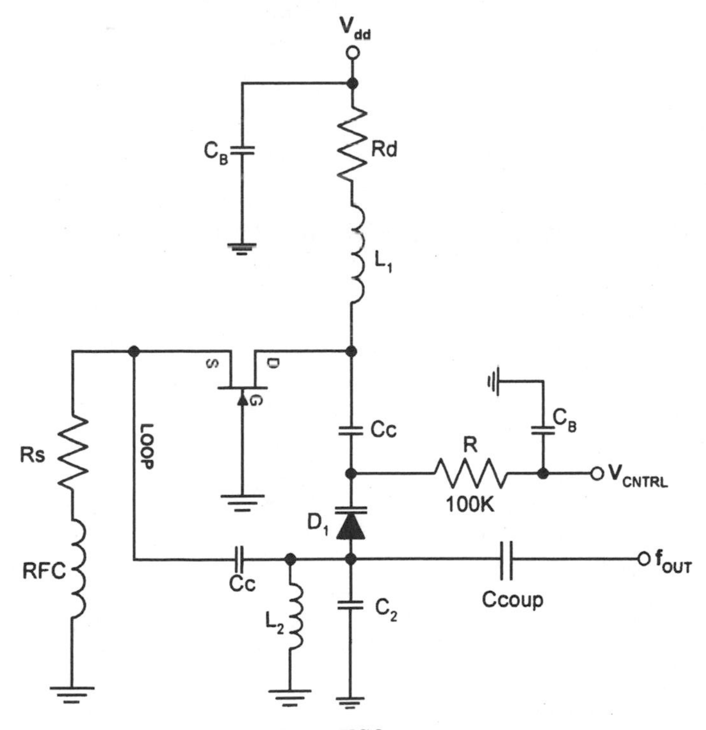

Figure 4.17 A JFET LC Colpitts VCO.

1. Choose a proper high-frequency transistor with an f_t that is much higher than the oscillation frequency (Q_1 is normally selected to have an f_t of about 5 times that of f_{OUT}).
2. Bias the active device Class A:
   a. Choose the supply voltage. Select a Q point for the transistor that is consistent with the available S-parameter file for I_C and V_C. Example: $I_C = 10$ mA; $V_C = 6$ V; $V_{CC} = 12$ V. Find the transistor's typical β, such as β = 50.
   b. Calculate $I_B = I_C/\beta$
   c. Calculate

$$R_b = \beta\,(V_C - 0.7/I_C)$$

   d. Calculate

$$R_C = \frac{V_{CC} - V_C}{I_B + I_C}$$

3. Calculate the component values for the LC resonator of the oscillator:

$$L = \frac{190}{2\pi f} \qquad C_1 = \frac{1}{48\pi f} \qquad C_2 = \frac{1}{48\pi f} \qquad C_C = C_B < 1\ ohm\ (X_C)$$

$$C_{D1} = \frac{1}{300\pi f}$$

4. Calculate

$$R_f = \frac{2500}{(0.025/I_C)}$$

   (R_f should be tweaked in the preliminary open-loop S-parameter analysis until both the input and output of the oscillator are close to 50 ohms on the simulator's Smith chart.
5. $C_{COUP} \approx$ 50 to 200 ohms (X_C) for a 50-ohm load. Find the necessary value of C_{COUP} by simulating the oscillator into a 50-ohm load, and use the lowest C_{COUP} reactance value that will still allow the oscillator to maintain a decent gain margin (>5). If a high input impedance buffer amplifier follows C_{COUP}, then $C_{COUP} = C_C$ (however, phase noise will go up).
6. Simulate and optimize as explained in Sec. 4.1, "Oscillator Simulation."

Notes. L and the capacitance of the varactor, D_1, are near series resonance, while C_1 and C_2 act as coupling capacitors to obtain 180 degree phase shift with L and D_1 (with a high Q). R_f, with its DC decoupling capacitor (C_C), feeds back some of the RF into the oscillator's input in order to diminish low-frequency gain for stabilization of the BJT, as well as to lower both the input and output impedance of the oscillator closer to 50 ohms. This not only makes it easier to simulate in a 50-ohm environment, but also in a real environment with a vector network analyzer (VNA).

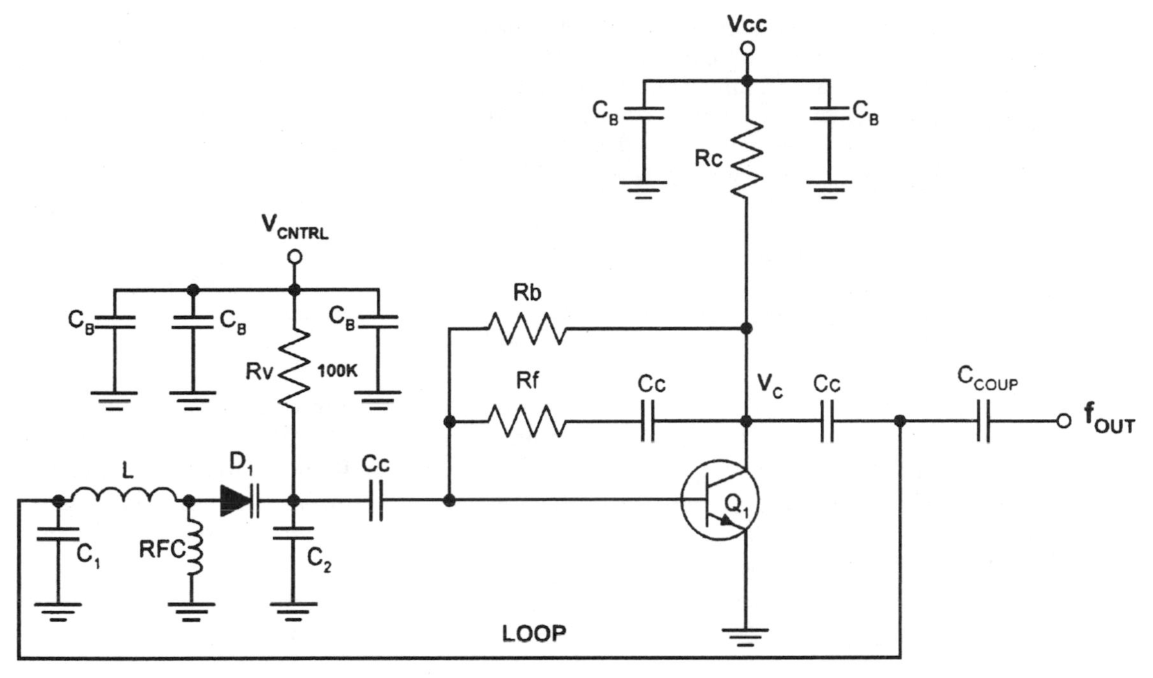

Figure 4.18 A complete bipolar VCO circuit.

R_b and R_c are the BJT's bias components, and the C_B's are bypass capacitors chosen to decouple all frequencies from 60 Hz all the way to f_{OUT}, and beyond, with an X_C of less than 1 ohm. This necessitates using various types of capacitors, such as electrolytic for the audio frequencies and two different value ceramic (or porcelain) capacitors for low and high RF.

The varactor bias voltage, chosen from the varactor's data sheet for the capacitance desired, employs the RFC and the R_V to block RF but pass the DC control voltage. Since a varactor is reversed biased, very little leakage current will flow through R_V, so the voltage dropped across this resistor will be quite small. The varactor is chosen to supply an appropriate range of capacitances for the VCO's tuning range, since the only component that will vary the frequency of oscillation will be the varactor's capacitance, which is controlled by V_{CNTRL}.

Q_1 will have an f_t that is 5 times or more above the f_{OUT} frequency so as to closely maintain the common emitter's 180 degree phase shift across the oscillator's entire tuning range.

The C_C's are placed to block DC, but to easily pass f_{OUT}. C_{COUP} will have a high X_C (50 to 200), and can readily be replaced by an equally high reactance inductor for harmonic suppression.

Depending on the frequencies of oscillation chosen, a varactor of sufficiently high value may not be available. This can be overcome by increasing L, which will allow D_1 to be decreased in value.

With the proper wide tuning varactor for D_1, a tuning bandwidth of 100 percent can be accomplished by employing a 10-to-1 capacitance hyperabrupt varactor, along with the proper tuning voltage range for V_{CNTRL}. However, when the VCO is used in this wideband mode, its output power will begin to decrease as the f_{OUT} increases. This is not a problem in less demanding VCO applications, or in a more narrowband (<50 percent tuning range) mode.

An MMIC version is shown in Fig. 4.19. It utilizes the same design equations for the resonator and varactor sections as does the above BJT design. The bias network and transistor is simply replaced by a high-frequency MMIC, along with its own bias network. The MMIC must, however, supply a 180 degree phase shift from its input to its output (which is the most common MMIC configuration).

Integrated *LC* and VCO oscillator (Fig. 4.20). Integrated circuits are now being manufactured that can be made to function as an LO or as a VCO by simply attaching a few external components. One such oscillator IC is the Maxim MAX2620, which can operate anywhere between 10 and 1050 MHz with low phase noise (−110 dBc/Hz at 25-kHz offset). It has a built-in integrated output buffer and a low-voltage power supply (+2.7 to +5.25V) for low power consumption (27 mW at a 3-V_{CC}). The MAX2620 can also be employed differentially to supply an IC mixer with an LO, or as an unbalanced oscillator to feed a double-balanced mixer.

Varactors and VCOs. When varactors are used in a VCO circuit, a high-value resistor (50 to 100 kilohms) is always placed in the varactor bias line between

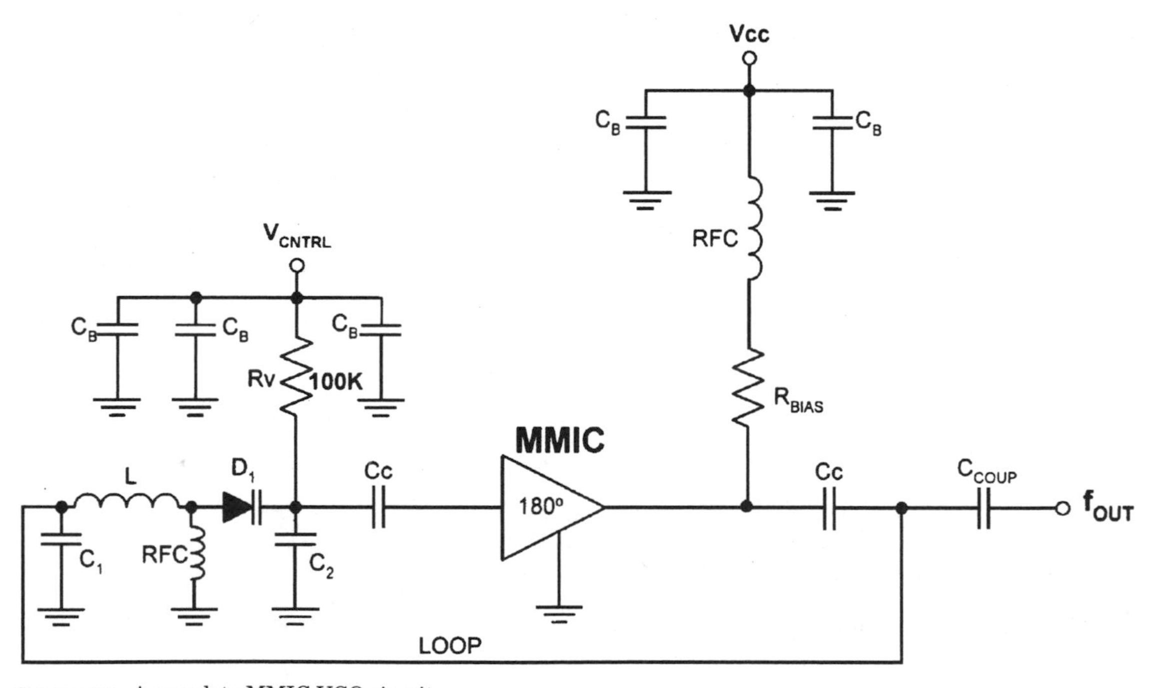

Figure 4.19 A complete MMIC VCO circuit.

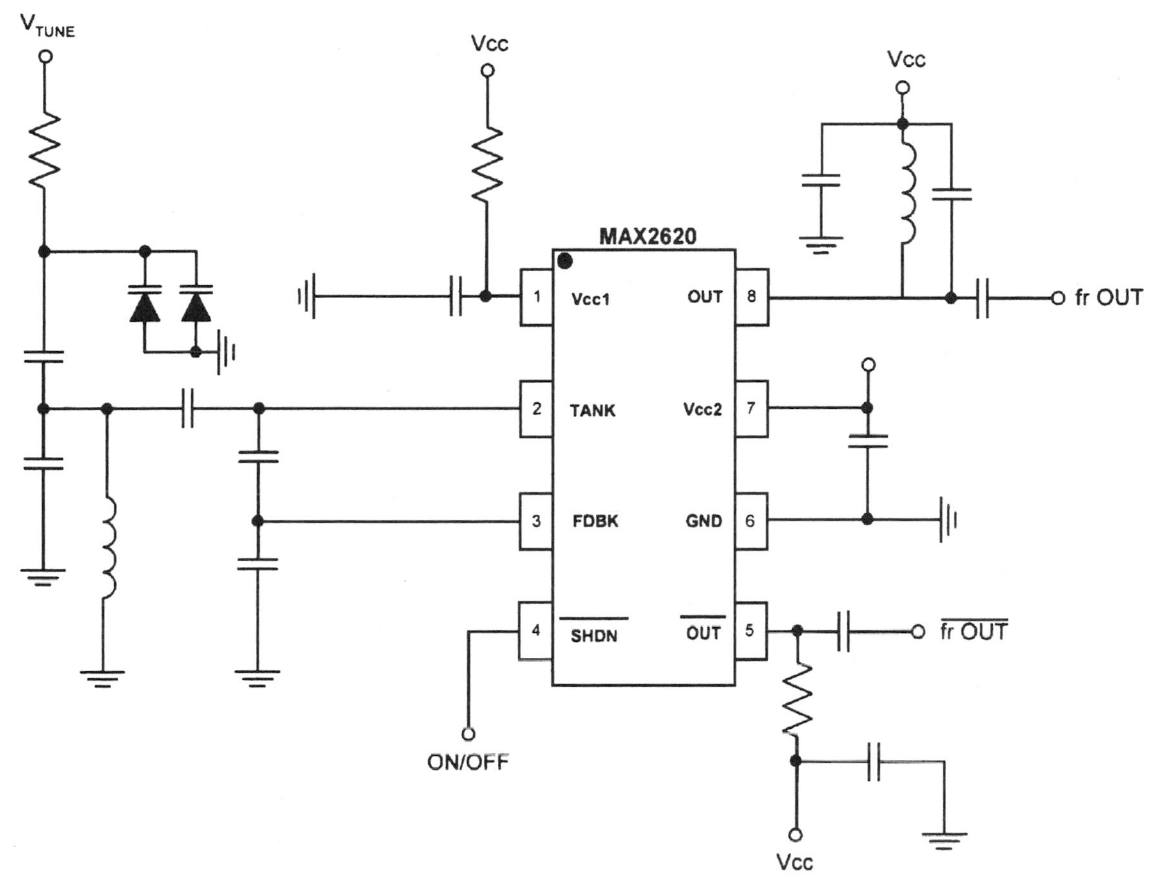

Figure 4.20 The Maxim oscillator integrated circuit with support components.

V_{CNTRL} and the varactor. The resistor is utilized to isolate the very small value of the varactor's capacitance and prevent it from being affected by the rest of the circuit's stray reactances, as well as the high value of any decoupling capacitor. This permits the varactor to maintain its desired capacitance value even while attached to a highly reactive control circuit.

An effect called *varactor modulation* can significantly degrade phase noise performance of the VCO. Varactor modulation is worsened when a more sensitive varactor is utilized, with more phase noise now contributed to the output of the VCO. This will not be as much of a problem in VCOs that tune only over a narrow range of frequencies, but can become significant with wideband VCOs employing highly sensitive varactors, such as the hyperabrupt types. In this wideband VCO application the varactor has, as it must, a wide capacitance shift when presented with a relatively low tuning voltage (*high sensitivity*), so any small noise or varactor control voltage variation will generate a large frequency modulation. This varactor modulation noise can be lessened by a less sensitive varactor, or by having a less wide-tuning VCO requirement.

Pulling—A parameter that specifies the shift of frequency and/or power when the VCO is in an output impedance mismatch condition.

Pushing—The variation in frequency or amplitude that occurs to a VCO due to changes in V_{CC}, is measured in MHz/V or dB/V, respectively.

Spurious signals—Undesired and nonharmonically related CW power output spikes present in a VCO's output.

4.3 Crystal Oscillators

4.3.1 Introduction

Since modern wireless communications equipment could not function properly with the extreme drift present in even the finest *LC* sinewave oscillator, it was necessary to develop crystal-controlled oscillators. In today's world of limited bandwidth and tight channel spacing, not only would an RF transmitter drift and interfere with adjacent channels, but its signal would be unreadable as it moved into and out of the passband of the receiver, creating changes in the volume, pitch, and distortion levels of an analog radio and degrading BER levels in a digital system. And the lack of any serious receiver frequency stability in the LOs would only contribute to this problem. Automatic frequency control (AFC) is one answer, as is the PLL, which, however, would not be able to function properly without a high-quality crystal oscillator as its reference. For a fixed-frequency RF source under a center frequency of 200 MHz, the crystal oscillator is the dominant choice.

A crystal oscillator requires only four things to precisely oscillate at a stable frequency and amplitude:

1. The loop gain must be +1 (but greater than 1 to start).
2. The oscillator circuit's impedance must be equal to its crystal's internal resistance.
3. The oscillator circuit must not drag down the Q of the crystal excessively.
4. The total oscillator circuit feedback phase must be zero degrees from output to input.

A crystal is the perfect choice for operation in an oscillator because it will vibrate at its own natural resonant frequency if an alternating signal at that same frequency is placed across the crystal, basically functioning as an ultra-high-Q series resonant circuit.

The most common crystal material used in oscillators is quartz because of its low cost, strength, and thermal stability. However, crystal-based oscillators can not easily change frequency, except by a few hundred hertz with a capacitive trimmer in parallel, or in series, with the crystal element.

The maximum frequency a crystal can reach on its fundamental is 200 MHz, and this only by using specialized inverted mesa methods. But most common crystals can reach only up to 20 MHz, yet can attain much higher frequencies on *harmonic* and/or *overtone* oscillator operation. For example, a *harmonic crystal oscillator*, with its output tank circuit tuned to one of these harmonics, can have an output at the crystal's second, third, fourth, etc., harmonic, yet the crystal itself is really only operating at its lower fundamental frequency. When operated as an *overtone crystal oscillator*, the crystal must actually vibrate at high harmonic (overtone) frequencies, and will work only at one of its odd harmonics—such as its third, fifth, or seventh—with the output tank circuit tuned to this chosen frequency. Overtone crystal oscillators normally require a special overtone crystal when operated in this mode.

As stated above, the crystal functions as a very high-Q series resonant circuit with high temperature stability and very narrow bandwidth (as the high-Q designation would indicate), with the crystal looking to the rest of the oscillator's circuit as shown in Fig. 4.9. R_s is the resistance of the crystal during series resonance, while L_m is the motional inductance, C_m the motional capacitance, and C_0 is the capacitance between the crystal's holder or, in a modern crystal, its plated electrodes. In fact, C_0 at VHF and above has so decreased in reactance that it has effectively shorted the output of the crystal to its input; this problem can be mitigated by resonating a small value coil in parallel with C_0.

Most oscillators will operate in series resonance mode, with the values of L_m and C_m governing the resonant frequency of the crystal. At series resonance the crystal is resistive with no reactances, since $X_L = X_C$, and can be described as:

$$f_s = \frac{1}{2\pi \sqrt{L_m C_m}}$$

As mentioned above, fundamental crystal operation usually peaks out at 20 MHz. This is due not only to the dangerously decreasing thickness of the crystal, but also to its decreasing R_S. The crystal's R_S can decrease to 10 ohms at 20 MHz on its fundamental, while the crystal in seventh overtone mode can reach 180 MHz with an R_S of 80 ohms. This demonstrates why many oscillators must run in overtone modes, which allow the crystal to be more easily impedance matched at higher operational frequencies.

L_m, whose value is based on the mechanical mass of the quartz crystal, can vary in inductance anywhere between 3600 mH at 1.5 MHz to 10 mH at 20 MHz. C_m, whose value is based on the actual stiffness of the quartz crystal, the size of the electrodes, and the size of the quartz, can vary anywhere between 0.007 pF at a fundamental frequency of 1.5 MHz to 0.02 pF at its fundamental frequency of 20 MHz. But when a crystal is operated on an overtone, the C_m will decrease. The designer can choose the value of the C_m desired, and the L_m will then be:

$$L_m = \frac{1}{4\pi^2 f_r C_m}$$

C_0 is a capacitance value that can be measured across a crystal at rest, and does not vary with frequency of operation nor with the number of its overtone, but only by the crystal's distance between its electrodes and the electrode's area. This value will normally be between 2 to 8 pF. The lower the value, the better for oscillator operation.

Various crystal specifications can greatly affect an oscillator's performance. The frequency accuracy of a crystal, while at room temperature in an oscillator test circuit, can vary from ±5 to ±100 ppm. The lower this value, the more accurate the oscillator's output frequency will be at 25°C, and the more costly the crystal. Frequency stability over some chosen temperature range is another specification, important to maintaining frequency accuracy under varying ambient and internal temperatures.

Aging affects the crystal's frequency accuracy over time, and can change this accuracy by as much as 6 ppm during a 12-month period—or as little as 0.75 ppm, depending on the type of package, crystal quality, crystal stresses, temperature, and frequency. However, the aging of a crystal will mostly occur within its first year; after which the rate will slow down to perhaps one-fifth its first year's value. For instance, a crystal might age 2 ppm over the first year and only 4 ppm over the next 10-year period.

4.3.2 Types of crystal oscillators

Since there are so many different circuit designs available, we will focus on only the most common crystal oscillators, such as the *Hartley crystal oscillator* of Fig. 4.23, the *Colpitts crystal oscillator* of Fig. 4.24, and the *Pierce crystal oscillator* of Fig. 4.25. As we can see, the crystals for the first two oscillators are placed in series in the transistor's feedback path, and, as a crystal has a very high Q (in excess of 75,000) and will thus have a very narrow bandwidth (BW $= f_r/Q$), only the tight band of frequencies within the crystal's natural resonance will actually pass onto the phase-shifting circuits, and will thus be in-phase at the oscillator's input. In fact, feedback that is off frequency by even the smallest amount will be rigorously attenuated, decreasing the level of the transistor's feedback, forcing the oscillator to return to its desired frequency. The phase-shifting network for the Hartley and Colpitts oscillators is the LC tank components, while the Pierce crystal oscillator employs a slightly different method of operation.

In the Pierce oscillator, a series resonant crystal has been substituted for the inductor of the Colpitts. Since at series resonance a crystal will display only a small, pure resistance to the oscillator's feedback from its output to its input, but will exhibit either a capacitive or an inductive element if not within this small window of series resonance, this will allow the 180 degree phase shift required for positive, oscillatory feedback. However, this point is usually shifted slightly higher in frequency than as marked on the series resonant crystal by about 50 ppm. In other words, the Pierce will oscillate by a very small amount higher in frequency than may be expected by the crystal's marked frequency (more on this below). This action forces the Pierce oscillator to stay accurately on frequency in low-power, medium-frequency applications.

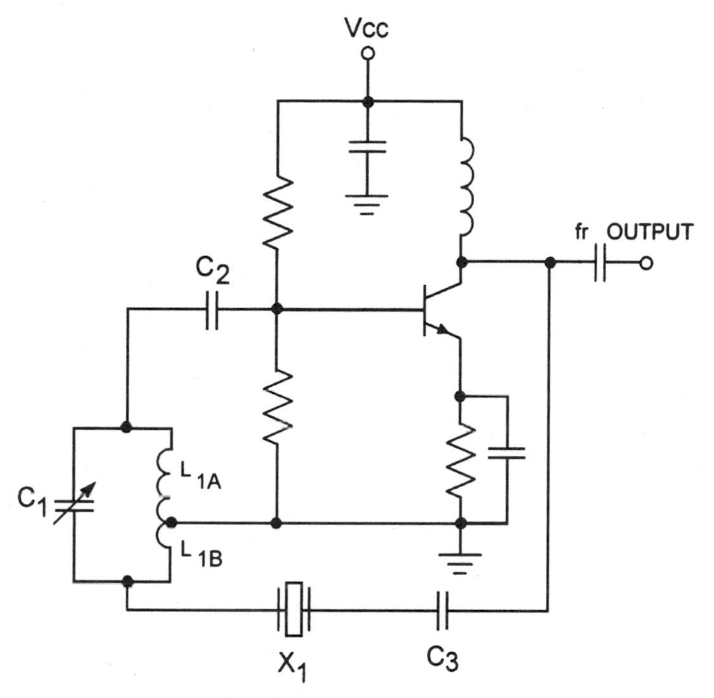

Figure 4.23 A type of Hartley crystal oscillator circuit.

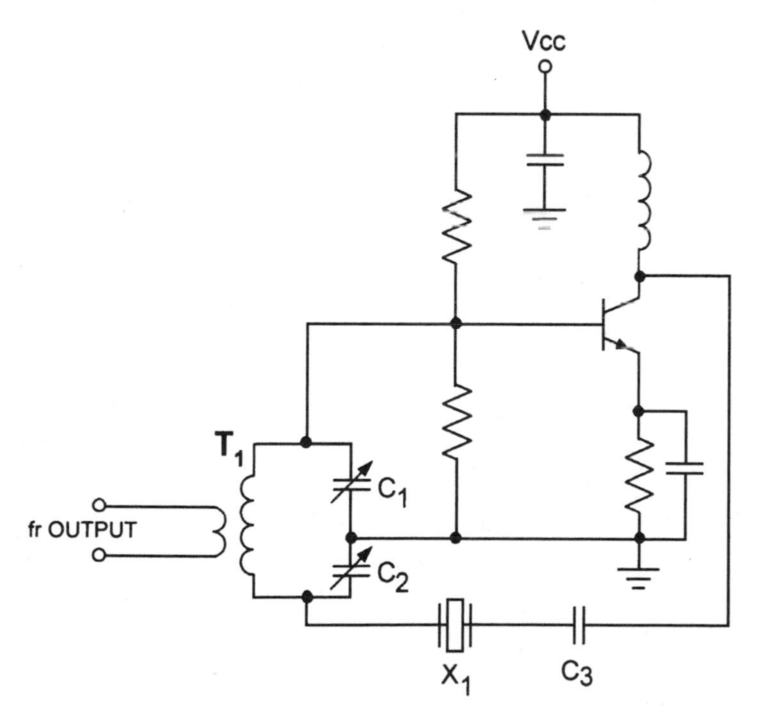

Figure 4.24 A type of Colpitts crystal oscillator circuit.

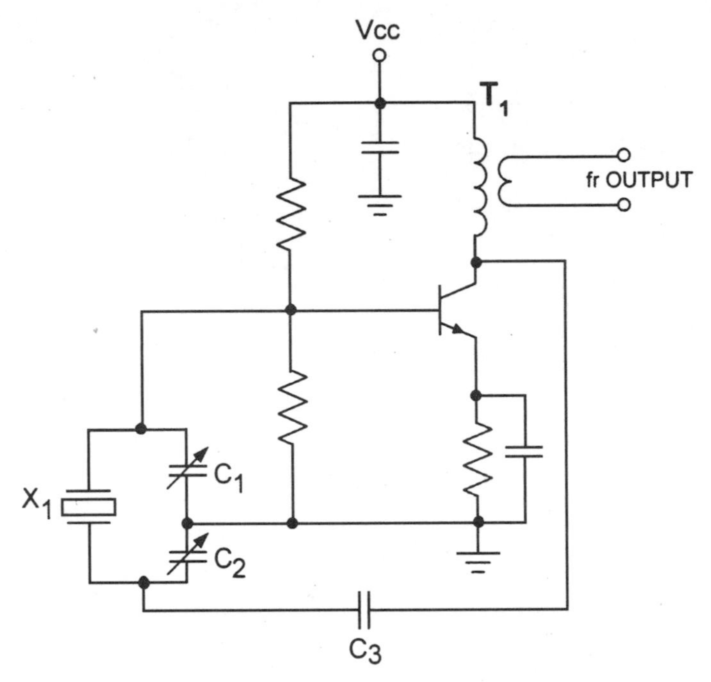

Figure 4.25 A type of Pierce crystal oscillator circuit.

4.3.3 Designing crystal oscillators

Designing crystal oscillators by the following design procedures, while verifying their operation as described in Sec. 4.1, "Oscillator Simulation," will permit the engineer to design and construct stable and reliable oscillator circuits for a variety of uses. Because of the low frequencies of some of these oscillators, we can employ a Spice simulator—instead of a linear simulator—if the transistor's *S*-parameter models are not available. The simulation technique is the same for either case.

Pierce crystal oscillator design (Fig. 4.26). The following design functions quite well from 600 kHz to 30 MHz, and uses a crystal that is in series resonance. However, since a Pierce oscillator will actually oscillate at a frequency that is 20 to 50 ppm above the crystal's marked series resonant frequency, the crystal itself can be specified to the crystal manufacturer as "parallel resonant." This will simply tell the manufacturer to build the crystal with a series resonant frequency that is approximately 90 to 100 ppm lower than a crystal specified at its series resonant frequency (both crystals are exactly the same except for this slight frequency modification). The parallel resonant designation will permit the Pierce to be slightly tweaked to operate at *exactly* the frequency marked on the crystal's can by the mere inclusion of a small variable capacitor shown as C_3 [or by a varactor circuit for a simple voltage-controlled crystal oscillator (VCXO)]. This adjustable capacitor, which is in series with the crystal, can also

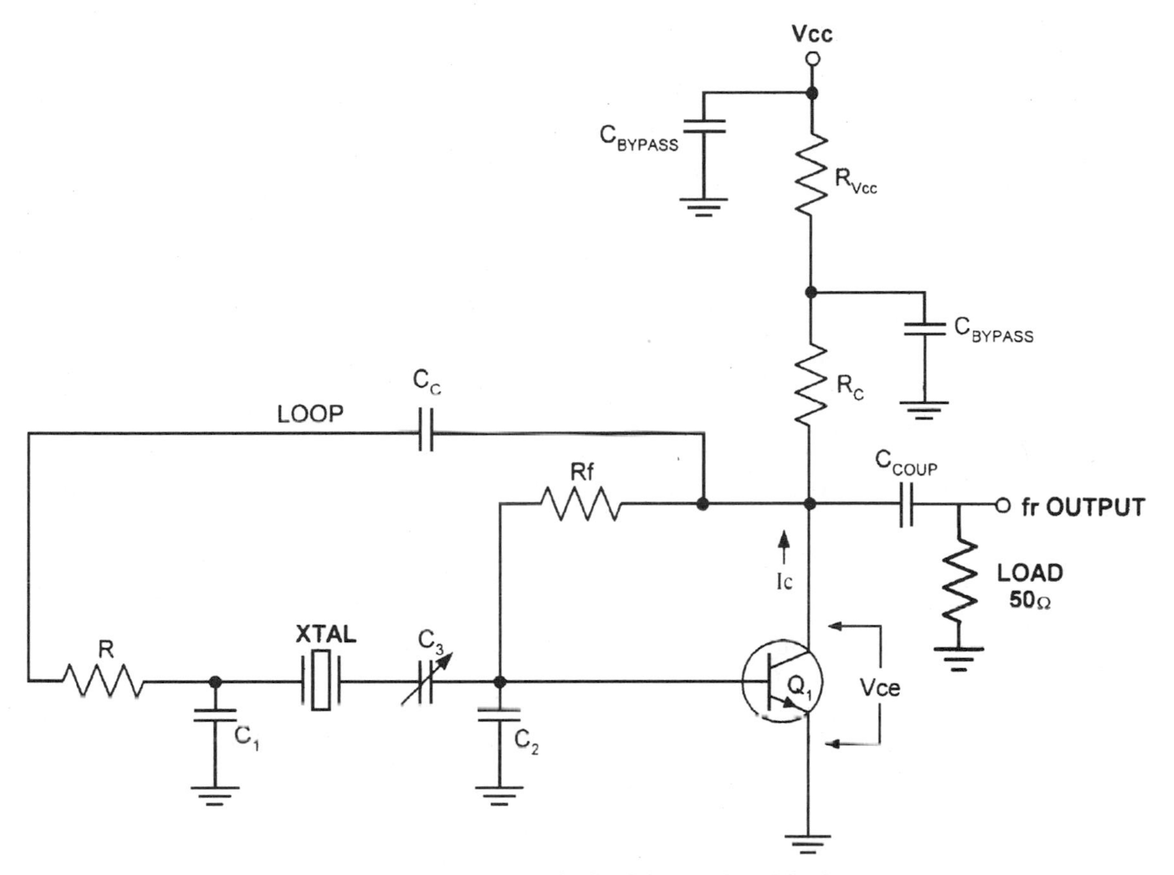

Figure 4.26 A Pierce crystal oscillator circuit showing feedback loop and load for design equations.

be utilized even if a *series* specified crystal is employed, but in this case it will vary the frequency of the oscillator from about 5 ppm to around 50 ppm *above* the marked series resonant frequency. However, since the impedance increases as the crystal gets closer to its true parallel resonant frequency (which is always above its series resonance frequency)—whether in series or "parallel" mode—the feedback gain will begin to decrease as the frequency of the oscillator is tweaked upward. Thus there will be a point reached where the feedback gain will decrease so much so that oscillations are no longer possible. This is why it is prudent to maintain the frequency of the oscillator as close to the actual series resonant frequency of the crystal as possible, without excessive tweaking; whether or not it is a series or "parallel" crystal.

If the oscillator's designer does not supply the crystal manufacturer with what is called the *load capacitance* (C_L) for a "parallel resonant" crystal, the manufacturer will assume it to be around 30 pF. The load capacitance is simply the load that the crystal will see when placed in the oscillator circuit, and slightly affects the accuracy of the "parallel" crystal's marked frequency. It can easily and more accurately be computed in a Pierce oscillator by:

$$C_L = \frac{(C_1 C_2)}{C_1 + C_2} + 5 \text{ pF}$$

Specifying the crystal in its series resonant mode will not require the above formula, and even a "parallel" crystal will be fine for most applications without this C_L specification—unless extreme frequency accuracy is required.

Obtain the crystal's motional capacitance (C_M), motional inductance (L_M), series resistance (R_S), and parallel plate capacitance (C_P) from the manufacturer for the crystal's frequency of operation, type of holder, and cut (typically AT). This will allow us to utilize the equivalent circuit of Fig. 4.9 to model the crystal in the linear circuit open-loop simulation program as a simple *LCR* circuit.

The Pierce oscillator is meant to work only on the crystal's fundamental-mode series resonance, but can function with overtone crystals if C_1 is replaced with a parallel resonant tank that is tuned midway between the desired overtone and the overtone just below it (Fig. 4.27). In this case, the crystal manufacturer must be told if the crystal is being run out of its fundamental mode.

Choose a transistor with a much higher f_t than required for the oscillation frequency ($5f_t$), and with a very high gain as well. The high f_t is required to assure as close to a 180 degree phase shift from the transistor's input to its output as possible, while the high gain is necessary because of this oscillator's rather high loop losses.

R, C_1, XTAL, and C_2 of Fig. 4.26 form a 180° phase-shift network, while R is also the feedback control element employed to place less stress on the crystal. As mentioned, the Pierce oscillates just slightly *above* the series resonant frequency of a series crystal, so C_3 is included to tune the oscillator *toward* the series XTAL frequency (but the oscillator can never quite reach it). By increasing C_3's capacitance, the frequency is lowered closer to the desired f_r of the oscillator, while decreasing C_3 increases the f_r further away from the series resonance of the crystal. R_{VCC} and C_{BYPASS} act in the decoupling role, while R_P, R_{VCC}, and R_C are the oscillator's bias resistors. C_{COUP} is used to couple power out of the oscillator into a 50-ohm load—without loading the oscillator down below a safe gain margin. If C_{COUP} is not of a high enough reactance value, the oscillator's feedback may become too low to maintain, or even begin, oscillations (see "Oscillator output coupling" in Sec. 4.2.5). Since the open-loop output and input impedance of a Pierce crystal oscillator are higher than 50 ohms, set the linear software simulator's termination impedances to about 300 ohms for more accurate results.

Follow these design equations to complete:

1. $C_{\text{BYPASS}} = C_C \leq 1 \text{ ohm}(X_C)$
2. $C_1 = C_2 = [2000 \text{ pF}/(10^{-6} \times f_r)] \times C_{\text{FACTOR}}$.
   ($C_{\text{FACTOR}} = 0.5 \leq 1$ MHz; $0.7 \leq 3$ MHz; $0.6 \leq 2$ MHz; $0.8 \leq 4$ MHz; $0.9 \leq 6$ MHz; $1 \geq 8$ MHz.)
3. $C_3 = 0$ to 6 pF trimmer

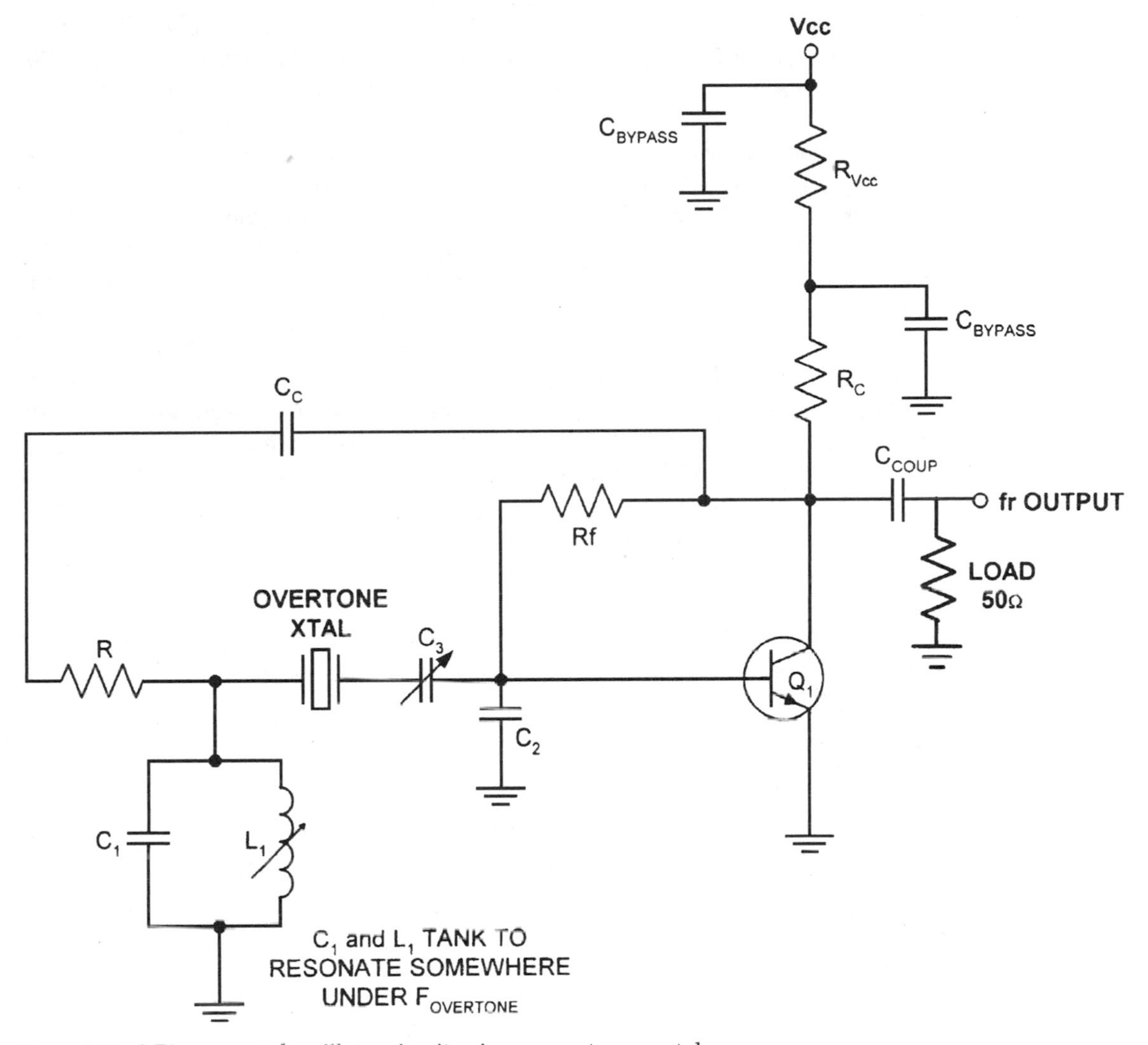

Figure 4.27 A Pierce crystal oscillator circuit using an overtone crystal.

4. $R = R_C = 3/2\pi f_r C_1$
5. V_{CE} = value in the S-parameter files for the transistor chosen, or as desired if in Spice
6. I_C = value in the S-parameter files for the transistor chosen, or as desired if in Spice
7. $R_{VCC} = \dfrac{(I_C R_C) + V_{CE}) - V_{CC}}{I_C}$
8. $R_f = \beta\,[(V_{CE} - 0.6)/I_C]$

Increase reactance value of C_{COUP} to decrease coupling power output to load; decrease reactance value to increase coupling. Start with 50- to 200-ohm X_C for a 50-ohm load, then simulate to confirm proper loop gain.

Pierce overtone crystal mode oscillator (Fig. 4.27). It may be necessary to employ a lower-frequency crystal than what is required in our design because of crystal cost or availability issues, or we may simply want a higher-frequency oscillator than can normally be obtained by running a crystal in its fundamental mode. A problem encountered in high-frequency crystal oscillator design is that as the frequency of the oscillator is increased, the crystal's internal resistance decreases, and a point is soon attained where it becomes troublesome to obtain a suitable impedance match for the crystal within the oscillator's circuit. This demands running a crystal at one of its overtone frequencies, which increases the crystal's series resistance, as well as more frequency accuracy and stability. However, we must always choose a crystal that has been cut specifically for overtone operation.

To design a Pierce overtone oscillator:

1. $C_{BYPASS} = C_C \leq 1$ ohm (X_C)
2. $C_2 = 2000$ pF/(1μ × f_r) (f_r is the actual overtone frequency we wish output from the oscillator, and not the crystal's much lower fundamental frequency).
3. $C_3 = 0$ to 6 pF trimmer.
4. $R = R_C = 3/2\pi f_r C_1$
5. V_{CE} = value in the S-parameter files for the transistor chosen.
6. I_C = value in the S-parameter files for the transistor chosen.
7. $$R_{VCC} = \frac{(I_C R_C + V_{CE}) - V_{CC}}{I_C}$$
8. $R_f = \beta\,[(V_{CE} - 0.6)/I_C]$
9. Increase the reactance value of C_{COUP} to decrease coupling; decrease the reactance value to increase coupling. Start with 100- to 200-ohm X_C for 50-ohm load and simulate.
10. Use proper overtone crystal (*odd modes only*), while modeling the crystal's equivalent circuit values according to the manufacturer's specifications for the overtone and frequency required in the linear software simulator program.
11. L_1 and C_1 will be a tank circuit resonant at:

$$f_{mid} = \frac{1}{2\pi\sqrt{L_1 C_1}}$$

with f_{mid} being a frequency that is approximately midway between the desired harmonic that we wish to output and the next lowest *odd* harmonic (of the fundamental) that is *below* this desired output harmonic.

Example: A frequency of 133 MHz is required. Design the Pierce overtone oscillator to function at 133 MHz with a 19-MHz crystal. In other words, the crystal will operate at its seventh overtone of 133 MHz, but has a fundamental resonant frequency of only 19 MHz. Choose L_1 and C_1 so that f_{mid} (in this case 114 MHz) is resonant or:

$$114 \text{ MHz} = \frac{1}{2\pi\sqrt{10 \text{ nH} \cdot 195 \text{ pF}}}$$

Now tune L_1 for the maximum power at the desired overtone. The C_1, L_1 tank will resonate at somewhere in the vicinity of 114 MHz, but the oscillator will output 133 MHz, or the seventh harmonic of 19 MHz.

4.3.4 Crystal oscillator issues

The different crystal oscillator circuit configurations employed in circuit design are required because of the various impedance levels found at different frequencies of oscillator operation. Since a crystal's internal series resistance can be as low as 20 ohms at 25 MHz, all the way up to 0.25 megohms at 500 Hz, special circuit designs are required to efficiently match and drive the crystal at these resistance values. The Pierce circuits above will be almost ideal for the majority of crystal oscillator needs in most wireless systems.

Any oscillator crystal in RF circuits should be calibrated to 5 or 6 decimal places in order to supply an accurate frequency for most LO applications. A crystal with less accuracy, especially at high frequencies, can result in an oscillator that can become unstable as a result of the huge frequency adjustments that must be made.

Oscillator start-up time is directly correlated to the Q of the oscillator's resonator, so the higher the Q the longer the start-up time. Crystal oscillators, with their ultrahigh Q, have prolonged start-up times up to, and sometimes surpassing, 100 mS. Start-up time will also be affected by the bias network of the oscillator's active device, since the bias network must reach its steady-state value before reliable oscillations will occur. Thus the RC time constant of the bias network can substantially slow down the onset of oscillations.

Obviously all passive and active components must be rated above the oscillator's frequency of operation, as well as the oscillator's voltage, current, or power. The inductors and capacitors must not have any series or parallel resonances that will interfere with oscillations, and the active element must have a gain that is more than sufficient to sustain oscillations at the frequency of operation.

Board layout is another critical aspect to proper oscillator operation (see Sec. 10.3, "Wireless board design").

SAW-based oscillators are becoming more popular in the VHF and above regions, and are similar in design and concept to crystal oscillators. However, surface acoustic wave (SAW) devices are limited in design usefulness; unless the center frequency is a common one, no SAW resonators are available without an expensive custom production run. And the initial frequency stability, temperature stability, Q, and aging characteristics are many times worse than those of an average crystal resonator. Nonetheless, replacing crystals with SAW resonators makes it possible to operate at very high frequencies of up to 2 GHz, and with powers of up to +22 dBm at 500 MHz.

Depending on the application, a crystal oscillator may require higher frequency accuracy over temperature than a normal noncompensated crystal oscillator (XO) can supply. This will demand that some type of compensated device be used, such as a *temperature-controlled crystal oscillator* (TCXO) or an *oven-controlled crystal oscillator* (OCXO). However, increased size, cost, current consumption, and complexity are the tradeoffs if such a compensated oscillator is to be adopted. And, depending on the angle of the cut for AT crystals, frequency stability over a desired temperature range can be optimized for an uncompensated crystal oscillator, sometimes making compensated oscillators unnecessary: Frequency stabilities of ±5 ppm from +25 to +70°C are possible with the appropriate AT cut angle in an XO. Wider temperature variations than this will quickly degrade an AT cut's frequency stability dramatically (down to ±20 ppm from −40 to +80°C), necessitating the use of a TCXO, or even an OCXO.

Most of the components making up any oscillator are temperature sensitive, especially important being the crystal and the ceramic capacitors of its resonator network. Even the finest crystal oscillator, if built with poor or inappropriate ceramic capacitors, may have unacceptable frequency drift. In a well-designed oscillator, the majority of the long and short-term frequency drift should originate only from within the crystal—and any circuit that adds more than double the drift of a lone crystal is improperly designed. The use of incorrect temperature-compensating ceramic capacitors, or capacitors with poor temperature tolerance versus capacitance, can destroy frequency stability of an otherwise good oscillator. However, if high frequency stability of better than a few ppm is required, then both the crystal and the entire oscillator circuit itself must be ovenized within an OCXO. The OCXO ovenizes not only the crystal, but all of the temperature-sensitive components. It has the highest stability commonly available in compensated crystal oscillators, with better than 0.001 ppm being common with SC-cut or AT-cut crystals over a wide temperature range. The oscillator itself is kept in a temperature-controlled oven that maintains the crystal and circuits at a temperature that is 10°C above the highest specified ambient temperature. The OCXO can even be tuned very slightly (by a few ppm) by a small screw located within the case. However, OCXOs are high in cost, consume much more current than a standard oscillator, have a certain warm-up period to reach full frequency accuracy, and may have poor aging characteristics because of the high heat that the crystal is constantly subjected to.

The TCXO is a highly temperature stable crystal oscillator, with better than 1 ppm stability. TCXO's make use of a temperature sensor, normally a *thermistor,* which generates a correction voltage to the TCXO's compensation network, a *varactor diode,* which then overpowers the frequency drifting effects of the changing temperature within the oscillator's circuits. TCXOs are lower in cost, with significantly less current requirements than an OCXO, and warm up nearly instantaneously.

4.3.5 Testing and optimizing crystal oscillators

To test and optimize the completed physical oscillator for start-up and proper functioning (frequency, amplitude, spurs, etc) under most real-world conditions as expected in normal operation, duplicate the following tests and optimization procedures.

1. Connect the closed-loop oscillator to a spectrum oscillator.
2. Confirm that the oscillator reliably starts at room temperature by turning it on and off a number of times. Cool the circuit with a canned cooler and repeat. Heat the circuit with a heat gun set on low and repeat.
3. Test for the expected output:
   a. Check that the oscillator frequency is stable at room temperature, and that it is stable during and after the above heat/cool test.
   b. Check for close-in spurious responses.
   c. Check the approximate phase noise level (see "Phase noise tests" on p. 94).
   d. Check that the amplitude is as expected, and that it is stable over time.
   e. Check for any wide-ranging spurious responses and excessive harmonic levels.

It will normally be necessary to improve the amplitude, stability, starting, and spectral purity of the oscillator by tweaking for optimum performance: Change the *L/C* or *R/C* ratio; vary the transistor's bias current; tweak any tuning capacitor in series with the crystal. Perform until an optimum point is reached that satisfies your required specifications. Always send the oscillator through the full complement of starting, frequency, amplitude, and spurious response tests after any tuning procedure is completed.

Chapter

5

Frequency Synthesizer Design

A method of combining the wide tunability of *LC* oscillators with the high frequency stability of crystal oscillators is a necessity in modern wireless communications design. We find both of these abilities in *frequency synthesis,* which is a method of generating a multitude of exceptionally accurate frequencies from a single, low-frequency crystal oscillator. It is the dominant technique for variable-frequency production in most receivers, transmitters, transceivers, and test equipment today.

By far the most widespread method of frequency synthesis is implemented by the *phase-locked loop* (PLL); but a newer technique, referred to as direct digital synthesis [DDS; sometimes called a *numerically controlled oscillator* (NCO)], is becoming increasingly prevalent in certain applications. We will concentrate on the PLL, which is easier to design, more versatile, and much higher in frequency.

5.1 Phase-Locked Loops

5.1.1 Introduction

The majority of frequency synthesis is derived from the phase-locked loop. Figure 5.1 demonstrates all of the vital circuits that make up a common single-loop PLL synthesizer: A low-frequency crystal oscillator feeds a *reference frequency* into the *R divider,* which decreases the reference frequency to equal the desired f_{COM} out of the *N adjustable frequency divider,* with the R divider allowing for different channel spacings. The reference frequency out of the R divider is then inserted into the *phase comparator,* which compares the phase of the R divider to that of the *N* adjustable frequency divider. The adjustable *N* frequency divider receives its own input frequency from the *VCO*'s output, dropping it down to a lower frequency that must be equal to the R divider's output. As the phase comparator is comparing the two frequencies at its input from the *N* and R dividers to see if they are of the same phase, it will produce a rectified

DC correction voltage at its output into the low-pass *PLL loop filter* if these two frequencies differ. This filter eliminates any AC variations and noise products, placing the now pure DC directly into the VCO's frequency control input. The all-important loop filter is required to filter powerful phase comparator constituents at the comparison frequency (f_{COM}) and its harmonics, since if these responses actually got through to the VCO they would adversely modify its stability. The varactor diode's bias within the VCO is affected by this DC control voltage, which immediately forces the VCO back on frequency if it has drifted off. These actions permit a frequency source to be adjustable over many discrete frequencies, but with the stability of the crystal oscillator reference.

The adjustable *N* frequency divider of the PLL is usually controlled by the operator through a front radio panel knob, or automatically by system commands. A microprocessor will normally supply digital control words to the PLL through a serial, but sometimes even a parallel, bus (see below) to change frequency. The microprocessor can also be employed to decode and drive display circuits to inform the radio operator of the exact channel of transmit or receive.

Premixing (Fig. 5.2) of a synthesizer can be used to obtain a higher frequency from a lower frequency PLL. The output of the PLL and a crystal oscillator can be fed into a mixer, and filtered by a bandpass filter to attain the sum of these two frequencies. Frequency multiplying of the PLL's output, at the expense of higher phase noise and degraded frequency channel resolution, can also be utilized to increase the PLL's output frequency.

A widespread manual PLL tuning scheme is shown in the circuit of Fig. 5.3. It has a shaft encoder, a microprocessor, the display with its driver, and a PLL chip with loop filter and VCO. Rotated by the radio operator, the tuning knob turns an optical or magnetic encoder that has two voltage outputs, A and B. The A output is a square wave in quadrature phase (90 degree phase shifted) to the B square wave output. The *A* output is connected directly to the microprocessor's interrupt line; when the A output from the encoder produces a falling edge, an interrupt will occur. The microprocessor will then immediately look at the *B* output to see if it is a 1. If it is, then the microprocessor considers that the knob has been rotated clockwise. The microprocessor will then increment the PLL's *N* divider by 1, increasing the output frequency, and update the frequency display appropriately. However, if *B* is a 0, then the microprocessor considers that the knob has been rotated counterclockwise, and the microprocessor decrements the PLL's *N* divider by 1, decreasing its output frequency.

Even though the PLL circuit in general obviously has many advantages, such as a much higher possible operating frequency than a crystal oscillator, is tunable in discrete steps, and is as stable as the reference source (which is usually a simple 10-MHz crystal oscillator), PLLs are far inferior to crystal sources when it comes to phase noise specifications. This can be a problem in digital wireless communications.

Most PLL chips are of the *charge-pump* type (Fig. 5.4). The charge pump outputs a current of steady amplitude, but with changeable duty cycle and

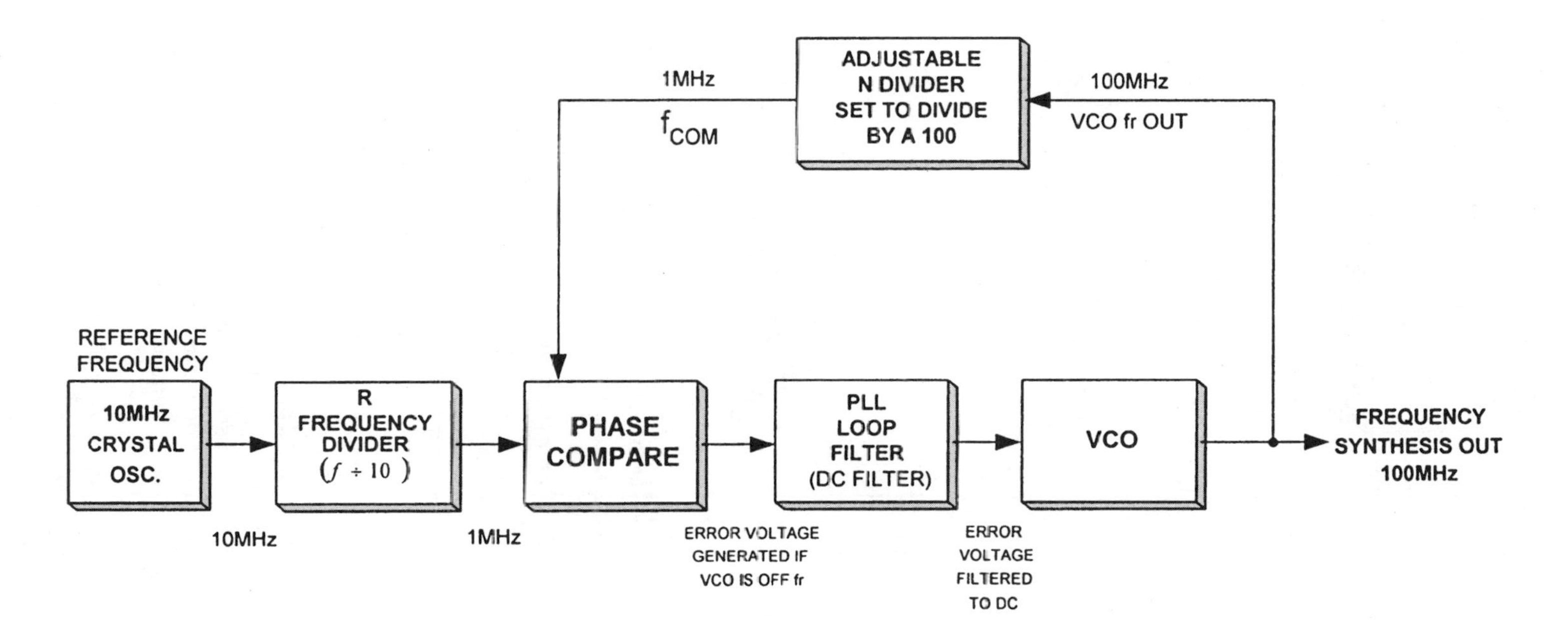

Figure 5.1 A simple phase-locked loop frequency synthesizer.

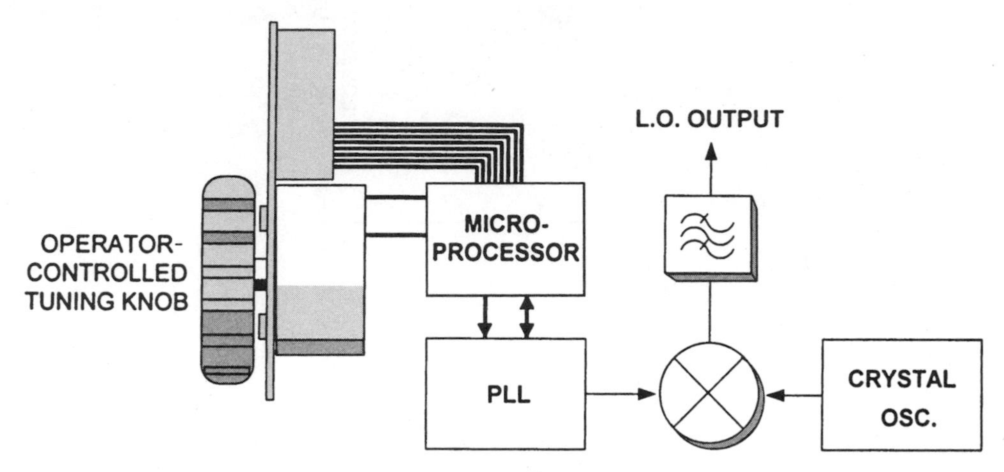

Figure 5.2 Premixing with a PLL and a crystal oscillator.

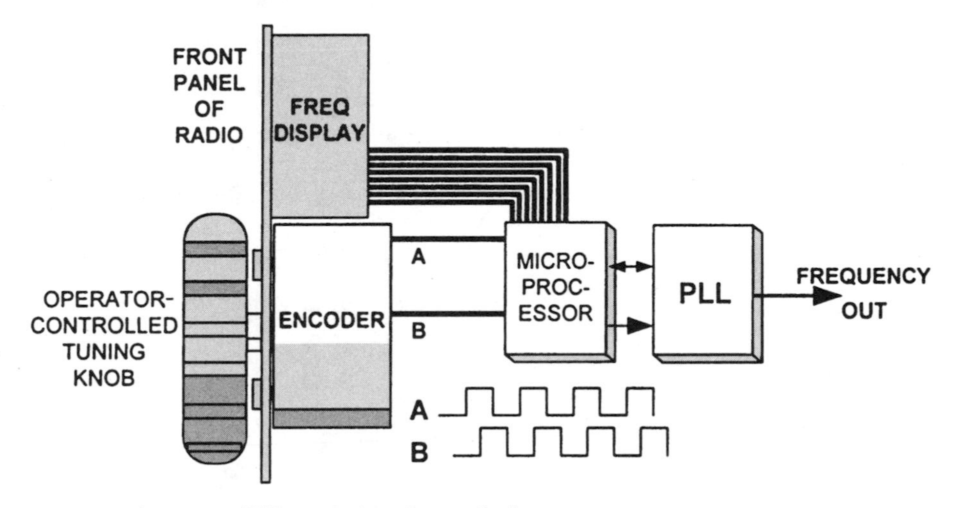

Figure 5.3 A common PLL manual tuning method.

polarity, into the PLL filter, which then converts this charge pump output into a DC control voltage for the VCO. The charge pump PLL permits the use of a passive filter, which is cheaper and adds little extra noise, unlike an active op-amp based loop filter.

A few words about *prescalers.* Prescalers (Fig. 5.5) take the high frequency of the VCO and divide it down to a more manageable lower frequency in the N divider section: The VCO frequency is fed into this prescaler, which divides the frequency down to $P + 1$, in which P stands for the size of the prescaler. At each of these $P + 1$ cycles, $A + B$ counters decrement by 1. This creates a count of $A(P + 1)$ and $(B - A)P$, which makes $N = A(P + 1) + (B - A)P$, or $P(B + A)$.

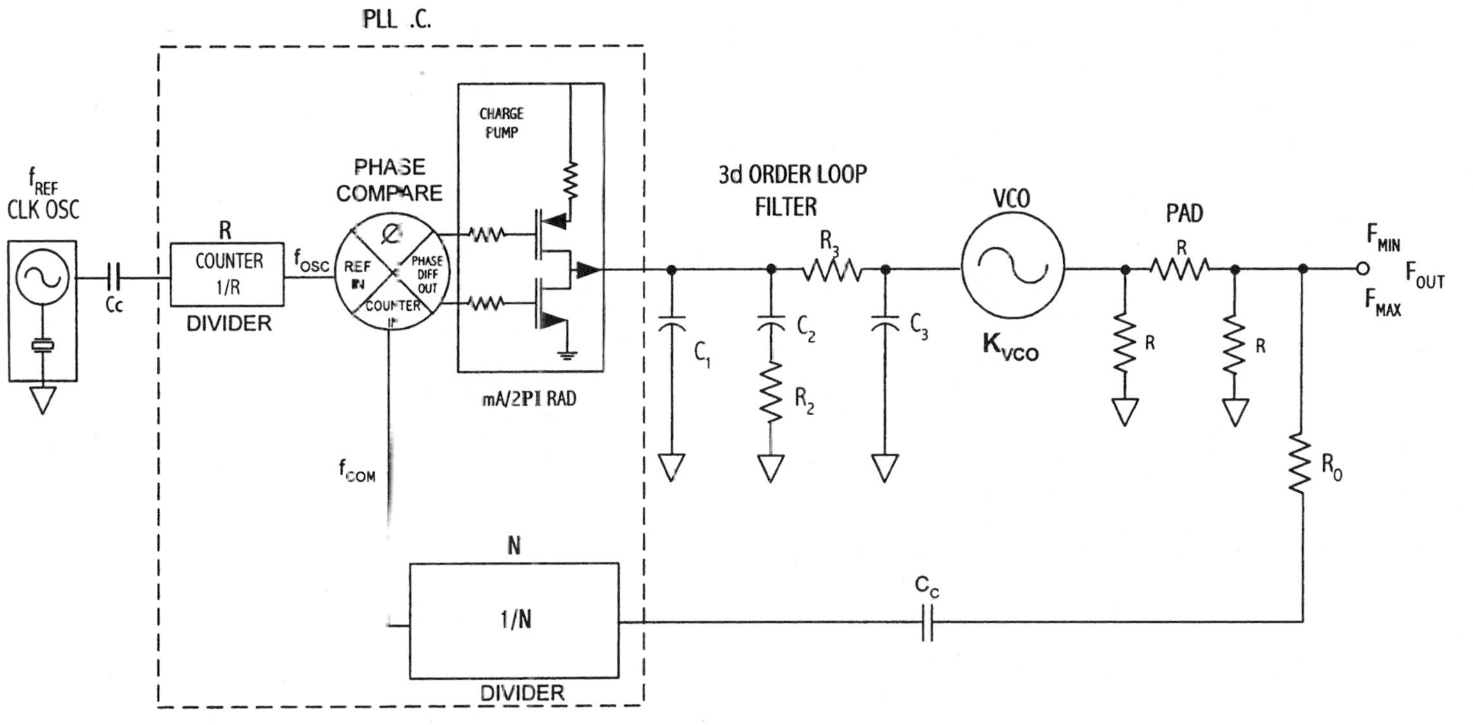

Figure 5.4 A complete phase-locked loop with filter, VCO, and reference.

N is not permitted to be less than $P(P-1)$, and if N *is* less than $P(P-1)$, then $B \geq A$, with:

$$B = \frac{N}{P} \qquad \text{and} \qquad A = N - BP \qquad \text{and} \qquad N = \frac{f_{\text{OUT}}}{f_{\text{COM}}}$$

The final outcome of using these dual-modulus prescalers, which are a part of the N divider, in PLLs is that it becomes possible to control the division ratio into the phase comparator in steps of 1(N), as opposed to the huge steps of 32 or 33 in a fixed-modulus 32 or 33 prescaler. This N value must always be an integer, with the largest N value being determined by the size of the B counter, since $N = P(B + A)$. Dual-modulus prescalers will, however, have certain illegal divide ratios, in which *specific* frequencies cannot be generated. If a particular N value results in a B register that is smaller than the A register, this will not be allowed, since B must be greater than or equal to A for a legal divide ratio. In other words, not all N values are allowed with a dual-modulus prescaler–equipped PLL. The tradeoff between having certain frequencies that are impossible to generate is that we can obtain better frequency resolution at the PLL's output than would normally be possible. However, if it is essential that certain frequencies be generated by the PLL dual-modulus prescaler (since N must equal $(P + 1)(A + P)(B - A)$ to be a legal divide ratio), then a legal divide ratio check should be performed by using National's *Easy PLL* or National's *Code Loader* program.

5.1.2 Designing phase-locked loops

The design of PLL frequency synthesizer circuits, until recently, was fraught with complications and uncertain results. However, PLL chip companies, primarily National Semiconductor, have released information that makes the design of a frequency synthesizer much more simplified than in the past. National Semiconductor has also released two different PLL design programs that almost completely automate the PLL design task. Two of these programs

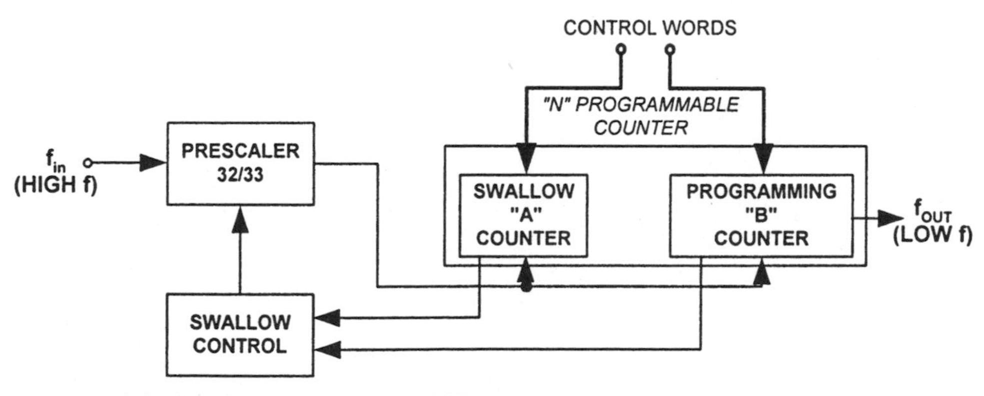

Figure 5.5 A dual-modulus prescaler for a PLL.

are described in the *Wireless Design Software* section, and both are located on the CD ROM disk included with this book.

The first place to begin in the design of a PLL frequency synthesizer circuit is in deciding on what we want our center frequency, frequency swing (minimum and maximum frequencies), speed (lock time), and channel resolution (spacing) to be, then select the appropriate PLL chip, VCO, and reference oscillator that can safely, economically, and repeatedly meet these criteria.

The following list should be completed so that the complete specifications of the PLL circuit of Fig. 5.4 can be clearly defined:

F_{MAX} = ______ HZ. The maximum output frequency required. Must never be more than the VCO was designed to handle, along with about 20 percent VCO frequency excess for a safety margin. Using a VCO with *less* tuning bandwidth decreases its phase noise.

F_{MIN} = ______ Hz. The minimum output frequency required. Must never be less than the VCO was designed to handle, along with about 20 percent VCO frequency excess for a safety margin. Using a VCO with *less* tuning bandwidth decreases its phase noise.

F_{OUT} = ______ MHz. Equal to $\sqrt{F_{MAX}F_{MIN}}$; must be a multiple of F_{COM}.

K_{VCO} = ______ MHz/V. Normally between 5 and 70 MHz/V. K_{VCO} is the VCO's gain (sensitivity) measured in megahertz per volt, and is the amount of frequency deviation, in megahertz, that the VCO will travel when 1 VDC is placed at its DC frequency control input.

$K\phi$ = ______ mA/2π. Charge pump gain ($K\phi$) is measured in milliamperes per 2π, and should be chosen at the highest value the PLL chip will allow in order to obtain the lowest phase noise at the VCO's output. This value will normally be either 1 or 5 mA/2π, and is typically selectable. A charge pump with a gain of 5 mA/2π would be preferred over a charge pump with 1 mA/2π if current consumption is not an issue. This charge pump gain value will also vary with the charge pump voltage supply.

F_{COM} = ______ kHz. Normally equal to the channel spacing. F_{OUT} and F_{REF} must be a multiple of F_{COM}. Sometimes it may be required to use an F_{COM} that is a certain fraction of the channel spacing because of the PLL's internal prescaler. The higher F_{COM} is, the better the phase noise of the PLL.

F_C = ______ kHz. The loop bandwidth of the PLL filter. F_C should be as narrow as possible to lessen spurious noise, but this will decrease switching speed. Normally F_C should be between 1 and 20 kHz, but must be at *least* 1/20 F_{COM} (National Semiconductor recommends a 2-kHz F_C if lock time does not matter). The choice of F_C will always be a compromise between reference sideband suppression and lock time—select a loop F_C to just meet the lock requirement, but with an acceptable margin.

ϕ (phase margin) = ______ degrees. Normally select a value between 30 and 70 degrees for the loop filter. The higher the phase margin, the higher the PLL's stability, but the slower will be its lock time. Choose a phase margin of

45 degrees, which is a good compromise between loop stability and loop response.

T3/T1 = ______ percent. Normally chosen to be 45 percent. *T3/T1* is the ratio, expressed as a percentage, of the poles of the loop filter. The higher this value (the closer to 100 percent) the more the reference spurs will be attenuated; but peaking will begin to occur within the filter's passband, and R_3 will increase in value, adding excessive thermal noise.

F_{REF} = ______ MHz. The frequency of the reference oscillator *before* the *R* divider. Must be a multiple of $F_{COM.}$ 10 MHz is a popular value, as applicable.

After filling out these required parameters, design the complete frequency synthesizer by performing the following calculations (or simply use the included National Semiconductor EasyPLL software):

1. $N = \dfrac{F_{OUT}}{F_{COMP}}$
2. $\omega_c = 2\pi F_C$
3. $T_1 = \dfrac{\left(\dfrac{1}{\cos\phi}\right) - \tan\phi}{W_C\left(\dfrac{T_3/T_1}{100} + 1\right)}$
4. $T_3 = \dfrac{T_3/T_1}{100} \cdot T_1$
5. $T_2 = \dfrac{1}{\omega_C^2\,(T_1 + T_3)}$
6. $C_1 = \dfrac{T_1}{T_2} \cdot \dfrac{K\phi K_{VCO}}{\omega_C^2 N} \cdot \left[\dfrac{1 + \omega_C^2 \cdot T_2^2}{(1 + \omega_C^2 T_1^2)(1 + \omega_C^2 T_3^2)}\right]^{1/2}$
7. $C_2 = C_1\left(\dfrac{T_2}{T_1} - 1\right)$
8. $C_3 = \dfrac{C_1}{10}$
9. $R_2 = \dfrac{T_2}{C_2}$
10. $R_3 = \dfrac{T_3}{C_3}$

If a broadband VCO is required in the synthesizer design, then more DC tuning voltage will also be needed, since very wideband VCOs may demand up to 20 or more tuning volts; but a typical narrowband PLL chip may be able to supply only 5 V or less. This increase in the necessary DC tuning voltage for a wideband VCO can be accomplished by employing a separate op-amp within the PLL filter as shown in Fig. 5.6. The VCO gain would then be:

$$\text{VCO}_{\text{GAIN}} = K_{\text{VCO}} A_v$$

where A_v = voltage gain of the op-amp and K_{VCO} = gain of the VCO in MHz/V. (The entire PLL design will still be the same as in steps 1 through 10 above, but now simply substitute K_{VCO} for VCO_{GAIN}.)

Another popular technique is to place a low-noise, high-supply-voltage op-amp at the DC tuning input of the VCO—with the loop filter's output placed into the input of the op-amp—and use the VCO_{GAIN} formula above to calculate the new gain of the VCO. The result of the VCO_{GAIN} calculation will be used as the new K_{VCO} in the above PLL formulas.

This completes the design of the most important part of a PLL synthesizer, the loop filter. The following will wrap up the total frequency synthesizer design by employing one of the most popular family of PLL chips in use today: the National LMX23XX (Fig. 5.7).

The complete National PLL chip's input and output pins are described in detail below for the widely used LMX2306 (which functions up to 550 MHz), the LMX2316 (functions up to 1.2 GHz), and the LMX2326 (functions up to 2.8 GHz):

1. Fl_o is an output pin that permits a parallel resistor to be attached between C_2 and R_2 of the PLL's loop filter. This will allow the PLL to obtain both a fast lock time and good phase noise specs by modifying the loop bandwidth on the fly. After the channel change occurs, loop bandwidth reverts back to normal.
2. CP_o is the output of the charge pump to which the loop filter is attached.

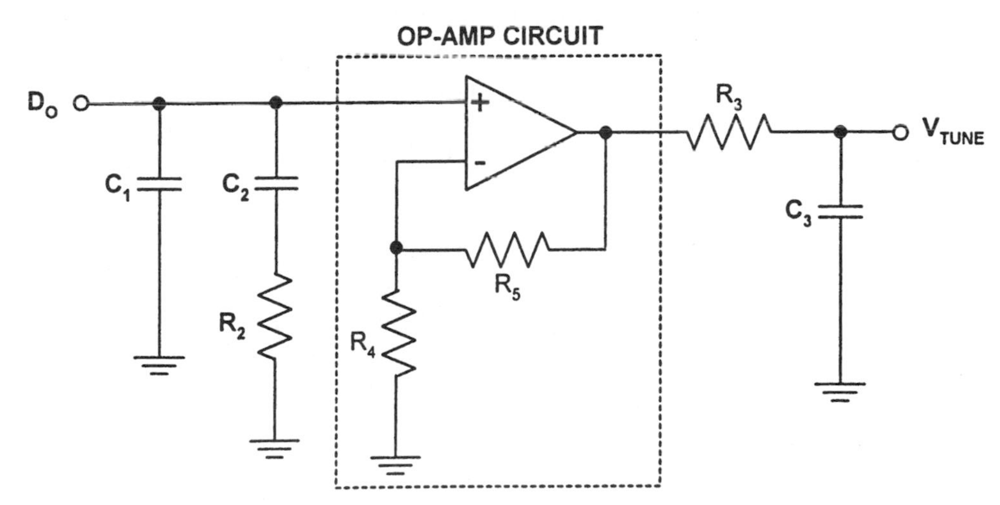

Figure 5.6 Circuit to increase the tuning voltage for wideband VCOs.

3, 4. GND at Pins 3 and 4 are for the charge pump and for the analog circuits respectively. Both can be short-circuited together and attached as directly as possible to system ground.

5. f_{IN} should be AC-shorted to ground through a 100-pF capacitor.

6. f_{IN} accepts the signal from the VCO's output through a series 20- to 200-ohm resistor. The series resistor, whose value depends on the VCO's output power, will lower the power into the preselector, allowing most of this energy to be delivered to the load.

7. $V_{cc}1$ pin is the heavily bypassed DC analog power supply voltage input. The input voltage for this particular chip may be anywhere between 2.3 and 5.5 V; but must equal pin 15's $V_{cc}2$ voltage.

8. OSC_{IN} is the reference oscillator input for a CMOS 100-kilohm output resistance clock oscillator. A clean crystal clock input is vital for a low-phase-noise PLL output.

9. GND is digital ground. It should reach system ground by as direct a route as possible.

10. CE is the chip enable pin when power is down for power-saving operation. It can be tied to V_{CC} if this feature is not required.

11. CLK is an input that accepts a CMOS clock signal from the channel select microcontroller for clocking data into pin 12.

12. DATA input accepts data from the microcontroller for the R counter, the N counter, and the function latch (which controls phase detector polarity, fast-lock modes, F_o/LD, counter reset, CP tristate, test modes, etc.), with the last two bits (control bits) informing the PLL as to whether the data should be sent to the R counter (0, 0), the N counter (1, 0), or the function latch (0,1) on command of pin 13, LE.

13. LE (load enable) pin controls when the PLL's registers will send data to the R, N, or function latches, depending on the control bits.

14. F_o/LD is an output pin that can typically be used as a lock detect (LD) output pin into a microprocessor, or into some out-of-lock alarm. A HIGH will be output when the PLL is in lock (on advanced PLL chips, such as with the National line, a trace may be taken from the lock detect (LD) pin back to the microprocessor. The pin, if digital lock detect is chosen by programming the proper PLL register, will output a HIGH as long as the VCO output frequency is locked. This HIGH or LOW signal can then be exploited by the microprocessor to indicate an unlocked condition by an LED warning on a display, or as an automatic shutdown of a runaway transmitter).

15. $V_{CC}2$ is the digital power supply voltage input pin, and should be tied to pin 7, which is the analog power supply input.

16. V_p is the power supply for the charge pump circuit, and must be greater than V_{CC}. (The DC control voltage into the VCO will always be a few

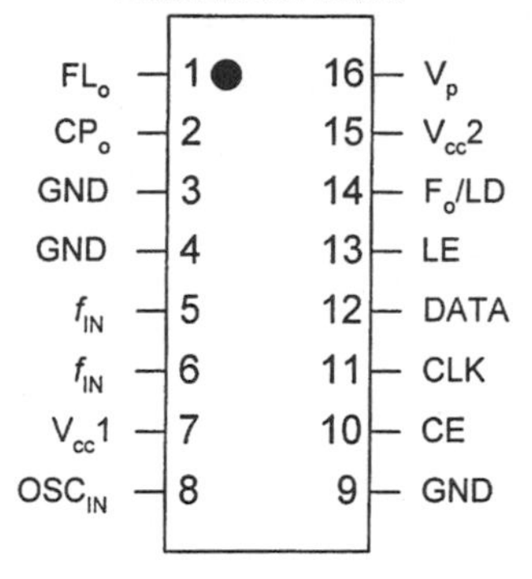

Figure 5.7 A popular high-quality PLL chip, the National LMX23XX series.

tenths of a volt less than V_p, so V_p must have the proper amplitude to fully drive the VCO's DC control input.)

After completing the filter's design calculations for the frequency synthesizer, the following final PLL design checks must be performed to confirm that the PLL will function as desired:

1. The loop bandwidth f_c should be *at least* 1/20 F_{COM}.
2. Make sure that C_0 is at least 5 times larger in value than the input capacitance of the VCO (which is usually around 20 pF for the average VCO input capacitance).
3. Since maximum PLL *phase detector* input frequencies normally are no higher than 10 MHz, make sure that F_{COM} is not above this amount.
4. R must generally be set to divide by at least 3 or more.
5. Check the completed PLL design to confirm that the *damping factor* is less than 1 by:

$$\frac{R_2 C_2}{2} \cdot \sqrt{\frac{K_\phi K_{VCO}}{N(C_1 + C_2 + C_3)}}$$

6. Check that $R_3/R_2 > 2$.
7. The *optimization index* can be checked by:

$$\frac{\dfrac{T_2}{1 + (\omega_C T_2)^2}}{\dfrac{T_1}{1 + (\omega_C T_1)^2} + \dfrac{T_3}{1 + (\omega_C T_3)^2}} \times 100$$

This formula confirms that the loop is stable with a fast lock time. Any value between 90 and 100 percent is considered stable.

8. A roughly estimated lock time can be found by:

$$\mathrm{LT} = \frac{-\ln\left(\frac{\mathrm{TOL}}{f_2 - f_1}\right)\sqrt{1-\zeta^2}}{\zeta\omega_N}$$

where
$$\omega_N = \sqrt{\frac{K\phi K_{\mathrm{VCO}}}{N(C_1 + C_2 + C_3)}}$$

$$\zeta = \frac{R_2 C_2}{2}\omega_N$$

LT = lock time, s
f_2 = higher frequency, Hz
f_1 = lower frequency, Hz
TOL = frequency tolerance (or acceptable frequency error at lock), Hz

9. To check for approximate phase noise (PN) of the PLL at 150 Hz from the center frequency: PN @ 150 Hz = 205 − 10 log F_{COM} + 20 log (N counter value), where 205 is the average value for a typical PLL chip, and is referred to as the PLL's 1 Hz normalized PN floor, or 1 Hz PNF. This formula does not take into account a noisy VCO.

Solving PLL problems. The most common issues found during the testing of a PLL after the design and construction phases are completed are noisy output, incorrect output frequency, spurious outputs, and an intermittent or continuous refusal to lock.

A noisy output can be caused by multiple problems, since in a well-designed PLL circuit the highest contributor to phase noise (Fig. 5.8) will be the PLL's own integral phase detector, but this internal self-generated noise can be swamped out by any of the following difficulties:

Poorly designed, noisy VCOs

A loop filter not wide enough to prevent the VCO from adding excess noise (VCO noise is tiny within the loop bandwidth, while outside the loop bandwidth VCO noise will be quite large)

A noisy or noncrystal reference source

A low charge pump current or voltage

Incorrect signal amplitude levels into the R or N dividers.

Reference spurs will also be encountered. These are spurious signals at frequencies that are located at an interval equal to the comparison frequency (F_{COM}) away from the carrier frequency (Fig. 5.9). These spurs may, as well,

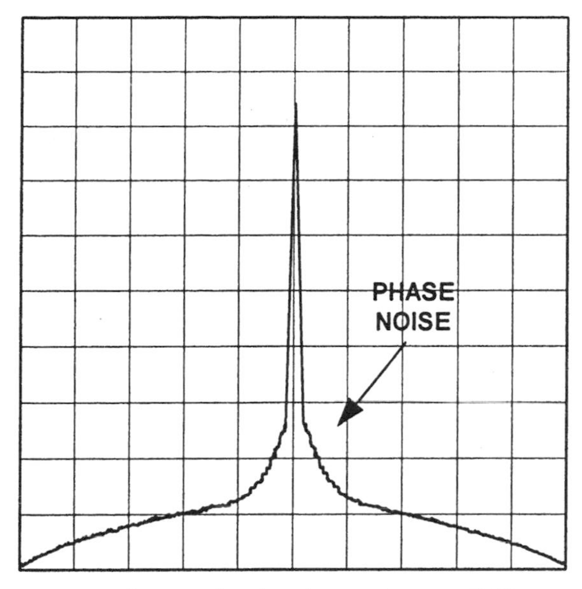

Figure 5.8 Spectra showing phase noise out of PLL.

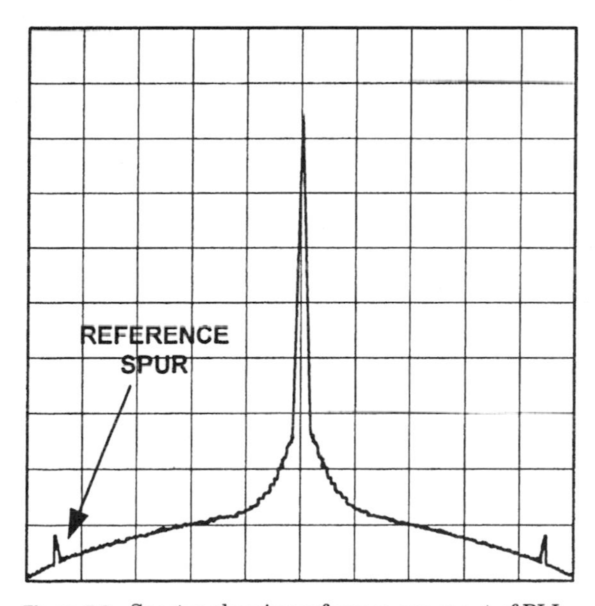

Figure 5.9 Spectra showing reference spurs out of PLL.

occur at harmonics of the reference frequency, and can be created by charge pump leakage and mismatch, PCB cross talk, improper decoupling of DC power into the PLL, and exterior noise and signal source ingress.

If the PLL circuit will not lock reliably, or has generally all-around inferior performance:

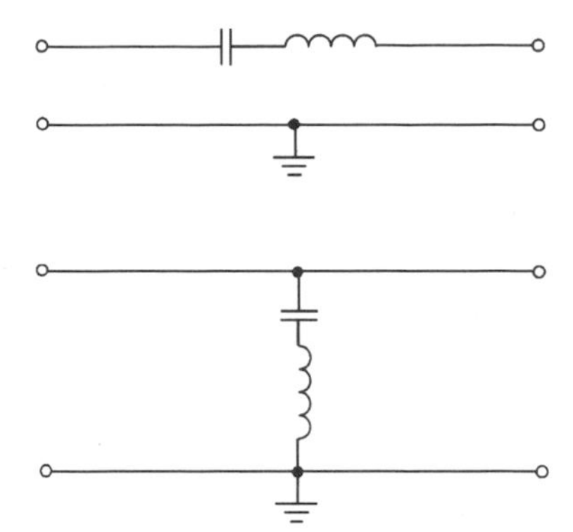

Figure 6.6 A basic series bandpass filter.

Figure 6.7 A basic series bandstop filter.

resonant circuit. This waste of power can be minimized by using only high-Q inductors in the tank. But because the inductor and capacitor currents in a parallel circuit are 180 degrees out of phase with each other, high circulating currents will always exist within the tank itself during resonance (Fig. 6.8). This circulating current is due to the two *LC* components exchanging current in a flywheeling manner. However, as these currents are completely out of phase, current flow *into* the tank is always at a minimum, and is dependent on the pure resistance within the tank caused by component Q limitations.

A simple bandpass filter (BPF) is shown in Fig. 6.9. Since it is an *LC* parallel filter in shunt with the output, all nonresonant frequencies will be sent to ground, while the bandwidth of interest will be passed on to the output because of the high impedance created by the tank at resonance. This creates a selective RF output voltage across the circuit so that it functions as a BPF. The bandstop of Fig. 6.10 has the parallel circuit in series with the output. Thus, it will pass all frequencies *except* resonant frequency, which is dropped across the high impedance of the tank. Since this will cause a decrease in the output amplitude at this single range of frequencies, it functions as a bandstop filter (BSF).

Most filters work by passing any frequencies within the passband with little attenuation while *reflecting*—not absorbing—most of the undesired signals within its stopband back toward the source. These reflections can become a serious problem in wireless systems design, as we shall soon see.

Filters must not only shape a signal, reject spurious frequencies, and choose one frequency band out of many, but they must also maintain a specific input and output impedance through much of their passband that is identical to the system's impedance (usually 50 or 75 ohms).

Different types of filters, such as *LC,* crystal, SAW, and distributed, will have various frequency bands in which they are most commonly employed because of size, price, and/or performance:

1. *LC* filters can be utilized from 1 kHz all the way up to 1.5 GHz. As the frequencies increase, however, so does the difficulty in implementation because of the distributed inductance and capacitance, which conspire to lower the frequency of the filter as designed, as well as distort its response.

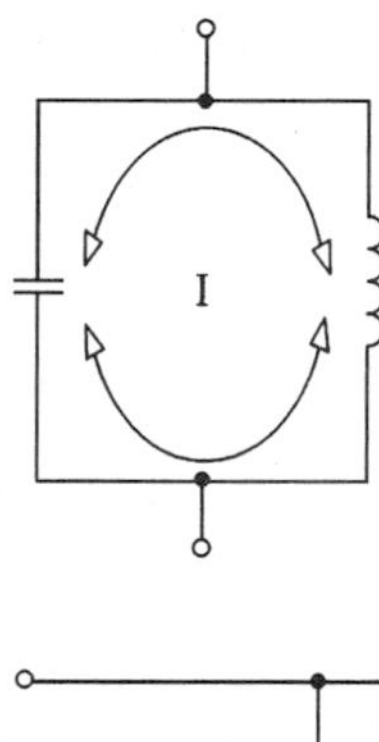

Figure 6.8 High internal circulating currents in a tank circuit at resonance.

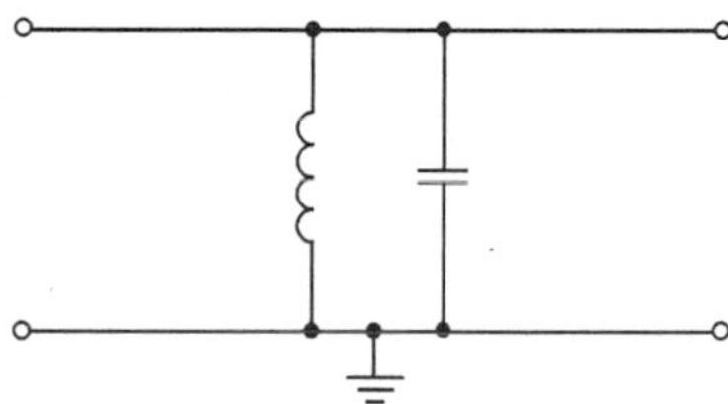

Figure 6.9 A basic parallel bandpass filter.

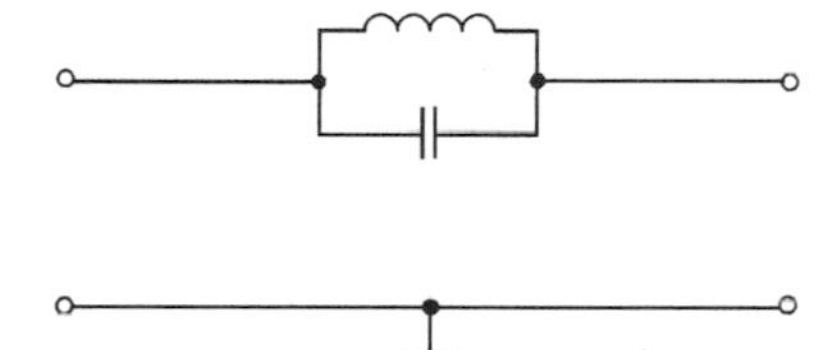

Figure 6.10 A basic parallel bandstop filter.

But *LC* filters are low in cost and can fit into a tight form factor (as opposed to distributed designs). They are not suitable for very tight bandwidth requirements that demand steep skirts.

2. Discrete crystal designs have been used from as low as 10 kHz to higher than 400 MHz, and can be employed in both narrow- and wideband applications. They are found in bandpass and bandstop applications.
3. Surface acoustic wave (SAW) filters use a piezoelectric crystal substrate with deposited gold electrodes. SAWs are capable of replacing *LC* filters in certain wideband applications between 20 MHz to 1 GHz, but will ordinarily experience a 6- to 25-dB insertion loss. Their filter skirts, or shape factor, are the sharpest of all the filter structures.
4. Distributed filters comprise copper strips placed on a dielectric substrate—a printed circuit board—that act as narrow- and wideband filter structures from 500 MHz up to 40 GHz (and beyond). They are low in cost and have high *Q* at high frequencies but, depending on frequency and design, can take up significant board space.

There are also special bandpass filter responses that are demanded for various requirements. For instance, the popular *Butterworth* response is adopted when no amplitude ripple to the signal is desired within the filter's passband; it has medium selectivity, medium group delay variations, and a good tolerance to component variations. (A filter that is sensitive to component tolerances will exhibit an undesired altered passband in S_{21} and S_{11} due to the normal variations in L and C values.) *Chebyshev*-type filters have a certain amount of passband ripple that will be forced on the input signal as it passes through to the filter's output. The "Cheby" does, however, have high selectivity, with high group delay variations being an unfortunate side effect of this innate selectivity. Amplitude ripple and high group delay variations, in digital signals, can cause an increased BER, so are undesired. Nonetheless, low-ripple Chebyshevs can easily be designed, and the group delay variations can be improved by widening the filter's passband—or using fewer poles. *Bessel* response filters have no ripple in their rounded passbands, and display very low group delay, but have extremely poor selectivity and poor tolerance to component variations.

There are many different types of *LC* circuit filter topologies that will furnish these responses of Butterworth, Chebyshev, and Bessel. The choice depends on the shape of the passband desired, the percent bandwidth required, the sensitivity to component tolerances and distributed reactances, and the ability to obtain easily realizable component values during design. More on this later.

There are numerous terms used when filters are discussed. Below are the most common:

Absolute attenuation—The maximum attenuation a filter is capable of at some chosen frequency in its stopband. Measured in dB.

Bandwidth—The width ($f_{LOW} - f_{HIGH}$) of the band of frequencies passed by a bandpass filter at its 3-dB-down points. Measured in Hz.

Center frequency (f_C or f_0)—The exact mathematical center of a bandpass filter. Measured in Hz.

Cutoff frequency—The point in a frequency response of a filter that is 3 dB below the average passband response, and which keeps on falling.

Decibels of attenuation per octave (dB / octave)—Filters can be designed as to how rapid their skirt slope falls. The decibels of attenuation per octave specification refers to a filter's steepness: If a filter is said to have a 15-dB/octave slope at a 1-GHz cutoff frequency, then the attenuation within the stopband will be 15 dB more at 2 GHz, while the filter's stopband attenuation at 4 GHz will be 30 dB. Knowing the required dB/octave fall of the filter's skirt assists the engineer in visualizing the attenuation as the frequency increases, or decreases, from its passband.

Differential delay—The group delay variation (GDV) between two particular frequencies, usually measured in nanoseconds. Delay equalization, either by discrete analog or by DSP techniques, can be used to almost nullify the effects of GDV, thus improving the BER of digital radios.

Group delay—The measurement, in nanoseconds, of the time retardation created by a filter or circuit to any discrete signal that passes through it. When the group delay varies across the passband significantly, as it does in tight Chebyshev filters (and to a certain extent in Butterworth filters), this will cause increased BER in digital radios. This variation in the group delay, called *group delay variations* (GDV), is especially severe at the edges of a filter's passband, so any filter that has steep skirts and/or a high pole count will have a high GDV. The problem can be lessened by widening the filter's passband beyond what is required by the signal; using fewer filter poles; or choosing a Butterworth type or, if selectivity specifications permit it, a Bessel filter response. GDV is directly equated to the *differential delay* specification above.

Insertion loss—The attenuation created in a filter within the center of its passband when the filter is terminated at its design impedance. Measured in dB.

Insertion loss linearity—The measure, in dB, in the variations in insertion loss with input power. Primarily a problem with crystal filters.

Passband—The band of frequencies from f_{LOW} to f_{HIGH} that a filter passes with 3 dB or less attenuation (f_{LOW} and f_{HIGH} are located at points where the insertion losses reach, typically, 3 dB). Measured in Hz.

Passband ripple—The amplitude fluctuations within a filter's passband. A ripple greater than 0.5 dB is usually considered unacceptable in digitally modulated radios. Chebyshev is the dominant filter topology that contains ripple within its passband. However, this ripple can be decreased to 0.1 dB or less by low-ripple Cheby designs.

Phase shift—The measurement of the variation of the phase of a signal as it moves through a filter from its input to its output.

Poles—Refers to the number of reactive components, inductors or capacitors, in a low-pass or high-pass filter; or the number of reactive pairs in a bandpass filter (for an *all-pole* filter). The filter's order matches its poles in an all-pole filter, and the number of poles governs the steepness of the filter's skirts.

Q (quality factor)—The ratio of the center frequency to its bandwidth at its 3-dB-down points. The narrower the bandwidth for the same center frequency, the higher the filter Q. Q also refers to the quality factor of the individual components that make up the filter. This is particularly important for inductors within the *LC* filter circuit, since the lower the individual Q's within each component, the higher the filter's insertion loss will be. The stopbands will also have a poorer stopband attenuation characteristic, and the filter's response at the band edges will be more rounded than with high-Q components.

Return loss—The measurement, expressed in decibels, of the difference between the signal power sent toward a filter's input and the strength of the RF signal power returned, or reflected, from the input back toward the

The fundamental building block in low-pass filter design is the *half-section,* as shown in Fig. 6.16. This half-section is composed of one series inductor and one shunt capacitor. By cascading an increasing amount of half-sections, we can obtain a filter with any filter skirt steepness desired. This is referred to as increasing the number of poles of a filter; where four poles equal one section in a low-pass filter. The half-sections may be joined either as pi sections (Fig. 6.17) or as T sections (Fig. 6.18). These sections will then combine with the next adjoining reactive element, creating a single inductor for the pi of Fig. 6.19. This is because there would be absolutely no reason to use two inductors in series with each other, or two capacitors in shunt with each other, when their values can simply be added to obtain the proper value (Fig. 6.20). Combining half-sections with either the pi or the T technique is equally valid, but the design that results in fewer inductors is usually preferred from a cost, Q, and size standpoint. A filter with either an even or an odd number of cascaded sections is also perfectly sound.

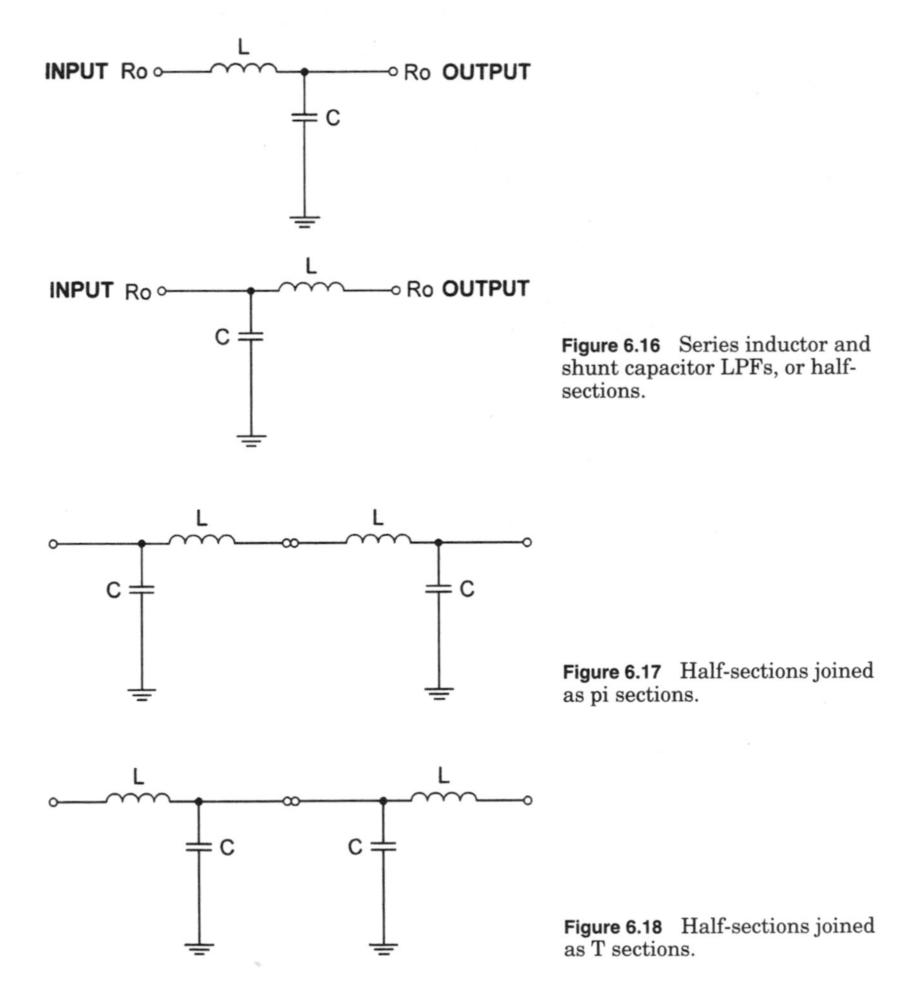

Figure 6.16 Series inductor and shunt capacitor LPFs, or half-sections.

Figure 6.17 Half-sections joined as pi sections.

Figure 6.18 Half-sections joined as T sections.

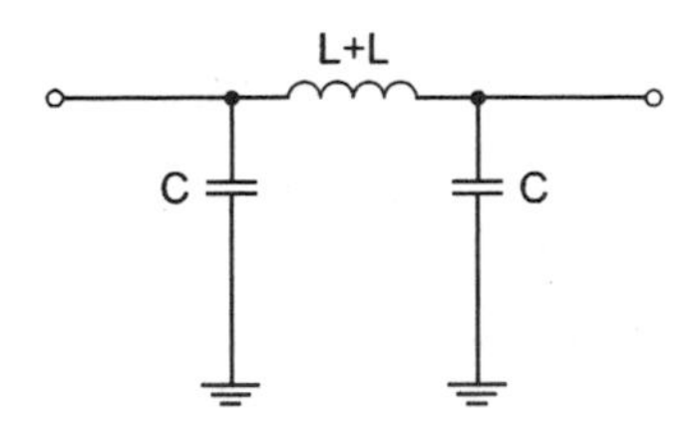

Figure 6.19 Joining two half-sections by adding series inductors.

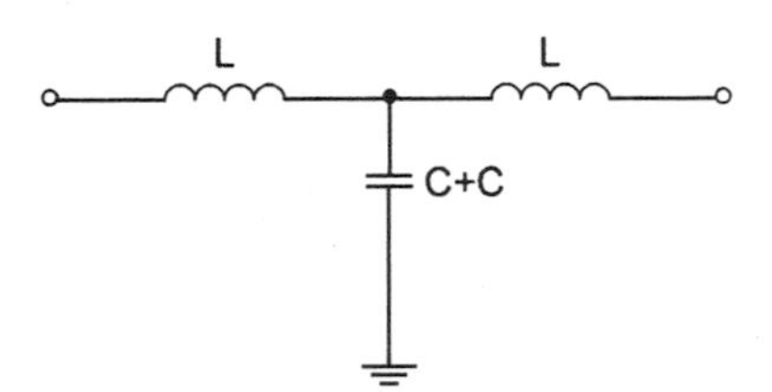

Figure 6.20 Joining two half-sections by adding shunt capacitors.

To design a low-pass filter, a half-section (as depicted in Fig. 6.16) can be calculated by:

$$L = \frac{\left(\frac{R_0}{\pi f_C}\right)}{2} \qquad \text{and} \qquad C = \frac{\left(\frac{1}{\pi f_C R_0}\right)}{2}$$

where R_0 = impedance at the filter's input and output and f_C = filter's 3-dB cutoff frequency.

After the L and C values are calculated, the actual number of poles can be increased from the present half-section's two, to any number desired. This is accomplished by combining half-sections as shown in Fig. 6.21, adding the capacitances, or by combining them as shown in Fig. 6.22 and adding the inductances.

However, combining half-sections as shown in Fig. 6.23 would be incorrect, and would result in a filter that not only has too many components, but also will not function as designed. This is because only shunt elements are allowed to "blend" with shunt elements, and only series elements can "blend" with series elements.

As the number of filter sections increases, the actual cutoff frequency will begin to decrease. This is referred to as the *cascade effect.* If this is not taken into account, the cutoff frequency will drop, especially when three or more sections are combined. Table 6.1 is a list of the correction factors that must be used to obtain an accurate cutoff frequency when cascading multiple sections.

To utilize this table in low-pass filter design: If f_C of the lowpass filter is to be 200 MHz, and a 2-section filter is required, then first multiply the adjustment factor of 1.15 times the f_C. This would be equal to 230 MHz. Now design the 2-section low-pass filter as if the f_C will be 230 MHz, and a filter with a true f_C at the desired frequency of 200 MHz will now be the result.

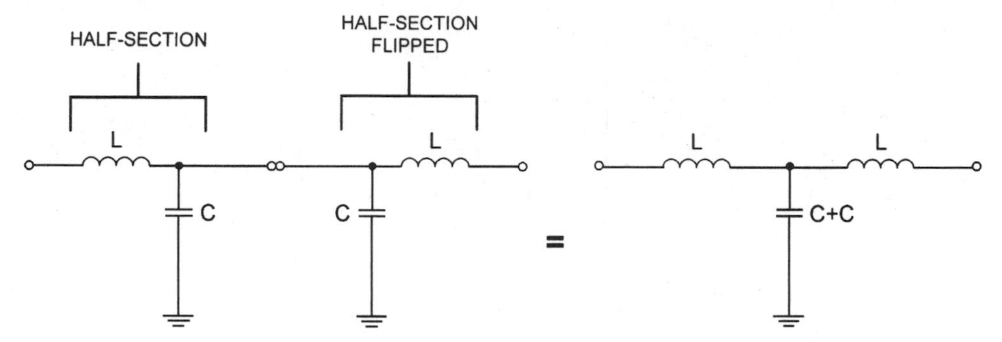

Figure 6.21 One way of properly combining half-sections.

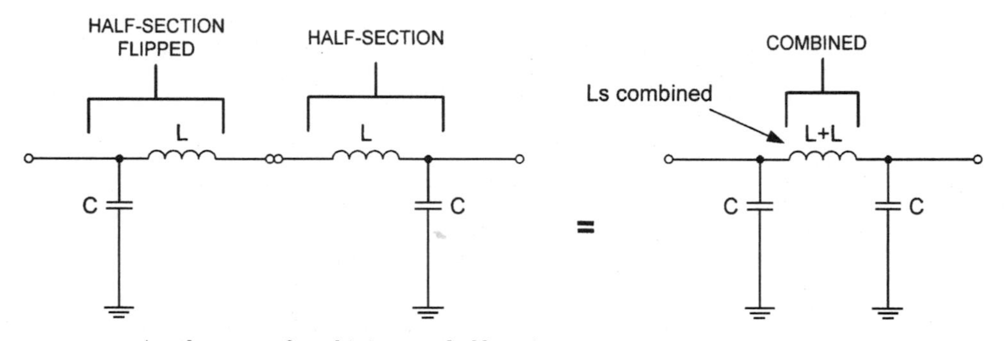

Figure 6.22 Another way of combining two half-sections.

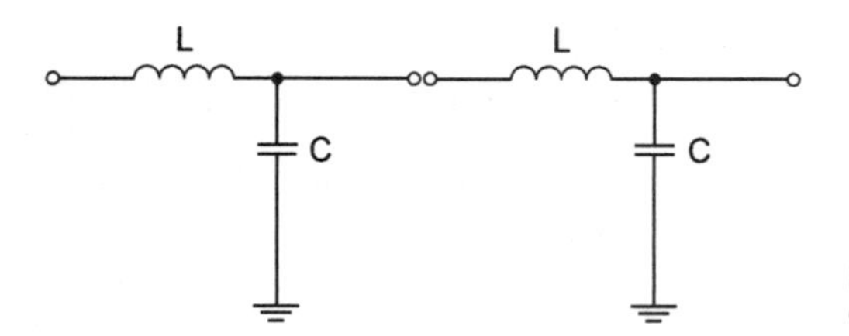

Figure 6.23 Incorrect way of combining two half-sections.

TABLE 6.1 Correction Factors for a Low-Pass Filter

Number of sections* ($^1/_2$ section = 2 poles = 1 L, 1 C	Adjustment factor
0.5	1
1.0	1.05
1.5	1.10
2.0	1.15
2.5	1.20
3.0	1.25

*Before combining of the half-sections.

If the low-pass filter is required to block DC, then a large-value capacitor with an X_C of 1 ohm at its lowest frequency of interest can be inserted at the filter's input. This will have no effect to the low-pass filter's response. The ability to block DC is especially valuable at the lower RF frequencies, where ferromagnetic cores are used in the filter's inductors (the DC can affect the permeability of the inductors' cores by saturating them, and thus changing the inductance of the coils, destroying the response of the filter).

The attenuation slope of the low-pass filter as presented can be approximated as an 18-dB increase in attenuation per octave for a 3-pole filter. In other words, if the filter were designed to have an f_C of 1 MHz, then the low-pass filter would have attenuated a 2-MHz signal by 18 dB. By 4 MHz, the signal would be down by 2 × 18 dB, or 36 dB. And if the low-pass filter were increased to 6 poles, it would have an attenuation slope that dropped by 36 dB per octave. With an f_C at 1 MHz, the signal is attenuated (has an insertion loss) of approximately 3 dB at 1 MHz, while at 2 MHz the signal is down to about 36 dB, and at 4 MHz it has dropped to 72 dB.

When multiple half-sections are combined with low-pass filter designs, the filter's attenuation response becomes that of the amount of the L and C components (poles) that result from the combination and joining of these half-sections. For instance, if a filter is created from three half-sections—which contain six reactive components (poles)—the filter would now contain only four reactive components (poles) after the appropriate components had been combined. The components left *after* the combination of the half-sections will be the indicator of the filter's attenuation response.

High-pass filters. Shown in Fig. 6.24 are a 2-pole high-pass filter that incorporates a series capacitor and a shunt inductor, and another that has a shunt inductor and a series capacitor. Both are half-sections. This is exactly the opposite component layout of that of the low-pass filter style above.

For the high-pass filter, as it was with the low-pass filter design, cascading these half-sections will produce a high-pass filter with as many poles as required.

To design a Butterworth high-pass filter, first calculate the L and C values of the single half-section of Fig. 6.24 by:

$$L = 2\left(\frac{R_0}{4\pi f_C}\right) \qquad \text{and} \qquad C = 2\left(\frac{1}{4\pi f_C R_0}\right)$$

where R_0 = impedance at the filter's input and output and f_C = filter's 3-dB cutoff frequency.

Take the half-section as calculated, and combine by dividing by two the combined inductors in Fig. 6.25, or by dividing the combined capacitors by two in Fig. 6.26. Unlike low-pass filter design, in which the combined components are added, a high-pass filter's combined components must be divided by 2. As for the similar situation of the low-pass filter above, the high-pass sections must not be combined as in Fig. 6.27.

and

$$CP = \frac{\left[\dfrac{1}{50\,(525\text{ MHz} - 475\text{ MHz})\,\pi}\right]}{2} = 63.6\text{ pF}$$

Transfer these values to Fig. 6.34, the bandpass filter's half-section. Begin adding and combining half-sections (Fig. 6.35). Continue combining half-sections until the 6-pole filter of Fig. 6.36 is obtained.

6.2 Distributed Filters

6.2.1 Introduction

Most distributed filters are constructed as microstrip structures. The actual materials to build them are low in cost, consisting of copper and a substrate. However, they can—depending on the passband frequency and the dielectric constant of the board material—take up much more space than lumped *LC* filters. Distributed filters are, nonetheless, usually the only economical and practical choice over 1.5 GHz for many applications. Unfortunately, distrib-

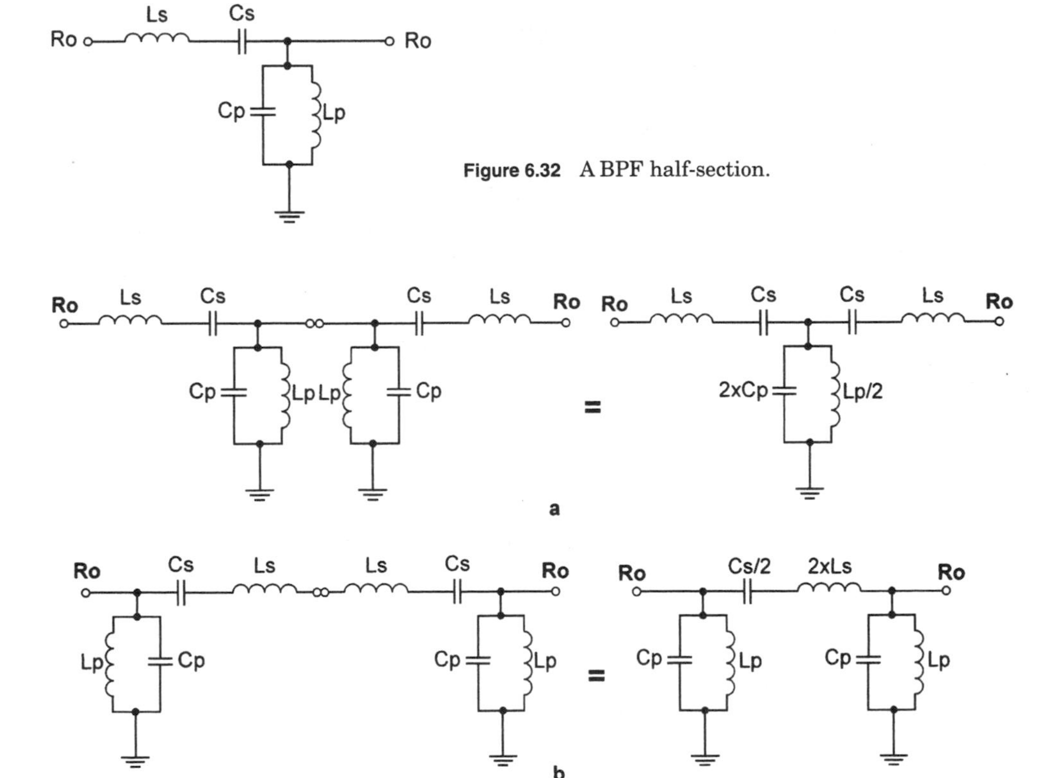

Figure 6.32 A BPF half-section.

Figure 6.33 Combining two BPF half-sections when placed (*a*) tank-to-tank and (*b*) series-to-series.

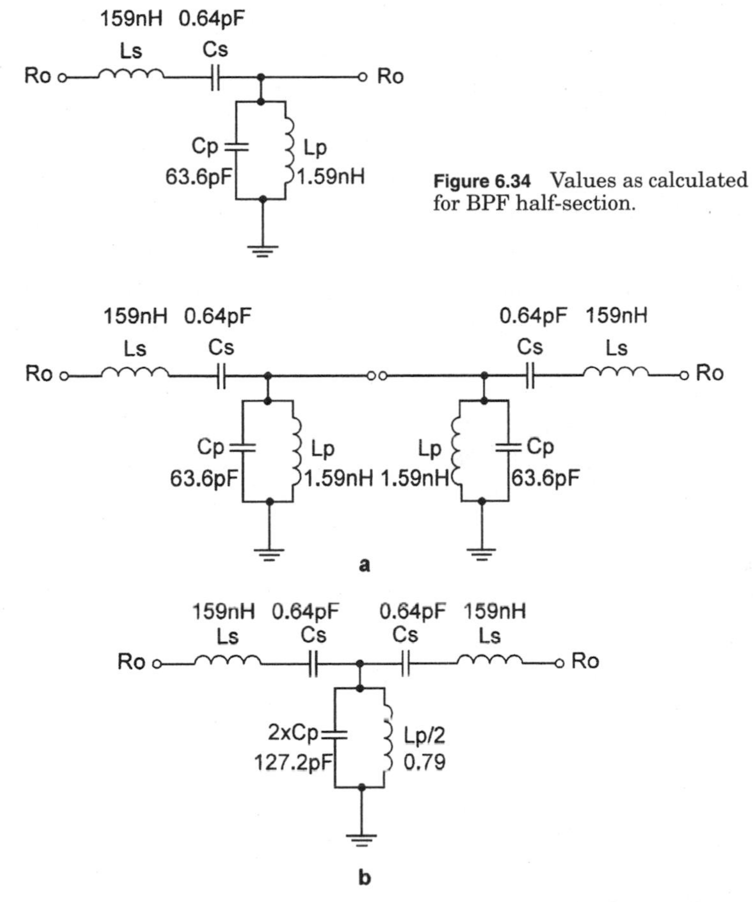

Figure 6.34 Values as calculated for BPF half-section.

Figure 6.35 (*a*) Two single half-sections; (*b*) combining the shunt tanks.

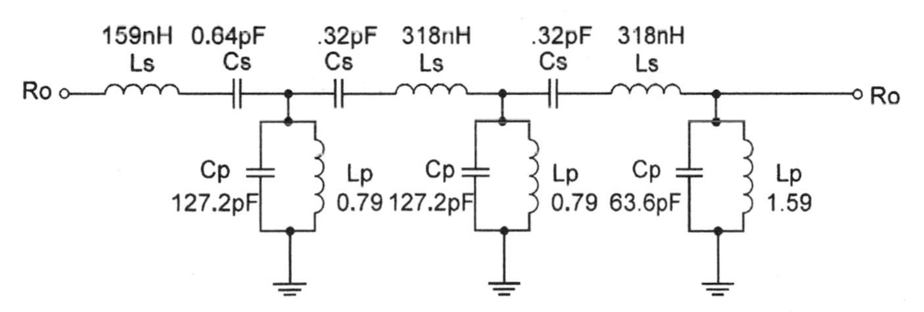

Figure 6.36 Completed 6-pole bandpass filter.

uted filters may display reentrant modes that will destroy their stopband attenuation at certain repeating harmonic frequencies.

Distributed bandpass filter structures cannot merely be a distributed copy of a lumped *LC* filter design. In order to obtain a narrow bandpass, with acceptable in-band ripple and insertion loss, specialized structures must be considered. Unhappily—unlike *LC* filters—the calculations to design such distributed filters are far too laborious. Most of these types of calculations, as well as the necessary optimization of such designs, are best left to a high-end software program such as Eagleware's M/Filters. However, a few easy-to-implement designs are presented for undemanding applications.

6.2.2 Types of distributed filters

Because of various requirements and specifications for microwave filters—such as stopband attenuation, narrow versus wideband, return loss requirements, and microstrip element lengths—a multitude of distributed bandpass filter structures have been created for microwave frequencies. Only the most common will be discussed here. Nearly all of the following structures can be designed with the various general passband shapes, such as Butterworth, Chebyshev, Bessel, as well as the less common *singly equalized transitional gaussian* (6 and 12 dB) and the *elliptic* types of the *Cauer-Cheby:*

1. *Edge-coupled bandpass filters* (Fig. 6.37) are effective in narrowband applications. But they can radiate—which necessitates shielding. They can also be an unusually long structure at the lower microwave frequencies: At 1 GHz, edge-coupled bandpass filters are up to 32 inches long in a seventh-order configuration (depending on board dielectric constant and thickness).
2. *Combline bandpass filters* (Fig. 6.38) employ both distributed and lumped (capacitor) elements, and are used in narrowband applications where size is at a premium: a seventh-order combline is only a little over 1 inch long at 1 GHz.
3. *Folded edge-coupled bandpass filters* (Fig. 6.39) are utilized in narrowband filtering applications, and are very similar to the edge-coupled bandpass filters above. However, they are considerably shorter: 8 inches at 1 GHz for the seventh-order type.
4. *Interdigital bandpass filters* (Fig. 6.40) are adopted in narrowband applications, and are quite compact: 2 inches long at 1 GHz for a seventh order.

It is especially important at high frequencies for the measurement of stub lengths to be accurate in distributed design, since the higher the frequency the shorter will be the stub, and the less room for error there will be. An effective

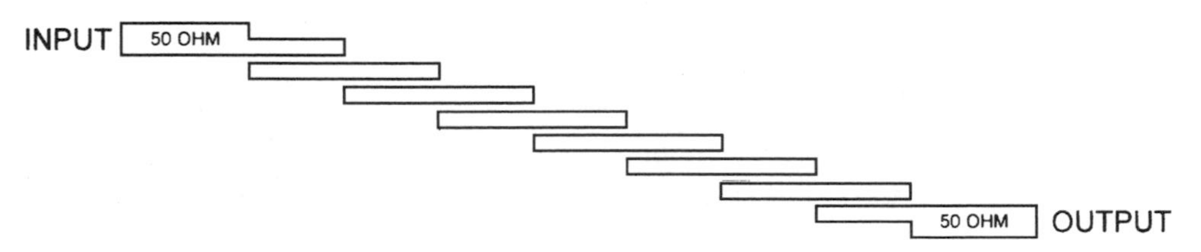

Figure 6.37 Sixth-order edge-coupled distributed BP filter.

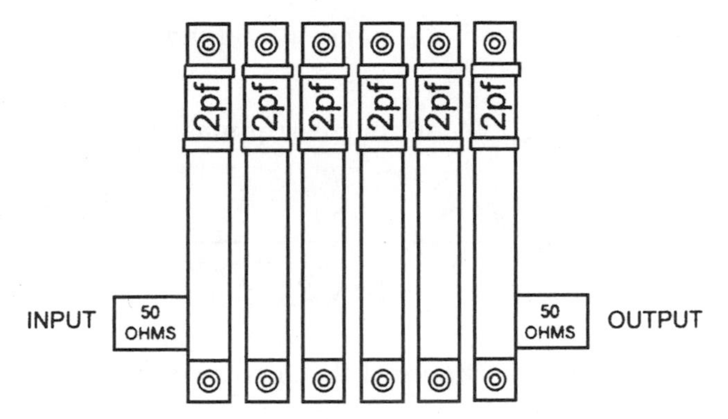

Figure 6.38 Sixth-order combline distributed BP filter.

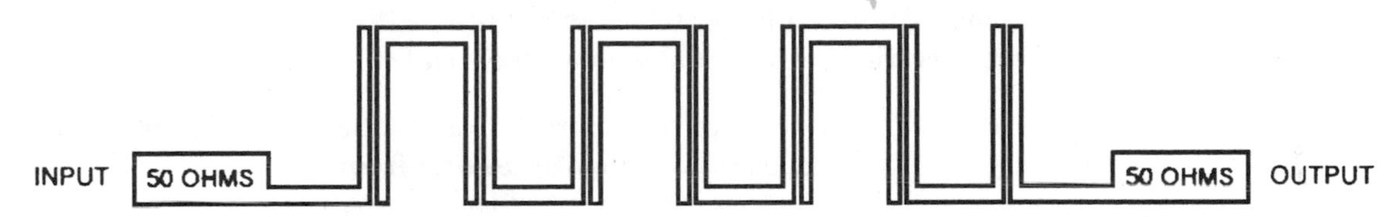

Figure 6.39 Sixth-order folded edge-coupled distributed BP filter.

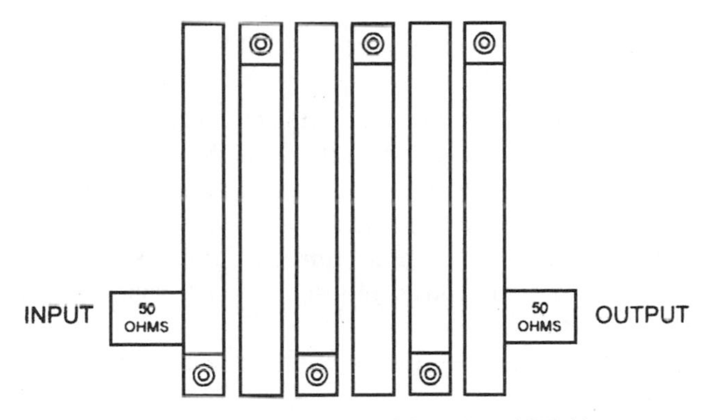

Figure 6.40 Sixth-order interdigital distributed BP filter.

technique for accurate grounded stub measurement and layout is shown in Fig. 6.41: The length of the stub is measured from the T junction to the center of the through-hole via.

6.2.3 Distributed filter design

A shorted-stub bandpass filter (Fig. 6.42). The width of the grounded stub is that of 100-ohm microstrip, while the length of the microstrip is exactly a quarter-wavelength, or 90 degrees; and is grounded by a via directly to the ground plane.

1. To calculate how wide a 100-ohm microstrip must be for the substrate in use:

$$Z_0 = \frac{377}{\left(\frac{W}{h} + 1\right)\sqrt{E_r + \sqrt{E_r}}}$$

where Z_0 = characteristic impedance of the microstrip, ohms (in this case it must be 100 ohms)
W = width of the microstrip conductor (use same units as h)
h = thickness of the substrate between the ground plane and the microstrip conductor (use same units as W)
E_r = dielectric constant of the board material

2. To calculate exactly how long the microstrip must be to be a quarter of a wavelength, with the chosen board material and frequency:
   a. Find the effective dielectric constant of the microstrip.

$$E_{EFF} = \frac{E_r + 1}{2} + \left(\frac{E_r - 1}{2} \cdot \frac{1}{\sqrt{1 + \left(\frac{12h}{W}\right)}}\right)$$

where E_{EFF} = *effective* dielectric constant that the microstrip transmission line actually sees as a result of the dielectric/air interface
E_r = rated dielectric constant of the PCB's substrate material (found on the PCB's data sheet)
h = thickness of the substrate material between the top conductor and the bottom ground plane of the microstrip; requires the same units as W
W = width of the top conductor of the microstrip; requires the same units as h

   b. After obtaining the effective dielectric constant of the microstrip substrate, find the propagation velocity of the microstrip transmission line:

$$V_P = \frac{1}{\sqrt{E_{EFF}}}$$

where V_P = fraction of the speed of light for the microstrip transmission line compared to light in a vacuum and E_{EFF} = effective dielectric constant as seen by the microstrip transmission line.

   c. Calculate the wavelength of the frequency of interest in a perfect vacuum:

$$\lambda = \frac{11{,}800}{f}$$

where λ = wavelength of the frequency of interest (f), mils
11,800 = the numerical value required to obtain a λ in mils while using an f in GHz
f = frequency of the signal of interest, GHz

d. Multiply the resultant velocity of propagation (V_P) by one-quarter the wavelength λ of the signal in order to arrive at the quarter-wavelength of the signal of interest (in mils) when it is placed into the microstrip:

$$\lambda/4 \text{ (in mils)} = V_P \times \lambda/4$$

An open-stub bandstop filter (Fig. 6.43). A bandstop filter, being basically a reversed bandpass filter, is also easy to design. Simply duplicate the above bandpass design procedures, but leave the stub open instead of grounding it through a via. Because it is now an open stub, however, the *end effect* will demand that the length of the stub be trimmed down by approximately 5 percent below the calculated length: Cut a small amount off the end of the open stub until the center frequency is as desired (not all electromagnetic microwave simulation software packages will take the end effect into account).

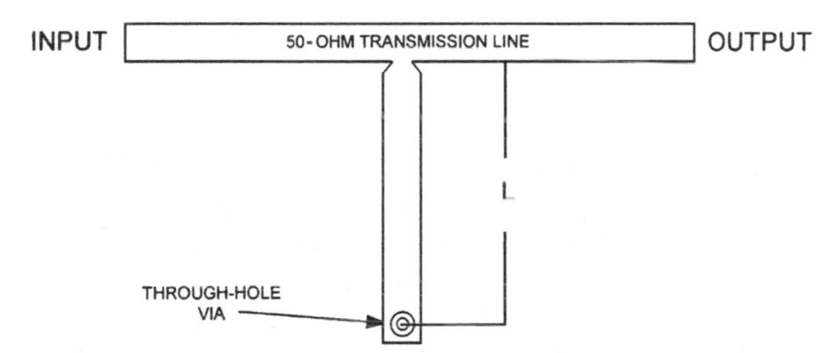

Figure 6.41 Proper distributed component length to ground via.

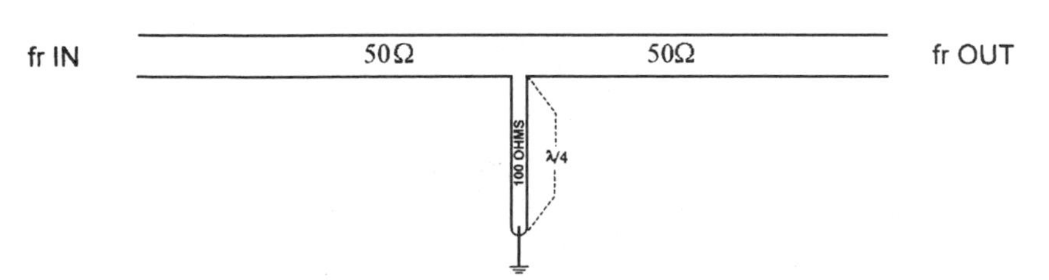

Figure 6.42 A microstrip layout for an equivalent lumped tank circuit for a high-impedance parallel tank.

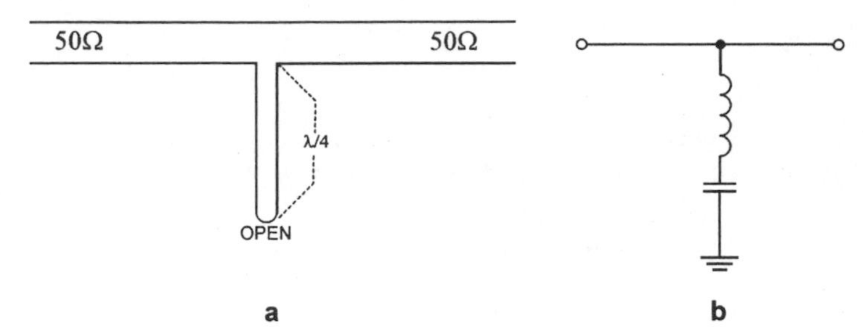

Figure 6.43 (*a*) A distributed bandstop filter; (*b*) with equivalent lumped series shunt circuit.

A 3-pole distributed microstrip Butterworth interdigital BP filter (Fig. 6.44). The following 3-pole BPF is fine for a simple distributed filter, but simulation will be required to optimize for proper performance. Specifically, adjust the spacing S between elements for improved S_{21} and S_{11} and to attain your desired bandwidth.

1. Compute the percentage of bandwidth required at the 3-dB-down points at the center frequency of interest:

$$\%\mathrm{BW}_{3\mathrm{db}} = \frac{(F_{u(3\mathrm{dB})} - F_{L(3\mathrm{dB})})}{f_{\mathrm{CENTER}}} \times 100$$

where $\%\mathrm{BW}_{3\mathrm{dB}}$ = percentage of the bandwidth at the 3 dB points
$F_{u(3\mathrm{dB})}$ = frequency of the upper 3-dB point, Hz
$F_{L(3\mathrm{dB})}$ = frequency of the lower 3-dB point, Hz
f_{CENTER} = filter's center frequency, Hz

2. The length of dimension L will be 90 degrees. Calculate the length L, in mils, of these 90 degree microstrip elements as in "Shorted-Stub Bandpass Filter," above.
3. Calculate the width W of the 60-ohm microstrip elements as in "Shorted-stub bandpass filter," above.
4. Calculate the length A of the two outer filter elements from the center of the microstrip input/output transmission lines to the ground via:

$$A = CF\left[L - \left(\frac{F_L}{F_u} L\right)\right]$$

where A = length to center of the input and output 50-ohm transmission line from the center of the ground via.
CF = correction factor; if $\%\mathrm{BW}_{3\mathrm{dB}}$ as calculated above is: 30% BW, CF = 1.30; 20% BW, CF = 1.35; 10% BW, CF = 1.70; 5% BW, CF = 2.0
L = length of the 90 degree stub, as calculated in step 2 above, from just below B section to the center of the ground via, in mils

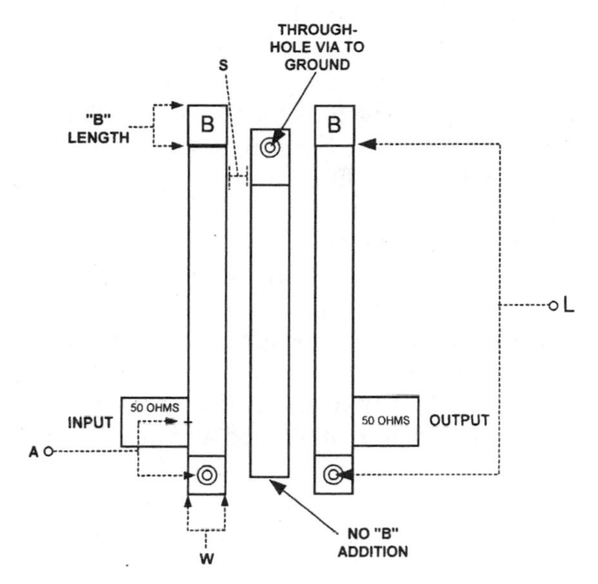

Figure 6.44 Third-order interdigital distributed BP filter design.

F_L = lower 3-dB filter frequency (must be the same as that used for the $\%BW_{3dB}$ calculations above), Hz
F_u = upper 3-dB filter frequency (must be the same as that used for the $\%BW_{3dB}$ calculations above), Hz

5. To calculate the length required of extension B on each of the outer elements (B is the same width W as the elements themselves, or 60 ohms):

$$B = ACF$$

where B = additional length to be added to both *end* stubs, mils
A = length to the center of the input and output 50-ohm transmission line from the center of the ground via as calculated above
CF = correction factor required for various 3-dB filter bandwidth percentages ($\%BW_{3dB}$) : 30% BW, CF = 0.20; 20% BW, CF = 0.14; 10% BW, CF = 0.05; 5% BW, CF = 0.01

6. To find the proper spacing between each grounded stub section:

$$S = ACF$$

where S = spacing between adjacent stubs, mils
A = length to the center of the input and output 50-ohm transmission line from the center of the ground via
CF = correction factor for various bandwidth percentages ($\%BW_{3dB}$) as calculated above: 30% BW, CF = 0.09; 20% BW, CF = 0.2; 10% BW, CF = 0.55; 5% BW, CF = 1.4

7. Ground each stub section directly to the ground plane through a via at the indicated end.
8. Optimize for good S_{21} and S_{11}, as well as for the desired bandwidth.

Low-pass filter (Fig. 6.45). A low-pass filter at microwave frequencies can be designed by using the distributed equivalent in microstrip. Design a lumped filter, and convert these lumped reactive values to distributed equivalent components (see Sec. 1.3.2, "Microstrip as equivalent components"). However, converting from a lumped LPF into an equivalent distributed LPF can be difficult because of the excessively low or high line impedances required to meet a calculated lumped filter's equivalent reactance, while remaining with a line length shorter than 30 degrees for decent filter realization. This can be overcome by designing a hybrid filter: a combination of lumped and distributed parts.

High-pass filter. A pure distributed high-pass filter is not easy to design. This is because of adjacent coupling when a simple equivalent circuit is employed, plus the added complication of having to use the series lumped capacitors that are needed in such a structure.

6.2.4 Distributed filter issues

Element collisions occur when the distributed microstrip elements become too close together in the filter's distributed design, and can actually touch or overlap. A different layout or dielectric constant would then be indicated.

Many distributed filters will have odd and even bandpass returns (*reentrance*) that will reduce the stopband attenuation to very poor levels at various frequencies. Some distributed filters that integrate lumped components, such as the combline microstrip/capacitor filter, will have less of these harmonic passbands within their stopbands. This same effect will also occur with low-pass and stopband filter structures. Reentrance with *half-wave* filter types occurs at every *other* harmonic.

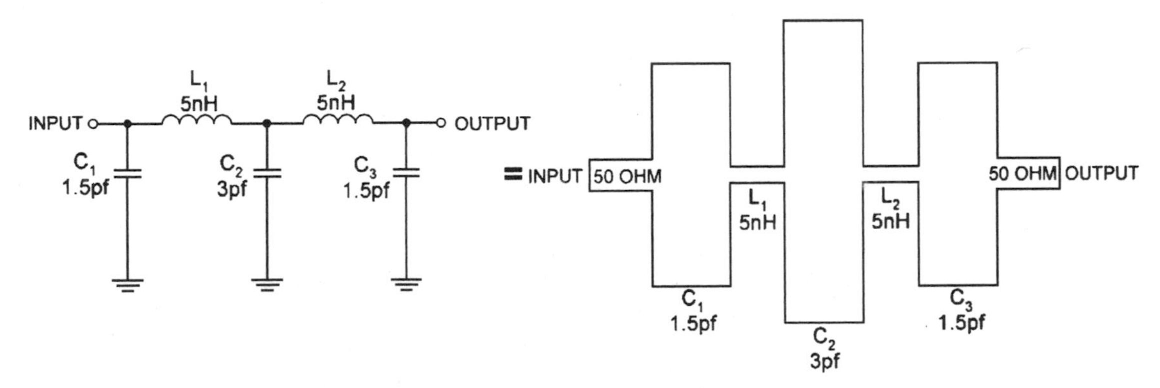

Figure 6.45 Lumped low-pass filter structure and distributed equivalent.

Both the bottom and top ground plane of a PCB should be connected together by through-hole vias in order to form a continuous ground plane, since most microwave circuits will have not only this bottom layer, but also a top ground plane around the circuit or filter structure itself, which covers most of the bare top surface area with copper. The top layer assists in reducing EMI and field coupling and produces better heat dissipation and grounding.

Distributed (and lumped) filters can become detuned if a hand is placed near to the circuit, or a cover or other conductive material is located closer than designed. This is due to *proximity effects,* and must be considered when a distributed design is synthesized or built. All high-end microwave simulator programs are able to take this into account by permitting the engineer to set the distance that the metal shield ("box") will be above and to the side of the circuit or filter.

6.3 Diplexer Filters

6.3.1 Introduction

Diplexers are two or more combined filters in a single package that are adopted to separate two or more different frequencies. This concept can be employed to separate the transmit from the receive frequency in a frequency division duplex (FDD) transceiver, in which application it is sometimes referred to as a *duplexer.* A diplexer also can be placed at the output of a mixer stage (Fig. 6.46), where it functions as an *absorptive* filter. In this mixer application, the first filter of the diplexer has a passband that corresponds to the undesired frequencies, so these pass right through and are then terminated into a 50-ohm load. These same undesired frequencies are blocked from entering the second filter by the filter's own stopband, but its passband passes the desired signals onto the IF sections of the receiver. Thus, the undesired signals through the first filter are absorbed, instead of reflected, because they are properly terminated into the 50-ohm resistive load. This will prevent any undesired frequency products—created by mixer nonlinearities—from being bounced off a standard filter's reflective stopband and returning to the mixer, causing increased IMD levels.

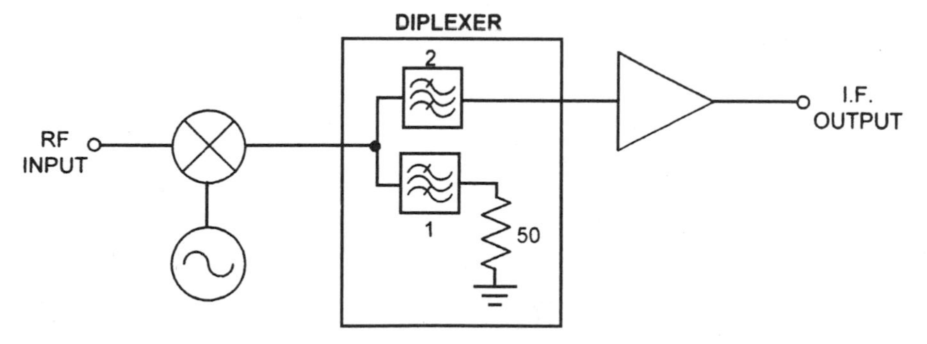

Figure 6.46 A BPF diplexer placed at the output of a conversion stage to decrease IMD.

There are many different combinations of two-filter diplexers: bandpass-bandpass, bandpass–low pass, low pass–high pass, etc., depending on the application.

6.3.2 Diplexer design

A diplexer must be designed as two *different*-frequency filters with nonoverlapping passbands. If their passbands are placed too close together, they will react adversely with each other. This deleterious effect will decrease return loss and increase insertion loss, while losing passband flatness and symmetry. In addition, each filter section of a bandpass diplexer should normally begin with a *series* resonant pole at the input, and not a shunt element, as a shunt element can short out the other filter's frequency of interest and destroy the return loss of the entire diplexer.

Both bandpass filters of the diplexer of Fig. 6.47 have a series resonant input pole instead of a shunt tank, and each filter has a 50-ohm input. (A 100-ohm input impedance is not required because each filter is resonant at a different frequency, f_{r1} and f_{r2}; this allows each filter to pass its frequency of interest, with the other filter looking like a complete open.)

A simple bandpass type of diplexer is shown in Fig. 6.48, and can be used in nondemanding applications:

1. $Q = f_r/\text{BW}$
2. $L_2 = \dfrac{Q \cdot 50}{6.28 f_r}$
3. $C_2 = \dfrac{1}{L_2(6.28\, f_r)^2}$
4. $L_1 = \dfrac{50}{Q(6.28 f_r)}$
5. $C_1 = \dfrac{1}{L_1(6.28\, f_r)^2}$

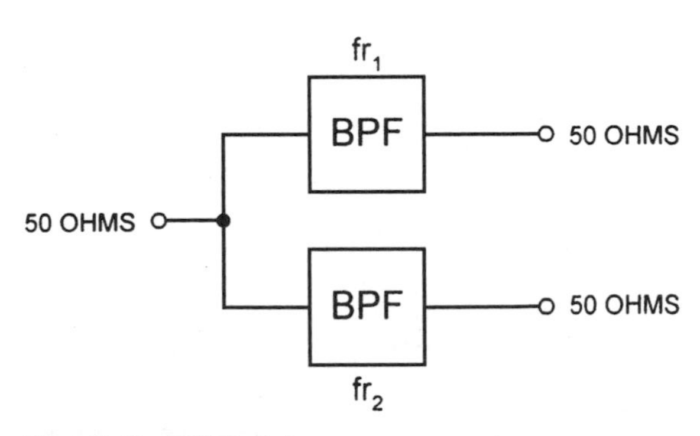

Figure 6.47 A BPF diplexer arrangement.

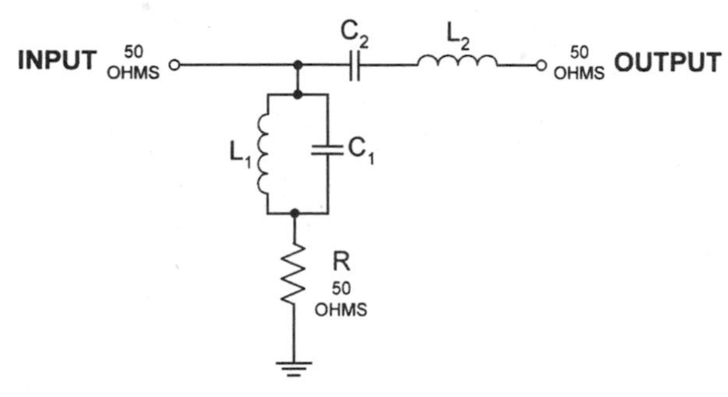

Figure 6.48 A type of BPF diplexer without a series input.

where BW = bandwidth of the desired output signal at 3 dB down, Hz, and f_r = frequency of the *desired* signal, Hz.

Other, much more selective diplexers (with more poles), can be rapidly designed by simply joining two standard 50-ohm filters, with nonoverlapping passbands and series input poles, together. Design the filters as presented in Sec. 6.1.3, "Image parameter design," or utilize any common filter design program.

6.4 Crystal and Saw Filters

6.4.1 Introduction

In certain low- and medium-frequency applications, crystal or ceramic components can be used in place of *LC* filters, especially for narrow bandwidth, tight-skirt filtering (however, special ceramic filters that are designed to function at up to 6 GHz are quite common). This is because of their superior *Q* and frequency stability compared to *LC* filters. Ceramics are much lower in *Q* than crystals, but are also lower in cost and more sturdy. Nonetheless, crystals and ceramics have almost the same general characteristics. Crystals, and to a lesser extent ceramics, basically function as ultrahigh-*Q* series (and parallel) resonant circuits, with an extremely low (or an extremely high) impedance at resonance over a very narrow bandwidth.

All crystals have a series and parallel resonant mode. The parallel mode is slightly higher in frequency than the crystal's series resonance, and is due to the parallel capacitance of the crystal's holder, C_{PLATE}, as shown in Fig. 6.49. At series resonance, *R*, which is a pure resistance of around 25 to 250 ohms [called the *equivalent series resistance* (ESR)], is the only impedance seen, since *L* and *C* will cancel each other. The resonant frequencies will depend not only on the thickness of the crystal, but also on the way it is cut, the crystal substance employed, and the holder capacitance.

However, additional crystal modes besides series and parallel can be used, such as the *overtone,* or *harmonic,* mode: A crystal is capable of being forced to resonate efficiently at odd harmonic intervals of its fundamental frequency, which are at the third, fifth, seventh, and up to the eleventh harmonic. To be

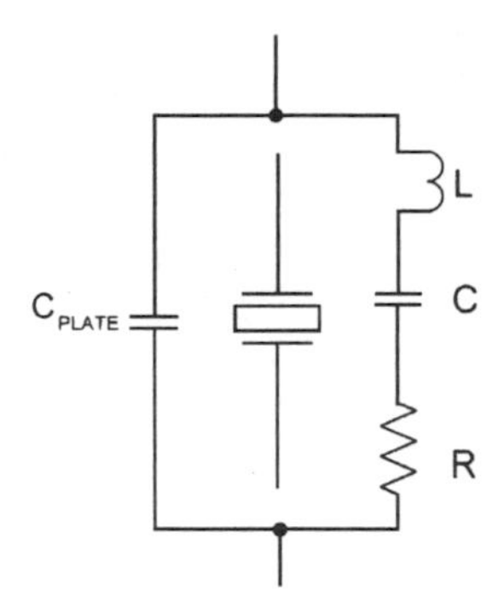

Figure 6.49 The equivalent circuit of a crystal in its holder.

completely correct, the overtone of a crystal is normally at a slightly different frequency than an exact odd-integer multiple (harmonic) of the fundamental, and is due to phase shifting within the crystal's structure. Nonetheless, these overtone modes will, unfortunately, force a crystal or ceramic filter to have reentrance modes at approximately odd-integer multiples of the series resonant frequency, causing decreased attenuation at specific points in the stopband. This can be overcome by special design procedures, specifically by adding an *LC* low-pass filter at the crystal filter's output to attenuate these extra passbands.

Crystals can be found singly, as well as in larger combinations, in RF filters. *Crystal-lattice filters* (Fig. 6.50) contain several crystals within a single package, and are adopted for use as a very sharp bandpass filter. The input and output employ RF transformers with shunt capacitors, while each set of crystals (Y_1 plus Y_2 and Y_3 plus Y_4) is cut to a different frequency (the matched set of Y_1 and Y_2 having a lower resonant frequency than the other matched set of Y_3 and Y_4). This is so we may attain the desired bandwidth and selectivity.

One form of the *ceramic ladder filter* is shown in Fig. 6.51. It contains a stack of ceramic filters, with coupling between the individual resonators being accomplished by capacitors. The coupling can also be done with shunt inductors. All the crystals in this filter are trimmed to the same series resonant frequency. Input and output impedance matching may be achieved with an *LC* network.

Today, crystal filters are normally made to order by various specialized companies, and will comprise resonators, transformers, and trimmer capacitors all within a single, small package.

6.4.2 Crystal and saw filter issues

SAWs are inherently capacitive in nature, and this capacitance must be tuned out. This can be accomplished by placing a series inductor at each SAW input to obtain a proper impedance match to a 50-ohm system, with some SAW matching circuits being a little more complex. When utilizing these matching inductors in a SAW filter circuit, use either *shielded* coils, or place a shield between the input and output of the SAW (called a septum), or place one coil

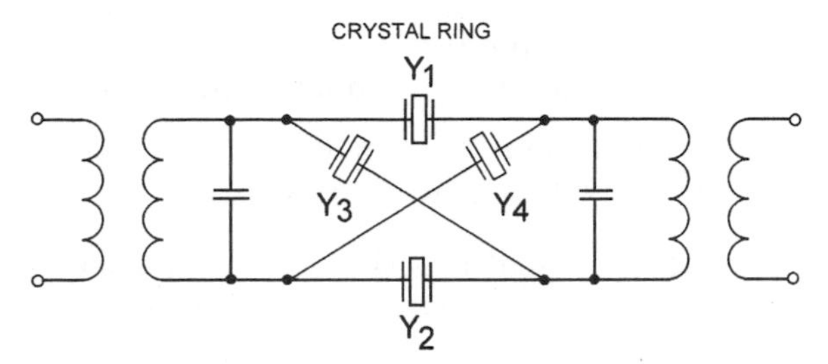

Figure 6.50 A crystal-lattice BPF circuit.

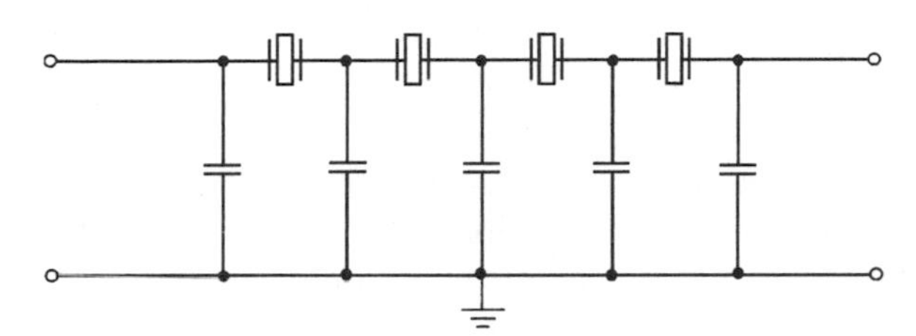

Figure 6.51 A ceramic ladder filter circuit.

on one side of the PC board and the other coil on the other side. These precautions are to prevent (or lessen) unwanted coupling between SAW ports, which would decrease SAW isolation and lower the effectiveness of the SAW filtering action, since nondesired frequencies could slip around the high-insertion-loss SAW filter.

There are a few limitations and problems with SAW filters that hinder their usefulness to some extent:

1. SAWs cannot have any of their filter specifications—such as center frequency, bandwidth, and insertion loss—changed after a production run, since they are manufactured in large quantities by a masking process that is similar to an integrated circuit process.
2. Off-the-shelf SAWs can be readily obtained only at certain common center frequencies and bandwidths. This is because of the expense of custom manufacturing these filters in smaller quantities.
3. SAWs can have very high insertion losses (up to 25 dB), especially at wider bandwidths. This usually must be corrected by an amplifier circuit placed after the SAW filter.
4. Variations in temperature across the SAW can cause increased BER in digitally modulated radios.
5. Unpredictable discharges of energy from the SAW structure can occur with some wideband SAW devices, causing destruction of other components within the circuit. This problem can be alleviated by placing shunt 5-kilohm resistors at the SAW's input and output.

6. SAW filters can have odd or even harmonic spurious responses. These spurious responses will degrade the SAW's stopband attenuation.
7. SAWs can have high time delays through the filter structure of up to 1μs or more, which are problematic in certain circuits.

SAWs do, however, have a few features that are hard to beat by any other filter type: They have as close to a brickwall filter response as can be obtained today, are physically small, do not need to be tuned, and are low cost when purchased in huge quantities.

6.5 Active Filters

6.5.1 Introduction

Active filters typically use an operational amplifier and an *RC* filter network to obtain low-pass, high-pass, and bandpass responses at low frequencies.

Passive *RC* networks themselves can be employed alone as a simple, nonresonant filter for certain audio applications, and can be utilized to attenuate RF, while passing only DC and the low-frequency AC. As an example, the basic RC filter of Fig. 6.52*a* works as a low-pass filter by a voltage divider action: The capacitor *C* will have a low reactance to higher frequencies, while the resistor is chosen to be of such a value as to be of a significantly higher resistance to higher frequencies than *C*. Thus, high-frequency signals are dropped across *R*, while little RF will be dropped across *C*. However, with lower frequencies, the reactance of *C* is higher than the resistance of *R*, so the low frequencies get dropped across *C*, and are then tapped from the output with low attenuation.

Reversing the resistor with the capacitor will create the opposite effect, producing a high-pass filter (Fig. 6.52*b*). Thus, any low frequencies will now be dropped across the high reactance of *C*, but not across the lower resistance of *R*. Higher frequencies will easily pass through the lower reactance of *C*, but be dropped across the higher resistance of *R*. Since the output is across *R*, a high-pass filter has now been formed.

A sharper filter response curve, along with an insertion gain instead of a loss, and filter buffering from the effects of the load, can be obtained if we insert a high-gain amplifier, such as an op-amp, within the above *RC* filter.

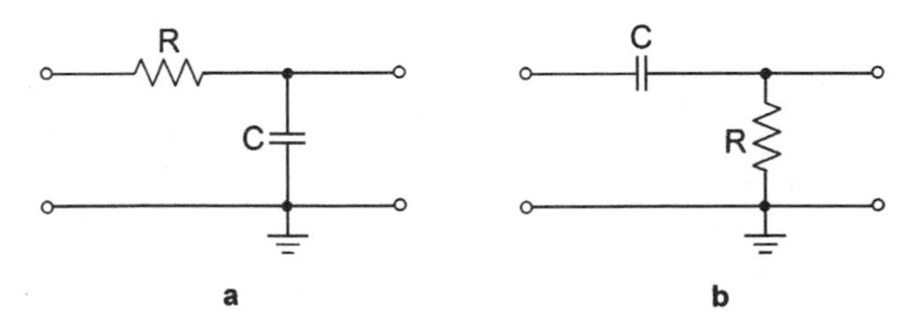

Figure 6.52 (*a*) A basic *RC* low-pass filter; (*b*) a basic *RC* high-pass filter.

Figure 6.53 is a common *RC* active low-pass filter, and functions so: C_2 passes the higher frequencies to ground, while C_1 sends a degenerative feedback to the noninverting input as the frequency increases. This is caused by the capacitor's decreasing capacitive reactance at increasing frequencies. Thus, C_1, C_2, R_1, R_2, and the op-amp efficiently form an active low-pass filter.

An active high-pass filter is shown in Fig. 6.54, with R_2, C_1, and C_2 forming a simple high-pass filter that is used to send the higher frequencies to the op-amp's input with little attenuation. At the lower frequencies, however, the increasing capacitive reactance of C_1 and C_2 attenuates, decreasing their signal at the filter's output.

The active bandpass filter of Fig. 6.55 employs a feedback network that readily passes all frequencies back to its input that are not within the filter's passband and, since this feedback is degenerative, all but a narrow passband of desired frequencies will be attenuated.

6.5.2 Active filter design

All of the active filters below have a Bessel response, and are meant to be driven from a low-impedance source that is significantly below R_1 in value.

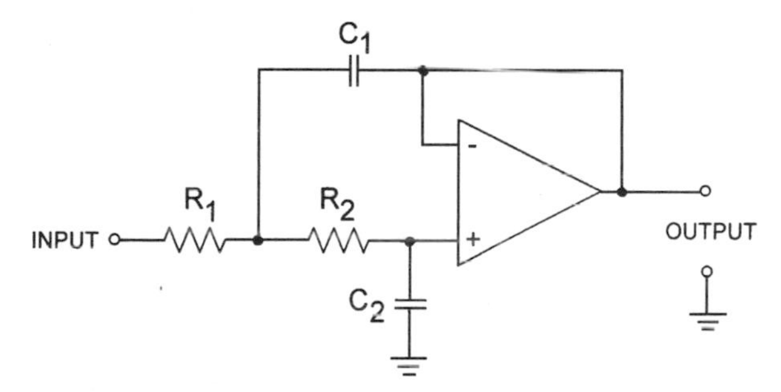

Figure 6.53 An active low-pass filter.

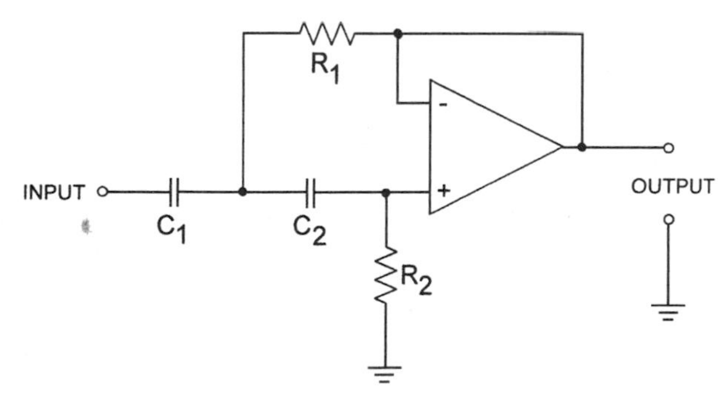

Figure 6.54 An active high-pass filter.

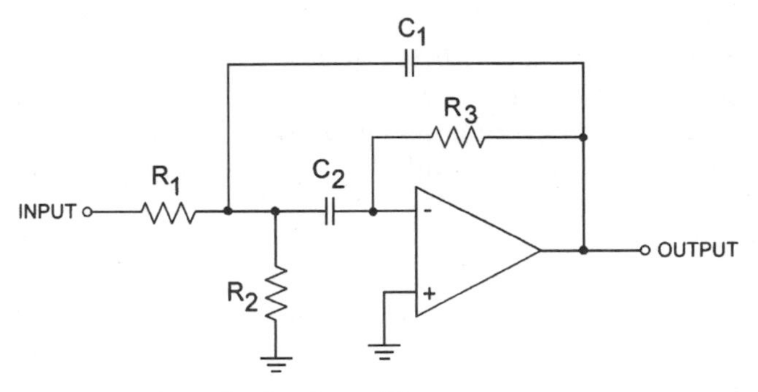

Figure 6.55 An active bandpass filter.

An active audio bandpass filter (Fig. 6.55).

1. Select the filter's desired voltage gain (A_v).
2. Select the filter's desired center frequency (f_c) and bandwidth (BW).
3. Calculate the required Q of the filter by $Q = f_c/\text{BW}$. (*Note:* $2Q^2$ must be greater than A_v, or the filter design will not function as expected.)
4. $C_1 = C_2 = 0.1\ \mu\text{f}$
5. $R_1 = \dfrac{Q}{A_v \cdot 6.28 \cdot f_c \cdot C_1}$
6. $R_2 = \dfrac{Q}{(2Q^2 - A_v) \cdot 6.28 \cdot f_c \cdot C_1}$
7. $R_3 = \dfrac{Q}{\left(\dfrac{6.28 \cdot f_c \cdot C_1 \cdot C_2}{C_1 + C_2}\right)}$

These active filters can be cascaded for more selectivity. However, the bandwidth will become increasingly more narrow as more active sections are added. The problem can be neutralized by anticipating the reaction, and designing the initial filter section *wider* than we would if employing just a single filter. As a very rough guide, we should double the bandwidth (BW) in the above equations for each section added in cascade. For example, if we require a three-stage bandpass filter with a bandwidth that will pass 500 Hz, then we would design the initial single stage to have a bandwidth of 1500 Hz, and then simply add two more of these duplicate stages. The three-stage cascaded audio filter will then end up with a final bandwidth of approximately 500 Hz.

An active audio low-pass filter (Fig. 6.53)

1. $A_v = 1$
2. $C_1 = C_2 = 0.022\mu\text{F}$

3. $R_1 = R_2 = \dfrac{1}{6.28 f_c \sqrt{C_1 C_2}}$

When cascading low-pass filter sections for added selectivity, we double the cutoff frequency (f_C) in the above equation for each filter we plan to cascade. For instance, if we are cascading two sections, and we desire a 1-kHz 3-dB cutoff frequency, then design each section to have a 3-dB f_C of 2 kHz. The final cascaded output will then have an approximately 1-kHz low-pass cutoff.

An active audio high-pass filter (Fig. 6.54)

1. $A_v = 1$
2. $C_1 = C_2 = 0.022\mu F$
3. $R_1 = R_2 = \dfrac{1}{6.28 f_c \sqrt{C_1 C_2}}$

When cascading high-pass filter sections for added selectivity, we cut in half the cutoff frequency (f_c) in the above equation for each filter we plan to cascade. For instance, if we are cascading two sections, and we desire a 1-kHz 3-dB cutoff frequency, then we design each section to have a 3-dB f_c of 500 Hz. The final cascaded output will then have an approximately 1-kHz high-pass cutoff.

6.6 Tunable Filters

6.6.1 Introduction

A filter that can change frequency with the application of a control voltage across varactor diodes can be an invaluable asset in the design of receivers and transmitters in today's packed spectrum. Certain ultrawideband transmitter frequency conversion stages can also benefit by allowing the selective filtering of excessive local oscillator feedthrough from entering the IF or RF stages.

Single varactors are capable of changing capacitance from 0.63 to 2.67 pF all the way up to 3.8 to 20 pF (and above) by the placement of a 0- and 20-V control voltage. This range of capacitances is perfect for tunable filter designs that are not in excess of the VHF and lower UHF regions.

Alpha Industries manufactures an entire range of low-resistance varactors that are ideal for operation in a high-Q, variable-frequency bandpass filter.

6.6.2 Tunable filter design

A half-octave range 50-ohm tunable bandpass filter (Fig. 6.56)

1. Design a basic *top capacitor-coupled bandpass filter* (Fig. 6.57) with any standard RF filter design program, such as AADE's low-cost *Filter Design* software (available at AADE.com). Select a center frequency for the top

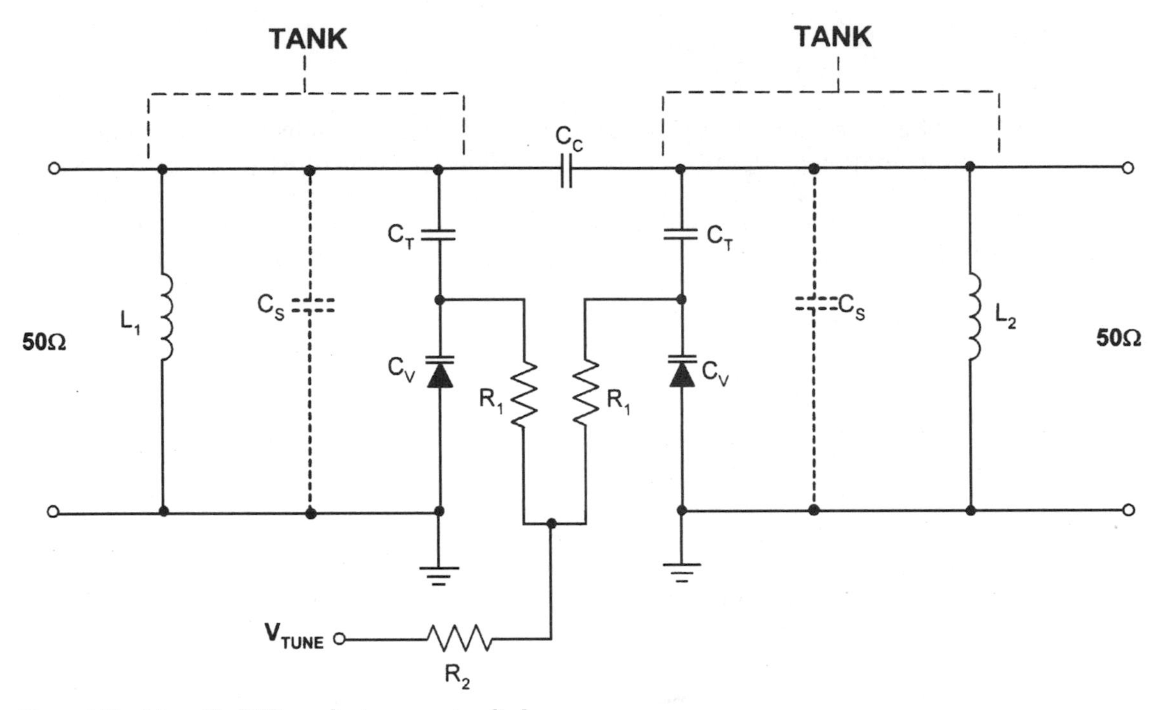

Figure 6.56 A tunable BPF employing varactor diodes.

capacitor-coupled bandpass filter at either the high end, low end, or middle of the tunable range of the desired bandpass frequencies, depending on the initial tuning voltages you plan to supply to the tuning varactors.

2. Now, remove C_1 and C_2 of Fig. 6.57, add the replacement components C_v (a varactor diode) and C_T, along with the bias resistors R_1 and R_2, as shown in Fig. 6.56. R_1 isolates the two varactors from the effects of each other and, with R_2, prevents a direct RF short-circuit to ground through V_{TUNE}; C_T blocks the DC inserted by V_{TUNE} from being short-circuited by L_1 or L_2; C_v supplies the variable tuning capacitance; C_C couples the two tank circuits consisting of L_1 and C_T/C_V, and L_2 and C_T/C_V. Select R_1 of 24 kilohms each and R_2 of 100 kilohms. If V_{TUNE} is to be located at 10 V with a 20-V varactor, choose a varactor diode (C_v) that has a value of C_1 in the center of the varactor's capacitance range, then select a C_T that is approximately 10 times this value. The capacitance of the series combination of C_T and C_V, in series, is:

$$\frac{C_T C_V}{C_T + C_V}$$

This is essentially the value of C_v alone because of the high capacitance of C_T. And since the value of C_1 and C_2 in Fig. 6.57 must equal the series combination of C_T and C_v of Fig. 6.56, then C_T is functioning only as a DC

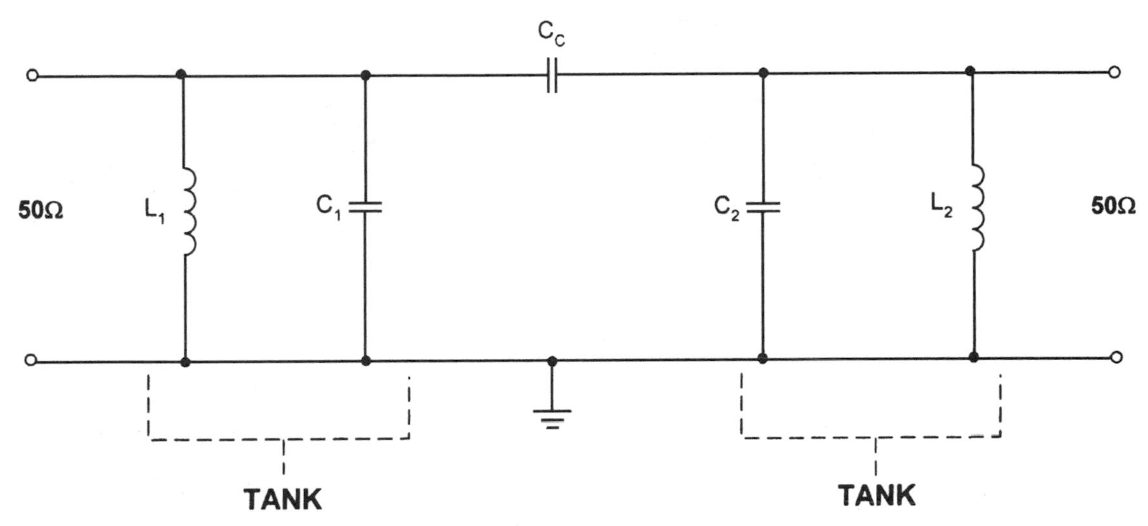

Figure 6.57 A capacitor-coupled BPF used as the basis for the tunable design.

blocking capacitor, while C_v, the varactor, is supplying *all* of the tuning capacitance for the filter's tanks.

3. We can apply a tuning voltage (V_{TUNE}) that will allow the varactor to linearly tune the filter to its maximum and minimum values, or, by supplying V_{TUNE} with discrete voltages, we can filter in discrete steps. Because of parasitic capacitance and inductance at these frequencies, the finished filter may have to be empirically optimized on the bench.

4. Since varactors are extremely limited in the value of their maximum capacitance, then as the frequency of the tunable filter is designed for operation below the VHF region, we would like some way of increasing the capacitance of the C_T and C_V combination (as C_1 and C_2 will naturally increase in value with decreasing frequency). The simple way of raising this overall capacitance level of the series C_T and C_V combination is to add a capacitor (C_S) in shunt with C_T and C_V, which will now increase the capacitance in each leg to:

$$C_S + \frac{C_T C_V}{C_T + C_V}$$

This will provide the larger required capacitance value dictated by the C_1 and C_2 capacitors, as calculated for Fig. 6.57. However, the relative tuning range will obviously decrease as the design frequency is decreased, since C_V will now have far less of an effect on altering the increased fixed capacitance of C_S (Fig. 6.58). In addition, the *minimum* value of capacitance will be set by the value of C_S. Nonetheless, more varactors can be paralleled to increase the tuning range, as required, and more varactor-tuned *LC* tanks can be added to increase the selectivity of this tunable filter.

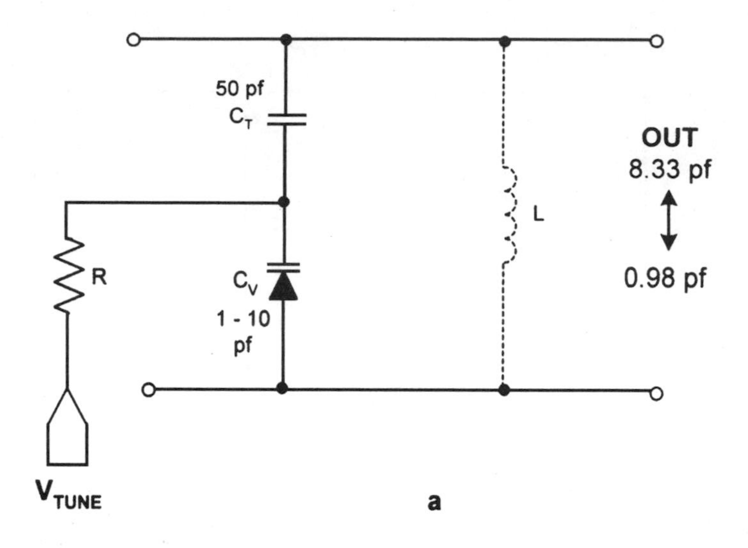

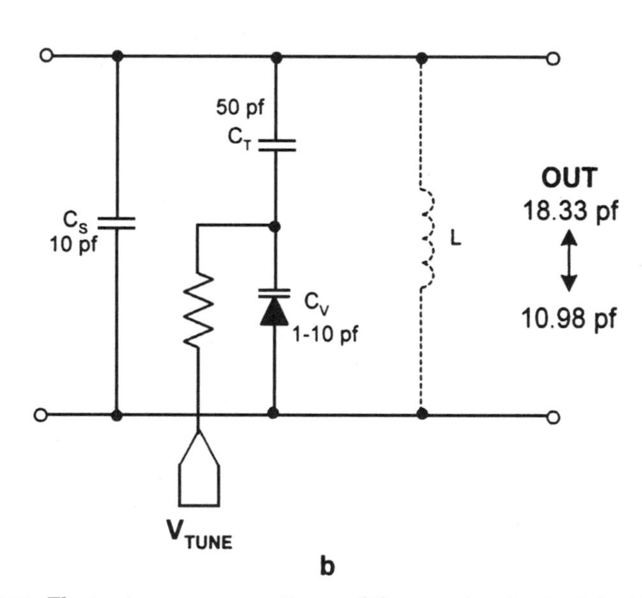

Figure 6.58 The tuning range capacitance of the varactor circuits (*a*) without C_s and (*b*) with C_s.

Over its entire tuning range, this type of filter cannot be expected to maintain its bandwidth, insertion loss, or return loss, but it should be quite acceptable for most variable-filtering applications.

A top-coupled inductor type tunable filter can likewise be built as a base for this tunable filter if upper frequency attenuation is more important than the superior, low-frequency attenuation of a top capacitor-coupled filter.

Always select high-Q inductors for any RF filter, but especially for these tunable varieties, to decrease any large variations in insertion loss over frequency. Since the filter as designed with software for the initial basic coupled capacitive filter of Fig. 6.57 will permit the choice of the tank inductor's value, select as low an inductance as the program allows. This is in order to decrease the filter's size and cost, and to increase the Q and the performance of the final design.

Chapter

7

Mixer Design

Mixers are 3-port active or passive devices. They are designed to yield both a sum and a difference frequency at a single output port when two distinct input frequencies are inserted into the other two ports. This process is called *frequency conversion* (or *heterodyning*), and is found in most communications gear. It is used so that we may increase or decrease a signal's frequency. The two signals inserted into the two input ports will normally be a continuous wave, produced within the radio by a local oscillator, and the incoming (for a receiver) or outgoing (for a transmitter) signal. If we want to produce an output frequency that is lower than the input signal frequency, then it is called *down-conversion*; if we want to produce an output signal that is at a higher frequency than the input signal, it is referred to as *up-conversion*. Indeed, most AM, SSB, and digital transmitters require mixers to convert up to a higher frequency for transmission into space, while superheterodyne receivers require a mixer to convert a received signal to a much lower frequency. This lower received frequency is then called the *intermediate frequency* (IF). Receivers must use this lower frequency signal, as it is much easier to efficiently amplify and filter with the IF stages tuned and optimized for a single, low band of frequencies, and the receiver's gain and selectivity are thus increased. The frequency conversion process within the nonlinear mixer produces the intermediate frequency by the heterodyning, or beating, of the input signal with the receiver's own internal LO. This mixer circuit can consist of a diode, BJT, or FET that is overdriven or biased to run within the nonlinear part of its operation. However, the beating of the two input signals yields not only the signal, the local oscillator, and the sum and difference frequencies of these two signals, but also many spurious frequencies at the mixer's output port. Nonetheless, most of these undesired frequencies will be filtered out within the receiver's IF, resulting in the new desired frequency (consisting of the carrier and any sidebands) now at the difference frequency. This new lower frequency will then be amplified and further filtered as it passes through the fixed tuned IF strip.

There are three basic classifications for both active and passive mixers. Briefly:

Unbalanced mixers have an IF output consisting of f_S, f_{LO}, $f_S - f_{LO}$, $f_S + f_{LO}$, and other spurious outputs. They will also exhibit little isolation between each of the mixer's three ports, resulting in undesired signal interactions and feedthrough to another port.

Single-balanced mixers will at least strongly attenuate either the original input signal *or* the LO—but not both—while sending less of the above mixing products on to its output than the unbalanced type.

Double-balanced mixers (DBMs) supply superior IF-RF-LO interport isolation, while outputting only the sum and difference frequencies of the input signal and the local oscillator, and significantly attenuating three-quarters of the possible mixer spurs. This makes the job of filtering and selecting a frequency plan a much easier task.

7.1 Passive Mixers

7.1.1 Introduction

Passive mixers permit a much higher RF input signal level than active mixers before severe distortion products within the output IF becomes unacceptable. These distortion products are in the form of intermodulation distortion (IMD), along with compression distortion. The IMD may fall in-band, or cause other signals to fall in-band, possibly swamping out or creating interference to the baseband signal. This causes additional noise, which will degrade system performance and BER.

Passive mixers also possess a lower noise figure than active mixers, which is very important for any stage within the front end of a low-noise receiver. However, passive mixers will have an insertion loss of around 7 dB, instead of the insertion gain that active mixers produce.

The passive mixer conversion losses are caused by the mixing diode's internal resistance, port impedance mismatches, mixer product generation, and the inevitable 3 dB that is wasted in the undesired sum or difference frequency. (This sum or difference frequency is removed by filtering, thus cutting a mixer's final output power in half).

Figure 7.1 is a common double-balanced mixer, which utilizes a *diode ring* to achieve frequency conversion of the input signal. The mixer's diodes are being constantly switched on and off within the ring by the LO, while the RF signal is alternately sent through the diodes, this mixes the two signals in a nonlinear manner, producing the IF output frequency. DBMs can function up to 8 GHz (and beyond) by using *hot-carrier* (*Schottky*) diodes, which possess low noise and high conversion efficiency.

DBMs made of discrete lumped components and placed on the wireless devices' PC board are seldom utilized today. Instead, double-balanced mixers

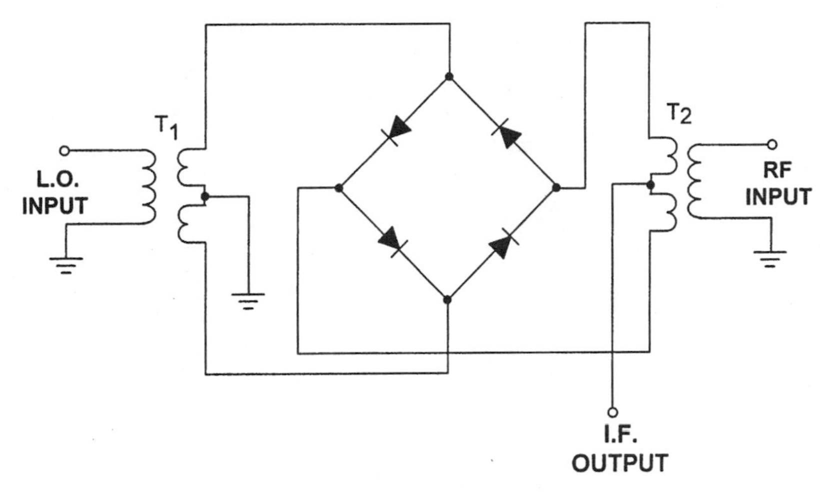

Figure 7.1 A double-balanced mixer stage.

are available in a module, with the diodes and transformers already balanced within a single, low-cost dual in-line package.

Lower-end passive mixers are available that employ either a single diode (Fig. 7.2*a*) or double diodes (Fig. 7.2*b*). These mixers are not double-balanced and, in contrast to an active mixer, must be supplied with a high-amplitude LO signal (but not as high as a DBM's). Nonetheless, they are cheap and require few components.

7.1.2 Types of passive mixers

There are several types of passive mixer designs available, depending on cost and performance levels required. Some of these passive diode mixers have already been introduced above, but will be further investigated in this section.

Figure 7.3 shows a one-diode, *single-ended* mixer. This type of device is found only in low-cost circuits, with the isolation between ports being supplied by bandpass and low-pass filters that are separated in frequency. The mixer circuit can also take advantage of the low level of LO power needed to drive the single-diode mixing element compared to the higher drive levels required of a DBM. The single-ended mixer shown, however, has a rather narrow bandwidth, poor port-to-port isolation, a low intercept point, and inferior intermodulation distortion suppression. If we would like to increase the specifications and overall quality of this device, we will need to increase the number of diodes. This will allow a higher LO drive input level, which automatically forces an increase in the mixer's 1-dB compression point because the P1dB is always about 10 dB below the LO for all diode mixers. Thus, the higher the LO drive that can be inserted into a mixer, the higher the P1dB possible. As we are then demanding a more powerful LO, this will unfortunately cost more and radiate a higher level of EMI.

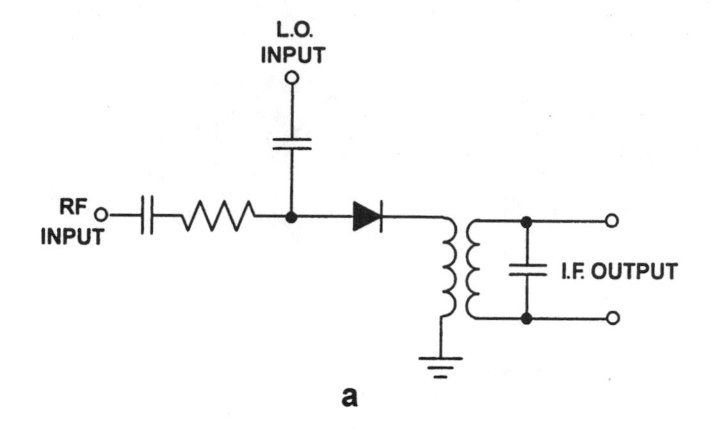

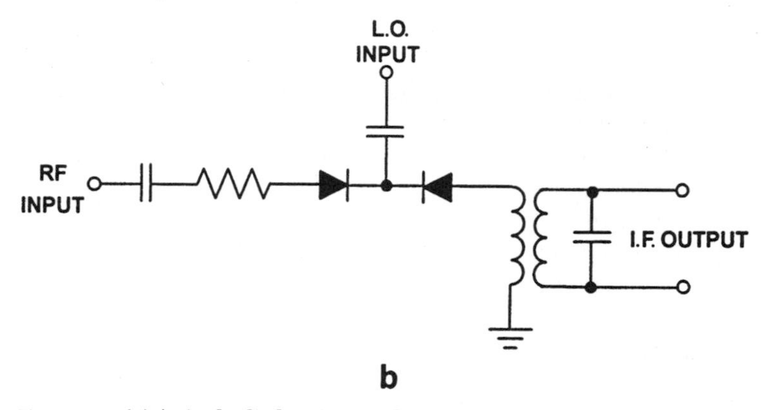

Figure 7.2 (*a*) A single-diode mixer and a (*b*) double-diode mixer stage.

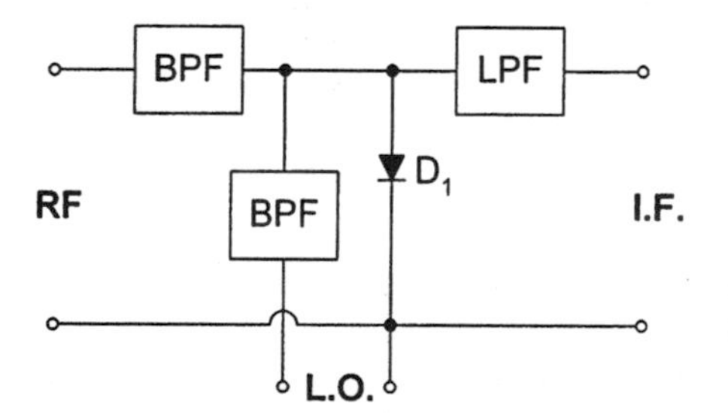

Figure 7.3 A basic single-diode mixer circuit with filtering.

Single-balanced mixers, as shown in Fig. 7.4, are composed of two balanced diodes, a balun (which converts the unbalanced LO output to a balanced mixer input, matches the diodes to the port's impedance, helps in port-to-port isolation, and balances the diodes), and two filters (one each at the RF and IF ports for improved isolation). These mixer types will balance out (cancel) and filter the LO power, preventing excessive LO feedthrough at the RF and IF

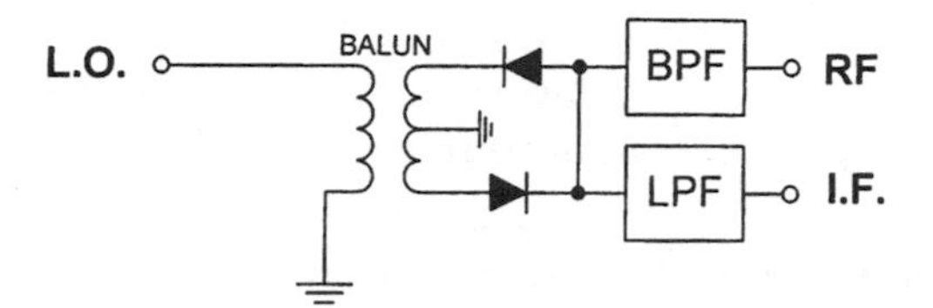

Figure 7.4 A single-balanced two-diode mixer.

ports. In fact, single-*balanced* mixers are superior to single-*ended* mixers in LO-to-IF and LO-to-RF isolation, as well as in their wider bandwidth operation. Furthermore, intermodulation distortion suppression is increased over the single-ended type because any IMDs made up of even harmonics will be suppressed by the balanced-circuit action, and since twice as many diodes are typically used with this circuit—along with higher LO power—the same RF amplitude levels inserted into the single-balanced mixer's input will create less IMD to be generated in the first place. Negative attributes (compared to a single-ended mixer) would be that the LO power must be higher, which necessitates a more expensive and power-hungry oscillator, and the parts count is increased, as a perfectly balanced balun and one more (matched) diode must be used, further increasing costs.

Single-balanced mixers' name comes from their single-balanced balun, while double-balanced mixers (Fig. 7.1) are so named because they employ two baluns. These double-balanced mixers will output only strong IMD products that are constructed of both odd RF and odd LO harmonics. This decreases the DBM's total output of mixer products to a quarter of the amount generated within any simple mixer. However, mixer products are suppressed to varying levels depending on the quality of the diode match and the accuracy of the balun balance. So, while a DBM requires twice the LO power as a single-balanced mixer, as well as double the number of balanced diodes and baluns, a DBM will have much better IMD suppression, a wider bandwidth, and a higher intercept point.

Triple-balanced mixers (*TBMs*), also known as *double double-balanced mixers* (DDMs), have baluns at all three ports, with two complete diode rings. They have increased intercept points for decreased mixer product generation and two-tone intermodulation distortion levels, as well as better port-to-port isolation and a wider possible IF bandwidth. However, TBMs need higher LO power and another matched diode ring and balanced balun above that demanded of a DBM. The price will also be higher.

7.1.3 Passive mixer design

As mentioned, DBM design and construction is best left to the mixer manufacturing companies. When utilizing DBMs in your wireless design, purchase a completed device from one of these companies, as it will be a cheaper solution than attempting to develop one from scratch. For pointers on selecting the correct passive mixer for your design, see Sec. 7.1.4, "Passive Mixer Issues."

Single-balanced mixers are a different story. The distributed passive single-balanced mixer presented next will be lower in cost to personally design and build than to purchase—depending on, of course, the quantities involved.

Distributed single-balanced microwave 90 degree narrowband hybrid mixer for UHF and above (Fig. 7.5). This mixer structure will have decent dynamic range, and requires approximately +5 dBm LO power, with satisfactory LO-to-IF and LO-to-RF isolation for most applications. It enjoys good IMD performance, with fair cancellation of even harmonic signals. However, the IF must be no higher in frequency than 50 MHz, since the difference between the LO and RF frequencies must be relatively small because of the mixer's resonant distributed design, which has to be able to react to *both* the RF and the LO frequency. For maximum LO rejection, design each microstrip section for the LO's output frequency.

Perform the mixer design by first computing the following microstrip lengths:

$A = (\lambda/4) \times V_P$ (at LO frequency) with 50-ohm microstrip.

$B = (\lambda/4) \times V_P$ (at LO frequency) with 35.5-ohm microstrip.

$C = (\lambda/4) \times V_P$ (at RF) with 50-ohm microstrip (C short-circuits RF to ground. Bends do not affect actual length, but are used for compactness).

$D = (\lambda/4) \times V_P$ (at LO frequency) with 50-ohm microstrip (D short-circuits the LO to ground. Bends do not affect actual length, but are used for compactness).

E = 50-ohm microstrip (the two 50-ohm microstrip traces must be of equal length).

RFC $= (\lambda/4) \times V_P$ (at LO frequency) with 100-ohm microstrip.

For the diodes, select the appropriate Schottky diodes for the frequency of operation and the application.

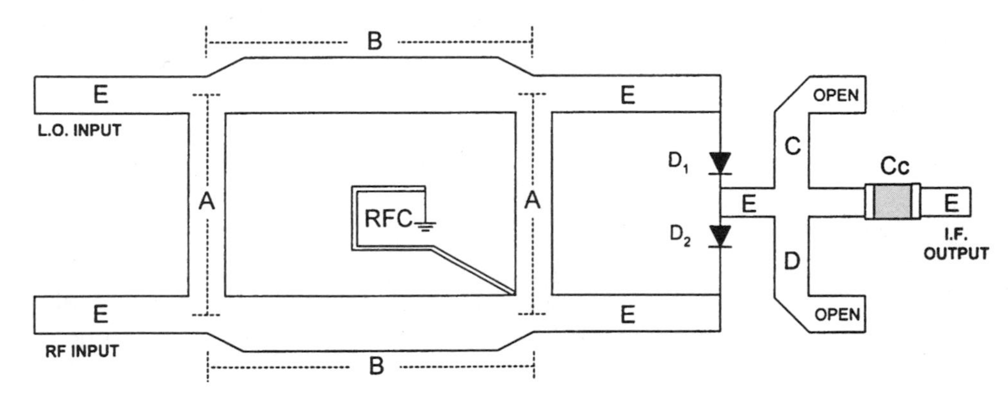

Figure 7.5 A narrowband microwave mixer for applications of UHF and above.

7.1.4 Passive mixer issues

The selection of a proper DBM for a particular receiver or transmitter application will depend on the required P1dB compression point, LO power, device cost, port isolation, and two-tone intermodulation and mixer-generated product suppression.

Undesired frequency product generation, and its suppression, is important in the entire heterodyning process, so we will delve into this subject a little further. Output mixer products (Fig. 7.6) are formed by the mixing, in the nonlinear diode elements of a DBM, of the incoming single-tone RF (and its resultant harmonics) with the single-tone LO (and its resultant harmonics). This creates high-order distortion products that are higher and lower in frequency than the desired product, which is normally either the sum or difference frequency of the LO and RF in a receiver, or the LO and IF in a transmitter.

Two-tone intermodulation products are created when *two tones* (f_1 and f_2) are placed at the RF input port of the receiver's DBM and, mixed with each other and the LO, give birth to high-order in-band spurious responses at the IF output port of the mixer. The higher the possible LO oscillator power, the lower the distortion products. Figure 7.7 demonstrates this point with three different level mixers (levels 7, 17, and 23), with each using its recommended LO input power of either 7, 17, or 23 dBm. In Fig. 7.7*a,* the level 7 mixer's IF output shows high third-, fifth-, and seventh-order two-tone IMD products for a 0 dBm RF input. In Fig. 7.7*b,* the level 17 mixer decreases the IMD products for the same 0 dBm RF input amplitude. In Fig. 7.7*c,* the level 23 mixer shows IMD products much further down than even the level 17 mixer, at approximately 65 dBc.

A further issue in mixer design is demonstrated in Fig. 7.8. Boosting a level 7 DBM's LO drive does not in itself drastically improve the IMD product suppression. This can be accomplished, in any significant way, only by increasing

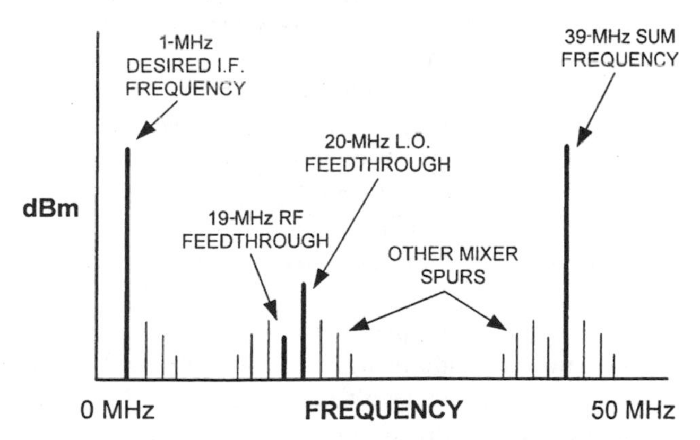

Figure 7.6 Various single-tone mixer spurs generated by the mixing of the RF (19 MHz) and the LO (20 MHz), and their harmonics.

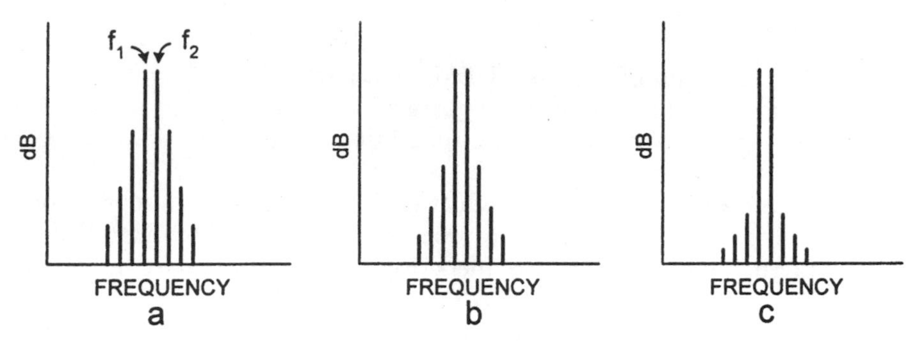

Figure 7.7 Different mixers levels and their two-tone output spectra: (*a*) Level 7; (*b*) Level 17; (*c*) Level 23.

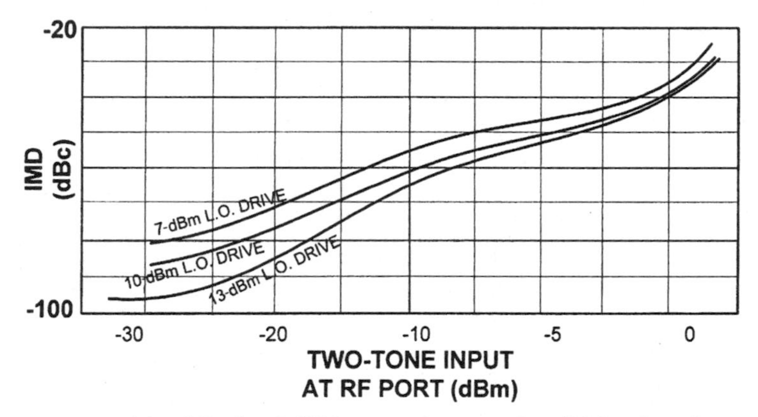

Figure 7.8 A level 7 mixer's IMD generation at various LO levels and two-tone input powers.

the number of diodes in each mixer leg from one to two in *series* (as well as with other techniques), and then increasing the LO drive, with a resultant improvement in IMD suppression.

The intercept point indicates the mixer's capability to suppress intermodulation distortion, typically referring to two-tone, third-order intermodulation products. A high intercept point decreases the undesirable generation of this IMD. In the world of DBMs, the intercept point and the 1-dB compression point do not directly correlate to each other; so choosing a mixer simply for its high 1-dB compression point as a guarantee of increased two-tone suppression could prove a mistake.

As stated, two-tone, third-order products can be reduced by increasing the "level" of the mixer and correspondingly increasing its LO drive and/or by decreasing the power of the input two-tone RF signal. But since we will generally use the recommended LO drive for our chosen level of mixer, which will be selected in consideration of the maximum LO power available within our design, as well as cost constraints, then decreasing the input RF level is normally the easiest and cheapest solution for two-tone, third-order product improvement: by decreasing the input two-tone RF signal by 1 dB, we will decrease the output two-tone, third-order products by 3 dB. However, the inverse is also true: Increasing the RF two-tone input by 1 dB will increase the output two-tone, third-order products by 3 dB. And as these third-order products are the most dangerous spurious signals because they can fall in band at the mixer's IF port, they must be attenuated to the lowest level the system requires.

To assure ourselves of decent intermodulation distortion performance and conversion loss variations, a Level 7 mixer should never be run with an RF input higher than −3 dBm (with the LO drive at the rated power level); a level 10 mixer never above 0 dBm; a level 13 mixer never above +3 dBm; and a level 17 mixer never higher than +7 dBm. In fact, decreasing these RF input levels to 20 dB below the LO drive is commonly done to reduce IMD generation to even lower amplitudes, while, unfortunately, also increasing the relative LO feedthrough.

It is possible to calculate the highest (in amplitude) two-tone, third-order spur level that is down from our desired signal by:

$$\mathrm{TOIM_{SUP}} = 2\,(\mathrm{TOIP} - \mathrm{RF_{IN}})$$

where $\mathrm{TOIM_{SUP}}$ = third-order intermodulation suppression down from the signal of interest, in dBc, at the mixer's output port
TOIP = third-order input intercept point of the mixer, dBm
$\mathrm{RF_{IN}}$ = power, in dBm, of the RF signal at the input to the mixer

Appropriate frequencies for the LO and IF should be selected during the frequency-planning stage that will minimize the number and strength of the mixer products present within the IF bandpass of the DBM. This is most conveniently performed by employing the appropriate software, such as Blattenberger's *RF Workbench,* or The Engineer's Club's *MixerSpur.* Both of these low-cost programs will graphically indicate if there are any dangerous mixer spurs within the IF passband.

As all mixers have not only a nominal, but also a minimum and a maximum LO drive level as recommended by the manufacturer for the particular DBM, we may sometimes desire a minimum drive level for two reasons: sufficient LO power may not be available and/or LO feedthrough must be minimized. However, two-tone IMD suppression, conversion losses, and return loss will all suffer as a result. Slightly increasing the LO drive level above the nominal value will end in a higher NF and a higher LO feedthrough; but will improve the mixer's two-tone IMD performance and mixer product suppression and decrease the conversion losses across the band. As indicated above, running the mixer at the recommended LO drive level is the best compromise for superior mixer performance.

In designing an *upconverting* superheterodyne receiver, the incoming RF signal should be placed at the passive mixer's IF port, while the now higher frequency IF output signal should be sent out of the mixer's RF port. This is also valid for transmitter design, since the mixer will also be performing up-conversion.

When viewing the mixer's *input intercept point* (IIP) specification on a data sheet [IIP is equal to the input RF power level, in dBm, in which the attenuation of intermodulation distortion will be at 0 dBc (0 dB below the carrier)], we sometimes may be required to convert from the input intercept to an *output* intercept point (OIP). This can be accomplished by:

$$\text{OIP} = \text{IIP} - \text{CL}$$

where OIP = mixer output intercept point, dBm
IIP = mixer input intercept point, dBm
CL = mixer conversion loss (usually 6 to 9 dB), dB

An *image reject mixer* in a superheterodyne receiver can be used to phase-cancel the offending image frequency and image noise instead of employing a filter for this purpose. One way this is accomplished is shown in Fig. 7.9. By using MIXER1 and MIXER2 to down-convert both the desired signal *and* the image to baseband by a 0 and +90 degree phase-shifted LO, the baseband *Q* leg of the signal is altered by 90 degrees, while the *I* leg is not phase-shifted at all. These two signals are then inserted into the COMBINER and added, which cancels the image frequency and adds the desired signal, doubling its amplitude. Image suppression is rarely better than 30 dB, however, so any high-amplitude signals present at the image frequency would still cause interference to be created in-channel.

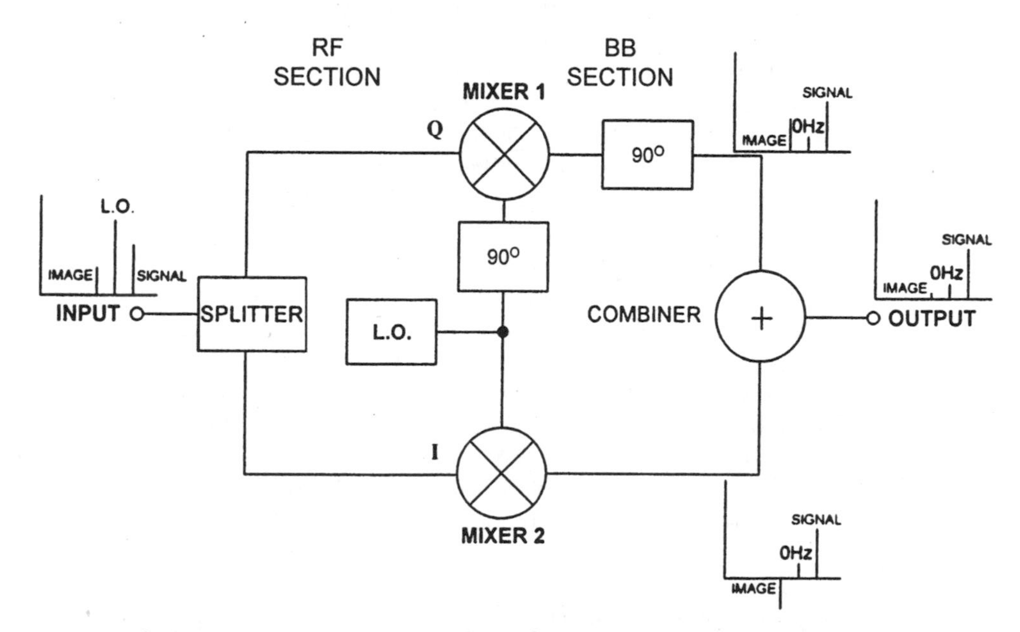

Figure 7.9 An image-reject mixer circuit and waveforms.

Operating a mixer in a *harmonic mode* allows a designer to use a much lower LO frequency than would normally be required. Ordinarily, only the sum or difference frequencies are employed at the IF output of the mixer, but any convenient mixing product may be utilized for this purpose: such as $f_{RF} - 3f_{LO}$, $f_{RF} - 5f_{LO}$, $f_{RF} + 3f_{LO}$, or the $f_{RF} + 5f_{LO}$ products. However, since these frequencies will be at a lower amplitude than the normal $f_{RF} - f_{LO}$ or $f_{RF} + f_{LO}$ products, supplemental amplification is required at the mixer's output, as is nonreflective filtering with a diplexer. The nonreflective filtering is necessary since the undesired signals and products are reflected back into the IF port of the mixer because of the reflective stopbands of a normal output IF filter, causing two-tone IMD performance to suffer (sometimes by as much as 25 dB).

Some common terminology used to specify a DBM:

Conversion compression—A specification that indicates the maximum value of the input RF signal level that will obtain a linear increase in IF output power. For example, level 7 mixers will usually have a conversion compression of +2 dBm.

Conversion loss—The rated signal level difference between the input and the output of the mixer at the rated LO input power. For instance, a level 7 (+7 dBm LO drive) mixer may have a loss in power from input to output of 8 dBm at midband. Decreasing the LO drive to 0 dBm may increase conversion loss by 0.5 dB or more.

Cross modulation—If two signals are present at the input port of the mixer—one modulated and the other CW—this term describes the undesired transfer of the modulation from one signal to the other.

High-side injection—Occurs when the LO frequency is higher than the RF frequency in a conversion stage.

Intercept point—Superior two-tone, third-order product suppression demands a high intercept point. This value is approximately 10 dB higher at the mixer's input than the *conversion compression* rating. *Cross-modulation* distortion and desensitization is also reduced with a high intercept point.

Interport isolation—The rating of the feedthrough between the mixer's LO, RF, and IF ports. This is the value, in dB, that one port's signal is attenuated at another port's input or output. The most important of these isolation specifications is the LO attenuation at the IF and RF ports, since LO feedthrough is a major problem in receiver and transmitter systems design, and the RF to LO isolation is normally of little concern because of the RF's low input levels. Typical LO-to-IF isolation is from 25 to 30 dB.

Low-side injection—Occurs when the LO frequency is lower than the incoming RF frequency in a conversion stage.

Noise figure (NF)—The noise added by the mixer. Equals the difference between the noise at the input of the mixer and the output of the mixer, in dB. When the mixer is driven with the proper LO drive level, the NF will equal the conversion loss.

7.2 Active Mixers

7.2.1 Introduction

Active mixers vary from passive diode mixers in that they can supply a conversion gain instead of a loss; they require far less LO power, are much less sensitive to port terminations, have better ultimate LO-to-IF isolation, and produce less mixer spurs. However, wide adoption of active mixers, such as the *Gilbert cell* type, has been dramatically hindered by a very poor IP3, high NF (around 15 dB), and the need for a DC supply voltage. The first two problems have limited the active mixer's role to the later stages of a receiver where NF matters little, and where the dynamic range of the signal is more under control by the AGC.

Active mixers are available up to RF frequencies of 5 GHz, with IF frequencies of 2 GHz, and are double balanced.

7.2.2 Types of active mixers

As with passive mixers, there are different types of active mixers.

The *single-ended FET* mixer of Fig. 7.10 consists of a JFET, some biasing components, and two tuned tanks. The RF input signal is dropped across the first tuned input tank and sent to the JFET's gate. An LO signal is inserted into the source lead, with the resultant converted signal removed from the JFET's drain and placed across the tuned output tank. This second tank is tuned to the desired IF output frequency, with most of the mixing, RF signal, and LO frequencies being severely attenuated by this circuit. The secondary circuit of the output transformer takes this signal, reduces the high output impedance, and places it into the IF amplifiers.

A *dual-gate MOSFET* mixer of the type shown in Fig. 7.11 employs a MOSFET, some biasing components, and a single tuned tank. The RF signal is sent through the coupling capacitor into the second gate, while the LO is inserted into the first gate, with the sum and difference frequency, along with the mixer products, being sent on to the tuned circuit. Since this output tank is tuned to exactly the desired IF, all other frequencies are attenuated, while the IF itself is dropped across the transformer's primary. The IF is then removed from the transformer's secondary and sent on to the IF strip for further amplification and filtering.

Another low-cost active mixer is the *single-ended transistor* mixer of Fig. 7.12. Both the signal and the LO are inserted into the base and mixed together by the nonlinear Class AB-biased transistor. Obviously, unless a diplexer is placed at the input, the RF and LO have no real isolation between their ports. The original RF signal and the LO frequency, as well as all mixing products, are present at the transistor's collector, but only the desired IF will be of any significant amplitude because of the primary and secondary tuned tank circuits.

A popular Gilbert cell mixer is shown in Fig. 7.13. The RF signal is inserted into the base of Q_1 of the modified emitter-coupled amplifier (composed of Q_1

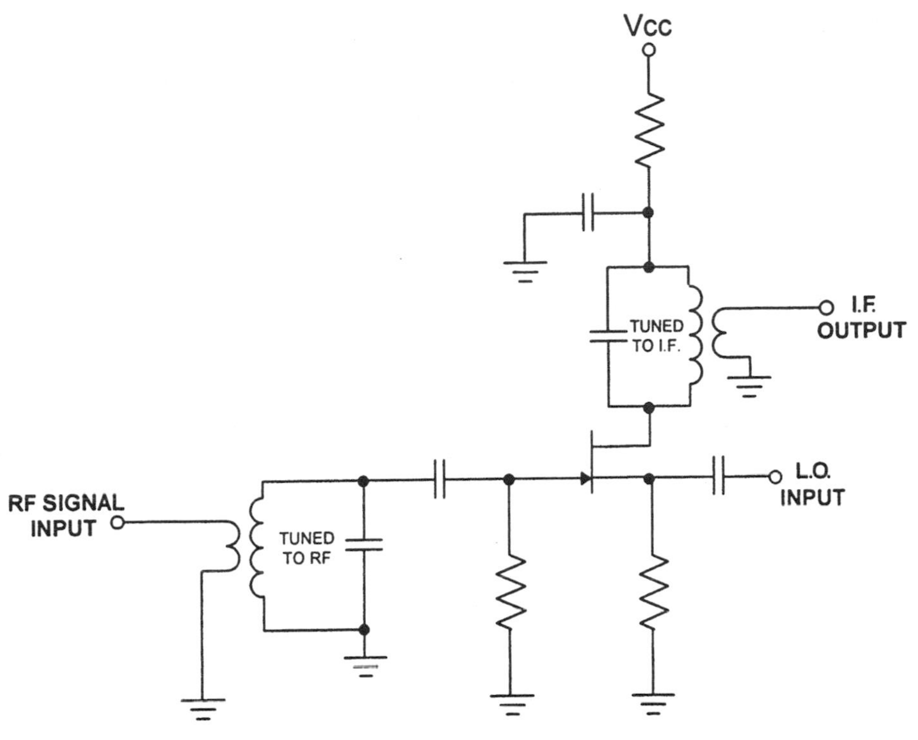

Figure 7.10 A single-ended JFET mixer circuit.

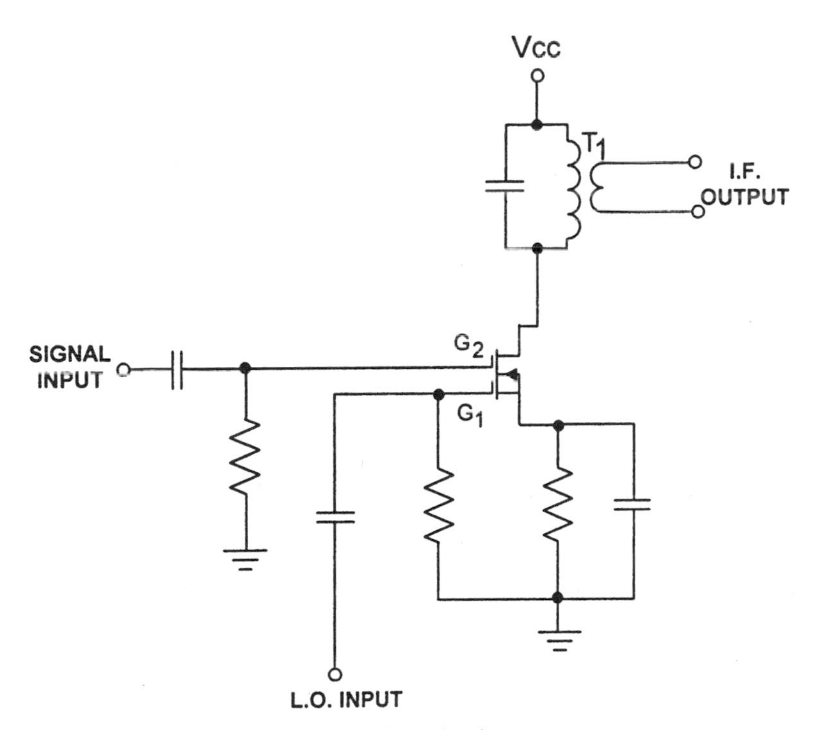

Figure 7.11 A dual-gate MOSFET mixer circuit.

minimize LO drive power, and output a relatively clean IF signal from the mixer's output port:

1. Z_1 and Z_9 are microstrip transmission lines, and will equal 50 ohms.
2. C_1 will block DC but pass the desired RF frequency with less than 1 ohm X_c.
3. Z_2 and Z_3 provide the proper input impedance match at the RF frequency for the GaAs FET device.
4. Z_4 acts as an RFC to the desired RF frequency, while Z_5 functions as a capacitor. They form bias decoupling for the negative V_{CC} supply.
5. R_1 functions as a low-frequency termination to maintain mixer stability. Values between 10 to 50 ohms should suffice.
6. C_2 is used to bypass the IF away from the RF port.
7. $-V_G$ should be adjusted from -5 to -1 V for best mixer operation.
8. A JFET must be selected that can operate at a frequency far above the expected RF input frequency.
9. Z_6 and Z_7 will match the S_{22} of the FET at the LO frequency.
10. Z_8 functions as an RFC to attenuate the LO from entering the bias supply ($+V_D$) or the IF output port, but allows the DC and IF to pass unhindered. C_4 passes the LO to ground, and acts as the RF ground for Z_8.
11. C_5 bypasses the IF to ground to decouple from $+V_D$.
12. $+V_D$ should initially be set to +5 V and then decreased for optimum performance.
13. L_1 and C_6 are chosen to match the IF frequency to the FET's output, while low-pass filtering the IF output for increased isolation.
14. C_7 is a DC block, but passes the IF with little attenuation. Could be series resonant to attenuate other frequencies besides the IF.
15. C_3 is a DC block, and should be chosen to operate at its series resonant frequency at the LO to assist in blocking the undesired IF, while increasing port isolation.

Integrated circuit double-balanced mixer (Fig. 7.17). Active mixers, such as the HP IAM-82028, are useful in non-noise-sensitive applications that require a low LO input power (0 dBm) and load-insensitive performance. The IAM-82028 Gilbert cell–based mixer operates with a flat RF-to-IF conversion gain of 15 dB over a wide RF input range of 0.05 to 5 GHz. This mixer also enjoys an IF output capability of DC to 2 GHz, an output P1dB of +12 dBm (which depends on V_{CC}: +7 V_{CC} yields +2 P1dBm; +12 V_{CC} yields +12 P1dBm), and will function with a V_{CC} of between 7 and 13 V. To design, simply add the coupling/decoupling components as shown, supply the proper grounds and V_{CC}, and the mixer is complete.

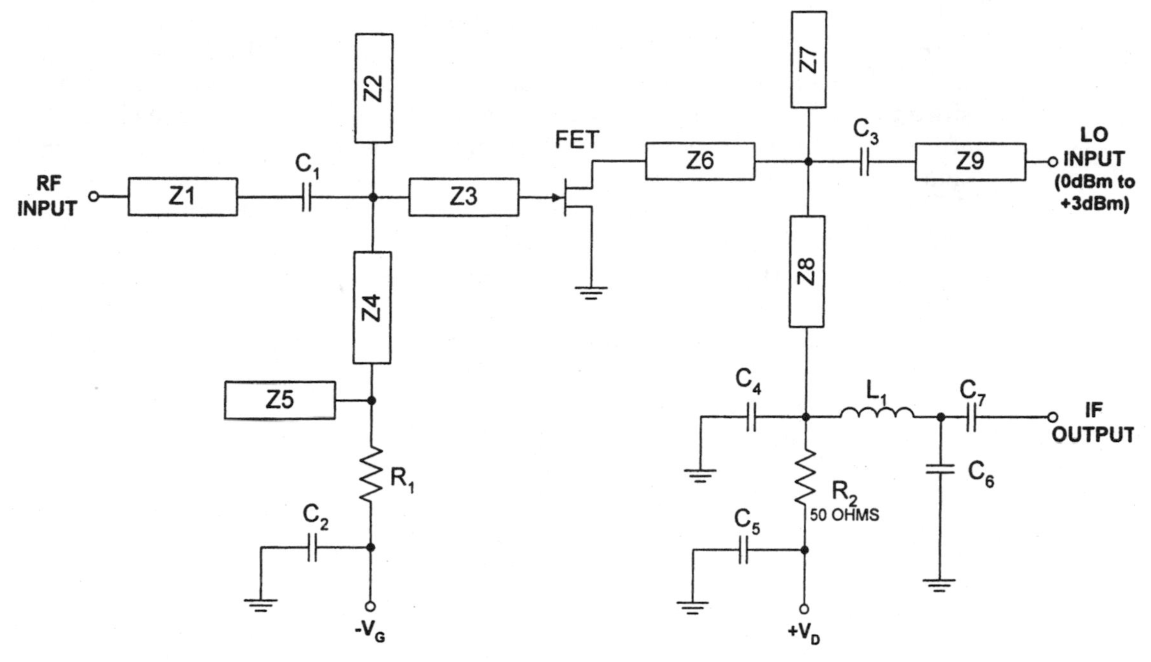

Figure 7.16 An active distributed mixer based on the JFET.

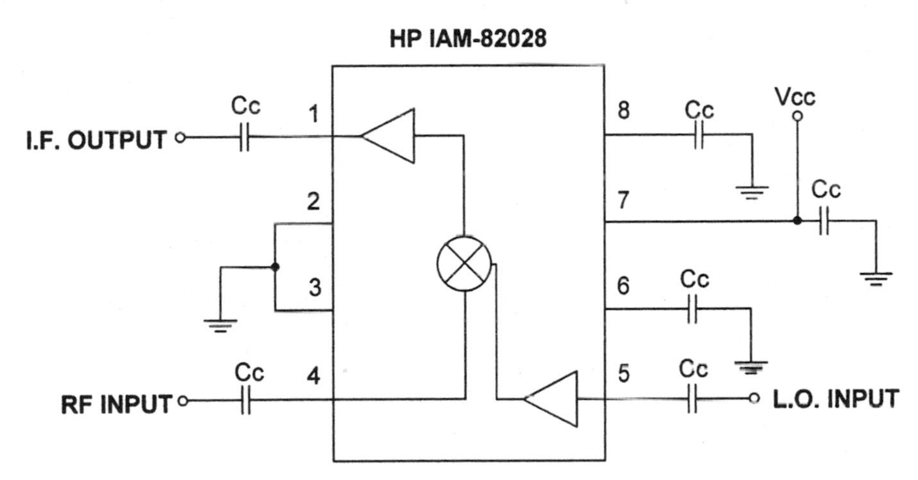

Figure 7.17 A popular IC active mixer.

7.2.4 Active mixer issues

When Gilbert cell IC active mixers (Fig. 7.13) are operated at low frequencies, a square-wave local oscillator must be used to decrease the steadily rising NF of the mixer caused by the longer off times of the mixer quad transistors (Q_3, Q_4, Q_5, Q_6). A square wave will minimize these off times.

When the active mixer is utilized in any up-conversion role, the input and output ports do not change as they would for a passive DBM. The input port will still be for the RF, while the output port will still be for the IF. However, since the IF port is generally capable of less than half (to as low as one-fifth) of the output frequency that the RF input port is capable of, then up-conversion (at least with any gain), will be limited to the rated output frequency of the IF port. Thus, most active IC mixers will be incapable of being used for up-converting a signal beyond 1 or 2 GHz.

Since most active MMIC mixers are much less sensitive to port mismatches than passive diode models, the LO input to the active mixer will normally not require an external buffer amplifier, nor will the IF port need a diplexer. But many IC active mixers will have DC voltages present at all ports, necessitating a series blocking capacitor at the RF, IF, and LO ports.

Chapter

8

Support Circuit Design

Most of today's support circuits are so useful that a modern transmitter or receiver could not satisfactorily function without them. Support circuits include electronic switches, attenuators, frequency multipliers, automatic gain control, the power supply, for example. Other circuits can only be called "bells and whistles." They are not essential for proper systems operation and are only present for added operator convenience; these circuits will not be discussed.

8.1 Frequency Multipliers

8.1.1 Introduction

Because sinusoidal crystal oscillators can rarely be designed to operate reliably at frequencies above 200 MHz, even on a crystal's overtone, frequency multipliers, or sometimes SAW oscillators, must be employed for this purpose. The frequency doublers and triplers used in FM transmitters to increase a signal's frequency, as well as in microwave local oscillator stages, are either a basic tuned-output nonlinear (Class B or C) amplifier or a diode multiplier circuit. These multipliers are not only able to increase an FM or CW signal's frequency, but also any FM deviation present. This is required in an FM transmitter system, as a carrier oscillator's frequency and its deviation may need to be multiplied by 30 or more times. For instance, a modulated RF carrier that began at 6 MHz, with an FM deviation of 150 Hz, could be altered, if fed through a 30 × chain of multipliers, into an output frequency of 180 MHz, with an FM deviation of 4500 Hz.

Because of their nonlinear nature, common Class C amplifiers (Fig. 8.1) work quite well as frequency multipliers, especially when run into their saturation curve. Since any distortion of a continuous wave produces harmonics of the fundamental, an output of such an amplifier/multiplier can be rich in harmonics. The input tank is tuned to the fundamental frequency that must be

frequency (as reactive components are), resistive multipliers will not be influenced by any variation in the input or output matching circuit with frequency changes, thus allowing the resistive multiplier to function over a very wide band of frequencies with high stability. Unfortunately, low efficiency prevents a Schottky diode multiplier from producing a high order of harmonics; doublers and triplers are the most common, with the maximum output power possible from this type of multiplier calculated by $P_{OUT} = P_{IN}/N^2$, with N equaling the harmonic and P_{IN} equaling the input power (in watts) to the multiplier. However, these multipliers can be used at very high frequencies of up to 100 GHz. And while an ideal Schottky diode would produce only odd harmonics, a real-life Schottky will generate both odd and even harmonics because of its small internal offset voltage.

If drive levels of over +10 dBm are expected into a Schottky diode shunt multiplier, such as the one in Fig. 8.4, then an adjustable DC bias circuit should be utilized for maximum multiplier efficiency (either a DC power supply or a simple resistor bias section can be added as shown).

SRD circuits can be replaced by transistor multipliers for frequencies between 500 MHz and 4 GHz (transistors are superior to SRDs below 2 GHz because of their lower cost, ease of tuning, better power output, and simpler design). Gunn and Impatt diodes can be used between 4 GHz and 16 GHz (Gunn multipliers are superior to an SRD in certain low-power applications because of their lower noise, lower cost, and ease of design implementation. Impatt diode multipliers are cheaper and easier to design than the SRD when multiplier frequencies must reach 8 GHz and over, but SRD multipliers will have a wider bandwidth and a lower noise generation). GaAs varactors can be utilized at up to 100 GHz (varactor diodes are lower in cost than SRDs and can reach higher frequencies, but cannot produce as many harmonic orders). PIN switching diodes can also be adopted in multiplier cir-

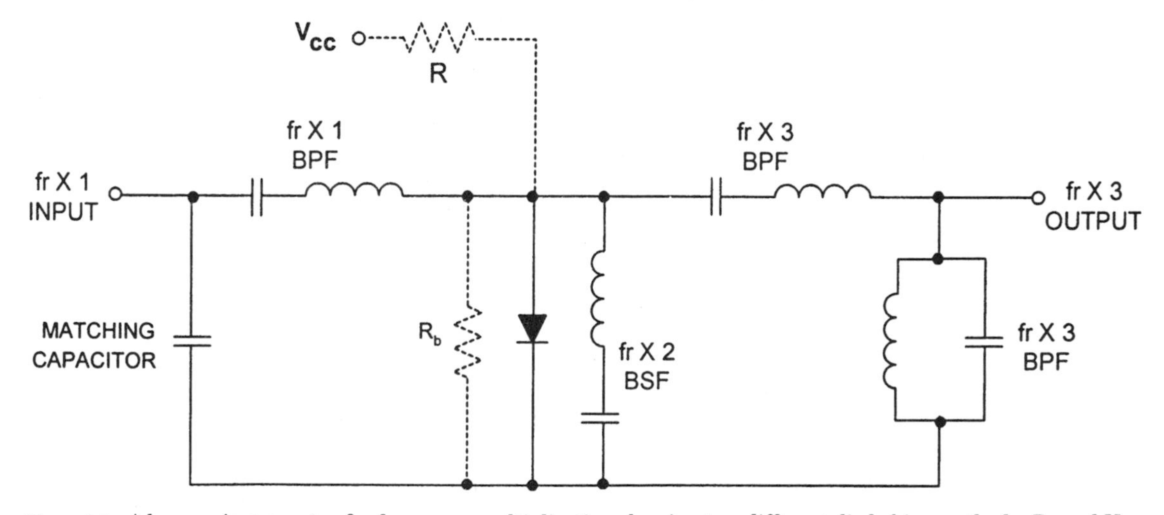

Figure 8.4 A harmonic generator for frequency multiplication showing two different diode bias methods, R_b and V_{cc}.

cuits instead of SRDs (PIN multipliers have similar characteristics to SRDs, but are much cheaper).

While multiplication is a viable and common method of increasing the frequency out of any oscillator, it should also be noted that phase noise will degrade a signal by a minimum of 6 dB per frequency doubling. Even PLL outputs can be multiplied to obtain almost any desired frequency, but at the expense of higher phase noise, as well as decreased frequency resolution. In fact, the phase noise of the frequency source feeding the input of the multiplier degrades by at least 20 log N at the multiplier's output, where N is the amount of multiplication.

8.1.2 Frequency multiplier design

As discussed above, there are numerous multiplier circuit topologies that are perfect for the particular frequency, cost, and multiplication desired. A few simple-to-design multiplier circuits are presented below (see also Sec. 10, "Wireless Design Software," for the program *Multfrq,* which can automatically design passive multipliers, and is available by free download from the Web).

MMIC multipliers. MMIC amplifiers can be adapted as odd/even harmonic frequency multipliers by overdriving the MMIC's input into saturation. To minimize the output MMIC's required drive level, it should have a low P1dB, and the harmonic of choice should not be above the 3-dB bandwidth of the device. Indeed, driving the MMIC into saturation—but not above its rated safe input level—while also increasing the DC bias current will materially assist in optimal harmonic generation. The DC bias current can be increased either by raising the MMIC's V_{CC} supply voltage or by decreasing its bias resistor value, R_{BIAS}. This will place a comb of harmonics at the MMIC's output, which may be picked off by a filter tuned to the desired harmonic. Proper spacing of the filter from the MMIC's output will maximize the exact harmonic of interest, with the spacing determined empirically.

By combining all of the above MMIC multiplication design techniques, a multiplication factor of up to 10 times the input frequency can be realized, with a loss of only 30 dB from the fundamental input signal.

Snap frequency multipliers (Fig. 8.5). Step-recovery, or "snap," diodes, function by switching between two impedance conditions: low and high. This change of state may occur in only 200 ps or less, discharging a narrow pulse that is quite rich in harmonics.

Snap-recovery diode (SRD) frequency multipliers are operated only in very high frequency circuits because of their high cost, the necessity to hand-tune each SRD circuit, the +17 dBm input power required, and other difficult implementation issues as compared to PLLs or active harmonic multiplication.

1. First, choose the correct SNAP:
   a. Lifetime rating on the SRD data sheet must be at least 10 times longer than the period (1/*f*) of the input frequency (given in nanoseconds).

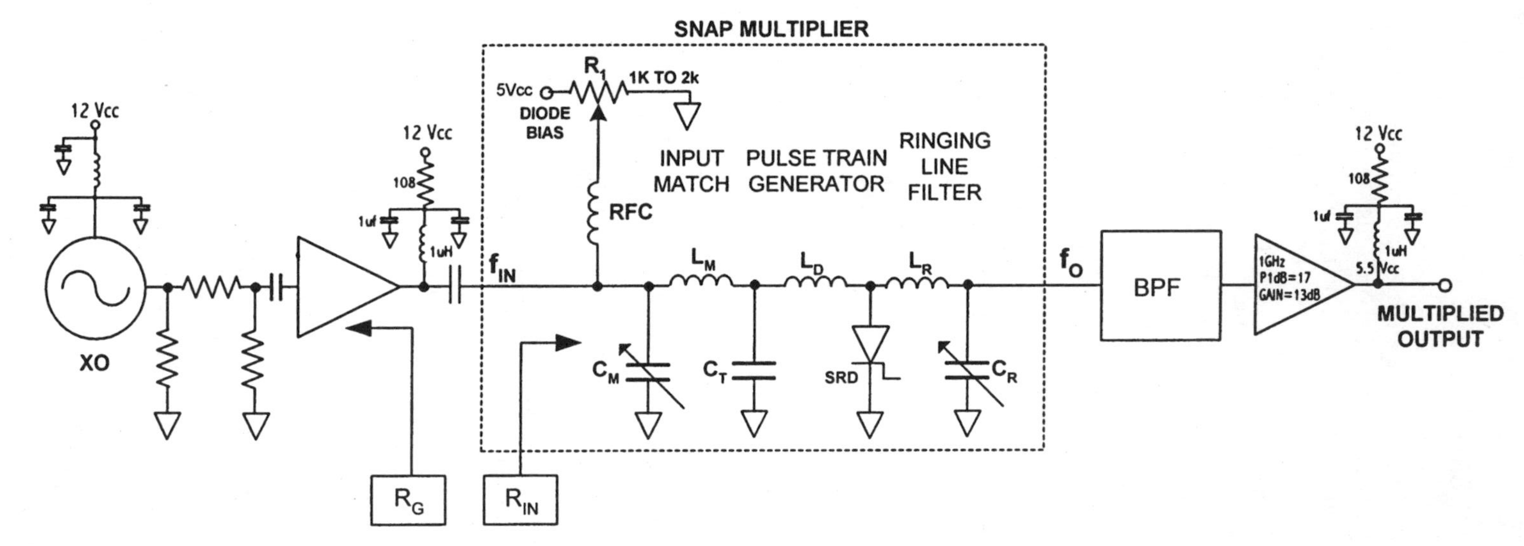

Figure 8.5 A complete snap multiplier with amplification.

b. For C_{vr} (or C_j), use nominal value on data sheet for L_D calculation.
c. Transition time (T_t) must be less than one period ($1/f_0$) of f_0 (the final output frequency) on data sheet (measured in picoseconds).

2. $T_p = {}^1\!/_2 f_0$.
3. $R_{IN} = 6.28\, f_{IN}\, L_D$.
4. $R_G = 50$ ohm.
5. RFC = 600-ohm impedance at f_{IN}, but capable of attenuating f_0 as well.
6. $C_m = \dfrac{1}{6.28\, f_{IN}\, \sqrt{R_G R_{IN}}}$.
7. $L_M = \dfrac{\sqrt{R_G R_{IDN.}}}{6.28\, f_{IN}}$
8. $C_T = \dfrac{C_{vr}}{(2 f_{IN} T_p)^2}$

   C_{vr} is provided on the diode's data sheet. It is sometimes given as C_j.

9. $L_D = \left(\dfrac{T_p}{\pi}\right)^2 \left(\dfrac{1}{C_{vr}}\right)$.
10. $L_R = L_D$.
11. $C_R = \dfrac{1}{(2\pi f_0)^2\, L_R}$.
12. $f_0 = f_{IN}\, N$.
13. Tune the snap circuit. Since the diode's output is a comb of frequencies, a spectrum analyzer is utilized to tune the SRD for maximum power output at the desired harmonic. Insert the proper frequency and power (+17 to +23 dBm) at the SRD circuit's input. Adjust C_M (input match), R_1 (diode bias), and C_R (line filter) until the desired output frequency is at maximum power, while confirming that all subharmonics and harmonics are properly attenuated.

Odd-order Schottky diode tripler multiplier (Fig. 8.6). This is one of the easier frequency multipliers to design for operation up to 600 MHz output, as presented by Wenzel. L_1 and C_1 form the input matching network, while blocking harmonics from entering the input of the multiplier, as well as increasing the input voltage to the diode (caused by this high-Q matching circuit transforming the impedance *and* the voltage).

D_1 and D_2 are the nonlinear harmonic-generating devices. L_4 is the DC ground for the diodes, with the DC being created by rectification of the AC of the input by D_1. L_2, C_2, L_3, and C_3 shunt the undesired frequencies to ground, while passing only the frequency of interest and matching the output (as with most multipliers, the diode switching times will be degraded unless these higher frequencies are shunted to ground). Input power should be approximately +10 dBm for a tripled output of around minus 5 dBm.

1. L_1 is close to series resonance with C_1 at $f_r \times 1$ (adjust C_1 for maximum third-harmonic power at the multiplier's output port, as well as for the best f_r return loss at the multiplier's input port).
2. D_1, D_2 are Schottky diodes (for low noise) of the *low-flicker* type.
3. L_4 at 30 MHz = 330 μH; 50 MHz = 100 μH; 75 MHz = 45 μH; 100 MHz = 25 μH; 125 MHz = 15 μH; 150 MHz = 10 μH; 175 MHz = 8 μH; 200 MHz = 6 μH; 250 MHz = 4 μH; 300 MHz = 2.8 μH; 600 MHz = 0.8 μH. (Example: If f_r is 10 MHz, and we require a 30-MHz output frequency, then we will need a tripler, since $f_r \times 3$ is 30 MHz, so we would choose an L_4 of 330 μH).
4. L_2 is at parallel resonance with C_2 at $f_r \times n$ (with n normally equaling 3).
5. L_3 is close to series resonance with C_3 at $f_r \times n$ (tune C_3 for maximum third-harmonic output power and return loss).
6. The resonant frequency for BPF is $f_r \times n$.

8.1.3 Frequency multiplier issues

Frequency multiplication up to high orders can be problematic for the stability of a circuit, as well as for the filtering out of all of the many subharmonics and harmonics (Fig. 8.7) at the multiplier's output. These subharmonics and harmonics of the fundamental may be spaced on either side of the frequency of interest by only a small amount, making it quite difficult to suppress them to high dBc levels.

It has been found that multiplying beyond a tripler with a normal silicon diode may add far more phase noise than the minimum of 20 log N that one would expect. This is due to high noise floors, as well as flicker noise at lower frequencies, created by the nonlinear device (the diode) employed in this

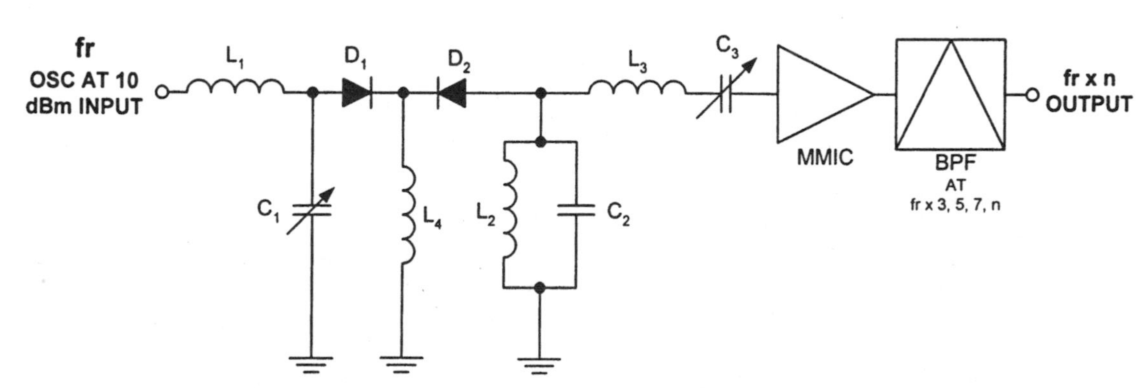

Figure 8.6 An odd-order frequency multiplier.

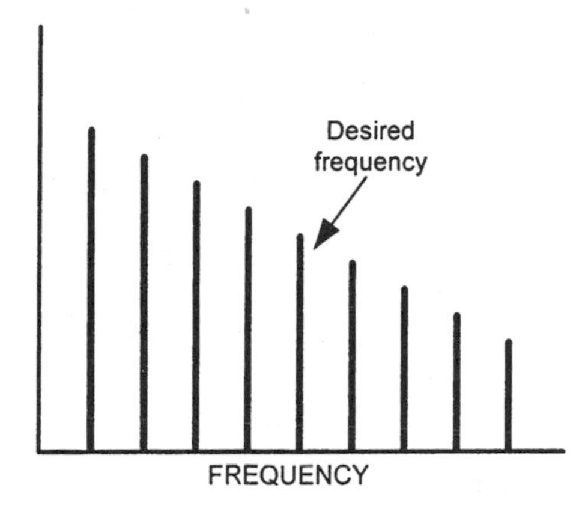

Figure 8.7 Subharmonics and harmonics of a multiplier stage.

action of multiplication. However, Schottky diodes will not have this increased noise problem: Doubler and tripler topologies using low-noise Schottky diodes can usually maintain a good 20 log N phase noise degradation, even up to ×7. Varactor multipliers (Fig. 8.8) can be delicate to tune for top performance, but they too add very low levels of phase and amplitude noise beyond the unavoidable 20 log N.

Diodes must have a certain *carrier lifetime* (τ) to function efficiently as a multiplier. This value must be equal to or greater than $\tau/(1/f_{IN}) = 10$. Longer carrier lifetime factors of up to 30 are more beneficial in that they permit the diode's reverse current to peak before the diode rapidly travels back to its high-impedance condition.

As mentioned, DC biasing a diode multiplier assists it in the frequency multiplication process. But because the diode of a simple frequency multiplier rectifies a certain amount of the RF, a resistor placed in shunt with the multiplier diode will furnish the necessary negative DC bias voltage. However, a small external negative DC bias voltage source can be used across the diode as required.

Transistor (Class C or B) frequency multipliers will have acceptable output noise levels, conversion gain, and bandwidth. Active multipliers can also consume less DC current than many diode multipliers because of the sometimes necessary inclusion of Class A amplifiers in a lossy diode multiplier chain to increase its output power. Additionally, any cascading of multiple varactor multiplier stages opens up a severe danger of instability (oscillations), which is not as much of a risk as with the active multiplier types. Still, a doubler is the safest active multiplier to construct, while cascading them for any further multiplication required. However, there are many low-phase-noise applications that demand the use of diode multipliers over the noisier active type—especially within digital communication radios.

In active multiplier design, it has been recommended by *Maas* and others that the duty cycle of a BJT or FET amplifier be adjusted to optimize the preferred output harmonic. The amplifier is biased close to cutoff to be on for 30 percent of the time for a doubler and 20 percent of the time for a tripler. In

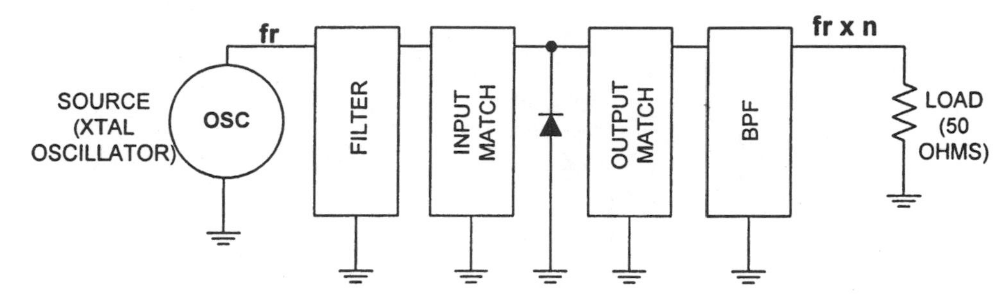

Figure 8.8 Frequency multiplication with a nonlinear element.

other words, the multiplier is acting as a Class C amplifier and rectifying the incoming RF signal that is to be multiplied. This action will output only pulses, creating distortion of the input sine wave, and thus harmonics. Obviously the Class C bias point can be varied to optimize the desired harmonic for the proper on period for the highest drain or collector current amplitude.

8.2 RF Switches

8.2.1 Introduction

The choice of which active or passive RF switch topology to adopt in a communications circuit or system will depend on many things, such as the expected input signal strengths, whether the switch needs to be reflective or absorptive, the necessary output intermodulation levels, bandwidths, frequencies, etc.

The basic transistor functions not only as an amplifier, but is also quite proficient at switching DC, AC, and RF and is especially important at frequencies below 10 MHz, where PIN diode switches have problems operating. An elementary circuit for switching RF currents is shown in Fig. 8.9; The NPN BJT will not conduct, and thus will not permit any of RF INPUT to proceed to the transistor's RF OUTPUT when its base is grounded through SW1. When the mechanical switch is flipped over to engage the positive supply, however, a base-current will start to flow through the base resistor, R1. When this occurs, the BJT will rapidly conduct, causing collector current to flow, and decrease the emitter-to-collector resistance. This allows the RF signal to pass on to RF OUTPUT. The RF switch can be utilized to remotely switch RF currents when long wires or traces would be problematic at high frequencies due to excessive losses, EMI, and other compelling reasons.

If required for the switching of amplifier supply voltages, DC may also be switched by adopting the BJT of Fig. 8.10: When SW1 is closed, a positive bias voltage is sent to the BJT's base, which switches on the transistor, permitting DC current to flow through the resistive load R_L. Turning the mechanical switch to off will stop base current flow, causing an almost zero collector current through R_L. However, instead of directly inserting a positive voltage on the BJT's base with the mechanical switch of SW1, a positive voltage can be

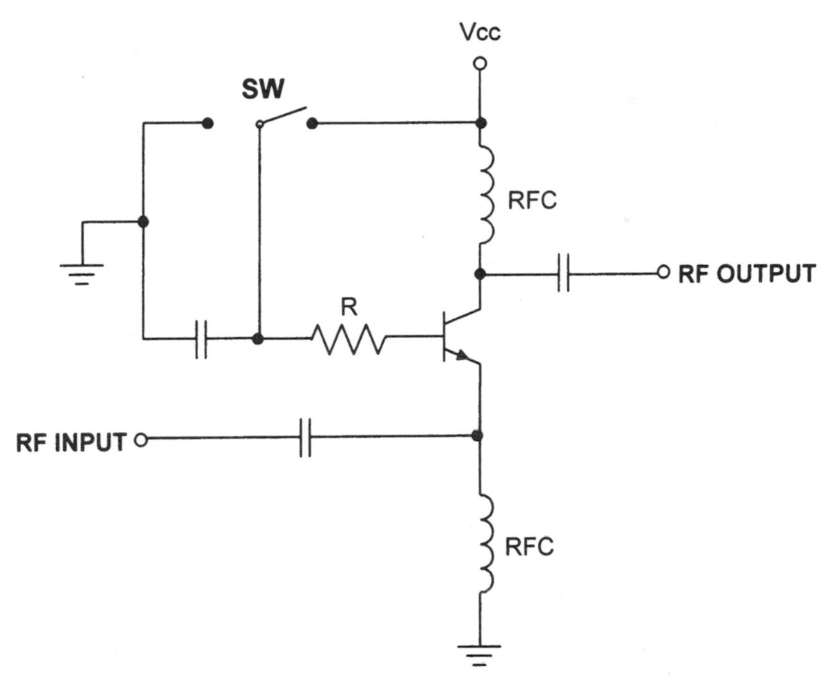

Figure 8.9 An RF switch using a transistor.

sent to the base in reply to some other system condition. An example would be during excessive reverse power as detected at the transmitter's output (which should shut down the system or supply a panel warning) or when a transceiver changes from receive to transmit (in which case the transmitter section, and not the receiver, requires DC power). The transistor switch is perfect for these applications, as is the passive *diode switch* below.

Diodes serve a vital function in switching applications in wireless communications and are used to switch in, or out, various values of crystals, filters, tuned tanks, subsystems, entire sections, etc., into or across the signal path. A simple diode switch is shown in Fig. 8.11. C_1 and C_2 block the DC bias, but easily pass the AC signal. When the mechanical switch (SW) is in the off position, a negative DC bias is placed across the diode, reverse-biasing D_1 and stopping the signal from passing on to the output. In order to forward-bias the diode and allow the RF signal to pass, the mechanical switch is turned on, placing a positive voltage that forward-biases D_1 into conduction. This permits the signal to cross the now low resistance of the diode to the output. R_1 is a current limiting resistor to ground for the DC bias voltage through the diode, while the RFC prevents the RF from entering the power supply.

8.2.2 RF switch design

The following are some easy-to-implement electronic RF switch circuit designs that will cover the majority of our wireless requirements.

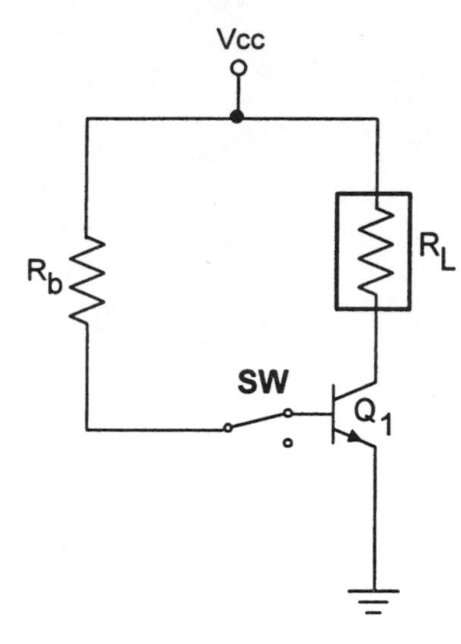

Figure 8.10 A switch for DC using a transistor.

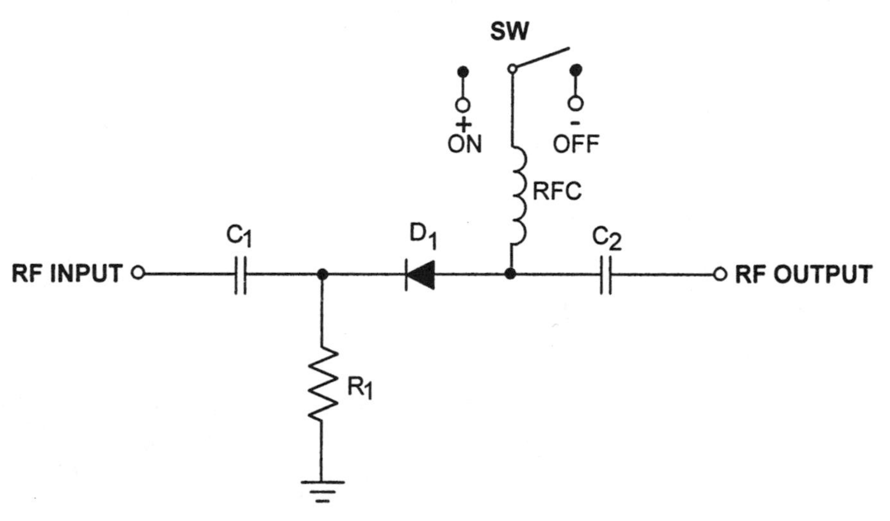

Figure 8.11 An RF switch using a diode.

Small-signal very high isolation PIN SPDT switch (Fig. 8.12). This switch can be designed with the following formulas, but requires both a forward and a reverse-bias voltage to turn ON and OFF.

1. RFC = 500 ohms (X_L) at f_r.
2. $C_B = C_C$ = 1 ohm (X_C) at f_r.
3. $R \approx \dfrac{\text{DC bias} - 0.9}{50 \text{ mA}}$

(The 50-mA value is PIN3's forward current draw, while PIN1 and PIN2 will each be on by 25 mA, for a low value of on resistance. DC bias is in volts.)

4. The design must have a positive voltage to turn on the series PINs and turn off the shunt PIN for an on switch condition; a negative voltage will switch off the series diodes and turn on the shunt diode, thus creating a high-isolation RF switch for the off condition. However, for low-level RF input signals, only a positive on bias is required, while off simply needs 0 V (if there is no negative bias, then high-level input signals can self-bias the PINs to on, which will lower the switch's off isolation drastically).

High isolation series/shunt RF electronic switch (Fig. 8.13). When this switch receives a $V_{CONTROL}$ of +10 V, the series RF_{IN} PIN1 becomes reversed biased (causing a high resistance), while the shunt PIN2 is forward biased (creating a low resistance), shunting any input to ground that may get through the series diode. When −10 V is placed at $V_{CONTROL}$, the series PIN1 is forward biased (giving a low resistance), while the shunt PIN2 is reverse biased (blocking the input signal from shunting to ground), allowing the RF IN to pass through to RF OUT.

Wideband diode RF switch (Fig. 8.14). This switch is capable of operating up to 1 GHz with low distortion and high isolation (up to 40 dB in the off condition).

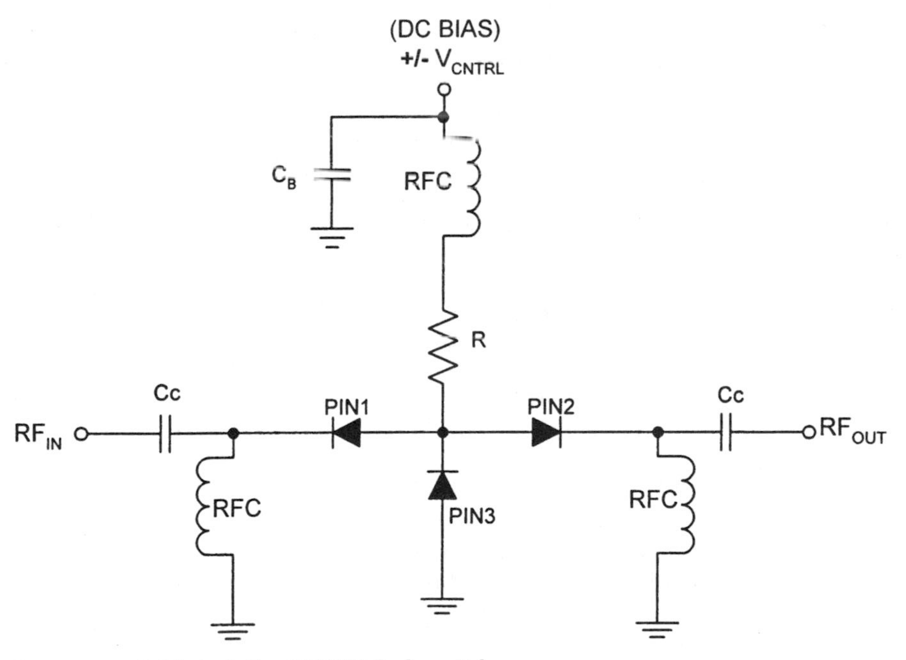

Figure 8.12 A high-isolation RF PIN diode switch.

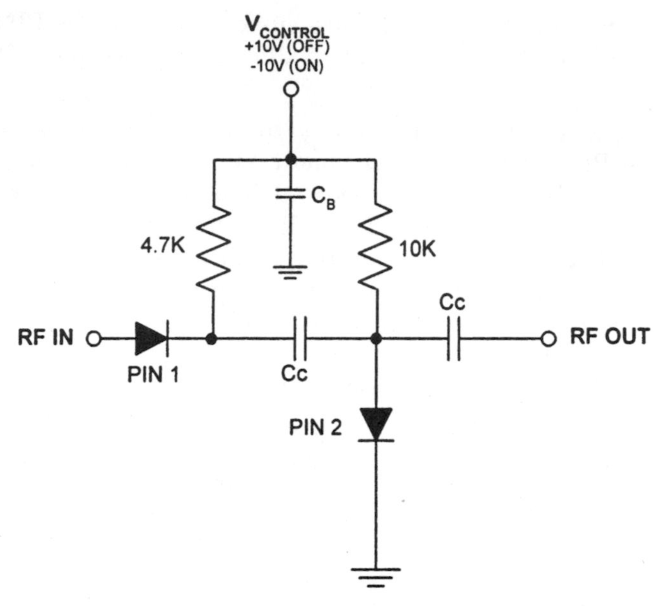

Figure 8.13 A high-isolation series/shunt PIN diode switch.

Its lower frequency is 50 MHz because of the decreased inductance of the RFCs as the frequency decreases, as well as the PIN diode's carrier lifetime limitations. A reverse bias on the off arm is not required, except to slightly improve harmonic distortion levels. It is recommended by HP (Agilent) that a 40-mA bias current be used to switch on the PINs in order to improve the output distortion levels. However, 15-mA forward bias is sufficient to switch the diodes on with minimal, but not optimal, distortion of the signal. For switching low-level amplitude input signals from a receiver, BIAS 1 and BIAS 2 need only be +5 and 0 V. For higher power signals, the bias should be +5 V (on) and −5 V (off). R is the current-limiting series PIN switch resistor for each diode pair.

1. RFC = 500 ohms (X_L) at f_r.
2. $C_B = C_C$ = 1 ohm (X_C) at f_r.
3. $R \approx \dfrac{\text{DC bias} - 0.9}{80 \text{ mA}}$

   DC bias is in positive volts.

Shunt PIN switch (Fig. 8.15). When this switch is in the on (0 V or negative voltage) condition, the insertion loss is low, while the return loss is high. When the switch is in the off (positive voltage) condition, the signal is shunted to ground before it reaches RF_{OUT}.

1. RFC = 500 ohms (X_L) at f_r.

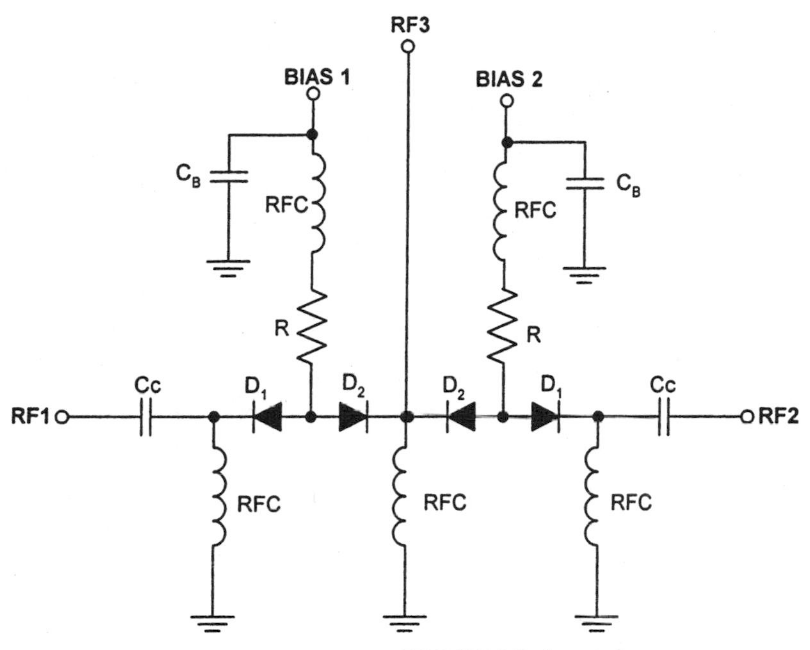

Figure 8.14 A wideband, low-distortion SPDT PIN diode switch.

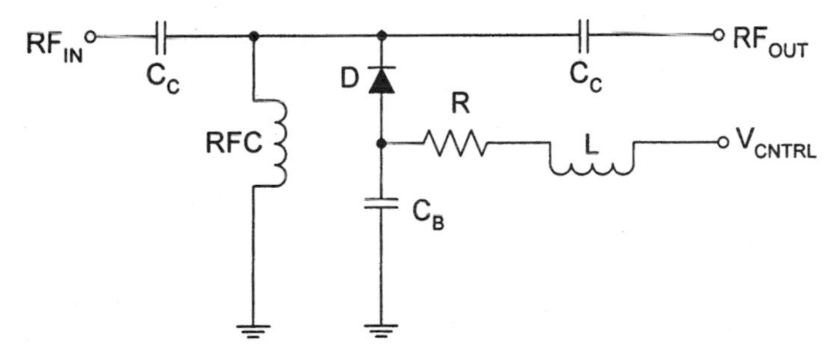

Figure 8.15 A shunt PIN diode switch.

2. $C_B = C_C = 1$ ohm (X_C) at f_r.

3. $R \approx \dfrac{\text{DC bias} - 0.9}{25 \text{ mA}}$

   DC bias is in volts.

Small-signal series PIN switch (Fig. 8.16). When this switch is in the on condition (+5 V or higher at V_{CNTRL}), the insertion loss is low, while the return loss is high. When the switch is in the off condition (0 V), the signal is severally attenuated by the diode, while also creating a low return loss.

1. RFC = 500 ohms (X_L) at f_r.

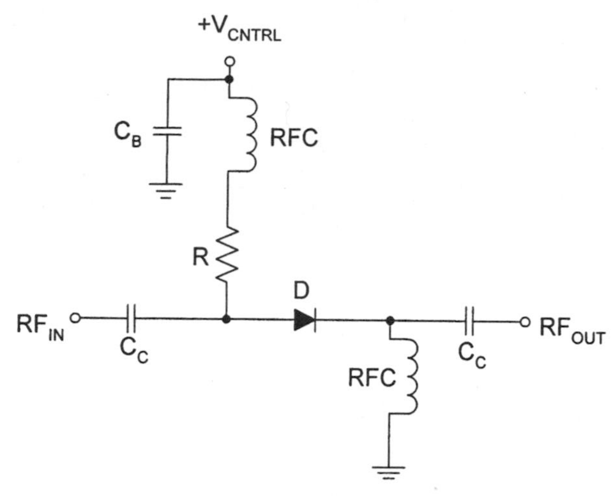

Figure 8.16 A low parts count PIN diode switch.

2. $C_B = C_C = 1$ ohm (X_C) at f_r.

3. $R \approx \dfrac{\text{DC bias} - 0.9}{50 \text{ mA}}$

Lower diode on bias currents than 50 mA can be used, but at the expense of gradually increasing distortion levels; 15 mA should be considered a minimum value. DC bias is in volts.

Series SPDT PIN diode switch (Fig. 8.17). A positive voltage at V_{CNTRL} will steer the INPUT to RF2, while a negative voltage will steer the INPUT to RF1.

1. RFC = 500 ohms (X_L) at f_r.
2. $C_B = C_C = 1$ ohm (X_C) at f_r.
3. $R \approx \dfrac{\text{DC bias} - 0.9}{50 \text{ mA}}$

(Lower diode on bias currents than 50 mA can be used, but at the expense of gradually increasing distortion levels; 15 mA should be considered a minimum value. DC bias is in volts.)

8.2.3 RF switch issues

As stated above, increasing the PIN's bias current decreases its IMD products: A bias current of 10 mA may work well for very low level signals, while a value of up to 60 mA may be required for higher-powered applications.

In some situations an electronic switch must be highly absorptive in the off position, or the high impedance of the switch will cause the power at its input to be reflected back to the source. This can severely degrade circuit perfor-

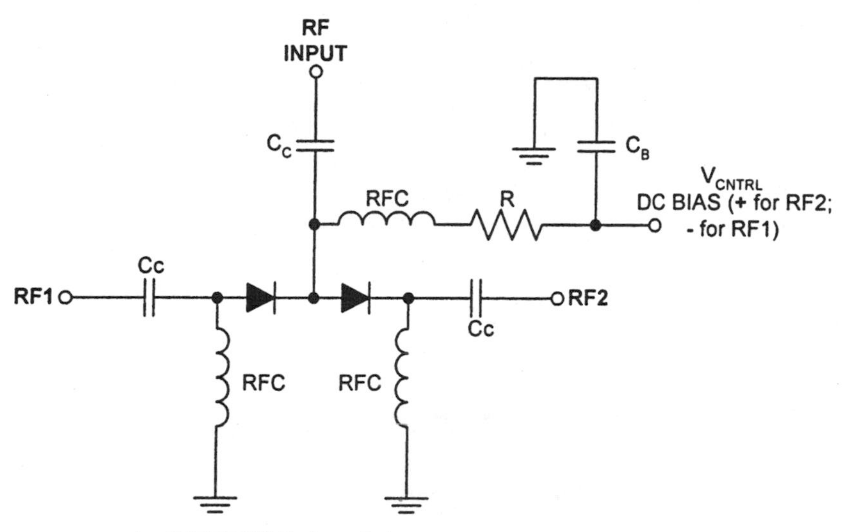

Figure 8.17 An SPDT PIN diode switch.

mance, and can actually damage RF stages. Designing a switch that switches to a 50-ohm termination in the off position will allow a desirable high return loss, minimizing reflections.

RF relays are less dependable than PIN switches and slower, larger, and more costly. Relays are, however, better for high-powered, wide bandwidth, low-IMD applications.

For basic switch requirements, a single or double diode configuration will serve most needs—except for switching reactive loads. The inductance of the load (such as an RF filter), has an undesired proclivity to resonate with the diode's off capacitance, which can give the switch very little of the off isolation that was expected.

8.3 Automatic Gain Control

8.3.1 Introduction

Automatic gain control (AGC) is found in almost all modern receivers. The popularity of this circuit is due to the necessity of increasing the usable dynamic range of a receiver, since without AGC powerful incoming signals would immediately saturate the receiver and create massive distortion, while feeble signals would go virtually undetected by the demodulator. Both would cause very poor BER in a digital system, or unreadable signals in an analog system.

Bias-based AGC networks function on a particular transistor characteristic: The gain of a transistor is increased when we raise the transistor's collector current; conversely, decreasing the collector current will also decrease the transistor's gain. Indeed, by raising the forward bias at the base we can easily increase the collector current, since increasing the base current will increase the collector current, and thus gain. As shown in Fig. 8.18, however,

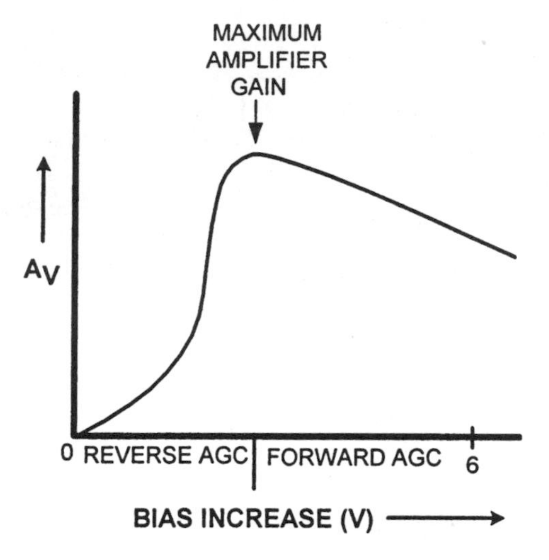

Figure 8.18 A base-bias voltage versus amplifier gain in an AGC.

a point will soon be reached in which this capability will not only level off, but the gain will actually start to decrease slowly with any increase in collector current.

The base current is created by the *DC bias voltage* that is impressed at the base of the transistor by the AGC circuit itself. In fact, many variable-gain amplifiers will depend only on this AGC voltage for their entire DC base bias.

Because of this capability of a transistor to increase and decrease gain when an external circuit increases or decreases the collector current, we see that there can be two methods of implementing AGC: *reverse* and *forward* AGC. Reverse AGC is by far the most popular, and will be found in the IF sections of many radios. Forward AGC may sometimes be designed into certain front-end RF amplifiers, but is undesirable for general applications because it wastes more collector current than reverse AGC, and has a much more gradual gain response.

With these DC bias–controlled amplifiers, care must be taken to confirm that severe distortion does not occur when the amplifier's gain is being varied by the AGC. Since the bias point is changed to decrease the gain through the amplifier, the stage can easily be biased into a nonlinear part of its operation, especially if the input signal is of a high amplitude. This is not as much of a consideration with AGC amplifiers that use voltage or current controlled diode *attenuators* at their input for this gain control function, since many newer AGC circuits will feed the detected and amplified control voltage to one or more variable attenuators, which are placed before fixed-gain amplifier stages (see Sec. 3.8, "VGA Amplifiers" and Sec. 8.4, "Attenuators").

The AC voltage needed to feed an AGC loop can be tapped off the last IF stage (Fig. 8.19) or, in some receivers, after detection by the detector. As shown in the figure, the IF signal is first tapped from the IF strip's output, RF amplified, rectified to DC, DC amplified, filtered to a steady DC, and sent to the base

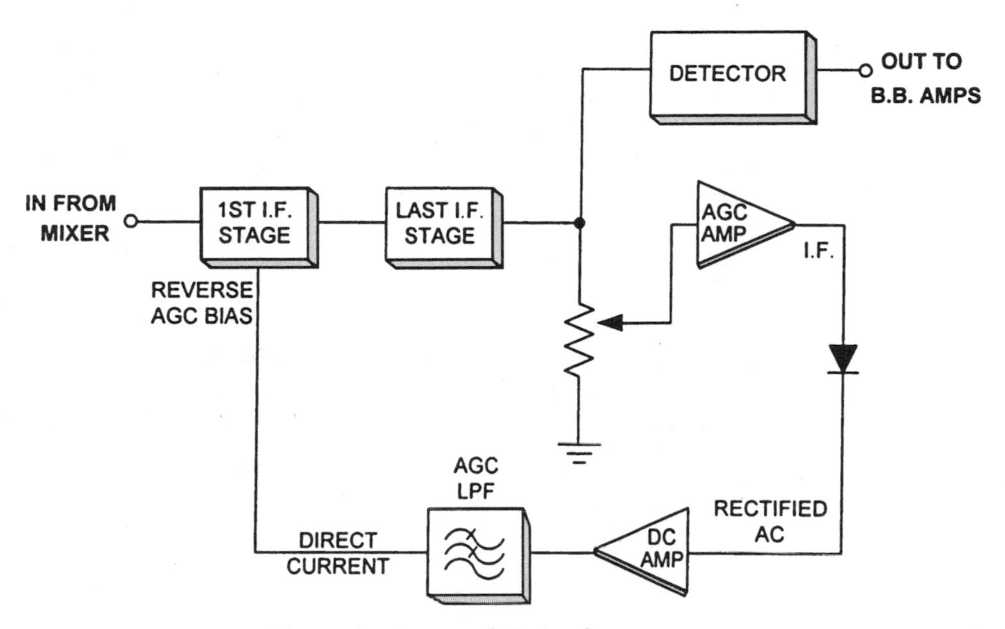

Figure 8.19 A type of AGC loop using the tapped IF signal.

of the first, second, and even third IF amplifiers through a trace on the PCB called an *AGC bias line.*

8.3.2 Automatic gain control design

The complete AGC circuit shown in Fig. 8.20 can be designed in various ways. However, the basics still do not change: The signal to be controlled must be sampled, detected, filtered, and placed into a variable-gain amplifier in order to change the stage gain according to input signal strength:

Sampling the signal. The signal to be controlled can be tapped from the IF by one of two ways. A large-value resistor that is much higher than the 50 ohms of the IF can be exploited to remove a small portion of the signal for feeding the AGC detector, or a directional coupler can be employed to remove a sample of the signal for AGC detection (see Sec. 8.8.2, "Directional Coupler Design").

Detecting the signal. Logarithmic amplifiers, or log amps, are adapted in some wireless applications to detect the peaks of the RF signal, then convert these peaks to a logarithmic DC output. Some high-frequency log amps can reach 2.5 GHz at their input, while still maintaining a high dynamic range of greater than 90 dB. These types of log amps are referred to as *demodulating log amps.* One such amp is shown in Fig. 8.21 as the AGC detector/amplifier in a receiver's automatic gain control feedback loop. R_{COUPLE} is a value significantly larger than

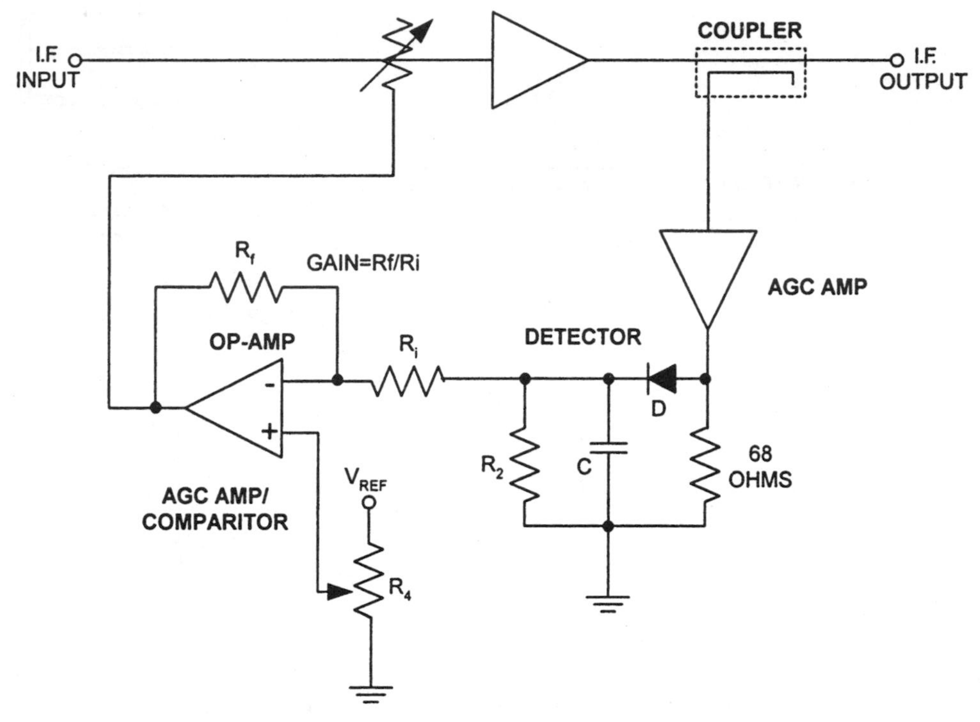

Figure 8.20 A common AGC circuit.

50 ohms to tap off a small portion of the IF signal into the log amp. As stated above, a directional coupler may also be used in this role. The log amp detects the peak RF, amplifies it, and then converts it to a log equivalent DC voltage output. The received signal strength indicator (RSSI) voltage is then placed into the *buffer amp,* and from there into the *integrator,* where the voltage is compared to V_{REF}. If the log amp voltage is below V_{REF} (a low input signal condition), then the integrator will output nearly 0 V to the attenuator. If, however, the voltage from the log amp is above V_{REF}, then a large negative voltage (near the op-amp's maximum power supply voltage) will be placed at the bias inputs to the IF attenuators. Figure 8.22 shows *input RF power* versus *DC output voltage* of a typical log amp.

If the attenuators, or a controlled bias VGA, required opposite voltages for gain control, then an inverting amplifier can be used, along with a positive supply voltage for the integrator.

An *AGC detector diode* is more commonly employed to detect the signal out of the coupler at the IF stages (Fig. 8.23). An unbiased detector (such as a self-biased or zero-biased, Schottky diode) can be used to convert the IF power to DC for the VGAs. C is chosen to have a low impedance to the RF in comparison to the diode (D_1) impedance. R_1, used in large-signal envelope detectors, presents a proper impedance match at the diode's relatively high input impedance for the 50-ohm coupler's output impedance, in addition to supplying a

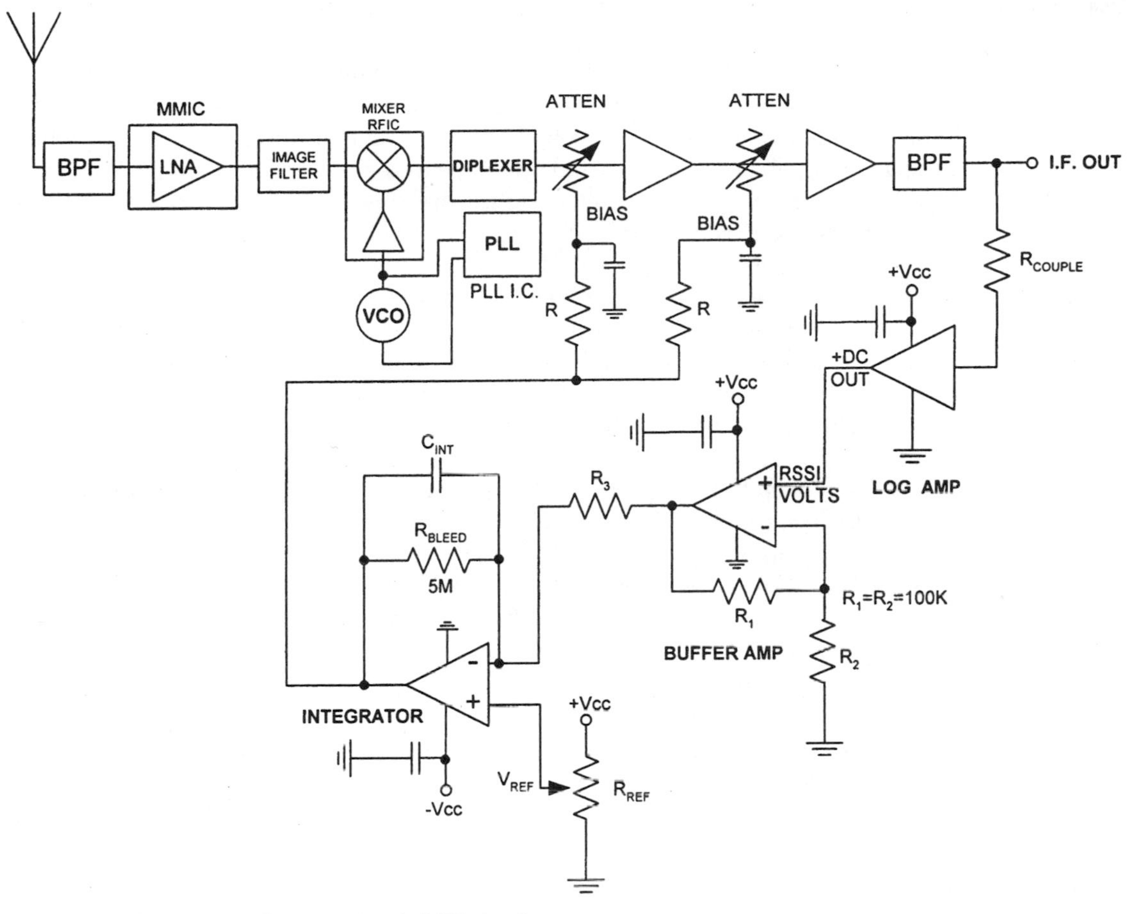

Figure 8.21 A log amp used in a receiver's AGC circuit.

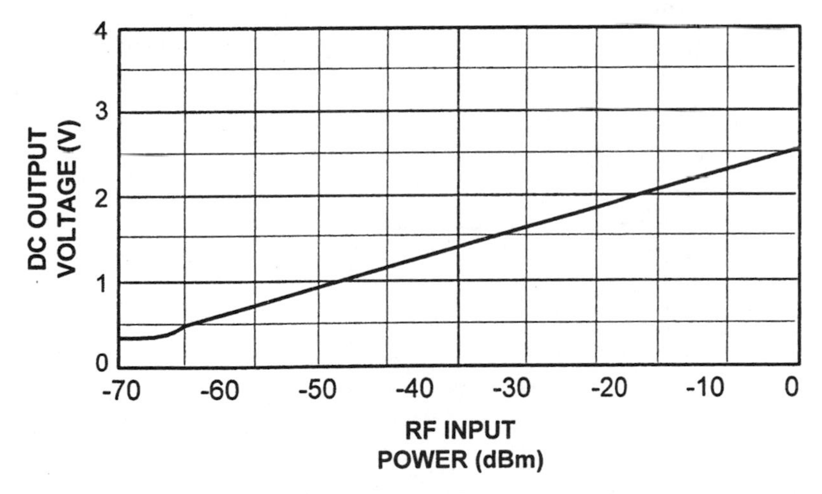

Figure 8.22 A log amp's DC output versus RF input power.

return for DC. R_1's value will normally be about 68 ohms. A small-signal square-law power detector circuit will replace R_1 with an RFC. R_2 serves as the load resistor for the *RC* time constant of the detector, and can be anywhere from 1 kilohm for the large-signal envelope detector to 5 kilohms for the small-signal square-law power detector (the difference between the two types of detectors is explained below). The DC OUTPUT should be placed into a high input impedance op-amp, or the DC output voltage will drop below expectations, since in designing an unbiased detector circuit (especially for square-law detector operation) the diode's junction resistance can be up to a few thousand ohms. Any Schottky detector's sensitivity will depend on the detector's load being much greater than the diode's internal resistance (R_J), so the input resistance of the next stage should be higher than 75 kilohms. Because of this high R_J value, input *LC* matching into the diode is rarely attempted.

The zero-bias Schottky diode should be chosen if small RF inputs are expected, since they will be much less sensitive to temperature variations than a *biased* diode arrangement (the detector's sensitivity is a measure of the input signal's amplitude compared to the output amplitude of the detector, measured in mV/μW, with higher values being better). Biased diodes have been employed to detect low signal levels in the past, as the detector circuit sensitivity can be dramatically improved by using a DC voltage to bias the diode to a point just before it begins to solidly conduct (0.7 V for silicon). This allows for less of an IF input amplitude before the diode strongly detects the signal. However, usually Schottky diodes need a *small* bias, even the "zero bias" types, when detecting very low amplitude RF signals. If not, then most of the RF signal's power will be dropped across the high junction resistance of the diode, and very little across the desired load resistance, unless the load resistance is made to be of a very high value.

As inferred above, Schottky diode detectors will detect power, since they operate in their square-law region at small signal input levels; high input levels will change the diode's response to *peak* voltage detection. So two basic types of detectors are utilized in RF circuits: the *envelope detector* and the

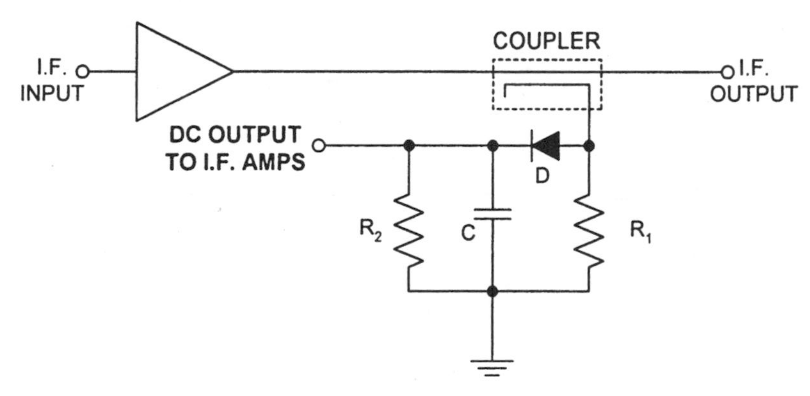

Figure 8.23 An AGC detector circuit.

square-law detector. The envelope detector detects the amplitude of the envelope of the RF signal, outputting a DC voltage that is equal to this value. The square-law detector detects the power of signal, outputting a voltage that is equivalent to this power. Both types of detectors use the same basic circuit arrangement shown in Fig. 8.23, except for the minor modifications mentioned above. But since a diode will act as a square-law device at low input powers (-60 to -30 dBm), and operate as a relatively accurate envelope-to-voltage converter at input powers greater than -15 dBm, it can be seen that both detectors can not only use almost the same circuit, but also the very same diode. Their type of operation will depend only on their input power.

Since the envelope detector iteration can operate with a much lower load than the square-law version and still maintain much of its sensitivity, the envelope detector is the most commonly accepted detector in AGC circuits (because of the lower internal resistance of the detector diode caused by the much higher RF input levels, the diode is now in almost full conduction).

Both kinds of detectors, especially the envelope type, will respond very quickly to an RF signal's increase in amplitude, but will usually have a much slower discharge time because of the requirement that C discharge through the large value of R_2. The charge-up is only through the RF source resistance of 50 ohms in series with the diode's on resistance, which can be relatively low during the IF signal's peaks. Decreasing the RC time constant of C and R_2 can alleviate this problem but, as discussed above, lowering R_2 also decreases detector sensitivity.

AGC amplifiers and integrators. DC amplifiers are normally needed to increase the AGC level of the DC signal that is fed into the VGA's gain adjust port. Normal RF amplifiers may also be required on the AGC's IF end to increase the signal into the detector. Raising the AGC's DC level can be accomplished with standard DC amplifiers, or with the circuit as shown in Fig. 8.24, a single-supply op-amp set to function as an *integrator.* In this circuit, a smaller C_{INT} or R_1, or a faster change in V_{IN}, will speed up the change in V_{OUT} into the VGA's gain adjust port. When the voltage at the inverting input is more positive than the voltage at the noninverting input, the output voltage will ramp down; if the voltage at the inverting input is more negative than at the noninverting input, the output will ramp up. The speed of the ramp-up/down will depend on the RC time constant of the RC components. A large-value resistor (R_{BLEED}) of approximately 2.2 megohms or higher is usually placed in parallel with the integrating capacitor, since all op-amps have a small input bias current that will quickly charge up this small-value capacitor: The resistor simply bleeds the current away. However, the resistor must be significantly higher in value than the input resistor R_1, or gain can be lowered excessively. Figure 8.21 demonstrates the integrator in a standard AGC circuit.

Variable-gain amplifiers. The variable-gain amplifiers in the IF strip can be of either the variable-attenuator type or the variable-bias type. Design of the appropriate attenuator and amplifier are explained in other sections of this book.

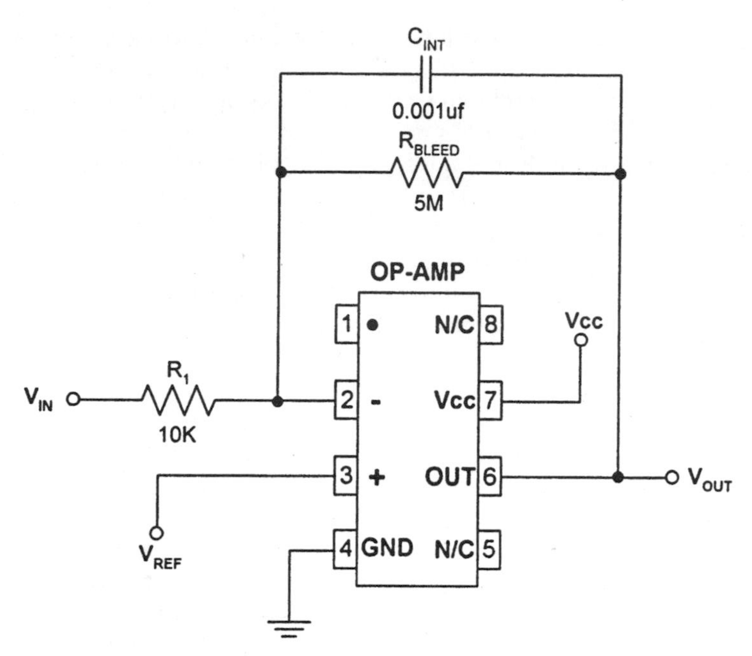

Figure 8.24 Single-supply op-amp designed to function as an integrator.

8.3.3 Automatic gain control issues

If there are any IF filters between the AGC output coupler and the gain-controlled IF amplifiers that have excessive group delay, this can cause time impediments in sensing that a signal has increased or decreased in amplitude, causing AGC loop instabilities. Another major cause of loop instability can be the desire to obtain excessively tight AGC control of the IF or baseband output amplitude. In most cases, an AGC circuit with less gain and less absolute control of the output amplitude will be much more stable. It is still possible, nonetheless, to safely design and construct an AGC loop that can control the output signal to within 2 dB (or better).

If the receiver's LNA is to be AGC controlled, it is almost always critical to maintain the noise figure of the receiver within reasonable limits. Unfortunately, any AGC action will naturally decrease the receiver's NF. By adding a *delay diode* in the AGC bias line back to any front-end amplifiers, the start of gain control can be postponed slightly. This will allow the LNA to maintain its NF, and that of the entire receiver, until it is absolutely required to decrease the gain at the front end. Even if there will not be an AGC connection to the LNA (or any other RF amplifier), a delay diode is still a good idea for the first IF gain-controlled amplifier in the IF strip, since this will help, in a small way, in maintaining a superior noise figure.

All AGC-controlled amplifiers should be decoupled from each other by a small value of in-line resistor (100 ohms) and a ceramic capacitor to ground at each VGA's DC gain control port. This will slow undesired interaction between gain-controlled stages.

In the early prototype stages of an AGC circuit it is a good idea to utilize trimmers in place of certain critical fixed resistors in order to allow the values of the AGC's loop to be empirically optimized.

8.4 Attenuators

8.4.1 Introduction

Attenuators are either fixed or variable circuits to reduce signal amplitudes and/or improve return loss, while maintaining the proper input and output impedance (normally 50 ohms), of the stages they are attached to. Attenuators are used extensively in wireless design.

Shown in Fig. 8.25 is a *step* (or *variable*) *attenuator* employed for testing of wireless circuits. Its attenuation can be varied in discrete steps by a manually turned knob, or by electronic control. Other variable attenuators are actually inserted between stages on a PCB, and can be either analog voltage or current controlled, with infinite attenuation resolution, while a digital step attenuator will have a limited number of discrete steps (2, 6, 12, etc.). All attenuators are rated for the maximum amount of attenuation they are capable of (15, 30, 45 dB, etc.), along with their maximum frequency and input signal strength levels.

SMA or BNC miniature coaxial in-line attenuators (Fig. 8.26) for testing or signal attenuation are available at various fixed values of up to 60 dB, with a maximum safe power dissipation of 25 W.

Integrated circuit solutions for variable and fixed attenuators are readily available, but their cost, performance, and size are usually inferior to the discrete designs. However, for small production runs and for low signal levels, either analog or digital IC variable attenuators can sometimes be the best choice.

8.4.2 Fixed-attenuator design

To design a 50-ohm pad for any attenuation value, first calculate the value of the attenuator's resistors with the following equations, then select the proper resistors for the maximum power dissipation expected.

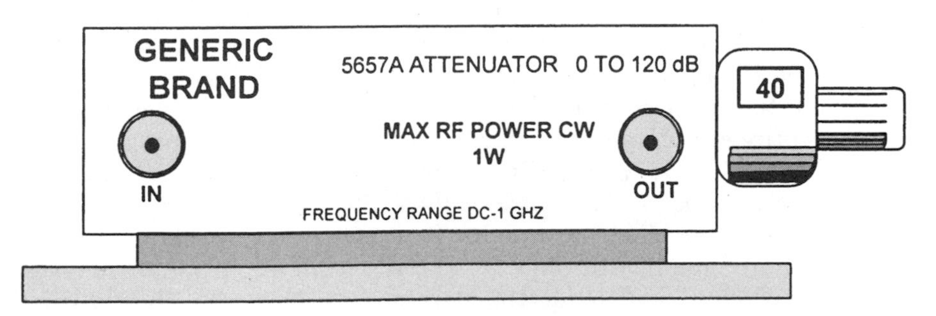

Figure 8.25 A common manually switched step attenuator.

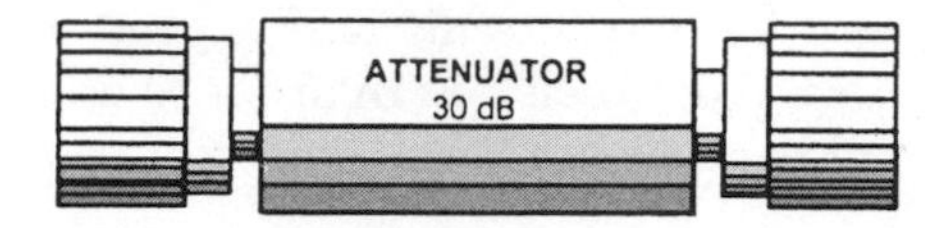

Figure 8.26 A coaxial in-line fixed attenuator with SMA connectors.

Fixed pi attenuator (Fig. 8.27)

1. $\chi = 10^{\text{Loss (in dB)/10}}$.
2. $R_3 = 0.5\,(\chi - 1)\sqrt{\dfrac{2500}{\chi}}$.
3. $R_1 = \dfrac{1}{\dfrac{\chi + 1}{50(\chi - 1)} - \dfrac{1}{R_3}}$.
4. $R_2 = R_1$.

Fixed T attenuator (Fig. 8.28)

1. $\chi = 10^{\text{Loss (in dB)/10}}$.
2. $R_3 = \dfrac{2\sqrt{2500\,\chi}}{\chi - 1}$.
3. $R_2 = \left(\dfrac{\chi + 1}{\chi - 1} \cdot 50\right) - R_3$.
4. $R_1 = R_2$.

8.4.3 Variable-attenuator design

It is quite difficult to iteratively design a quality *absorptive RF* attenuator (the electronic switches shown in Sec. 8.2.2, "RF Switch Design," can be adopted as *reflective* attenuators, in undemanding applications, by employing different levels of on/off bias to control the level of attenuation). It is easier to use a popular and proved absorptive variable-attenuator design. A frequently utilized voltage-variable attenuator is shown in Fig. 8.29, as described by Agilent. This attenuator functions quite well over a frequency range of 1 MHz to 3 GHz with the HP HSMP-3810 series of Agilent PIN diodes, and will have very low distortion at low signal levels. It is economical and has a good return loss over its entire attenuation and frequency range (greater than 11 dB over a control voltage of 0 to 15 V from 10 to 3000 MHz). Figure 8.30 is a graph of the circuit's attenuation versus control voltage V_{CONTROL}.

This four-diode attenuator functions this way: The DC returns for D_2 and D_3 are supplied through R_1 and R_2, while R_3, R_4, and R_6 furnish the proper impedance match for the particular PIN diodes chosen (in this case, the

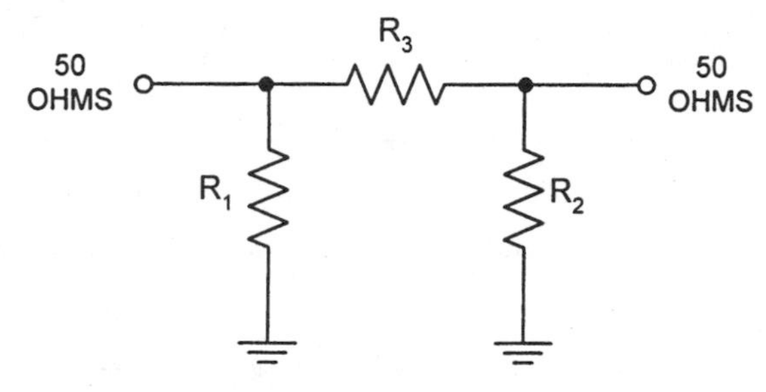

Figure 8.27 A 50-ohm pi attenuator.

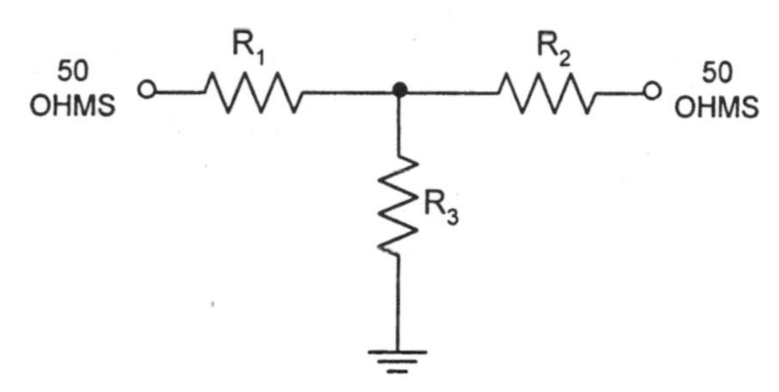

Figure 8.28 A 50-ohm 'T' attenuator.

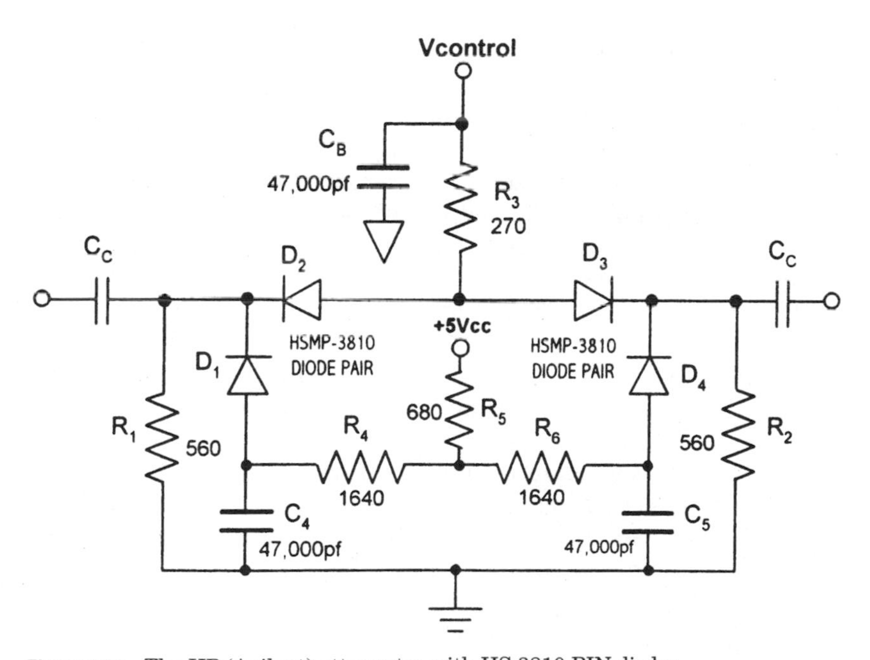

Figure 8.29 The HP (Agilent) attenuator with HS 3810 PIN diodes.

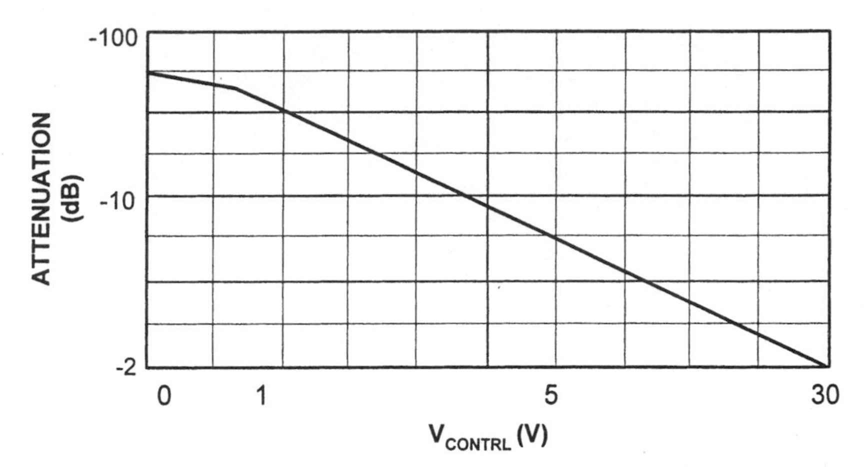

Figure 8.30 The HP (Agilent) attenuator's approximate attenuation versus control voltage (V_{CONTRL}) at 100 MHz.

HSMP-3810). The values for R_1, R_2, R_3, R_4, and R_6 were chosen experimentally by Agilent to match their diodes in this absorptive, single-power-supply attenuator.

8.5 Baluns

8.5.1 Introduction

A balun transforms a balanced line into an unbalanced line, since it is sometimes necessary to have a balanced output from an unbalanced amplifier for interfacing with other stages. Impedance matching may also be required.

A balanced stage's input or output consists of two parallel conductors with two input lines, one with a 0 degree signal, the other line having the same amplitude signal but 180 degrees phase shifted, with each conductor having equal currents flowing in opposing directions (Fig. 8.31*a*). An unbalanced stage's output has a single conductor for the current, with a second conductor for the ground return, and is the dominant technique found in contemporary RF design (Fig. 8.31*b*). But when a balanced source must be converted to unbalanced, this demands that the two differential signals be mixed (combined) so that they are in phase in order to output an unbalanced signal.

For antenna use, baluns may be purchased in connectorized weather-resistant packages, in which they are utilized for placing an unbalanced signal from the coax transmission lines into a balanced dipole antenna, while also matching any impedance variations. If a balun was not used in this situation, RF currents on the center conductor of the coax would pass to one leg of the dipole, while the RF current on the ground conductor would pass to the other leg of the dipole; this would result in RF radiating from the coax's ground shield, causing EMI. These balun structures, at HF frequencies, can be as simple as a wideband, untuned transformer.

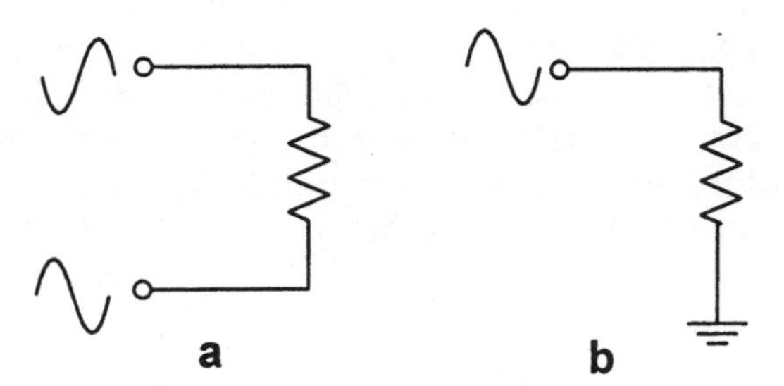

Figure 8.31 (*a*) Balanced input; (*b*) Unbalanced input.

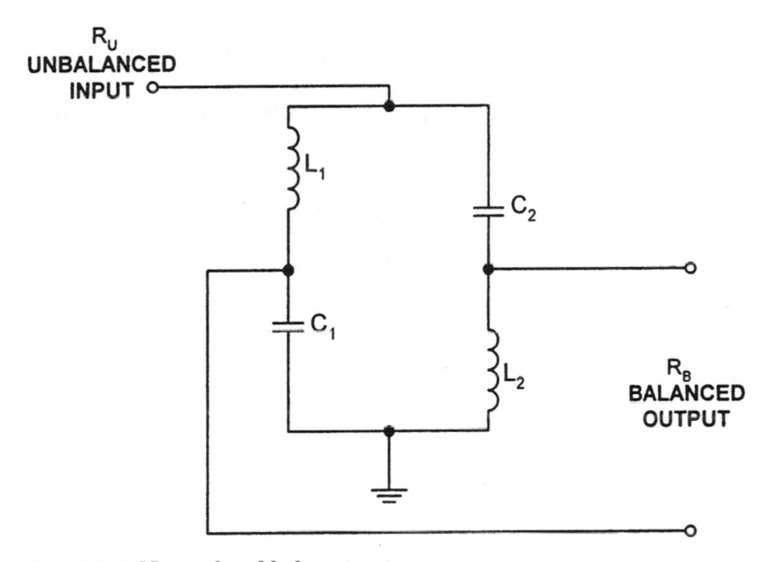

Figure 8.32 Narrowband balun structure.

8.5.2 Balun design

Presented are two designs—one lumped balun for VHF and under, and another for microwave and above as a distributed circuit.

Narrowband lumped balun (Fig. 8.32)

1. $L_1 = \sqrt{\dfrac{R_U R_B}{6.28 f}}$.
2. $L_2 = L_1$.
3. $C_1 = \dfrac{1}{(6.28 f)(\sqrt{R_U R_B})}$.
4. $C_2 = C_1$.

where R_U = unbalanced source or load resistance, ohms
R_B = balanced source or load resistance, ohms
f = center frequency, Hz.

Microwave distributed balun (Fig. 8.33). This planar structure is relatively broadband (30 percent bandwidth when six sections are used), and is simple to design and implement. With the addition of more sections, wider bandwidths are possible, while maintaining the required 180 degree phase shift between ports 2 and 3 for balanced operation:

1. $A = \frac{\lambda}{4} \times V_P$.

2. $B = \frac{\lambda}{2} \times V_P$.

where A = length of 50-ohm microstrip (calculate width as described in Sec. 1.4, "Transmission Lines")
B = length of 50-ohm microstrip (calculate width as described in Sec. 1.4)
V_P = velocity of propagation as a fraction of the speed of light (depends on E_r of PCB; calculate as described in Sec. 1.4) .

8.6 Splitters/Combiners

8.6.1 Introduction

A splitter/combiner is a 50-ohm circuit used to "mix" different signals in a linear manner, leaving them unchanged, with no new signals created. For a splitter, a signal is placed at the input, and two or more signals, usually of equal amplitude and phase, are removed from two or more separate ports at the output. For a combiner, two or more signals of equal phase are placed at the input, and a single signal is removed from the one output port that is equivalent to their vector sum. A splitter and a combiner are exactly the same circuit; to turn the splitter into a combiner, the circuit is merely reversed, with the input becoming the output.

Complex splitter/combiners can be utilized to split or combine the signals of various RF power amplifiers into a single high-power output.

For two in-phase, same-frequency signals placed at the input to a combiner, the insertion loss will be quite small (less than 0.4 dB). However, if these two signals are at different frequencies, the insertion loss for a two-way (three total ports) combiner will be 3 dB. With the same circuit configured as a two-way splitter, the insertion loss will be approximately 3.5 dB—with 3 dB of the loss due to the signal being split into the two output ports.

8.6.2 Splitter and combiner design

50-Ohm *LC* Power 0 degree splitter/combiner (Fig. 8.34). This *LC* device will have a lower insertion loss (3.5 dB) than a resistive splitter, while maintaining a high 20-dB isolation between ports. However, this circuit will have a

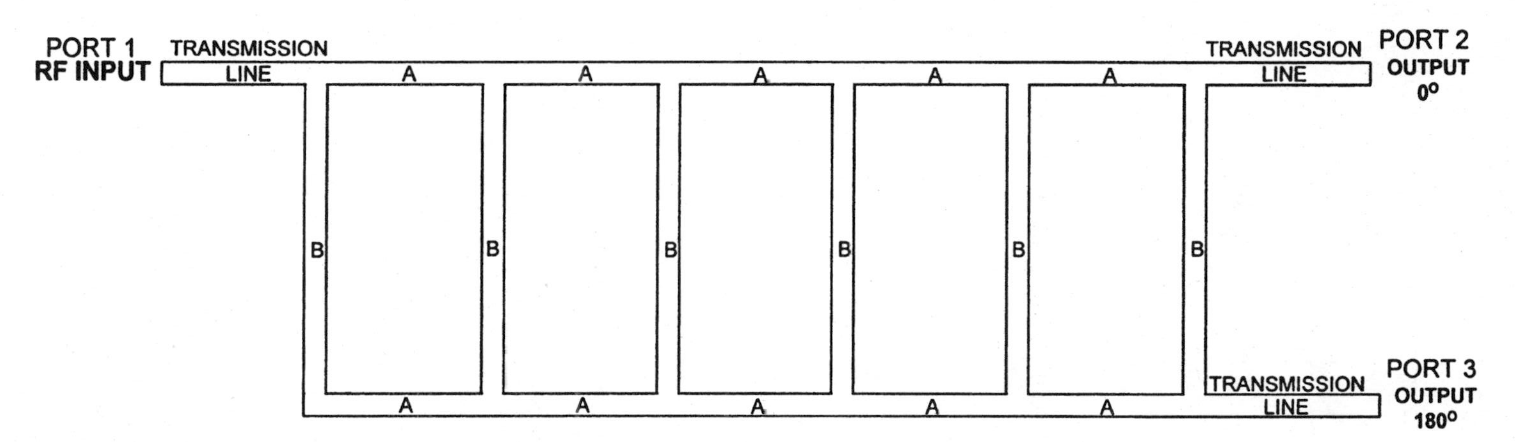

Figure 8.33 A distributed microwave balun.

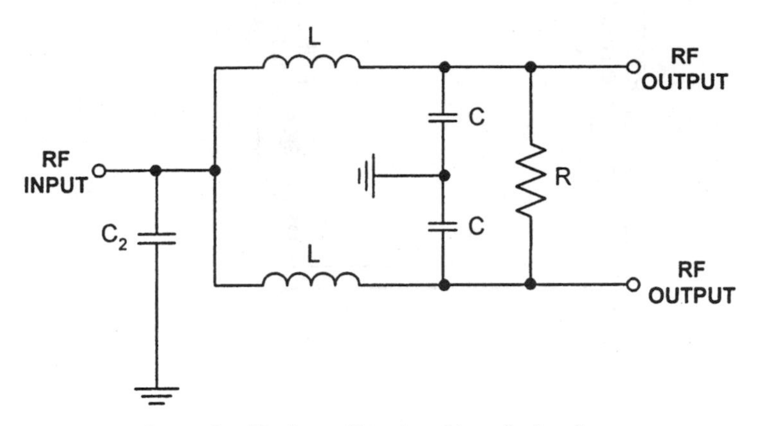

Figure 8.34 A reactive 50-ohm splitter/combiner for low loss.

significantly lower bandwidth (around 20 percent) than a resistive design, since it is reactive.

1. $L = \dfrac{50}{4.4\,f}$.
2. $C = \dfrac{1}{445\,f}$.
3. $C_2 = 2\,C$.
4. $R = 2\,Z_{IN} = 100$ ohms.

90 degree *LC* power splitter (Fig. 8.35). This circuit is similar to the above splitter, but has a 90 degree phase difference between its two output ports. It also possesses a 20 percent bandwidth.

1. $R = 50$ ohms.
2. $L = \dfrac{R}{2\pi f}$.
3. $C_1 = \dfrac{1}{2\pi fR}$.
4. $C_2 = 0.5\left(\dfrac{1}{2\pi fR}\right)$

Resistive splitter/combiner (Fig. 8.36). This simple resistive splitter is extremely wideband, but has a high insertion loss (6 to 7 dB) and low isolation between ports (same level as the insertion loss).

1. $R = \dfrac{50}{3}$.

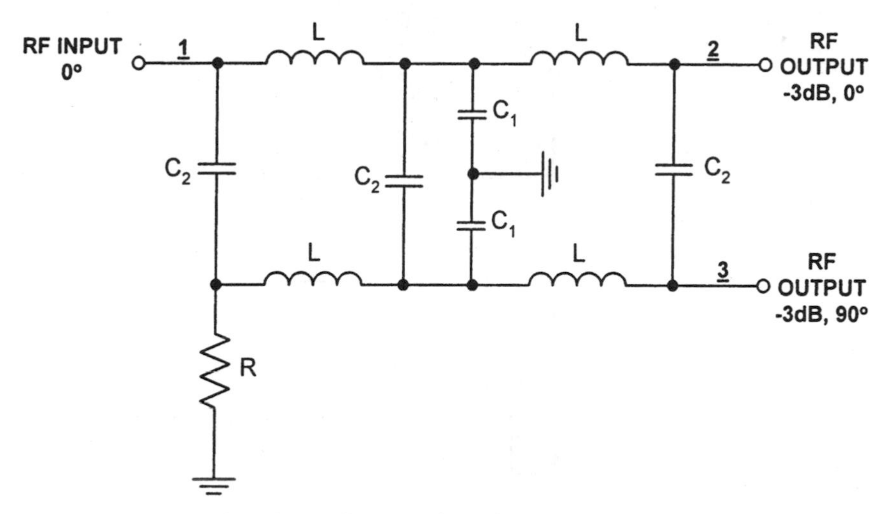

Figure 8.35 A lumped 90 degree directional coupler.

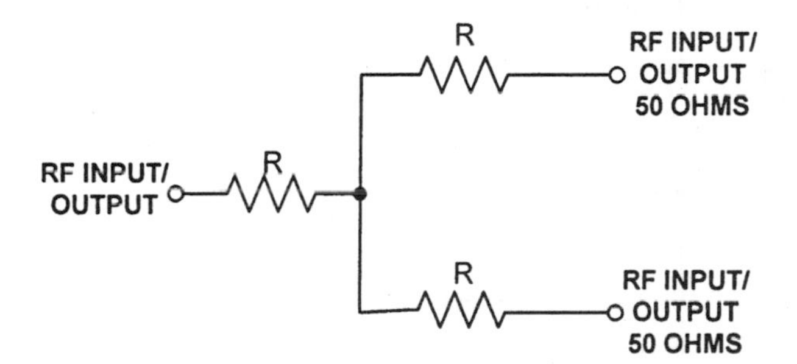

Figure 8.36 A 50-ohm resistive wideband splitter/combiner.

8.7 Power Supplies

8.7.1 Introduction

Most wireless communications equipment runs off DC power supplies that obtain their energy from the AC mains. This is because batteries will furnish current only for a limited period of time, and so are reserved for equipment that will not normally be near AC—such as most portable devices. AC main voltage is converted by a power supply into the required DC levels needed by the system in order to provide power at a constant and regulated voltage.

The basic power supply is shown in Fig. 8.37. It consists of a two- or three-pronged plug, a transformer, a rectifier, a low-pass filter, and a regulator. The transformer changes the AC main voltage of 120 VRMS into any desired voltage, either up or down; the bridge rectifier circuit turns the AC into a pulsating DC; the low-pass filter converts the varying DC into a steady DC; the *three-terminal regulator* maintains the output voltage within tight specifications. C_3 suppresses any output oscillations, and helps in regulation, with R_B

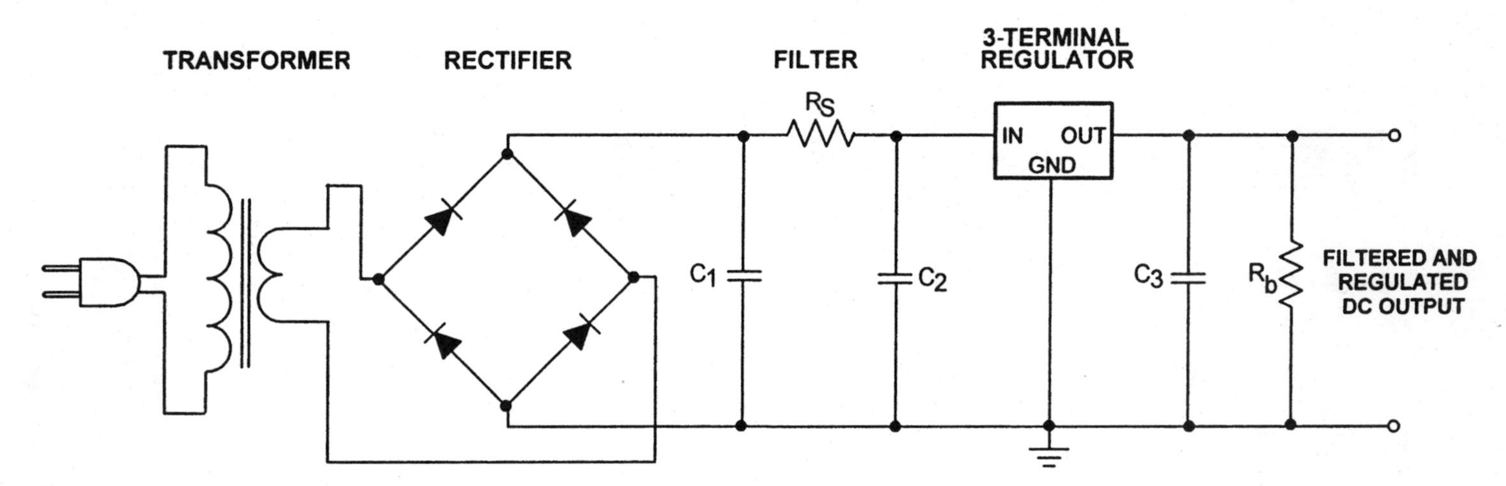

Figure 8.37 A complete linear power supply.

being a bleeder resistor that drains a fixed current from the regulator to help stabilize the output voltage, as well as drain hazardous voltage levels from the filter capacitors when the power supply is shut off. More on each of these circuits below.

Transformer. Since a transformer conveys AC energy from one circuit to another by electromagnetic induction, we can increase or decrease the current or voltage by changing the ratio of the windings between the primary and the secondary. A low-frequency transformer is made up of a primary coil, which obtains energy from an alternating current source. The primary's expanding and contracting magnetic flux lines flow through a core made of steel plates, which concentrates this flux with the least amount of losses. The primary's flux lines cut the secondary coil, inducing an AC voltage and producing a current that flows through the transformer's load.

Rectification. Rectification is the first step in obtaining a smooth DC output voltage. AC power can be changed into pulsating DC by employing one of three general rectification circuits.

The most basic technique is *half-wave* rectification (Fig. 8.38), which has a peak voltage that is almost equal to the input AC peak voltage and demands few components (a single diode). However, this method gives us a troublesome-to-filter 60-Hz output.

The second method, *full-wave rectification* (Fig. 8.39), has a simple-to-filter 120-Hz output. Unfortunately, only half of the input's peak AC voltage is available to the load because of the transformer's center tap.

The dominant method in modern quality power supplies is *bridge rectification* (Fig. 8.40), which not only furnishes us with an easy-to-filter 120 Hz, but also the full input AC peak voltage levels at the output.

Filtering. A low-pass filter is necessary in order to smooth out the pulsating DC power that results from rectification, since such amplitude variations would be unacceptable for many electronic circuits. Filtering is used to eliminate this pulsating component, while giving us a constant, unchanging current output.

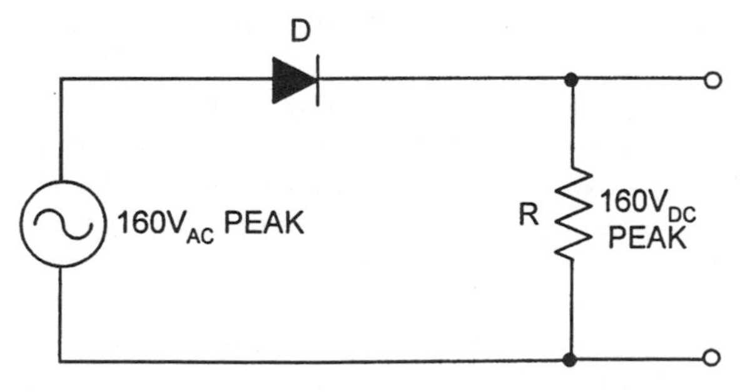

Figure 8.38 A half-wave rectifier circuit.

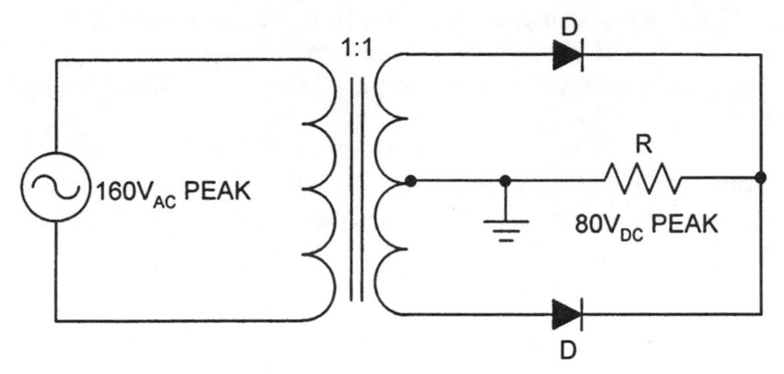

Figure 8.39 A full-wave rectifier circuit.

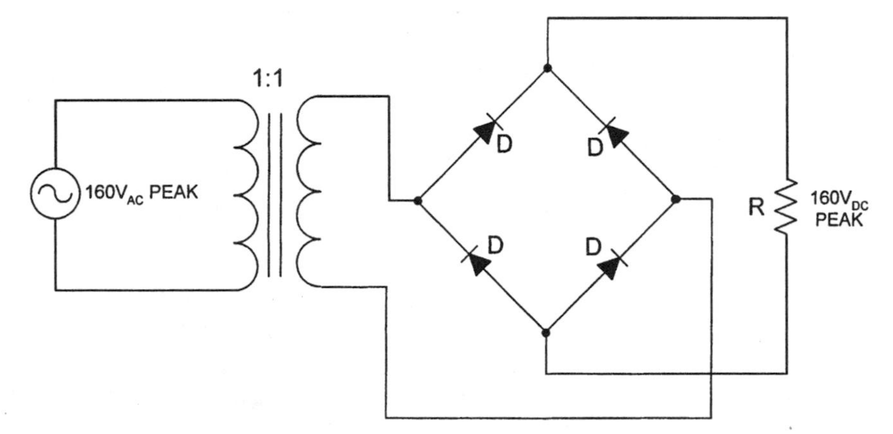

Figure 8.40 A bridge rectifier circuit.

One such filter that will remove any AC component riding on the rectified DC output is shown in Fig. 8.41. C_1 filters the majority of the ripple, with R_S and C_2 functioning as an AC voltage divider. The small amount of ripple left over from C_1 will be dropped across R_S, with very little across C_1. This is due to R_S's increased resistance over C_2's decreased reactance to the relatively high ripple frequency. Swapping out R_S with an inductor (Fig. 8.42) would allow the filter to continue to function properly at high current drains.

Regulation. Modern equipment and circuits will perform reliably only when they are furnished with a constant supply voltage. If we consider that power supplies without any regulation will shift their output voltage when the voltage varies at the mains, or even when the resistance of the load changes, we can see that regulators are required to avert, or at least decrease, these undesired effects. Regulators will also smooth the output voltage, thus assisting the power supply's filter section. Figure 8.43 illustrates one of the most common, and easy-to-implement, regulators available—the three terminal (3-T) type. Much more on regulators will follow.

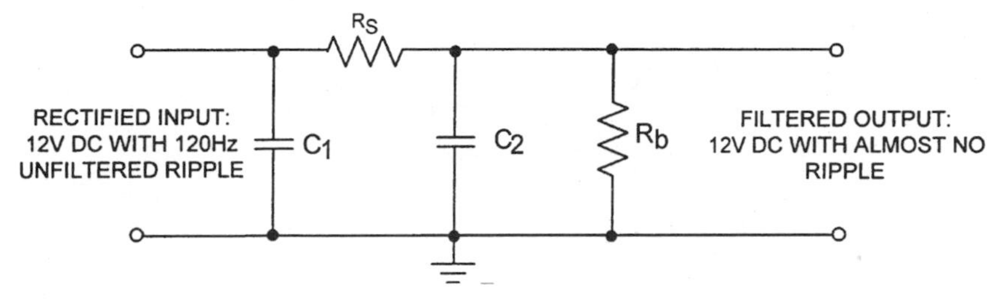

Figure 8.41 A type of power supply filter using a series resistor.

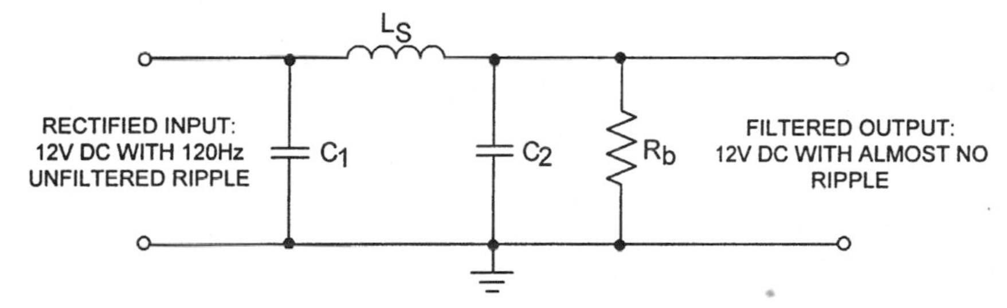

Figure 8.42 A type of power supply filter using a series inductor.

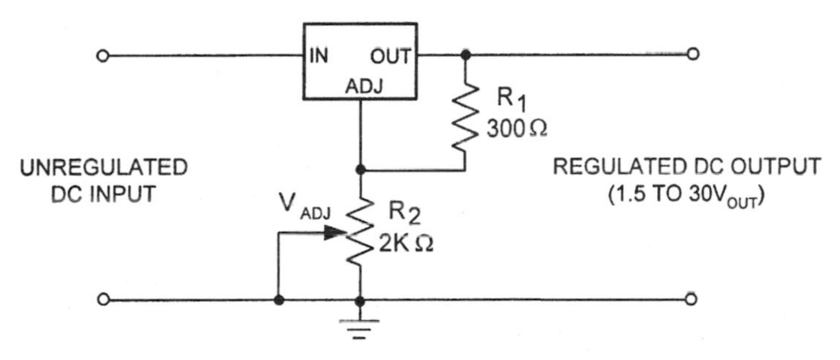

Figure 8.43 The common three-terminal (3-T) regulator with adjustable output voltage.

Switching-mode power supplies. Figure 8.44 demonstrates one widespread method for implementing a complete switching-mode power supply (SMPS): The main inserts a 120-V RMS AC signal into the SMPS' input, where high-amplitude transients that attempt to enter the supply and cause damage will be shorted to ground through the *metal oxide varistor* (MOV), thus imparting limited protection from lightning strokes or surges. The bridge rectifier, consisting of the four diodes, rectifies the 60-Hz main AC. Thermistors TH1 and TH2 limit inrush current by forcing the current, when the thermistors are cold, to slowly flow through their high resistance. After the thermistors have been warmed up by this current flow, their resistance falls, allowing almost complete current flow. Capacitor C_1 filters much of the rectified AC to DC. The

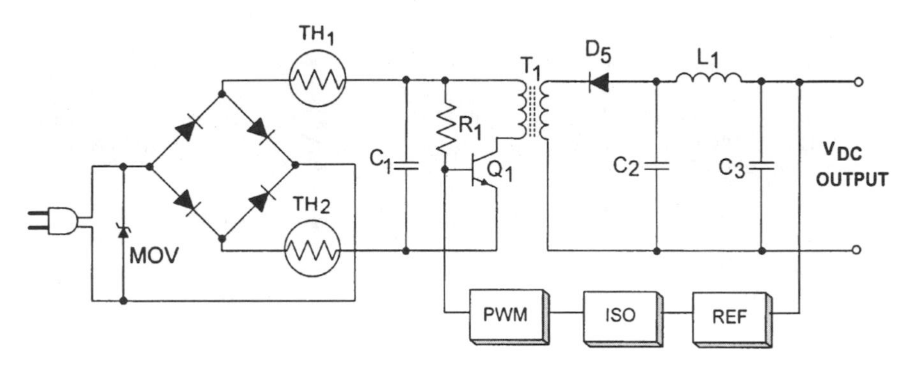

Figure 8.44 A switching-mode power supply.

chopper transistor Q_1 is switched on and off by the pulse width modulator (PWM), which, when turned on, will change the width of its output pulses. The pulse's width is dependent on the required amplitude of V_{DC}, which is governed by the REF voltage. In other words, V_{DC} must not fall below this REF voltage, or the PWM increases its duty cycle to compensate. R_1 is the Q_1 start-up resistor, while ISO (the *isolator,* usually an *optoisolator*) supplies isolation between the low-voltage secondary and the higher-voltage primary. The chopped-up (by Q_1) direct current is sent through transformer T_1, rectified by D_5, low-pass filtered by C_2, L_1, and C_3, and is then placed at V_{DC} OUT as a regulated DC.

As a caution, never run a switching-mode power supply without a load attached at its output, or it may become damaged or run improperly. And care is always warranted when probing the circuits of any SMPS, since very high voltages exist within some portions of this circuit.

8.7.2 Types of power supply regulators

The 3-T linear regulator, like the switching regulator, has become a common type in electronics today. However, for certain applications and for cost constraints, there are many different voltage regulator designs available. For instance, the cheapest regulator circuit of them all is the simple *zener shunt voltage regulator* of Fig. 8.45. Because of reverse-diode action when the zener hits its avalanche knee voltage, which assures that the reverse voltage across the device will change very little for a large increase in zener current, the load is in parallel with the diode and is therefore well regulated. And as the current rises, the circuit forces the surplus voltage that is not dropped across the diode to be dropped across R_S, since the sum of the zener voltage and the voltage drop across R_S must equal the input voltage. In essence, the zener alters its resistance as the current changes in order to keep its zener voltage (V_Z) constant with the parallel load. (V_{IN} must be a little higher than V_Z for this type of zener regulator to remain in regulation.) However, since the zener is in shunt with the load, the current through the diode can be considered wasted. In many applications, this is unacceptable. The following regulators solve this problem.

Another low-cost, but higher-performance, regulator, is the *series-pass transistor regulator* of Fig. 8.46. If the voltage across the load attempts to increase for any reason, the regulator's output voltage will stay nearly steady. This is due to the following action: The voltage across the load R_L must equal the voltage drop across the zener minus the voltage drop across the E-B junction of the series-pass transistor Q_1. The transistor's base voltage is set by the zener and, since the voltage dropped across the zener cannot change, then any rise in the regulator's output voltage will force Q_1's emitter to be that much more positive than the base, which is the equivalent of making the base *less* positive. This ends in a smaller base-to-emitter voltage, resulting in less emitter current through the transistor, and thus an increased voltage drop across Q_1. Any attempted rise in voltage across the load R_L is substantially lowered in value, and a steady output voltage is maintained for undemanding applications.

Figure 8.47 shows a *series-pass regulator with feedback,* which maintains a far more consistent and steady output voltage than the two regulators discussed above. This circuit contains R_1, R_2, and R_3, which monitor the output voltage as a voltage divider network. And since these resistors do form a voltage divider, we can also set the required output voltage to a wide range of values by simply moving the wiper of the adjustable R_2 all the way from a little above the zener's V_Z to just under the unregulated supply voltage. For instance, if we wish to increase the output voltage, we can move the potentiometer's wiper downward, and the voltage on Q_2's base decreases, lowering

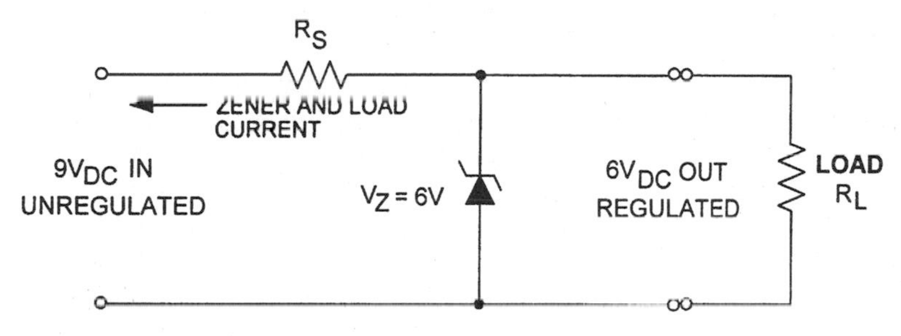

Figure 8.45 A simple zener shunt regulator.

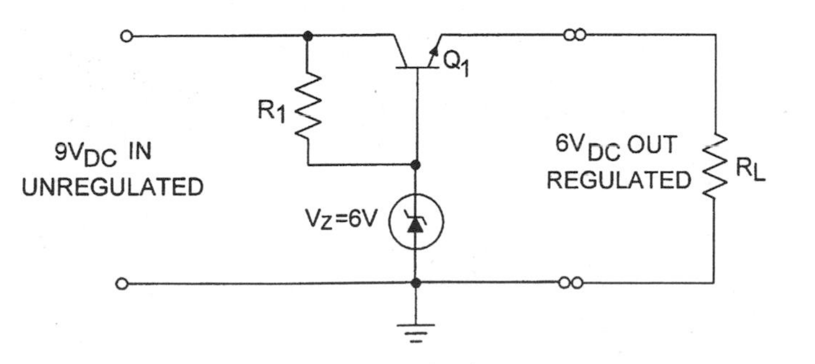

Figure 8.46 A series-pass transistor regulator.

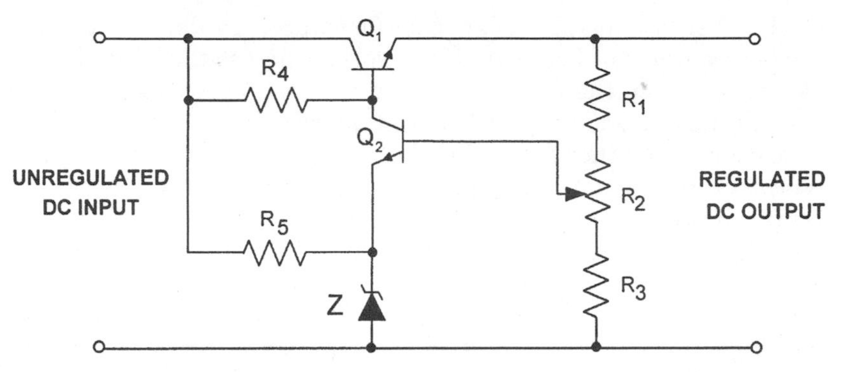

Figure 8.47 The series-pass regulator with feedback.

its forward bias so that this error detector/amplifier transistor will now conduct less. This forces Q_2's collector voltage to increase, which then forces the base voltage of Q_1 to rise, making this series-pass transistor conduct harder. Now, increased current will flow through any load placed at the regulator's output, thus increasing the output voltage. Q_2's emitter is clamped at a reference voltage by the zener, with R_5 setting the diode's idling current, while the collector resistor for Q_2 and the base bias resistor for Q_1 is supplied by R_4.

Three-terminal regulators (shown in Fig. 8.43) are, because of their low cost, size, and weight, high efficiency, and great simplicity, the most common type of regulator for almost any circuit or system requiring 3 A or less (much higher currents are possible with the addition of external components). These devices are integrated circuit regulators that include full internal current limiting and thermal protection circuits, so if the IC's internal power dissipation rises excessively, the 3-T will shut itself down before burning out. As shown in the figure, the 3-T circuit employs R_1 and R_2 to fix the output voltage level. R_2 can also be varied in order to change this output voltage for almost any requirement.

Very efficient *switching regulators* are found in an increasing number of power supplies. These devices regulate by outputting a variable-duty-cycle pulse into a low-pass filter in step with the output load requirements, and perform this switching with speeds of between 20 and 500 kHz. Because of the fast rise and fall times of these waveforms, however, strong switching noise constituents will be produced within the regulator. These must be heavily filtered to reduce excessive hash output into sensitive radio gear. Thus, their use is still limited in certain communications gear, where designers may favor linear power supply regulators, such as the 3-T. If selected, switching power supplies require not only filtering of all the input and output leads, but also shielding and very short trace runs to minimize EMI generation.

Figure 8.48 is a common architecture for a switching regulator. Q_1, a switching pass transistor, receives pulses into its base from a VCO that is controlled by a comparator. Q_1 then outputs a switching voltage in step with these commands into the lowpass filter of L_1 and C_1, which filter the pulses into DC. By increasing or decreasing the pulse's duty cycle into this *LC* filter, the average DC output voltage will increase or decrease. D_1 protects Q_1 from the inductive kickback of the stored charge within L_1, which would normally produce a very high spike of voltage into the emitter during switching. The voltage divider of R_1 and R_2 program the desired output voltage of the switching regulator by using the voltage dropped across R_2 as a comparison to the zener's reference voltage, which then turns on or off the comparator, and thus controls the VCO, all depending on the voltage level across R_2. Most of the above circuit is found in IC form, with built-in current-limiting and thermal protection.

A widely used integrated version, with typical support components, is displayed in Fig. 8.49. This is a common regulator arrangement in many production switching-mode power supplies. As above, R_1 and R_2 program the desired output voltage, with the voltage across R_2 being fed back by the *feedback line* to the internal comparator circuits. The low-pass filter is composed of L_1 and C_{OUT}, with D_1 shunting the inductive kickback of a discharging L_1 to ground (when the IC switches off), instead of across the IC. C_1 is used to give the regulator stability at higher current draws. Further filtering at the regulator's output may be required if undesirable ripple amplitudes are still present.

8.7.3 Regulator design

Three-terminal regulators are the easiest type of regulator to design and implement, and are also the most prevalent. The only parts needed for a

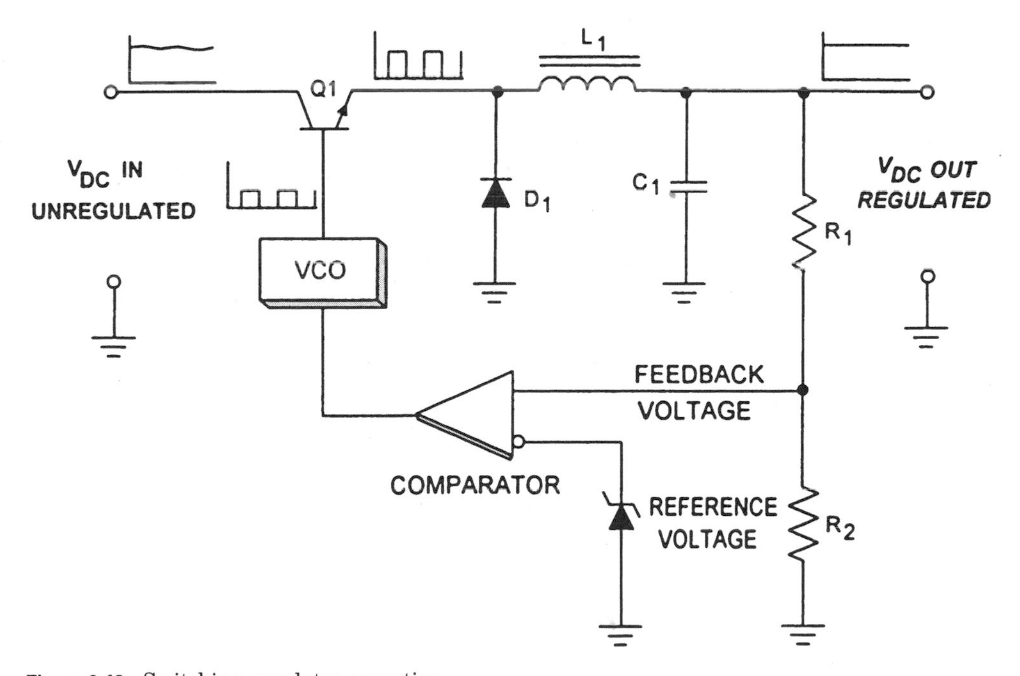

Figure 8.48 Switching regulator operation.

8.8.2 Directional coupler design

A lumped directional coupler (Fig. 8.56) can be designed as follows:

1. $R = 50$ ohms. Replacing R with a 50-ohm amplifier, or placing a high-impedance circuit in shunt with R, will permit reverse power to be sensed.
2. $L = \dfrac{50}{2\pi f}$.
3. $C_1 = \dfrac{1}{2\pi f 50}$.
4. $C_2 = \dfrac{10^{(CF/20)}}{2\pi f 50}$.

The C_2 value *must* be less than $0.18/2\pi f50$. CF = coupling, in dB, desired at port 3 (it must be less than −15 dB for this type of coupler), and it must be in the form of $-X$ dB in the equation, not as $+X$ dB or X dB. The CF value will remain within specifications over a bandwidth of only 10 percent.

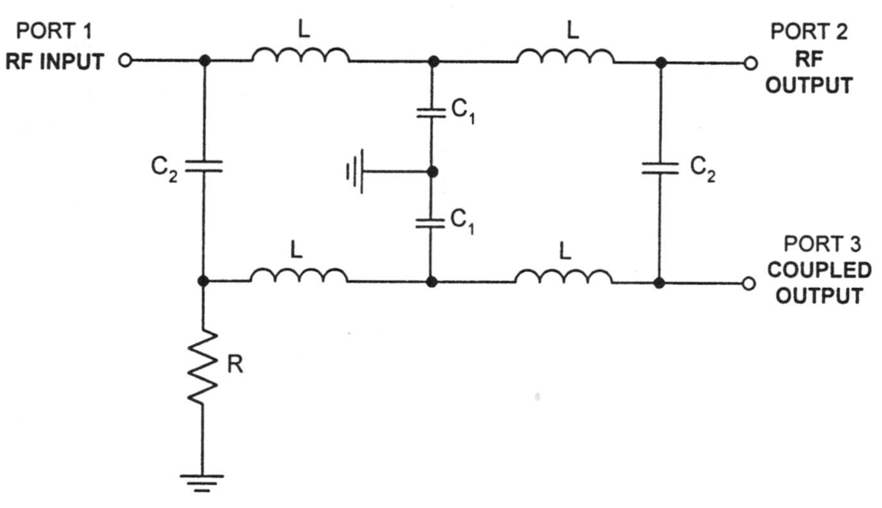

Figure 8.56 A lumped directional coupler circuit.

Chapter

9

Communications System Design

Without a solid understanding of the complete communications system all the way from the transmitter's modulator input to the receiver's demodulator output—including everything in between—and how the selection of various components, circuits, and specifications can make or break an entire system, any wireless design will surely fail. The interrelationships of the transmitter, the receiver, the antenna, the air-interface link, the type of digital or analog modulation, adjacent channel and cochannel considerations, etc., are critical to a dependable link for a high data rate at the required BER for digital radio, or with the expected voice quality for analog radio.

9.1 Receivers

9.1.1 Introduction

The most difficult-to-design element in any communications system is the receiver. A receiver must have a low noise figure (at VHF and above), low group delay variations and IMD, high dynamic range, stable AGC, appropriate RF and IF gain, good frequency stability, satisfactory gain flatness across multiple channels, low phase noise, negligible in-band spurs, sufficient selectivity, suitable BER and—sometimes the most critical specification of all—be within certain cost constraints.

An important concern of any superheterodyne receiver is the *image frequency,* in that any signal received within this image band will be amplified by the receiver's IF stages and then be unavoidably transferred on to the demodulator to be output as interference. This image frequency can be eliminated only at the front end of a receiver, before down-conversion, by a filter that blocks the interfering frequency: the *image filter.* When the local oscillator is higher in frequency than the incoming RF signal (high-side injection), the image is any frequency that is at twice the IF plus the desired RF signal frequency [(2 × IF) + RF], or at the IF plus the LO frequency (LO + IF). If the local oscillator is lower in frequency than the incoming RF signal (low-side injection), then we can find the image frequency by (2 × IF) − RF or LO − IF. Taking the first case, high-side injection, the image is any signal (or even noise at that frequency) that differs from the LO by the amount of the IF—just as the desired signal does—but is higher instead of lower than the local oscillator. Subtracting the desired signal from the local oscillator frequency will give the IF, which will, of course, easily pass through the receiver's IF amplifiers. But any frequency (the image) that is *higher* than the LO by exactly the same amount that the signal is *below* the LO will give us the same frequency. This same frequency will also easily pass through the IF amplifiers, and create interference and a decrease in the SNR.

As mentioned, the dominant technique for attenuating the image frequency is by front-end filtration. The filtering can be further assisted by using as high an IF as possible to move the image as far away from the desired frequencies as possible. This will make the filtering of the image a much easier task, with decreased risk of excessive group delay variations caused by a tight filter. Maintaining this first image far from the desired frequency is assisted by utilizing double- or triple-conversion receiver designs. With these multiple-conversion receivers, the first IF is at a high frequency, while the second and third IFs are much lower. These lower IFs will supply most of the selectivity and gain, since the lower the IF, the more simple, stable, sensitive, and selective the amplifiers will be. This is mainly due to the circuit's decreased stray capacitances and inductances at these frequencies, along with less of a requirement to employ special high-frequency components.

We will address typical receiver and transmitter system design issues, the communications link and its impairments, and the communications system as a whole.

9.1.2 Receiver design

Most receivers are of the *down-conversion* type, which takes the RF input and immediately begins to convert it either to a single lower IF or down to two or more increasingly lower IFs. The other type of receiver, called the *up-converting* superhet, is operated in wide-tuning-range applications, and is especially dominant in HF SSB ham radios. It takes the incoming RF and converts it to some higher frequency—typically about twice the highest expected receive frequency—to distance the image frequency from the LO to assure simple RF front-end filtering. Up-converting is rarely seen at VHF and above.

A standard double down-conversion superhet receiver block diagram is shown in Fig. 9.1. First, the antenna picks up the electromagnetic waves from the environment and, because of its natural passive gain, amplifies any signals that are within its bandwidth. The inductor L_1 short-circuits static buildup on the antenna to ground to prevent it from entering, and possibly damaging, the delicate LNA of the receiver's front end. L_1 can also be an inductor within BPF1. The signal is amplified by the antenna and sent into the input of the receiver's BPF1. BPF1 is a filter sometimes called a *preselector*, and is utilized to select a band of frequencies, while rejecting out-of-band signals, thereby minimizing IMD products. However, this filter must not be so tight that insertion loss (IL), noise figure (NF), and group delay variations (GDV) are increased excessively. BPF1 will also help reject the first LO frequency from radiating its CW signal back through BPF2 and the LNA. The LNA, because of its reverse isolation, can significantly attenuate this reradiation by itself, selecting the proper mixer with an adequate amount of port isolation. BPF1 will further assist BPF2 in the attenuation of image frequencies and image noise located at $(2 \times \text{LO}) - \text{RF}$, or $\text{LO} - \text{IF}$. (Normally image noise at a second or third mixer stage can be ignored because of the predominant noise contribution of these first stages of the receiver.) In fact, the frequency of the IF is chosen to permit a receiver to reject this first RF image frequency without requiring an excessively expensive, complex, and tight image filter before the first down-mixer, MIXER1. The next stage, the LNA, will set most of the receiver's noise figure and IMD performance. The LNA normally gives us approximately 20 dB of front-end RF gain, with a low NF of less than 2 dB. BPF2 attenuates any harmonics created by the nonlinearity of the LNA, as well as the image noise caused by the LNA itself (since no amplifier is noiseless). BPF2 will further reject some of the out-of-band signals and LO feedthrough.

In receivers with a single filter in the receiver's front-end, the BPF1 would match the antenna to the input of the LNA, decrease the amplitude of the out-of-band signals from overloading this amplifier, and provide a certain amount of image filtering. But considering that filter insertion loss *before* the LNA is directly translated into increased NF, an image filter (BPF2) is usually placed between the LNA and the first mixer to decrease IL at the BPF1 preselector. The two-filter method decreases the NF, while assisting the preselector in rejecting undesired signals as well as any image signal and the ever-present image noise.

As stated above, but well worth repeating: Low noise figure for a receiver means utilizing a low NF and high-gain amplifier in the receiver's front end, since any losses before the receiver's LNA will correlate dB for dB in an increased NF; 3 dB of filter loss before the LNA equals 3 dB more NF, which translates into a less sensitive receiver. Unfortunately, there is a slight compromise that must be made in LNA design: For the highest front-end gain, the LNA must be matched to the receiver's front-end filters. But this matching can increase the noise, since an optimum NF match will rarely coincide with a high return loss, or low mismatch loss, match. So, considering that the first stage of a receiver is required to be designed for the lowest NF and the highest gain, we will normally match for the best NF that the amplifier can provide, while providing acceptable gain.

We can see the importance of LNA gain by glancing at the formula below, as the second stage of the receiver will only add to the overall noise figure of the receiver by the second stage's NF *divided* by this first stage of gain, or:

$$\text{NF}_{\text{OVERALL}} = \frac{\text{second-stage NF}}{\text{first-stage gain}}$$

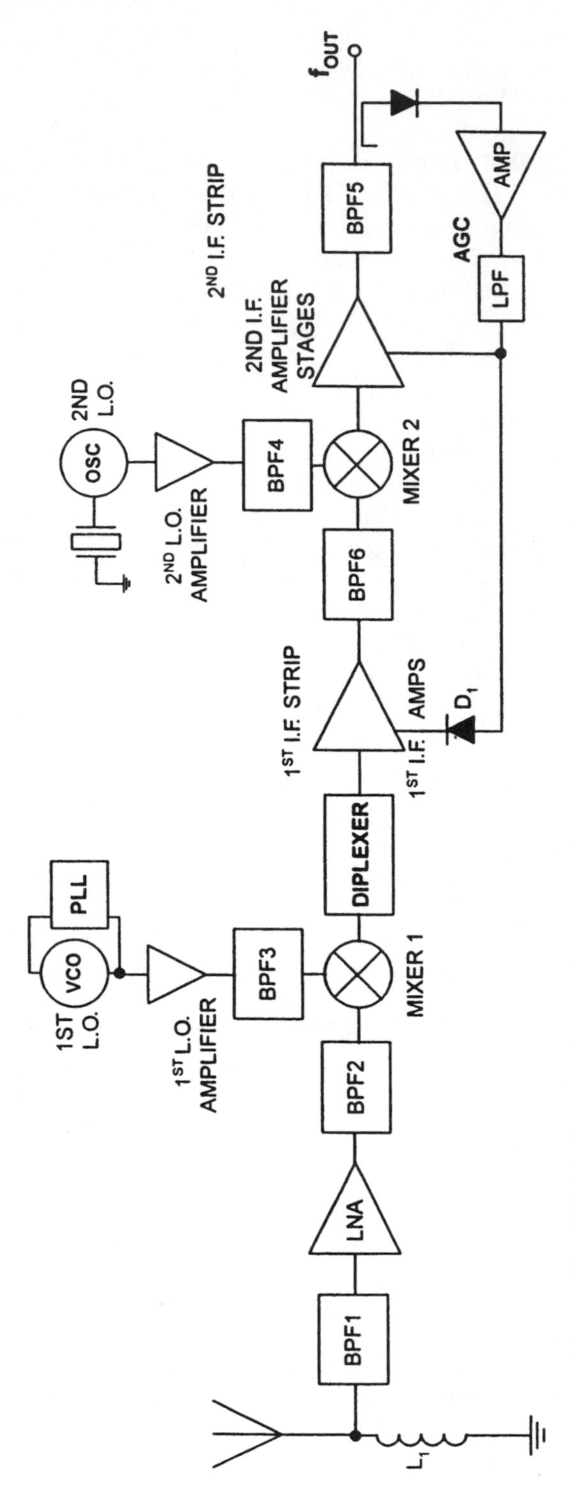

Figure 9.1 A heterodyne VHF dual-conversion receiver.

If the receiver is required to operate only within the HF region or below, then both the LNA and BPF2 could be dropped from the receiver design, as noise figure is not as important in this region because of massive natural and artificial noise generation. This will decrease sensitivity, but will give more IMD immunity in a crowded HF spectrum.

MIXER1 will normally be subjected to high amplitude input signals, so a high compression point is mandatory to decrease mixer-generated intermodulation distortion. This will generally demand a high-level diode double-balanced mixer (DBM) instead of an active mixer stage in this sensitive location. And for decreased IMD generation, the RF signal into the DBM should be at least 10 dB less than that injected into the DBM's LO port. Thus, a level 10 mixer could accept a maximum of 0 dBm at its RF input port before excessive IMDs became a problem; 15 to 20 dBm less RF would be even better, and would be needed for higher-quality, lower-distortion receivers. DBMs have the added advantage that they suppress even-order LO and RF mixer-generated harmonic products, as well as the RF and LO frequencies themselves, at the IF output. Attenuation of these frequencies is far from complete, however.

The diplexer, placed after the MIXER1 stage, will filter and pass the desired IF but, unlike other filters, it will stop other frequencies from entering the IF's bandwidth by *absorption* rather than *reflection*. Reflection of the undesired frequencies—such as LO harmonics, the sum of the RF and LO, and the IMD products—would cause RF power to be sent back into the mixer, which would unbalance its diode ring (causing increased IMD), as well as adversely affecting the mixer's dynamic range and conversion loss. Indeed, many viable receiver architectures may simply pad the output of the mixer so that these reflections are attenuated not only as they enter the pad, but also as they are reflected back into the mixer's IF port. The pads lower the input/output VSWR by supplying the mixer with an almost pure $50 + j0$ ohm termination (pads placed at the RF port will, however, reproduce thermal noise, which would have normally been removed by the image filter, at the image frequency). A wideband, high-isolation amplifier may also be used at the mixer's output, as shown for the second IF strip, since this will permit all of the mixing products to pass through this amplifier, and, after filtering from a normal reflective IF filter (such as BPF6), the BPF will "bounce" much of the undesired signals back toward the sensitive IF port of the mixer. However, these reflected signals will have been significantly attenuated by the reverse isolation of this wideband amplifier. Nonetheless, the amplifier must have a high P1dB to linearly accept the sometimes high-powered out-of-channel signals that can occur, without producing significant distortion.

The LOs must generate low phase noise in order to mitigate BER and adjacent channel selectivity degradation. This can be difficult in phase-locked loop–based LOs, but is relatively easy in properly designed high-Q crystal oscillators (see Chap. 4, "Oscillator Design"). All oscillator-generated spurious signals must also be as low as possible to minimize receiver spurious responses. The LO amplifiers are broadband types used to buffer the local oscillator from the mixer ports to minimize VSWR, and thus mixer-generated IMD, and to increase the oscillator's output power to the nominal levels to maintain the mixer's NF and conversion losses. In design of this LO chain, the LO amplifiers can be overdriven for maximum power output and flatness, since the LO signal involves only a single frequency. This means that no IMD will be generated, only harmonics, which are easily filtered out.

Both BPF3 and BPF4 are employed to reduce wideband noise, harmonics, and possibly any subharmonics that will be present in the oscillator's outputs. Reduction of generated wideband noise improves the mixer's NF and the receiver's sensitivity, while harmonic suppression prevents a decrease in the mixer *second-order intercept point* (IP2) and the subsequent generation of increased LO second harmonics.

The first IF chain of the receiver will furnish some gain and filtering (certain first IFs may supply only filtering), along with delayed AGC of any amplifiers present, before injection into MIXER2. If desired, this first IF filtering of the DIPLEXER and BPF6 will also remove the second image frequency from reaching MIXER2, but the filtering itself must not be so tight as to introduce excessive group delay variations, causing increased BER. The filtering is present mainly to provide some channel selectivity and the rejection of spurious signals, since image noise into MIXER2 is not a concern if relatively high gain stages are adopted in the receiver's front end, which is where the NF is set (the noise floor at MIXER2 will be mainly noise amplified by the LNA).

The second IF amplifiers and filter stages supply most of the receiver's gain and selectivity, since lower-frequency circuits are cheaper and more stable. However, any such high-gain IF stages should be shielded against EMI that may have been emitted from other internal stages, or from external emissions, in order to prevent amplifier oscillations and receiver interference. The second IF strip will also increase the receiver's total dynamic range by the use of variable-gain amplifiers (VGAs) within an AGC loop. This automatic gain control will not only control the gain of the second IF stages but, with many high-end receivers, it will also set the gain of the first IF strip and the LNA. The second IF strip will output f_{OUT} into an external modem, or into an internal detector stage.

Considerations. When deciding on the number of filter poles required for a receiver's (or transmitter's) RF or IF, calculate the amplitude of all undesired signals (such as adjacent channels, mixer products, and LO feedthrough) and how much they must be attenuated to meet specs. A program such as Kirt Blattenberger's *RF Workbench,* and to a certain extent The Engineer's Club's *MxrSpur,* can be invaluable in this regard. Run simulations for the chosen filter type (Butterworth, Chebychev, etc.), along with the preliminary number of poles (3, 5, 7, etc.). Add some extra attenuation margin (approximately 10 percent) to this figure. Confirm that the filters will now properly attenuate all undesired signals, without adding excessive group delay variation, insertion loss, sideband cutting, cost, or ripple to the design.

To discover the receiver's AGC range needed for the IF strip, and in some cases for the RF stages, find the difference in dB between the lowest RF signal expected that will still be able to supply the desired amplitude at the receiver's output, and the highest RF signal expected. This will be the range, in dB, that the AGC must *lower* the gain of the receiver to maintain the nominal output signal level—even as the RF signal amplitude increases drastically.

The complete RF and IF gain for the entire receiver, all the way from its front-end input for the antenna to the output of the last IF, will typically be in the neighborhood of 125 dB. A minimum value for such gain would be 90 dB, but in some specialized applications of down-conversion directly into a modem, the gain may be as little as 18 dB. The receiver's gain figure includes gains, which are supplied by the amplifiers, and losses, which are caused by filters, switches, pads, and mixers. Both antenna and receiver gain are finally combined to confirm that there will be a signal of sufficient amplitude at the receiver's output to drive the detector or modem at the lowest signal levels expected, and with sufficient *fade margin*.

Reciprocal mixing is a common problem in receiver design that places the noise sidebands of the local oscillator into the IF of the receiver, and is caused by the heterodyning of out-of-band interfering signals and the LO noise. It significantly decreases the receiver's SNR if this interfering signal—or the LO noise—is not attenuated. This mixing action takes place within the receiver's mixer, with the interfering signal usually being quite close to the desired signal:

$$f_i + f_N \text{ or } f_i - f_N = f_{\text{IF}}$$

where f_i = interfering signal, which is close to the desired signal
f_N = frequency of the LO noise, which is close to the frequency of the LO
f_{IF} = frequency of the receiver's IF

Frequency plan. The receiver's internal frequency plan can make or break a design. The proper RF and IF bandwidth and the exact LO and IF frequencies must all be selected carefully, or serious interference and mixer-generated spurious responses can greatly decrease the expected receiver specifications, sometimes rendering an entire design almost useless (see Chap. 7, "Mixer Design"). In order to catch most mixer-generated spurs, such as $f_{\text{RF}} - 3f_{\text{LO}}$, $f_{\text{RF}} - 5f_{\text{LO}}$, $f_{\text{RF}} + 3f_{\text{LO}}$, $f_{\text{RF}} + 5f_{\text{LO}}$, it will be necessary to display these spurious output frequencies, and their amplitudes, by employing the proper RF system simulation software (see Sec. 10.10, "Wireless Design Software"). If it is found through simulation that the specifications cannot be met with the current design, then more IF filtering, new LO and IF frequencies, and/or a new mixer topology may be required. This is because we do not want any strong spurious frequencies, due to the many nRF ± nLO mixer products, to fall within the bandwidth of the receiver's IF. Such spurs would cause the BER to be degraded in a digital receiver, and in an analog receiver, even the reception of undesired signals and interference would be possible within our desired channel.

High-side injection occurs when the LO frequency is *greater* than the RF frequency in a conversion stage, while low-side injection occurs when the LO frequency is actually *lower* than the incoming RF frequency in a mixer stage. The choice of whether to operate a superheterodyne receiver with a high- or a low-side injection will depend on the system design and whether the final demodulated signal needs to have the sidebands with an inverted frequency spectrum. In digital communications, this frequency inversion can be a major consideration, since the output of the IF is typically sent straight into a modem, which may or may not require inversion. But even if we do not have a choice as to whether we can frequency-invert or not, we may still select our high- or low-side injections throughout both the transmitter and receiver stages, and base our high/low preferences on the availability of oscillators, multipliers, or PLLs, as well as their cost and design complexity, along with the location of our undesired image frequency and the undesired mixer spurs.

General receiver calculations. There are several important design formulas to assist the designer in calculating the receiver's specifications, and these are presented below. However, it will be far easier and more accurate to obtain these important receiver specifications by using the included *AppCad* program by Agilent. This software will, within seconds, compute total receiver gain, NF, SNR, MDS, sensitivity, noise floor, input/output IP3, dynamic range, etc.

1. The total gain required of a receiver can be calculated by finding out what the lowest expected RF signal level will be after the antenna (into the receiver), and deciding on what the minimum receiver output signal requirement is into the modem or detector:

$$G_{\mathrm{dB}} = P_{\mathrm{OUT}} - P_{\mathrm{IN}}$$

where G_{dB} = required gain of the receiver, dB
P_{OUT} = lowest acceptable signal output level of the receiver, dBm
P_{IN} = lowest expected RF signal level into the front end of the receiver, after the antenna, dBm.

2. Minimum discernible signal (MDS) is a sensitivity rating for receivers, and is the lowest signal detectable. This can be at 0 dB above the receiver's noise floor, and can be calculated by:

$$\mathrm{MDS\ (dBm)} = -174\ \mathrm{dBm} + 10 \log_{10} \mathrm{BW} + \mathrm{NF}$$

where BW = noise bandwidth of the receiver, or approximately the 6-dB-down bandwidth (instead of the typical 3-dB bandwidth), and NF = receiver's noise figure, dB.

3. The *third-order intercept point* (IP3) is approximately 10 to 15 dB above the P1dB compression point, and is the location where, if the gain slope of the receiver could continue, the undesired output third-order frequency products would be at the same amplitude as the output two-tone fundamental test signals that had been originally placed at the receiver's input. To compute the total cascaded input IP3 ($\mathrm{IP}_{\mathrm{TOT}}$) of multiple stages of a receiver (Fig. 9.2), use the formula:

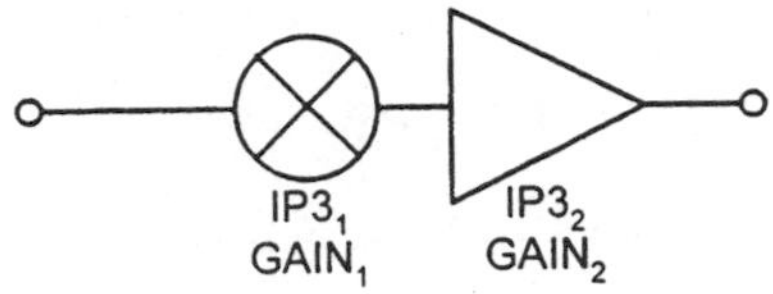

Figure 9.2 Cascaded stages for IP3 calculations.

$$\mathrm{IP_{TOT}} = \cfrac{1}{\cfrac{1}{\mathrm{IP3}_1} + \cfrac{\mathrm{GAIN}_1}{\mathrm{IP3}_2} + \cfrac{\mathrm{GAIN}_1 \times \mathrm{GAIN}_2}{\mathrm{IP3}_3}}$$

where $\mathrm{IP_{TOT}}$ = receiver's total IP3 from its input to its output and $\mathrm{IP3}_n$ and GAIN_n are linear terms, not decibels.

4. The importance of a low NF in a receiver is demonstrated by its linear relationship with SNR. For instance, if a receiver has an NF of 12 dB, and the SNR at its output is 20 dB, we can improve the SNR to 30 dB (using the same input signal power and receiver gain) by decreasing the receiver's NF to 2 dB. This is a 10-dB improvement in NF *and* in SNR. Thus, if a certain signal power is placed at the input of a 5-dB NF receiver, and this creates an output with an SNR of 20 dB, then the actual *input* SNR to the receiver will have been 25 dB. In other words, the receiver added 5 dB to the noise at its output: the receiver's NF directly correlated into a decrease in the output signal's SNR. This also affects the power that the transmitter must send to the receiver's antenna to make up for the decreased SNR caused by a receiver's poor NF, with a relationship that is also dB for dB.

The NF for a receiver can be calculated by using *Friis's equation,* with the noise figure itself not needing to be referenced to any particular bandwidth, since it is a ratio between the input and the output of a receiver (or amplifier) over exactly the same bandwidth:

$$\mathrm{NF_{TOTAL}} = 10 \log \left[(10^{\mathrm{NF}_1/10}) + \frac{(10^{\mathrm{NF}_2/10}) - 1}{(10^{G_1/10})} + \frac{(10^{\mathrm{NF}_3/10}) - 1}{(10^{G_1/10})(10^{G_2/10})} + \frac{(10^{\mathrm{NF}_n/10}) - 1}{(10^{G_1/10})(10^{G_2/10})(10^{G_n/10})} \right]$$

where $\mathrm{NF_{TOTAL}}$ = total NF for the entire receiver, dB
NF_1 = noise figure for the first stage, dB
NF_2 = NF for the second stage, dB
NF_n = NF for the *n*th stage, dB
G_1 = gain for the first stage, dB
G_2 = gain for the second stage, dB
G_n = gain for the *n*th stage, dB.

Alternatively, we can calculate the required NF for a receiver if we wish a desired receiver output SNR. These calculations need the receiver's planned bandwidth:

1. Calculate the thermal noise power within a certain receiver's bandwidth:

$$KTB = -174 + 10 \log BW$$

where KTB = thermal noise power within the specified BW, dBm, and BW = receiver's bandwidth, Hz.

2. Calculate the lowest sensitivity that would produce a signal at the specified SNR at the receiver's output:

$$SENS_{LOW} = (-KTB + SNR)$$

where SNR = receiver's desired output signal-to-noise ratio, dB, and KTB = thermal noise power, dBm.

3. The NF required to *just* meet the sensitivity specifications:

$$NF_{MAX} = SENS - (-SENS_{LOW})$$

where SENS = required sensitivity of the receiver as specified and $SENS_{LOW}$ = the sensitivity of the receiver at 0 dB NF (for the specified SNR). NF_{MAX} = maximum noise figure that the receiver can have and still satisfy the sensitivity requirement of SENS

Example. If a receiver under design has been specified to have a sensitivity of −100 dBm, what is the maximum NF that we can permit the receiver to have (with zero sensitivity margin) if the receiver's bandwidth is 100 kHz?

First, calculate KTB = −174 + 10 log 100,000 = −124 dBm.

Now calculate what the sensitivity of the receiver would be with 0 dB NF. Let's say the modulation chosen to be received will need a minimum of 18 dB SNR for a certain desired BER: −124 + 18 = −106

What is the maximum NF allowed for this receiver with zero margin?

The answer is: −100 − (−106) = 6 dB.

In reality, a safety margin would have to be added to the above calculation (the reasons for this are presented later). To add a safety margin, the NF of the receiver would have to be reduced even further. To increase the sensitivity margin by 3 dB, the NF would have to be *lowered* to 3 dB (6 dB − 3 dB = 3 dB). Depending on the radio service, more margin may be required, especially if the *implementation margin* is taken into account (the implementation margin is the loss in system specifications that occurs from the design stage to the actual implementation stage, as well as from mass production variances.

9.1.3 Receiver issues

As stated, the receiver's NF, which is dominated by its first stages, determines the sensitivity of the receiver at VHF and above. But since externally generated noise and interference in metropolitan areas can reach high levels, an LNA with an NF of 2.5 dB is usually more than sufficient. This metropolitan environment will also mean that there will be many other radios broadcasting—some at very high power—and some will have fundamental and harmonic frequencies quite close to our own receiver's frequency. Moreover, since our receiver may be of a mobile type, the fixed hub or cell site transmitter our receiver is linked with may be transmitting from a distance that is anywhere from 35 miles to 35 feet away. This can place extreme stress on the receiver's dynamic range, causing high levels of odd-order IMD (especially the third order) to drop into our receiver's IF passband, producing poor BER performance in a digital radio, or a decrease in fidelity in an analog radio. We can lessen this problem by adopting an LNA and first mixer with a high IP3 and/or, in certain circumstances, a front-end attenuator controlled by AGC. The front-end attenuator is used only to reduce the desired input signal levels to an amplitude that will not overdrive the receiver's LNA or first mixer; this solution would be unacceptable for attenuation of undesired adjacent channels, since the frequency of interest—especially if at a low amplitude—would also be attenuated. The noise figure will, of course, increase as attenuation is added.

Half-IF spurs can be a problem when a receiver possesses a *low*-frequency IF relative to its RF. This type of spur is created by the nonlinearities within the mixer permitting spurs in the receiver's IF by:

$$\text{Half-IF (Hz)} = [\,(0.5 \times \text{IF}) + \text{RF}]\,2 - (2 \times \text{LO})$$

This formula means that any RF that is at half the IF *plus* the RF, and that substantially gets past the first RF filters that are before the first mixer, will be able to cause IF in-band interference. Half-IF interference is due to the second harmonic of the half-IF plus RF mixing with the LO's second harmonic, with the resulting difference falling dead in the IF band. To mitigate, any RF signals that are at half the IF *plus* the desired RF should be properly attenuated before they exit the first front-end receive filter.

9.2 Transmitters

9.2.1 Introduction

Proper transmitter design is critical, since it is a device that radiates an electromagnetic signal. This signal can, if the transmitter is improperly designed or constructed, interfere not only with other wireless communications, but also with many different types of non-RF electronic equipment.

Harmonic and spurious outputs, wideband noise and phase noise, frequency and amplitude stability, and the signal's peak and average output powers are but a few of the critical parameters that must be addressed before any transmitter design can begin. Spurious signals generated by the mixing of the LO and IF (along with their harmonics) are of particular concern, as are two-tone intermodulation products created by two or more frequency components mixing together in any nonlinearities—at least up to the fifth order ($3F_1 \pm 2F_2$

or $3F_2 \pm 2F_1$). Other undesired output signals, such as harmonics of the desired RF carrier signal and the feedthrough of the LO and IF, can all cause interference. Transmitted noise (especially in a multipoint environment) will raise the noise floor of the receiver at the other end of the link, lowering its SNR, which will decrease the distance the communication link can reach; so any Class A or Class AB power amplifier should be specifically designed to output a minimal amount of additive wideband noise. Depending on the frequency, power, band, modulation, and service, certain frequency-stability requirements are mandated by law, or are simply required for proper demodulation at the receiver and/or to prevent adjacent channel interference.

9.2.2 Transmitter design

The generic linear double-conversion transmitter of Fig. 9.3 could just as easily be a single-conversion unit. The choice as to whether to employ single or double conversion is based on the transmit RF frequency and the much lower input frequency. At higher RF operation, double conversion is required to properly suppress, through low-cost IF filtering, the close-in sum or difference frequencies of the input signal mixing with the LO, as well as the LO feedthrough, while also suppressing spurious mixer responses caused by the limitations of a real-life IF filter's limited realizable percentage of bandwidth. In other words, if we desire a high transmitter RF output frequency with only a single conversion stage—yet we have a low input frequency that must be converted to this much higher carrier frequency—we would need a filter with an impossibly narrow bandwidth and ultrasteep skirts, along with the escalating problem of the inevitably high group delay variations that would severely distort the output signal.

Taking the first element of Fig. 9.3, the antenna, we see that it is at DC ground through the inductor to protect the RF output filter (RF BPF) and the *solid-state power amplifier* (SSPA) against static buildup discharge damage. The RF BPF suppresses much of the transmitter-generated harmonics, wideband noise, IMD products, and out-of-band conversion frequencies. As an added consideration in FM service, most of the SSPAs will run saturated for maximum efficiency, which will create large harmonic output levels; these harmonics of the fundamental must be sufficiently attenuated by this last RF output filter. In fact, since the output filter is typically reflective in the stopbands, most of these undesired harmonic frequencies are actually reflected back into the SSPA, which will then create significantly higher-than-expected harmonic output from the transmitter. This annoying effect will necessitate an increase in the rated attenuation of the RF output filter by approximately 15 to 20 dB.

If the SSPA is to be operated at less than saturation for digital or SSB voice communications, it must be designed to maintain the desired output power with low distortion levels for the RF signal. This means we may have to run the power amplifier at up to 10 dB (or more) under its maximum output power rating. To put it another way, the SSPA has been "backed-off" in power by up to 10 dB in order to maintain the required linear operation that a particular modulation technique demands for decreased spectral regrowth (a form of

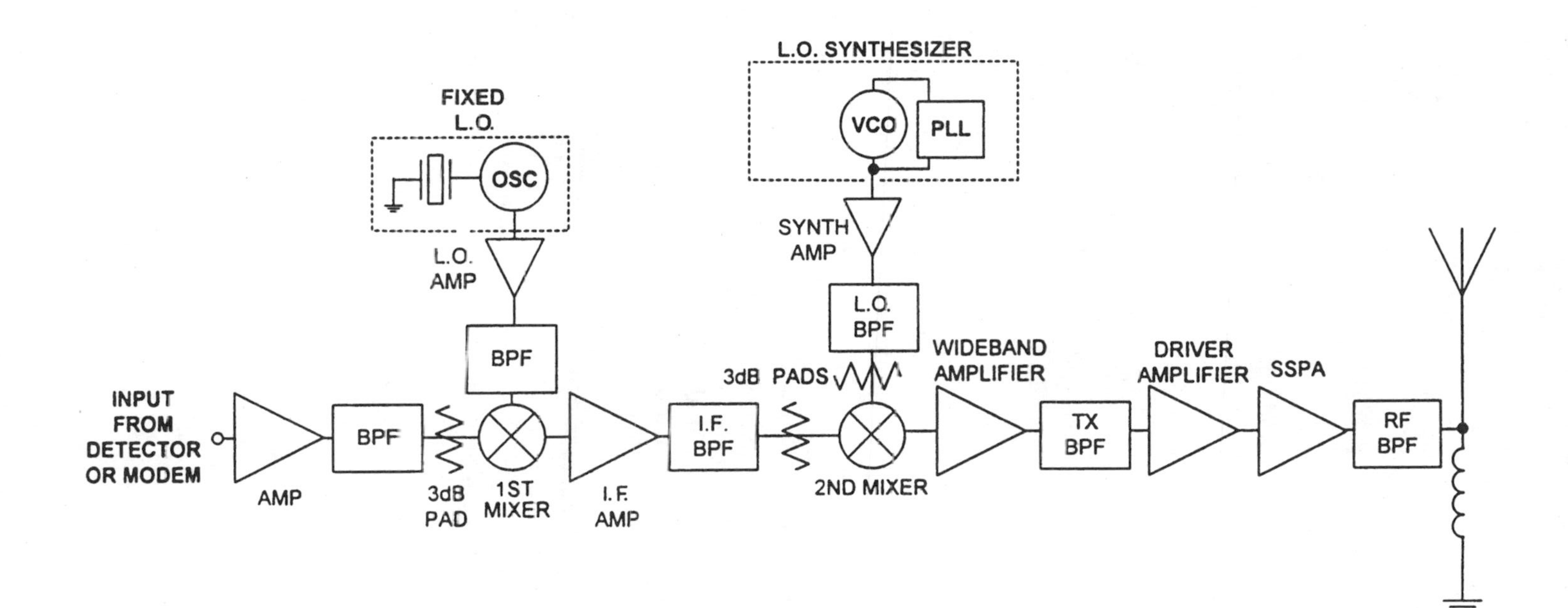

Figure 9.3 A standard linear transmitter.

IMD) and to maintain a good BER. The SSPA must also be exceptionally stable, and not begin oscillating nor decrease power with the wide impedance variations encountered in a mobile or portable operation. The impedance variations are caused by the antenna being constantly presented with many conducting structures that are passing nearby within the mobile environment. As well, the driver amplifier must be able to supply the necessary input signal amplitude to the SSPA, and without significant distortion levels.

The TX BPF must be tight enough to attenuate the LO feedthrough, the undesired sum or difference frequency, and other mixer products—but not contribute significantly to the signal's group delay variations nor amplitude ripple. The wideband amplifier is not always required, but will permit all the undesired mixer products to pass through to the TX BPF, where they are reflected back toward the mixer stage. The wideband amplifier, because of its high isolation, will attenuate most of these reflected signals so that they do not create increased IMD levels within the mixer. Many transmitter designs may suppress reflections from the TX BPF's stopbands by use of a diplexer or attenuator pads, or may simply dispense with all of the above and place the TX BPF directly at the output of the mixer port. This choice will depend on the output power from the mixer; the less output power, the lower in amplitude the reflections, and the less the requirement for their suppression. (These mixer products reflect back into the diode ring of a passive mixer stage, causing mixer diode imbalance; with the resultant increase in spurious outputs and decreased third-order intercept points.)

The second mixer itself can be a double-balanced type, of a level 10 (10 dBm LO power) or higher, to suppress IMD. Typically, the IF input port to the mixer should never have a signal higher than 10 to 15 dB less than that at the LO port, or excessive intermodulation products will result. As an example, if the LO port is at +10dBm, then the IF input must be at or lower than 0 dBm. The 3-dB pad located at the LO port helps to present a 50-ohm impedance to the sensitive LO BPF. This LO BPF suppresses LO harmonics, which lowers mixer IMD, as well as wideband noise, which improves the mixer's NF. The synthesizer amplifier, SYNTH AMP, buffers the output of the synthesizer, minimizing VSWR, as well as increasing the synthesizer's output power to the mixer's nominal level to maintain its rated noise figure, IMD, and conversion loss specs. The synthesizer stage itself should have high stability and low phase noise so as not to degrade the downstream receiver's SNR and BER. The 3-dB pad at the input to the second mixer may be used to maintain a more even 50-ohm input impedance for the IF BPF, which will shift in characteristics if not presented with its design impedance. The IF BPF should be tight enough to reject the first mixer's products, its sum (or difference) frequency, the LO feedthrough, and excess noise (since increased bandwidth also increases input noise), yet loose enough to minimize group delay variations and sideband cutting. Selection of the proper filter topology, each with its own positive and negative attributes, is vital (see Chap. 6, "Filter Design").

The wideband IF amplifier stage, IF AMP, not only does amplification chores, but will also have high reverse isolation to prevent the mixer products from the first mixer from reentering its output port. Moreover, the IF AMP will provide a relatively decent 50-ohm termination for the IF BPF at the mixer port because most mixers, because of their continuous switching action, have difficulty maintaining anything close to $50 + j0$ ohms. The first mixer converts the low input signal up to the IF, while the LO BPF filters wideband noise, harmonics, and spurs created by the LO and/or its LO AMP. The fixed LO is a high-Q, and thus low-phase-noise and high-stability, crystal oscillator. The BPF and/or the AMP may or may not be present, depending on the modulation source and its requirements (usually a modem in commercial data communications equipment, or an I/Q modulator).

The transmitter's output may also be tapped by a directional coupler in order to feed the signal's amplitude information to a microprocessor to maintain proper transmitted output levels by an automatic level control (ALC) circuit. A temperature sensor and reverse power level signals may also be output to a controller to prevent SSPA damage or destruction.

The PA stage of a digital transmitter should have an SNR of greater than 65 dBc so as not to degrade the overall system's signal-to-noise ratio, while the phase noise of the LOs should be better than 95 dB/Hz at 10 kHz for a typical QAM transmitter. The digital transmitter's IF and RF filters should also pass the entire signal with no passband cutting, with an amplitude tilt of less than 2 dB, and with in-band ripple of less than ±0.5 dB. Spurious signal outputs into adjacent channels will normally need to be better than 65 dBc in most of the radio services.

9.3 Link Budgets

9.3.1 Introduction

Before any piece of hardware is designed, a *link budget analysis* must be performed. This will tell us how much NF and gain the receiver requires, and how much power the transmitter must output, in order to reach our desired range at a specified BER and SNR. Performing a basic link budget analysis is simply calculating the final SNR and signal strength at the output of a wireless receiver after the signal is sent from the transmitter, across the entire transmission path, and through the receiver (Fig. 9.4).

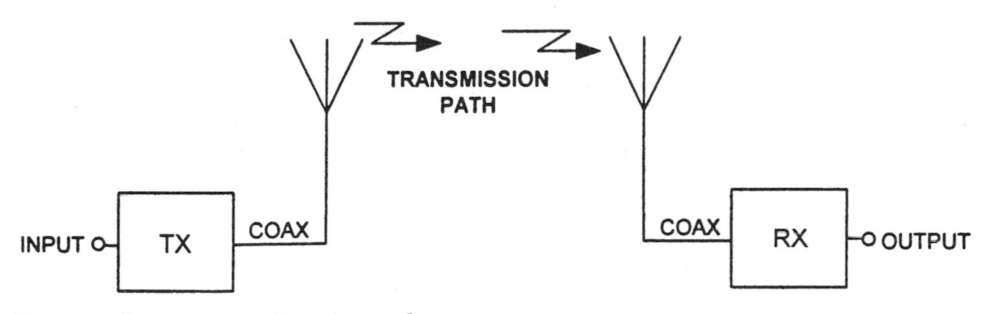

Figure 9.4 Transmitter and receiver path.

Because of the vagaries of wireless communications caused by atmospheric and multipath fading, a certain amount of *fade margin* will be necessary to make sure that the link will remain up—even under infrequent but severe weather conditions—for a certain percentage of the year. This fade margin is a safety allowance of excess receiver NF and gain, transmitter power, or antenna gain placed in our link budget to assure a dependable wireless connection, with a certain amount of permitted downtime (in seconds) per year. For instance, a 20-dB fade margin for digital communications systems can be added to the link budget, which will not only cover atmospheric anomalies and multipath, but also equipment aging and repairs. A common figure for dependable operability that could be expected for certain digital systems, and with this 20 dB of fade margin, might be 99.99 percent with a BER of 10^{-8} throughout a year-long period.

It is important when calculating two-way links that *each* direction of the link has the required link budget to function at a desired BER, considering that all duplex links may not be power-, bandwidth-, SNR-, frequency-, or even modulation-symmetrical.

9.3.2 Link budget design

After we find the range (in kilometers or miles) over which the communications link must reliably transmit information, we must then calculate the wireless link's *free-space path* loss. This is the loss, in dB, that occurs to an RF signal at a specific frequency over a specific range—but it does not account for *any* impairments.

Another figure we must obtain is the fade margin, as mentioned above. This will be required to assure the link of reception reliability during any unexpected, but persistent, atmospheric anomalies and multipath effects.

We would also like to find the minimum power that must be available at the receiver's output for proper demodulation within the modem or demodulator, and what SNR is required for the particular modulation and error correction in use.

In the following example case, it is being assumed that the RF designer will be told what the maximum transmitter equivalent isotropically radiated power (EIRP) will be. But in order to obtain the necessary BER at the receiver's output, a compromise will normally have to be made between transmitter output power (which must generally be minimized to the lowest level that assures reliable communications), and the receiver's NF and gain versus IMD generation. However, higher-gain antennas and lower-loss coax cable on either the transmit and receive side, or both, can sometimes be an easy way to increase the fade margin.

To perform a link budget analysis:

1. Calculate the free-space path loss over the desired link distance. As mentioned, free-space path loss does *not* include any losses caused by the atmosphere or by multipath, but merely accounts for the inverse square law signal spreading of the RF wavefront as it leaves the transmitting antenna

(for instance, if the distance from the transmitter source is doubled, the receiver will receive only a quarter of the energy it would have received at half this distance). This means that no matter how directional, and thus high-gain, an antenna may be, the power will still drop off commensurate to the inverse square law, and will decrease in field strength over distance. Thus, if a transmitter is sending at equal powers into two antennas—one low gain, the other high gain—the high-gain directional antenna merely *started off* with a higher field strength than would have been possible with the use of the low gain, omnidirectional antenna. By following the formula below, and working it out for various distances, we can obviously see that for every time the distance between the receiver and transmitter is doubled, a further 6 dB of path loss occurs:

$$L_P = 32.4 + 20 \log f + 20 \log d$$

where L_P = path loss, dB
f = frequency, MHz
d = distance, km

2. Decide on the fade margin. The higher the frequency, the longer the link path, and the greater the desired reliability, the more fade margin is required. Fade margins of 10 to 20 dB in 20-km digital microwave links are quite common, while higher frequencies and longer links may require up to a 30-dB margin.

3. Since the magnitude of the noise floor of a receiver will set the minimum signal level that will still be easily detectable above this noise—or a signal that will have a *positive* SNR—we can calculate the signal strength required at a receiver's antenna inputs, at a receiver's particular NF, to obtain a desired SNR at the receiver's output:

$$S_{\text{dBm}} = -174 + \text{SNR} + \text{NF} + (10 \log_{10} \text{BW}_{\text{N}})$$

where S_{dBm} = signal strength, in dBm, needed at the receiver's *antenna inputs* (at the receiver's front end), to obtain an output IF signal at a desired SNR in a 50-ohm system
SNR = required signal-to-noise ratio, in dB, for the type of modulation used, at the output IF (i.e., at the detector or modem input)
NF = noise figure, in dB, of the receiver
BW_N = The 6-dB bandwidth of the IF, in Hz (or, more accurately, the *noise bandwidth*)

4. Calculate the *power* that will be present at the receiver's input *after* its own antenna, and after its trip across the link from the transmitter.

$$P_r = P_t + G_t + G_r - \text{PL}_{\text{dB}}$$

where P_r = power, in dBm, present at the receiver's input

P_t = power, in dBm, delivered by the transmitter *into* its own antenna
G_t = gain, in dB, of the transmitter's antenna
G_r = gain, in dB, of the receiver's antenna
PL_{dB} = free-space path loss, in dB, between the transmitter and receiver antennas

5. Now, confirm that the transmitter is outputting enough power to overcome the free-space path losses and to account for the desired fade margin, and/or that the receiver will have a low enough NF for the desired SNR and enough gain for proper output signal strength into the modem or detector:

$$P_{OUT} = P_t - L_p + G_t + G_r - L_t - L_r + RX_{dB}$$

where P_{OUT} = power at the receiver's output into the modem or detector, dBm
P_t = transmitter's power output into its own antenna coax, dBm
L_p = free-space path loss as calculated above, dB
G_t = transmitter's isotropic antenna gain, dBi
G_r = receiver's isotropic antenna gain, dBi
L_t = loss of the transmitter's coax, dB
L_r = loss of the receiver's coax, dB
RX_{dB} = gain, dB, of the receiver section itself (including conversion and filter losses, and the front end and IF amplifier gains)

The following formula will tell the designer the signal strength output of the receiver stage when the signal strength at the receiver's antenna is known:

$$P_{OUT} = P_{SIG} + G_{dB;} - L_r + RX_{dB}$$

where P_{OUT} = signal strength, in dBm, available from the receiver's output into the modem or detector input
P_{SIG} = power, in dBm, picked up by the receiver's antenna from the transmitter's antenna
G_{dBi} = gain, in dBi, of the receiver's antenna
L_r = loss of the coax cable, in dB, between the receiver's antenna and its front end
RX_{dB} = gain, in dB, of the entire receiver stage—including filters, conversions, and amplifiers—from the front end to the output into the modem or detector stage

The following formula will give the designer the signal-to-noise ratio at the output of the receiver stage:

$$SNR = P_{OUT} - (-174 + 10 \log (BW_N) + NF + L_r + RX_{dB})$$

where SNR = signal-to-noise ratio present at the output of the receiver, dB

P_{OUT} = signal strength, in dBm, available from the receiver's output into the modem or detector input

RX_{dB} = gain, in dB, of the entire receiver stage—including filters, conversions, and amplifiers—from the front end to the output into the modem or detector stage

L_r = loss of the coax cable, in dB, between the receiver's antenna and its front end (presented as a *positive* number only)

NF = noise figure of the receiver, dB

BW_N = 6-dB bandwidth of the IF (or, more accurately, the noise bandwidth), Hz

The above calculations should give us an indication of the signal power, noise figure, and gain required across a particular link.

Real-life path losses. Since free-space path loss in a microwave link will not be the only loss encountered, other impairments must be added to this to establish a worst-case scenario within your link.

The path losses attributed to rain begin to dramatically increase at frequencies above 10 GHz. A heavy rain shower can attenuate a 10-GHz signal by 2 dB per kilometer (or more). As these frequencies increase, so do the losses. It is, however, only the space between the transmitter's and receiver's antennas that actually has this rainfall that will force such severe attenuation, so much of a transmission path may in fact be clear. A wet snow has a similar attenuating effect, while a dry snow will have little significance, even at the higher microwave frequencies. Dense fog will attenuate a 10-GHz signal by up to 1 dB per kilometer, with much smaller attenuation levels as the frequencies decrease. Water vapor and atmospheric oxygen absorption create attenuation at 18 GHz and above, with various high attenuation peaks at several microwave frequencies. *Fresnel zone* clearance inadequacies, atmospheric reflection, and scattering will also conspire to increase path losses. But not all impairments are continuous attenuation mechanisms, especially the atmospheric effects causing reflection. Nonetheless, all such losses can be compensated for by increasing the power from the transmitter, decreasing the receiver's NF and increasing its gain, and raising the transmitter and receiver antenna gain. The end effect we want is to make sure that there is enough signal amplitude out of the IF of the receiver to drive the modem or demodulator, and that that signal has the SNR required for reliable demodulation at a low BER.

9.3.3 Will it work?

Consider a simple link budget analysis of the communications system and path of Fig. 9.5.

A 40-km link, with no obstructions, is needed to operate dependably at 2.4 GHz with a transmitter/antenna combination that can output a +47 dBm (50 W) EIRP signal with our modulation of choice. We find that the receiver must, through all atmospheric conditions, maintain a signal into the detector

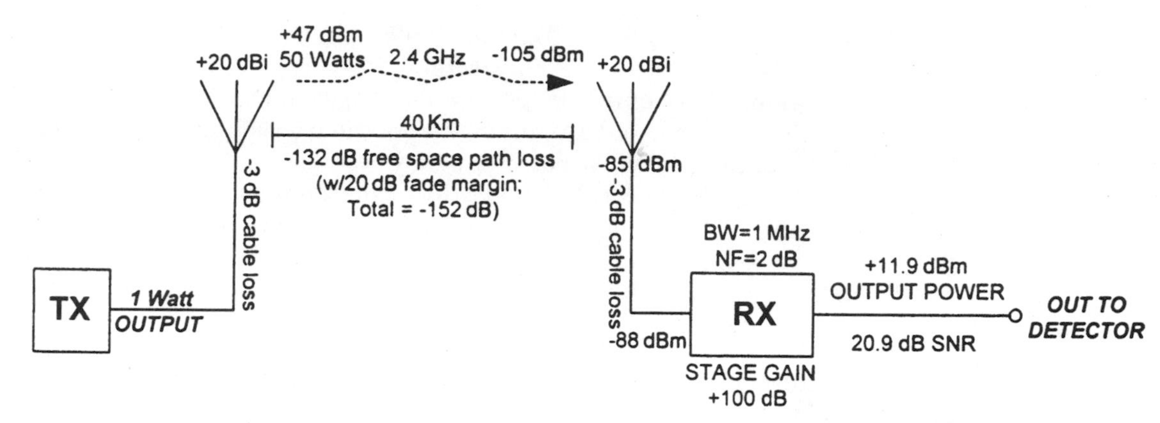

Figure 9.5 Basic link budget analysis.

or modem of +10 dBm, with a minimum SNR of 20 dB, at a bandwidth of 1 MHz. What would be the specifications of the receiver necessary to meet this requirement?

First, we must calculate the free-space path loss between the transmit and receive antennas for a frequency of 2.4 GHz, then add 20 dB for the link margin to be assured of reliable operation. We can now subtract the calculated path loss, with link margin, from the EIRP of the transmitter system to arrive at the power level that will be present at the receive antenna under a worst-case situation of −20 dB over and above the free-space path loss. We find this figure to be −105 dBm. The receive antenna, with its 20 dB of gain, will take the −105 dBm signal and amplify it by 20 dB to −85 dBm, while the 3 dB of cable losses will drop the −85 dBm signal level to −88 dBm. Thus, −88 dBm of signal is finally placed at the receiver's front-end input from the original transmitter signal.

We can see that a total gain of 100 dB will be required of the receiver to meet or exceed the +10 dBm output power requirement into the modem or detector, or 100 dB + (−88 dBm). We have left ourselves a small gain implementation budget of about +2 dBm to cover design-to-realization (real-world) losses, for a total signal level out of the receiver of +12 dBm. We then employ the SNR formula as shown above to find that we have a good SNR of over 20 dB at the receiver's output and a small 0.9 dB for the SNR implementation budget, for a total of 20.9 dB. We have found that the link will be quite dependable over almost all atmospheric conditions and during most parts of the year.

The final value of the communications receiver's output signal above the noise depends on the receiver's bandwidth, its NF (which is why receiver NF is so important, since a 1-dB improvement in NF translates to a 1-dB improvement in SNR at the detector), antenna temperature, antenna gain, cable losses, and transmitted EIRP. However, the actual signal *amplitude* at the receiver's detector input depends only on the receiver's antenna gain and cable losses, receiver RF and IF gain, and the transmitted power.

9.3.4 Link budget issues

Fresnel zone clearance must be considered when setting up a microwave link across the earth. This is because optical line of sight between the transmitting and the receiving antenna is only one factor in most microwave links, the other being the first Fresnel zone clearance.

The Fresnel zone is referred to as the *radio line of sight,* and will need more clearance that the optical line-of-sight since a radio wave, when it passes near an object (such as a building or a mountain), will become defracted or bent, causing the radio wave to become degraded, even if the obstruction is many feet below (or to the side) of the radio wave.

This is why we must confirm mathematically, and not optically, that we have sufficient Fresnel zone clearance above any mountain or building to avoid attenuation of the transmitted signal at the receiving station. The following formula can be used in conjunction with Fig. 9.6 to verify this:

$$h = 72.1 \sqrt{\frac{d_1 d_2}{f(d_1 + d_2)}}$$

where h = clearance between the top of any obstruction and the direct line of sight that is required for zero attenuation to the transmitted signal, feet

d_1 = distance between the transmitter and the obstruction, miles.

d_2 = distance between the obstruction and the receiver, miles.

f = frequency of the transmitted signal, GHz

If the answer is required in meters:

$$h = 17.3 \sqrt{\frac{d_1 d_2}{f(d_1 + d_2)}}$$

where the terms are as defined above except that h is in meters and d_1 and d_2 are in kilometers.

We could ignore the Fresnel zone clearance entirely but, depending on the frequency and the geometry of the obstruction and whether the obstruction is close to the optical-line-of-sight, the additional path losses could be anywhere between 6 and 20 dB.

As we have discovered, the noise floor of a receiver system is a critical issue. To calculate this, including the *antenna's* NF as well as the receiver's NF, we can use the formula below. This formula works quite accurately for terrestrial total receiver system NF calculations, but assumes a receiving antenna's noise temperature to be 290 K, which is a reliable antenna temperature estimate for all antennas in an earthbound link environment. And since the formula calculates the signal strength, in dBm, required at the input of the receiving antenna for the receiver to output at a 0-dB SNR, we would have to confirm that the transmitter on the other end of the link has the power to permit us,

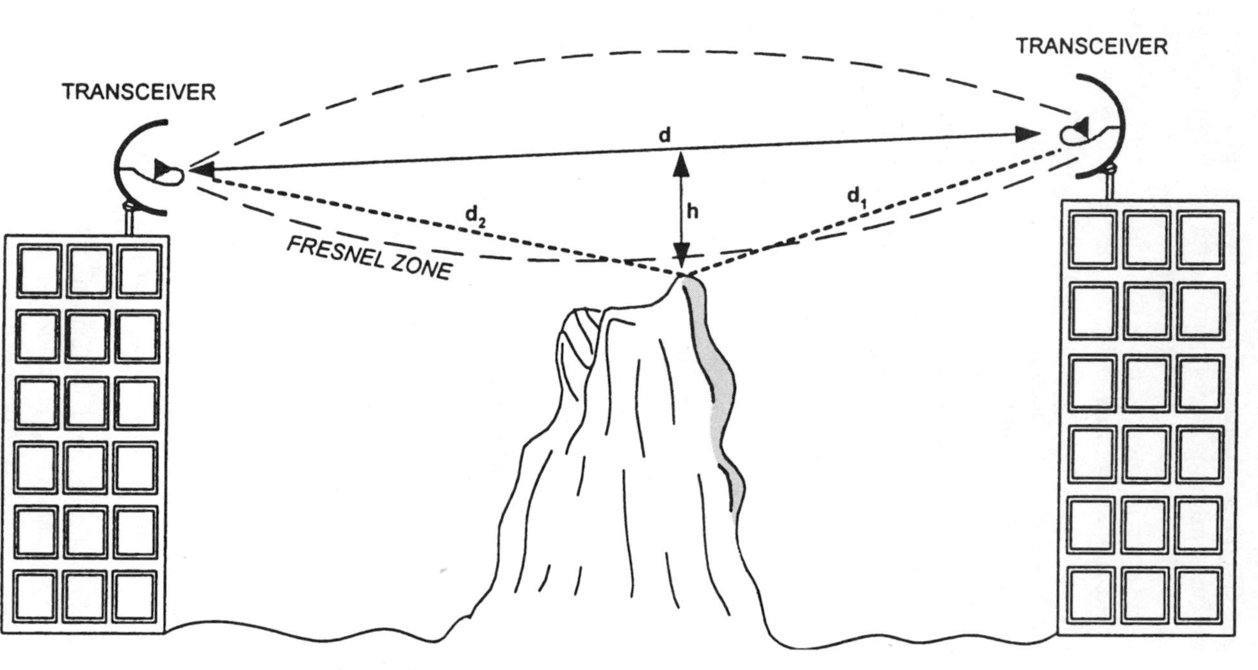

Figure 9.6 Measurement of Fresnel zone clearance.

with sufficient fade margin, to obtain this SNR (which is the value as required by the receiver's modem or detector in use at the output of the IF):

$$S = 10 \log \text{BW} + 10 \log [290 (10^{\text{NF}/10} - 1)] - 198.6 - G_A$$

where S = entire receiver's sensitivity, in dBm, with a 0 dB SNR at its output

BW = system's IF bandwidth, Hz

NF = receiver's noise figure, dB

G_A = receiver's antenna gain, dB

Another factor that affects the noise of the receiver system is the antenna's orientation. If the receiving antenna is pointed more toward the sky (but *not* toward the sun) in order to receive a signal from a transmitter that is placed on a mountain top, as an example, then its noise temperature, and thus its NF, will be less than if it were pointed at the ground (the ground has an approximate noise temperature of the above-mentioned 290 K). However, even if the receiver antenna were pointed at the coldest region of space (to communicate with a satellite transmitter), its sidelobe reception would still increase its noise temperature.

9.4 The Complete System

9.4.1 Introduction

Complete communications systems design must take into account many variables, such as the link itself, general transmitter and receiver specifications, type of modulation, data rate, and BER. Most of these issues have been or will be covered in other sections of this book. This section, however, will concentrate on the overall technical issues that affect systems performance.

9.4.2 Wireless system design

At the beginning of any wireless design project, certain system parameters must be established. In a digital data link, we would need answers to some of the most basic of questions: What will be the frequency of operation? Will the system need to be full duplex or half duplex? Will transmit and receive be separated in frequency [frequency division duplex (FDD)] or in time [time division duplex (TDD)]? What is the bandwidth, modulation, fade margin, gain, BER, SNR, phase noise, group delay, transmit power, and receiver NF? What number of conversion stages will we need for the transmitter and receiver sections? What are the system's required channels, frequency plans, frequency stabilities, dynamic ranges, third-order intercept points, and frequency inversions? Most of these specifications are interrelated, and must be answered, since digitally modulated radio systems have to be especially designed for low levels of phase noise, group delay variations, IMD levels, amplitude ripple and shape, and frequency variations so as not to adversely affect the BER of a phase/amplitude-modulated digital signal.

After all specifications and requirements have been plotted for the communications system, a detailed block diagram showing the gain, frequency, bandwidth, and signal levels (in dBm) should be drawn for the receiver at both its highest expected input signal level (to confirm that no section is being overdriven), and its lowest expected input signal level (to confirm that the output power into the detector is adequate); as well as for the transmitter section. These diagrams assist in verifying that the gain and bandwidth distribution is appropriate and that spurious and harmonic suppression will be effective. After this is accomplished, the actual circuit design for each stage can begin.

A generic wireless system. Figure 9.7 shows a generic digital linear *time division duplexed* (TDD) transceiver with its own built-in modulator and demodulator. (Other digital transceiver systems may drop internal modulation and demodulation entirely, and begin and end with the IF. A modem would be placed at the transceiver's IF ports, which would feed the transmitter section with a modulated IF signal and demodulate the IF from the radio's receiver section.)

On the transmit side, the baseband digital data is input into the DAC, which converts it into serial analog data, which is filtered and sent into the I and Q inputs of the modulator. The I/Q modulator takes the I and Q signals and mixes them in their own DBM, with one DBM fed an in-phase LO and the other a 90 degree out-of-phase signal. This action attenuates the indeterminate frequencies produced in this multiplication process. The outputs are added in a linear mixer and output as a single IF signal—along with attenuated sidebands and the undesired carrier that were not fully suppressed because of phase and/or amplitude mismatching between the I and Q legs. The IF is then amplified, filtered, up-converted, and amplified again before being sent out of the antenna. (In some lower-frequency systems, the signal directly out of the modulator can be used for transmission into space, and would need to be simply filtered and amplified before being transmitted from the antenna.)

On the receive side the transmit/receive switch (T/R) is switched to route the incoming signal through the initially low-insertion-loss front-end attenuator (employed to lessen very high-level input signals), into the LNA (to increase the signal power and lower the receiver's NF), through the image filter (to reduce image noise and interference), into the down-conversion stage (to heterodyne the RF to the lower IF), through the diplexer (to decrease reflection-generated mixer IMD), into the IF strip with AGC (to decrease/increase the signal level and supply selectivity), and into the demodulator stage (or modem), thus outputting a digital data stream.

The following is another example of a good general systems design breakdown, this one of an ordinary FM transceiver.

In the full duplex radio of Fig. 9.8, a duplexer is placed at the transceiver's front end to allow simultaneous TX/RX on different frequencies while employing the same antenna. This duplexer must have a high enough level

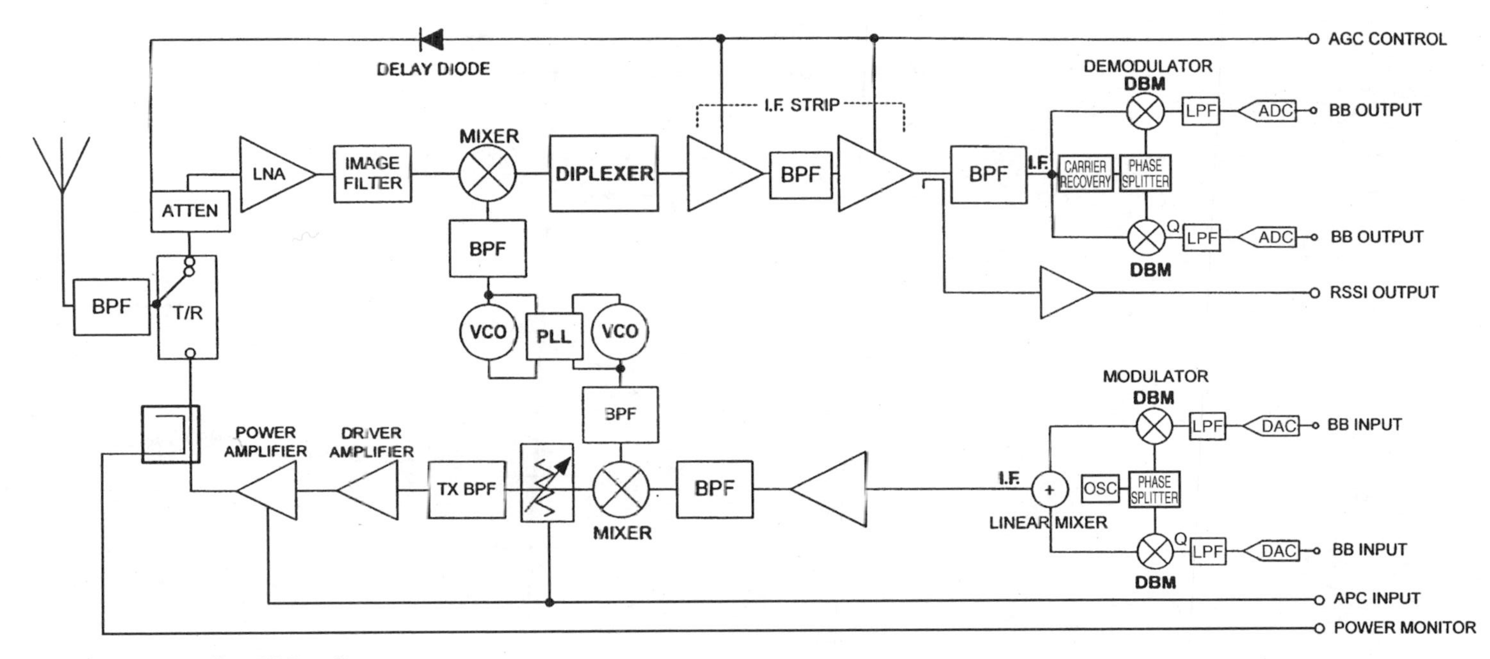

Figure 9.7 A complete TDD radio.

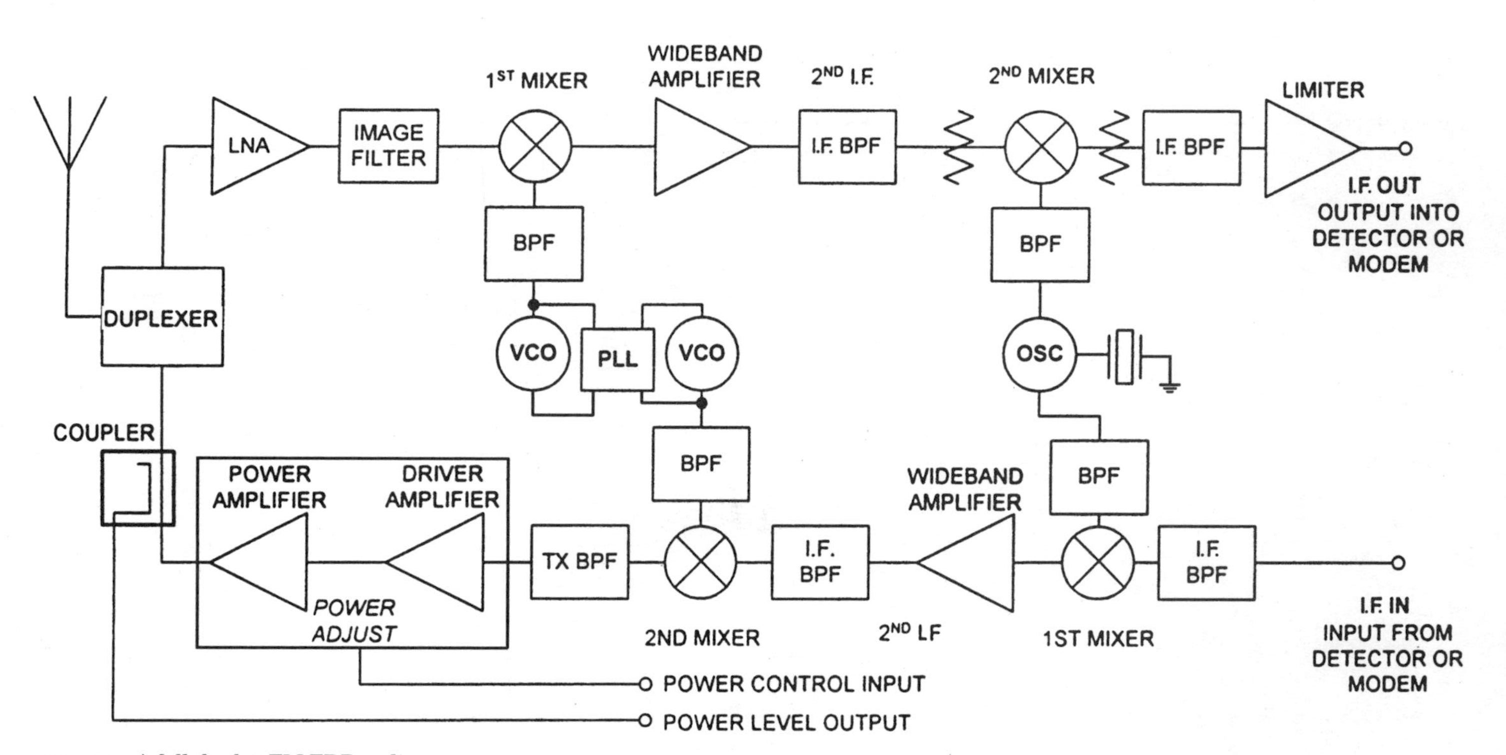

Figure 9.8 A full-duplex FM FDD radio.

of attenuation to prevent the high-power transmit frequency and power amplifier (PA) noise from negatively affecting the very sensitive receiver, while also attenuating transmitter harmonics and some of the receiver's image frequency, and at the same time not adding excessive group delay variations.

The LNA will set most of the receiver's NF, and thus sensitivity, and must be high in gain and low in internal noise generation. However, just deciding to design an LNA with the lowest NF and the highest gain will result in a receiver with poor intermodulation performance, since the mixer's third-order intercept point (IP3) will be reduced by the gain of the LNA (this would be true even if the LNA itself had an infinite IP3). This makes the first mixer, and thus the entire receiver, very intolerant of strong input signals—the mixer is basically predetermining the entire receiver's IP3. As the first mixer is such a critical element in receiver IP3 performance, DBM diode mixers are usually employed for this purpose, since they have a far higher IP3 than most active mixers. The DBM thus permits more gain in the LNA stage, which lowers the effect of noise contributions of the following stages—including the high-NF first mixer stage. Obviously then, LNA gain and first mixer IP3 must be selected with care to maximize receiver IP3 and minimize NF so that we may obtain a highly linear and sensitive receiver. Most sensitive receivers will have an image filter placed just before this mixer, which will further assist the diplexer's filtering action, and will almost entirely eliminate the LNA's own self-generated noise within the undesired image band from adversely affecting the receiver's SNR.

The wideband amplifier that follows the first mixer has a high reverse isolation to prevent the large mixer-generated sum frequency from reflecting off the reflective stopbands of the IF BPF and reentering the first mixer, which would increase IMD. The IF BPF rejects all signals that are not on channel, which is an important task, as the gain of the IF strip is quite high (usually at least 90 dB). In some radios, rejection of the second image noise may be necessary, and the IF BPF provides this capability as well. The attenuators into/out of the second mixer stage help to increase the return loss of the conversion stage, thereby decreasing IMD. This is especially vital at the IF port, where reflections off of the filter's stopbands will be quite severe. In this FM receiver case, a limiter sets the IF_{OUT} amplitude into a discriminator, with IF_{OUT} then flowing into signal-processing integrated circuits and to an output speaker.

On the transmitter side, the modulated IF is placed into IF BPF at the IF in port, converted up in frequency by the first mixer, amplified and filtered, then mixed up by the second mixer to RF. A wideband, high-isolation amplifier at the first mixer's output port is used to dampen reflections back into the mixer, with attenuators performing a similar function. The second mixer's TX BPF suppresses wideband transmitter noise, as well as harmonics and mixer products. These two frequency-conversion stages are necessary so that we may economically filter out the close-in sum (or difference) frequency, as well as the inevitable LO feedthrough, from exiting the transmitter in strength. The driver amplifies the signal to an input level that is acceptable to properly drive the power amplifier (PA). In this FM case, the PA will run saturated, which demands that the driver has enough gain to keep it so. Some of the signal is tapped by the coupler for output power confirmation to a microprocessor, with the majority of the signal sent to and filtered by the duplexer, and broadcast out of the antenna.

The power amplifier stage can be the most difficult part of the transmitter to design. The PA must be very tolerant of low output return losses, which are caused by the large impedance variations created by the mobile antenna and its rapidly changing physical environment, as well as reflections off the duplexer's reflective stopbands. Not only must the PA not be destroyed by these VSWR variations, but must also not degrade total performance specifications. In fact, many FM services demand that the PA be able to vary its power output over some specified range, be highly efficient (for increased battery life), generate minimal harmonic levels, and have enough gain for efficient saturated output.

9.4.3 System design with RFICs

Most wireless system designs (Fig. 9.9) now rely heavily on the ever-increasing use of *radio-frequency integrated circuits* (RFICs). RFICs may contain the complete LNA/mixer stages, the entire IF stages, or even the total transceiver. However, a complete transceiver on a single RFIC chip is usually available only for low-data-rate or low-cost voice applications—at least for the foreseeable future, and with the single exception of the upcoming one-chip solution for *Bluetooth* devices (for further information on this important wireless technology, please go to *Bluetooth.com*). Nonetheless, higher integration levels are slowly becoming a reality as more and more companies attempt to make a complete wireless system on a chip.

Depending on the frequencies and specifications required, adopting RFICs can significantly lower design and production costs, as well as the physical size, of the complete radio system. In fact, multiple RFICs have already replaced many discrete systems, as high levels of integration can allow one RFIC to replace dozens, if not hundreds, of components. This not only decreases the cost of designing the communications system in the first place, but also simplifies the actual construction of the systems themselves.

RFICs are available for each part of a radio, with the level of the desired integration depending on the design flexibility required. The design of Fig. 9.9 demonstrates low levels of RFIC integration, but still allows for a far more rapid time to market than a completely discrete design. Most of the chips for this, and similar radio designs, are available from MAXIM, RFMD, Mini-Circuits, National, Motorola, etc.

Many designs must still rely on discrete, or at least individual MMIC, amplifiers and mixers when RFICs are not compatible with a particular design goal, while other wireless systems may have to employ an almost entirely discrete design for price/performance reasons. Indeed, the RFIC solution is compatible with a particular wireless design only if the chip already

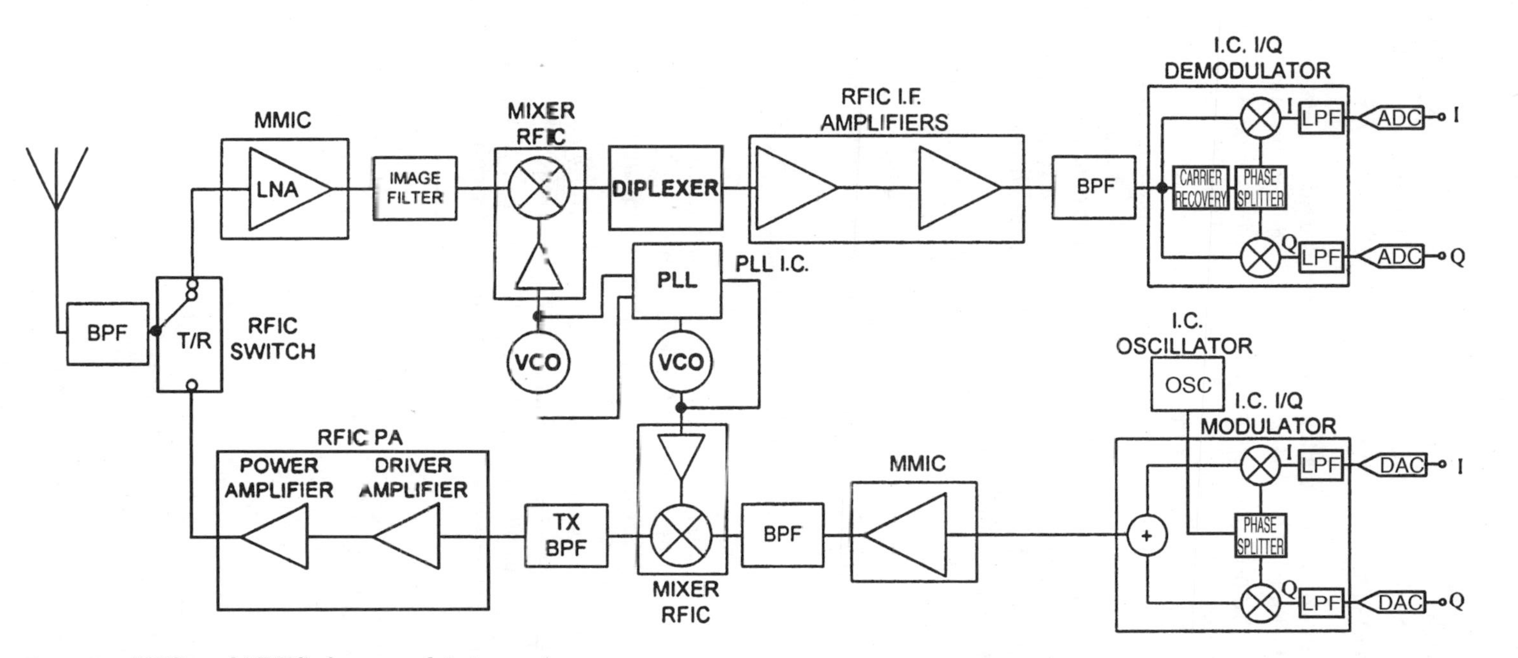

Figure 9.9 RFICs and MMICs for a complete transceiver.

exists to perform the required functions, or if volume production makes it economically viable to actually design and manufacture the RFIC itself—an extremely expensive option.

An RFIC design consideration. For some portable wireless devices, the receiver's front-end RFIC (consisting of a conversion stage and RF and IF amplifiers) may have high-impedance outputs, which interface with external discrete IF filters. These high impedances are adopted to decrease current drain for the chip, which can become quite heavy because the RFIC's built-in IF impedance matching output buffer is attached to a normal 50-ohm circuit. Thus, the RFIC's output may be connected to an external high-impedance IF filter (of a few thousand ohms) in order to decrease this drain, or with the internal buffer bypassed altogether and directly attached to a high impedance.

To explain further: For the lowest current drain possible in any amplifier stage, there is a nominal output impedance, which will depend on the voltage of the power supply and the RFIC's output IP3 requirements. (This is exactly the same idea behind increasing the efficiency in any power amplifier). In other words, the RFIC buffer's output intercept point will be governed by the impedance of its load (R_L), and the buffer's own current consumption (I_Q), or:

$$\text{OIP3} \approx 40 + 10 \log_{10}\left(\frac{R_L \times I_Q^{\,2}}{2}\right)$$

where OIP3 = third output intercept point, dBm
R_L = impedance of the RFIC buffer's load
I_Q = RFIC buffer's current consumption

This demonstrates that the higher the buffer's current consumption and/or the higher the load resistance, the higher the OIP3. By increasing the load resistance, it can be seen that the buffer current can be safely decreased while maintaining the desired OIP3—an important consideration in most portable applications.

As an example of the current savings that can be realized with the above technique, a front-end receiver chip from RFMD, the RF2418 (Fig. 9.10), which combines an LNA, mixer, and IF buffer on a single chip, will draw 15 mA from a +3 V DC supply with a 3-dB NF, a −7 dBm IP3, and a +18-dB gain into a 50-ohm filter. However, if the IF output of this buffered chip is connected to a 500-ohm load, the gain will not only increase to 24 dB, but the RFIC's current draw will drop to 6.5 mA.

Why do RFIC designers even bother adding this problematic buffer stage? Because these engineers are able to efficiently match only the naturally high output impedance of the RFIC to the normally low input impedance of an external filter with this active buffer stage—instead of applying bulky passive *LC* matching. However, as stated above, the matching buffer will consume a lot of power if a low output impedance is used, which is why some low-power RFICs will have these high-impedance output ports instead of the normal 50-

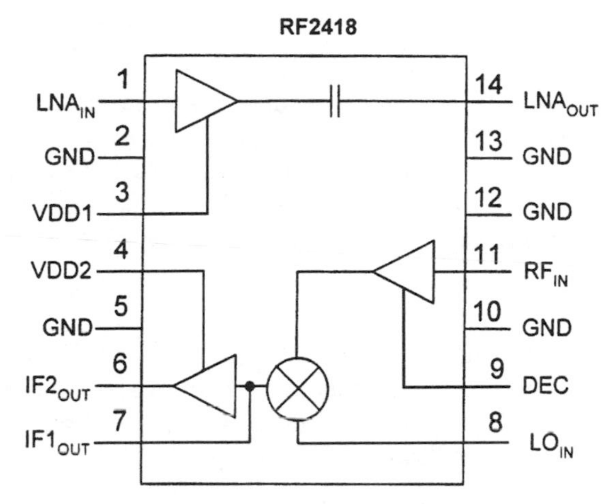

Figure 9.10 The RF2418 IC from RFMD.

ohm matched outputs. And in the case of the RF2418 RFIC, its buffer may be completely bypassed for an even larger power consumption savings, with the IF output now going into the input of a high-impedance filter stage.

9.4.4 System issues

Within communications systems there are certain inherent general design problems that the engineer should be aware of:

1. The Class A or Class AB power amplifier of a microwave transmitter in multipoint service (where many clients are served by a single central hub transceiver site) should be designed to transmit minimal *additive wideband noise.* This wideband noise effectively lowers the SNR at the central hub receiver when in this multipoint environment, since there may be hundreds of client transmitters that are on at the same time—whether modulated or not.
2. *Adjacent channel interference* (ACI) problems can be alleviated by employing different polarizations or different frequencies in adjacent sectors in a multisectorized hub antenna environment, by using sectorized hub antennas with proper side and back lobe rejection, by avoiding excessive hub transmitter power level outputs.
3. *Amplitude ripple* and *tilt* across a wideband digital channel must be minimized to a level not to exceed 0.5 dB for QAM and QPSK so as to decrease BER degradation.
4. Such esoteric impairments as *in-channel tones, transient bursts,* or *TDMA timing issues* can create crippling cochannel interference or decreased BER. In-channel tones and transient bursts can be lessened by redirecting the client antenna (if possible), by using a more directional antenna, or by

tracking down and eliminating the interferers (if possible). Timing is a control issue, and will normally not be your concern.

5. *Frequency stability* is quite critical for the upstream channel, especially when wideband cable modems are employed as the modulator/demodulators. In these cases, stability must be better than ±12 kHz to catch a standard cable modem's preamble (or the link will not lock up), as well as to maintain the BER over time with no automatic reboots. A longer preamble that can tolerate a frequency spec of ±25 kHz is possible with some cable modems.
6. *Group delay ripple* (GDR) must be no more than 75 ns for QAM-64, and less than 200 ns for QPSK, as increased GDR increases ISI. Simple analog design considerations (mainly for the system's bandpass filters) and digital adaptive equalizing will keep GDR in spec.
7. Increased system *linearity,* through hardware design and/or the appropriate SSPA back-off at the hub and client transmitter, will lessen BER degradation of a digital signal by decreasing intermodulation distortion levels.
8. *Multipath* causes phase cancellation, which creates both amplitude and phase distortions of the desired signal. This decreases the received signal's strength, thus decreasing SNR and increasing BER. It also creates undesirable amplitude notches or slopes, causing increased ISI. Increasing the gain, and thus the directionality, of the antennas; proper location of the hub transceiver (such as not placing it in front of a tall building or mountain); and using equalizers will all mitigate, but cannot eliminate, multipath problems.
9. *Near/far* receiver issues can mean some compromises in transmitter and receiver design. Depending on how close a client is to the hub, a close-in receiver may become saturated by the constant power output of the hub transmitter in a multiclient environment. This can be significantly reduced by using receiver front-end attenuators, utilizing the natural shadowing effect created by a high-mounted hub antenna (allowing the close-in receiver to be under the main lobe of the transmitter's antenna), employing different gain antennas for the close-in client receiver to that of the farther out client receiver, or using a switchable front end that will permit the LNA to be bypassed if the client receiver is too close to the hub transceiver.
10. *Phase noise* created by the nonperfect nature of a real LO will degrade the BER and increase the ISI. A digital wideband client receiver, for example, should possess a phase noise spec −85 dBc/Hz at 10 kHz, or better, to lessen BER degradation of the incoming signal.

Assorted issues. The hub antenna in a multipoint system environment must be located on as high an area and/or tower as practical. This will allow the coverage of the largest amount of client transceivers, since most nonbusiness antennas will normally be located at a single-dwelling roof level—and may be up to 15 to 20 miles away. This permits clearance for the Fresnel zone and increased mitigation of hub multipath problems, with the antenna as the highest local structure.

The *implementation margin* (IM) is another important aspect of systems design. IM is the decrease in SNR, and the corresponding increase in BER, that occurs in a system from the design to the actual building of the radio. The IM losses must be accounted for by increasing the required SNR of the radio to compensate for this effect during the design phase. When all of these modem and radio impairments in an imperfect practical wireless link are added—excluding the fade margin—an IM of up to 6 dB is common in high-data-rate radios. These hardware impairments can be caused by excessive phase noise, amplitude errors, noisy carrier recovery, jitter, group delay variations, noise through excess bandwidth, nonlinearities, thermal noise, adjacent channel interference, frequency instabilities, etc.

The choice as to whether we should use *frequency division duplex* (FDD) or *time division duplex* (TDD) in a radio system should obviously be addressed early in the design cycle. FDD radios operate with separate transmit and receive frequencies—isolated from each other by a duplexer filter—to allow the radio to transmit and receive during the same time period. A TDD radio utilizes the same frequencies for both transmit and receive, but employs a high-isolation switch in the transceiver's front end, near the antenna, to switch between transmit and receive during different time periods. Normally the choice as to which one to exploit will be dictated by the *systems engineer* on the project, as will most of the other system specifications.

Chapter

10

Wireless Issues

There are many important topics in wireless communications that must be addressed, not only for a full understanding of RF design, but also to be able to successfully implement complex circuits. Noise, EMI, PCB layout, prototyping, FCC rules, etc., are all issues of significance in the world of wireless.

10.1 Noise in Components and Systems

Noise is of crucial concern in radio, since the higher a signal is above the noise, the higher will be its SNR and the farther away it can be detected with a desired BER. There are two primary classifications of noise: *circuit generated* and *externally generated.* Both are unavoidable and limit the possible gain of any receiver's amplifiers. However, noise can be minimized by careful and cautious circuit and systems design. For instance, if we employ LNAs and tight filtering in the front end of receivers, and decrease the noise contributions from local oscillators and the image frequency, and use proper shielding and layout techniques, we can significantly improve on the noise floor amplitude.

Noise manufactured within circuits and systems produces a haphazard and fluctuating voltage that varies widely in frequency. *White noise* (also referred to as *Johnson* or *thermal noise*) is created by a component's electrons randomly moving around under the influence of thermal energy. *Shot noise,* because of its characteristics, can also be considered another type of white noise, but is caused by electrons entering the collector or drain of a transistor, and by the haphazard movement of electrons across any semiconductor junction.

Zener diodes are especially problematic as a noise contributor because of their shot noise effects. In fact, all zeners will add shot noise, but the diode's designers will have minimized other, more complex, noise contributors in "low-noise" zener types.

External noise is caused by not only artificial sources of electromagnetic interference, such as dimmer switches, car ignitions, and electric motors, but also by natural sources, such as the static created by atmospheric lightning. External noise also includes *space noise,* produced by solar flares and sunspots, and *cosmic noise,* induced by the stars radiating interfering signals in all directions.

Circuit-generated noise power, in watts, can be calculated by the formula shown below. This simple formula states that the two contributors to noise in a circuit are the temperature and the bandwidth of the circuit; the lower the temperature and the lower the bandwidth, the lower the noise contribution. The actual carrier frequency of the signal itself has absolutely no effect on the production of this noise:

$$P_N = KTB$$

where P_N = noise power, W
K = Boltzmann's constant, 1.38×10^{-23}
T = circuit temperature, K
B = circuit bandwidth, Hz

A related noise contributor is from an outside origin, such as a signal source, and is created by the same mechanism as above. It is referred to as *source noise,* and can be calculated by:

$$\text{NF} = 10 \log_{10} \left(\frac{P_{NO}}{P_{NI}} \right) \qquad \text{at 290 K}$$

where NF = noise figure, dB
P_{NO} = output noise power, W
P_{NI} = input noise power, W
290 K = the reference temperature used in most measurements, in kelvins.

10.2 Electromagnetic Interference

10.2.1 Introduction

Attenuating *electromagnetic interference* (EMI) and *radio-frequency interference* (RFI) to the lowest levels possible is a major requirement of most wireless designs. Undesired electromagnetic radiation escaping from a radio's enclosure will not only interfere with neighboring wireless equipment, but small RF levels flowing *within* the radio cabinet itself can destroy the proper operation of an otherwise solid design.

10.2.2 Designing for EMI suppression

Analog EMI suppression. Suppression of EMI is necessary for all designs—both wireless and nonwireless—because of rigid European and strict FCC regulations, as well as for protection of your design against improper operation.

Metallic shielding of synthesizers and oscillators to prevent coupling energy into other circuits, or to prevent other circuits from injecting energy into the frequency sources, is vital. Any such EMI propagating out of a frequency source can decrease the isolation between stages and/or cause harmonic or mixing products to form in the rest of the radio's stages, while EMI entering a synthesizer or VCO can create undesirable spurs. Traces leaving a frequency source should be shielded by placing them at a lower PCB layer (in stripline form), which is especially important from high-power frequency sources, such as those used to feed diode mixer conversion stages. However, when shielding of the top layer is employed to prevent microstrip traces from radiating, the characteristic impedance of the microstrip can be greatly affected by the top and side shield enclosure if it is placed too close to the track.

Temperature-reducing perforated shielding should be utilized with caution—and never when a design is to operate above 2 GHz—as harmonics of the fundamental signal frequency can easily escape such an enclosure.

Lack of shielding or proper layout can cause a filter to become virtually useless if the signal to be filtered is able to propagate around the filter itself. This can be especially problematic with SAW devices because of their high initial insertion loss, which allows EMI to pass almost unfiltered.

Traces for transistors, inductors, mixers, and RFIC's, and even the components themselves, can all radiate; so self-shielded parts should be used whenever required.

Tough EMI radiation problems can be mitigated by applying RF-absorbing foam or rubber over an offending circuit (at 400 MHz and above), or placed within the shield itself to prevent the *waveguide effect.* (The waveguide effect is undesirable low-loss coupling through a long shield structure that creates a waveguide-like transmission line.)

In most wireless PCB situations it is prudent to ground all external cases, shields, and septums straight to the ground plane in order to reduce EMI emissions.

Even though placing shielding over a radiating circuit—or shielding a circuit that is being adversely affected—will help EMI substantially, any such electromagnetic emissions can still travel through the dielectric of the PCB and influence susceptible circuits. A method of attenuating this type of EMI is to utilize a "fence" of through-hole vias, which are located around the affected circuit, from the top of the PCB to the bottom ground plane. At very high frequencies, or if high-amplitude harmonics of the fundamental are present, then vias will have to be placed in a relatively dense pattern.

Preventing EMI from corrupting the DC power supply, and thus the entire wireless device, is critical. A generous amount of decoupling capacitors that are capable of shunting the entire spectrum of possible EMI frequencies—along with a high-frequency choke—should be placed close to all active devices. This will prevent unwanted oscillations and/or contamination of our desired signal (see "Coupling/Decoupling of Amplifiers"), and is especially important for VCOs and LNAs, which must be heavily decoupled from noise-producing conducted emissions that can take place through the common power supply.

Digital EMI suppression. EMI created in digital circuits within the radio system is quite similar to that in the above analog circuits. However, because of certain design variances, and the huge amount of harmonics generated by the square-like waveforms of a digital circuit, there are some added considerations.

Traces on a PCB can transmit EMI at high frequencies—almost as if they were small monopole or loop antennas—if the traces are unterminated, or terminated into a high impedance. If the trace is not microstrip, but of some unknown or uncontrolled impedance, then the electromagnetic fields will leak from the trace and couple into other neighboring ones, causing cross talk between circuits. In fact, the EMI radiation itself may not even be at the fundamental of the actual frequency that is running through the trace, but may be at one of its many harmonics. These harmonics are more easily radiated because of their shorter wavelength. Indeed, as the length of the trace gets closer to one-eighth the wavelength of the signal of interest, then *transmission line effects* must be considered to preserve signal integrity, such as confirming proper line impedances, using shorter line lengths, and employing proper terminations. (Transmission line effects in general infer that after a trace becomes longer than one-eighth the wavelength, the signal's response to the trace must be considered.) And as the frequency of operation increases, reflections will become more of a problem on a simple trace, causing extreme signal degradation of the original digital waveform.

Just as in analog RF design, 50 ohms is accepted as the characteristic impedance of transmission lines and terminations of most high-speed digital designs, as inordinate $\Delta I/\Delta T$ cross talk and EMI (and increased power consumption) are created with reduced impedance values.

For increased EMI suppression in digital circuits within wireless systems, these considerations must be kept in mind during high-speed design and layout:

1. Keep traces away from the board edge. This can decrease PCB edge emissions by up to 20 dB.
2. Traces should be routed at 90 degree angles to the next layer to lessen cross talk.
3. Most modern boards have extremely high component and trace density, so the coupling of EM energy (cross talk) from one trace to another is a large concern. This can be partially mitigated by slowing pulse rise times, increasing the distance between traces, and utilizing shorter traces.
4. Heavy decoupling in the high $\Delta I/\Delta T$ atmosphere of digital logic should always be employed to prevent noise and signals running into undesired areas.
5. Splitting a common analog/digital ground plane into two partitions prevents, or at least reduces, undesirable EMI signals in one section from interacting with the other.

10.3 Wireless Board Design

10.3.1 Introduction

In the world of microwaves and RF—unlike most other disciplines of electronics—the PCB's layout, construction, and materials are almost as important as the circuit design itself. In fact, a perfect wireless design of even the most basic oscillator or filter can be ruined by improper board layout.

10.3.2 Board materials

At microwave frequencies insertion losses become a concern in choosing a PCB board material. A majority of these losses comprise both conductor and dielectric losses. When dielectric materials with low *loss tangent* (also called *dissipation factor, tan delta,* and *TAND*) are used at lower frequencies (under 2 GHz), the losses in the copper conductors can overwhelm any losses in the board material itself. At higher frequencies, however, this may not be the case: A board material with a dielectric constant (also called E_r, K', and *relative permittivity*) of 3.5 will slowly begin to display an almost equal loss in the dielectric and in the conductor as 20 GHz is reached. Higher-dielectric-constant materials will also increase the loss in the dielectric, and an increase in the conductor loss slightly more, over that of lower dielectric constant materials. Thus, it can be demonstrated that both the dielectric and the conductor losses (overall insertion losses) increase with increased loss tangent *and* dielectric constant—especially at higher frequencies. In general, it is recommended that a maximum E_r of 10 must be chosen for up to about 4 to 5 GHz, an E_r of 6 for up to 6 to 7 GHz, an E_r of 4 up to 13 to 14 GHz, and an E_r of 3 up to 30 GHz. These values will be *maximum* recommended dielectric constants in order to minimize insertion losses. However, dielectric constants of smaller values can be employed to control the size of the distributed circuit elements.

At microwave frequencies, high-dielectric-constant materials can force the microstrip circuit topologies to become too small to be realized, both economically and physically. Increasing the substrate's thickness, to a certain extent, will ameliorate some of this effect, but at the cost of possibly confronting different undesired modes of signal propagation, along with increased board via inductance. Nonetheless, thinner substrates will increase insertion losses, while increased thickness will decrease these losses. (Major compromises in RF must always be considered in every design.)

Since the ubiquitous FR-4 circuit board material, which is made up of epoxy resin/glass laminates, exhibits excessive loss and dielectric constant variations at approximately 1.5 GHz and above, other board materials must be chosen for operation in excess of this frequency. FR-4 has a loss tangent of 0.008 at VHF, and approximately 0.02 at higher frequencies, making the loss tangent too high for most microwave work. The majority of the higher-frequency board materials are not only far more expensive than FR-4 type materials, but are also disturbingly soft structures, and may bend quite easily; an example is Teflon™

(called *Duroid*). Still, these softer board materials can be backed by a thick metal ground plane, called a *carrier,* that adds the required rigidity. The carrier is typically aluminum plate, and can be up to a quarter of an inch thick.

All of these microwave circuit board materials can be chosen to exhibit different dielectric constants (2 to 11), loss tangents, temperature effects, and dielectric and dimensional tolerances. Unfortunately, most demand special processing to fabricate a complete printed circuit, so companies that process FR-4 may be unable to work with the more specialized materials. These processing steps will also increase the price of making the printed circuits and vias on the microwave PCB substrate. Nonetheless, a relatively new board material from *Rogers Corporation* (the largest creator of high-frequency PCB substrates in the world) gives any board house the ability to manufacture a complete microwave PCB. This rigid, high-frequency laminate material is the Rogers RO-4000 series. While the cost of processing a layout with this substrate is comparable in cost to FR-4, the board material itself is, of course, slightly more costly, so FR-4 should be used whenever possible for large production runs. Operation of the RO-4000 material is viable up to 20 GHz, with good dielectric constant variations over temperature, as well as very even thickness tolerances across the PC board.

Rogers Corporation conducted a test (utilizing Agilent's microwave simulation software *EEsof*) to demonstrate the interactions that can occur when the dielectric constant, loss tangent, or dielectric thickness of a PCB is varied (while the other two properties are left unchanged). This is an important example of how board material can affect the outcome of a design, sometimes quite drastically. Figure 10.1 displays a distributed edge-coupled bandpass filter centered at 1 GHz with a bandwidth of 10 percent, exploiting a 50-mil Duroid material. Figure 10.2 is a frequency domain graph showing all substrate parameters nominal for this bandpass filter. Figure 10.3 shows what occurs to the filter's center frequency and passband when the dielectric constant is varied both upward and downward. A slight increase in E_r shifts the passband lower in frequency, while a slight decrease in E_r shifts the passband higher. Variations in the E_r of the board material will affect the passband frequency of a filter because any change in E_r changes the electrical length of each of its elements by altering the velocity of propagation through the dielectric.

Both manufacturing and temperature variations will influence the dielectric constant, necessitating a board material with a tight initial E_r tolerance and a

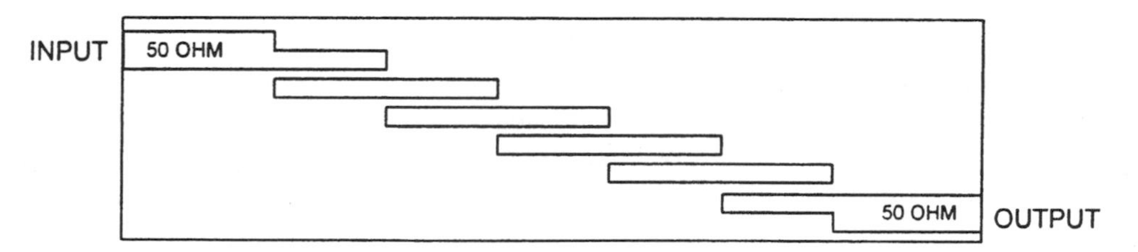

Figure 10.1 Sixth-order edge-coupled distributed BP filter.

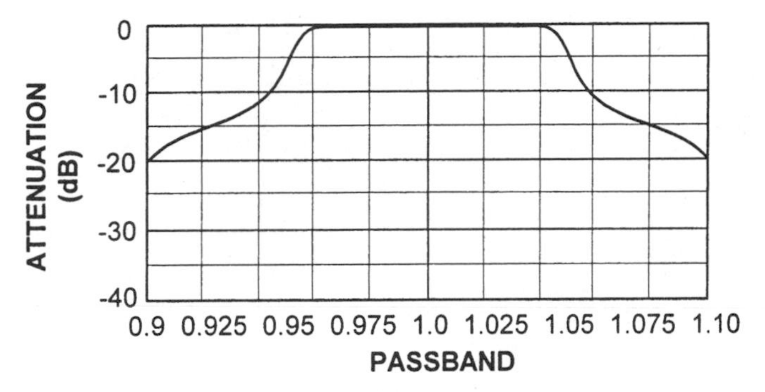

Figure 10.2 Dielectric constant at a nominal value of 10.20; distributed filter's center frequency as expected.

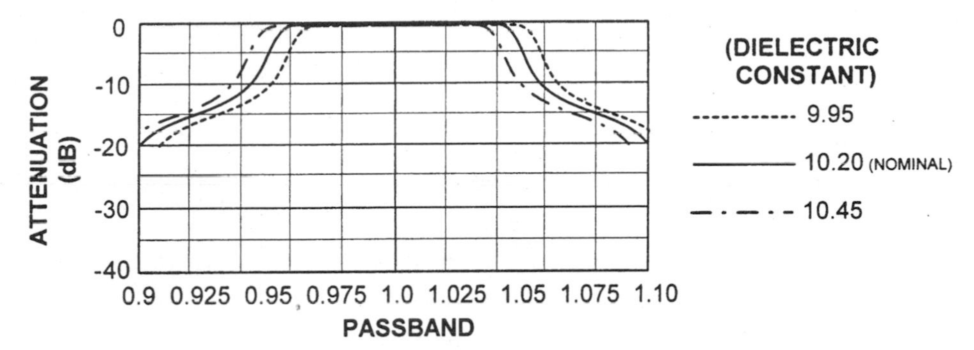

Figure 10.3 Variations in center frequency of a distributed filter as the dielectric constant is varied.

low *temperature coefficient of dielectric constant* (TCK). Figure 10.4 reveals the increased signal attenuation as the loss tangent is increased, causing a slightly lower filter output than expected. (Most quality microwave board materials will have low loss tangents, so perhaps a more adverse effect on a signal at microwave frequencies can be an improperly cleaned PCB, or a solder mask that is placed over the entire PCB board.) Figure 10.5 demonstrates that thickness deviations can change the bandwidth of the filter: Thicker than expected substrates will increase the bandwidth, while a thinner substrate will decrease the bandwidth.

Lumped filters will show little noticeable effect with almost any modern substrate material, while distributed filters are very sensitive to dielectric constant, loss tangent, and thickness variations of the PCB's substrate.

10.3.3 Board layout

Both high-frequency analog and digital board layouts must be approached with caution in order to obtain proper circuit operation. The following are the most important suggestions and precautions.

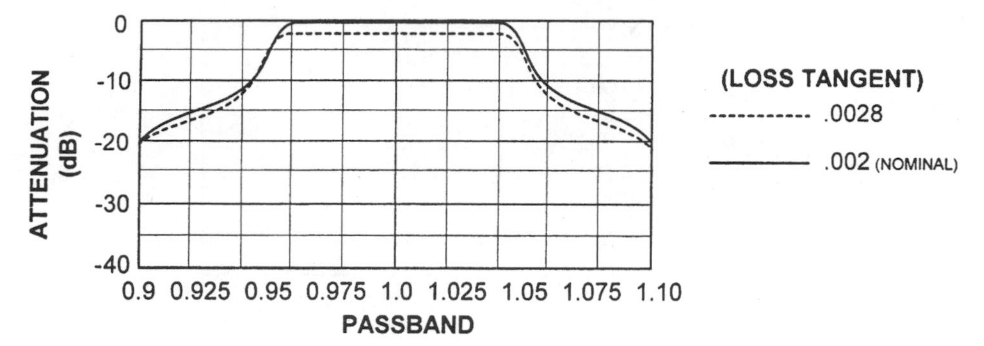

Figure 10.4 Increase in insertion loss of distributed filter as loss tangent is increased.

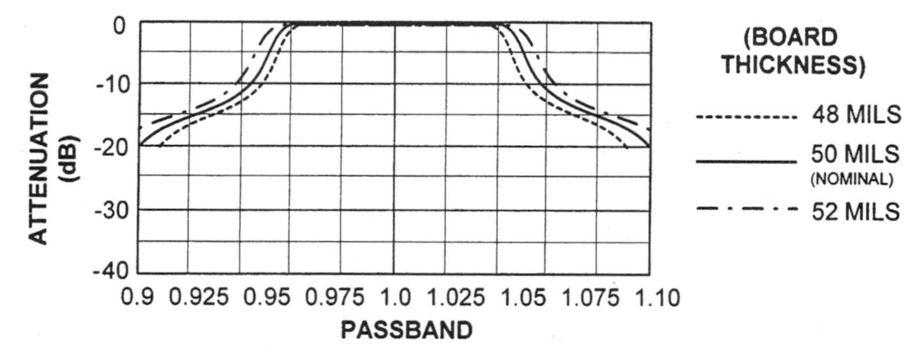

Figure 10.5 Variations in the bandwidth of a distributed filter as the board thickness is varied.

High-frequency analog board design. The lengths of all component leads must be minimized to decrease losses and abnormal circuit operation caused by the added lead inductance. Transmission lines, usually microstrip, must be employed to maintain 50-ohm constant impedances, thus decreasing mismatch losses and reflections caused by impedance discontinuities. Some lower-frequency circuits can exploit simple traces of unknown impedances, but these would still have to be kept very short so that transmission line effects did not disturb circuit operation (even a short 1-cm-long wire can have an inductance of approximately 10 nH, or 63 ohms at 1 GHz, forming an almost pure inductor). All bends in microstrip traces should be mitered or rounded to prevent radiation into adjoining circuits. Microstrip into or out of an active device that has narrow leads should be tapered for a decreased impedance bump (Fig. 10.6). All ground returns from all discrete active and passive devices, as well as ICs, must be sent to the board's ground plane by the shortest route possible, normally through a via. This is to lower the return path's inductance to ground (Fig. 10.7), and is especially important for proper stability and gain of a discrete transistor in its emitter return circuit. The entire board layout should also be as physically condensed as possible to minimize losses and radiation (EMI), but not so much as to increase undesired coupling. The decoupling components for

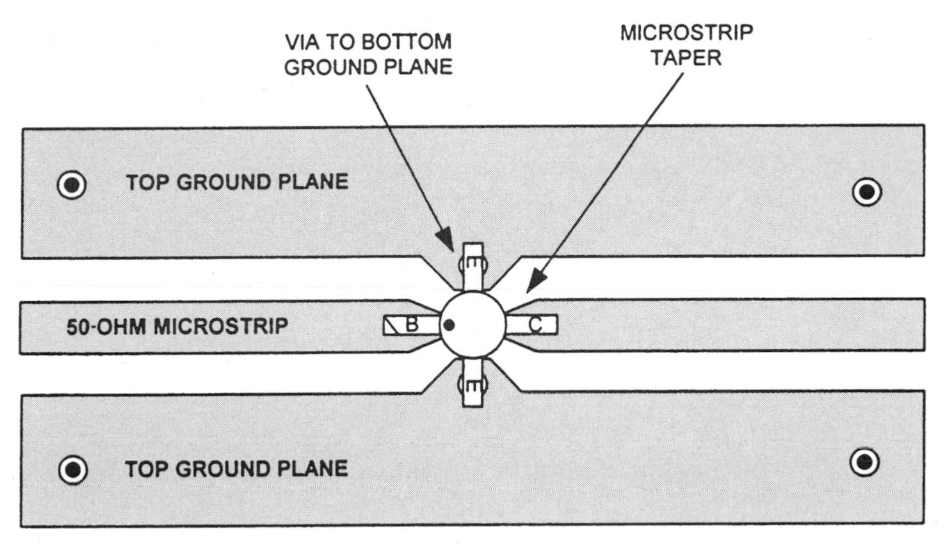

Figure 10.6 Proper board layout for a transistor as seen from top of PCB.

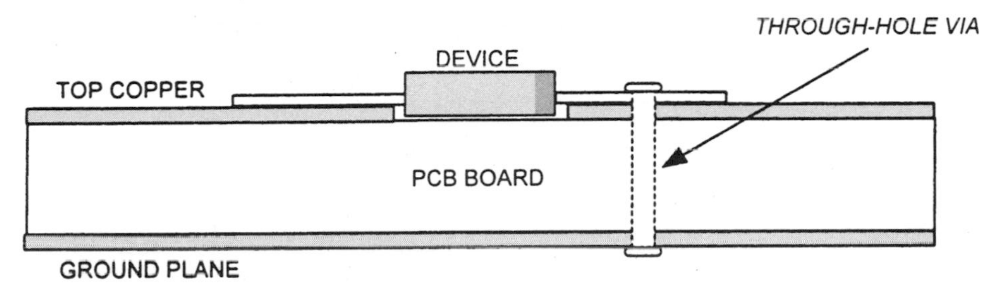

Figure 10.7 Side view of a through-hole via to ground plane.

the power supply (PS) should have a low capacitive reactance to ground—and a high inductive reactance—at *all* frequencies for *all* devices being supplied (Fig. 10.8). This is to block any signals from entering, and contaminating, the power supply. The decoupling function is accomplished by capacitors (and/or inductors) that are optimized for different frequencies; such as electrolytics for low frequencies and ceramics or porcelain for the much higher frequencies. Since vias, as stated, will have a certain inductance, just as any conductor will, this value can be calculated by:

$$L = 5.08h\left[\ln\left(\frac{4h}{d}\right) + 1\right]$$

where L = inductance of the via, nH
h = length of the via, inches
d = diameter of the via, inches

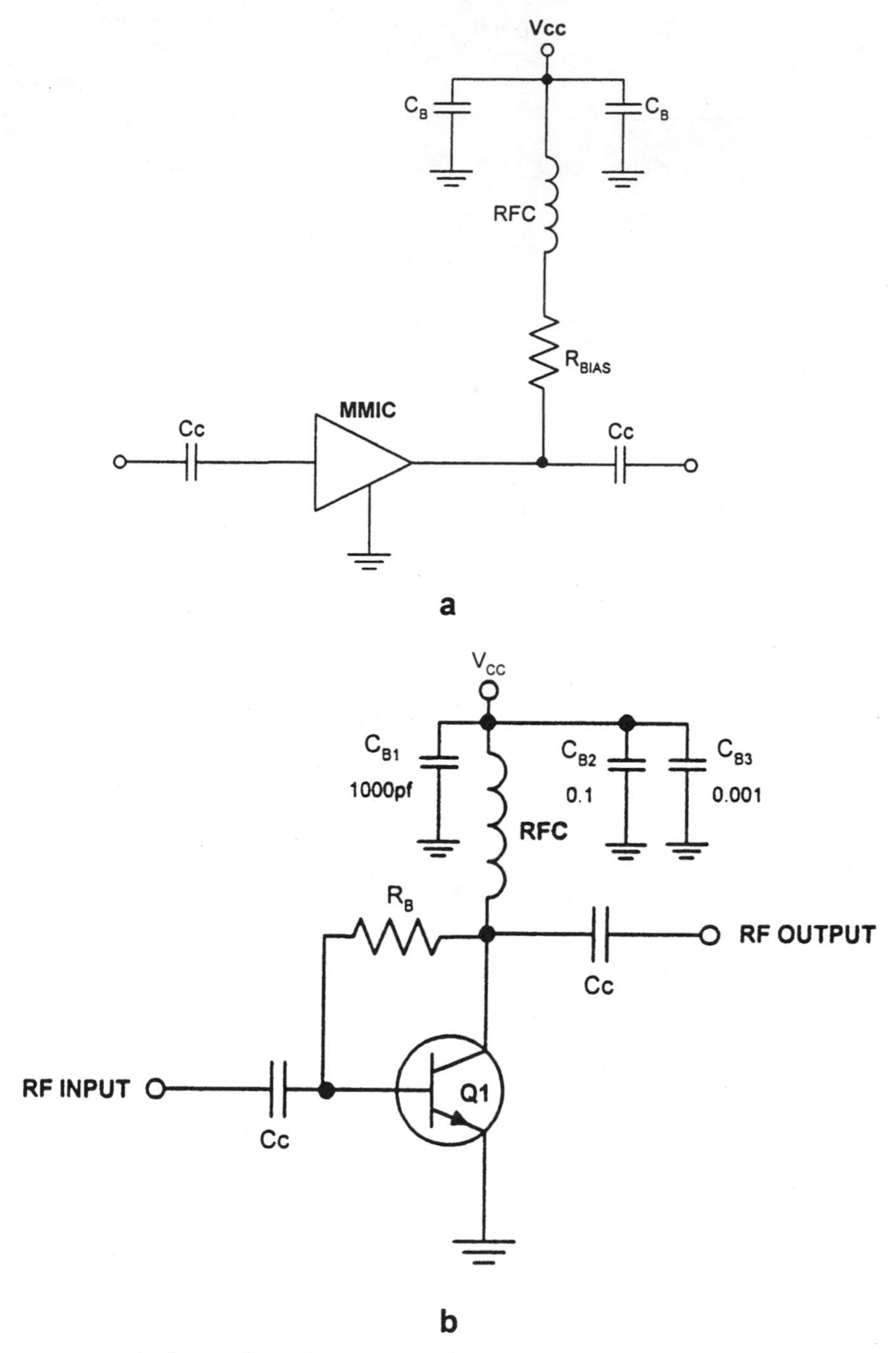

Figure 10.8 (*a*) Proper decoupling for a MMIC; (*b*) proper DC bias RF capacitor decoupling for a transistor.

This formula infers that varying the diameter of a via will not do much to change the via's inductance, while altering its length has a profound effect, which is why all vias (and conductors in general) to ground must be kept short, or the inductive reactance can become significant at higher frequencies.

The printed copper board's traces will have a certain amount of resistance, and this resistance can adversely affect a circuit's operation. Since a trace has this resistance, it will mean that any two points along a trace will *not* have the same voltage, which can cause problems as to what the actual ground reference level is (it *should* be zero volts). At RF, this resistance is further increased by the *skin effect.*

Vias should be placed at regular intervals of a quarter-wavelength or less through the top ground plane down to the true bottom ground plane in any RF circuit, as the only real ground plane of a two-sided board (one substrate, one upper and one lower copper sheet) is considered to be the continuous bottom copper layer.

All components at RF frequencies will have some reactive and resistive parasitic effects, so only resistors, inductors, and capacitors that are rated at the frequency of operation—or above—should be used in an RF circuit. Depending on the application (coupling, decoupling, filtering, matching, etc.), running into a component's series or parallel resonance unexpectedly can destroy proper functioning of the wireless circuit.

Mutual inductance (coupling) of traces, components, and wires must be accounted for in any design. This undesired coupling of energy can be alleviated by:

1. Keeping traces that are carrying RF currents separated by distance
2. Employing shields
3. Reducing the area of the current-carrying loops
4. Using right angles between traces

A concept similar to tapering the microstrip into an active component to lessen the impedance bump is shown in Fig. 10.9. To decrease impedance variations and lower VSWR when a RF signal encounters a passive component, such as a coupling capacitor, the component should ideally be of the same width as the microstrip and the solder fillet itself should be smooth so as not to disturb the signal flow.

As the return currents of a microstrip ground plane are flowing directly under the microstrip that is carrying the signal currents (Fig. 10.10), the ground plane must never be broken, or an unexpected impedance discontinuity will result within the microstrip.

Always terminate all microstrip transmission lines with their characteristic impedance to avoid unpredictable reactive effects.

High-speed digital design. A digital signal's rise and fall time, and not its frequency, govern the signal's speed as it relates to PCB design. The traces for a

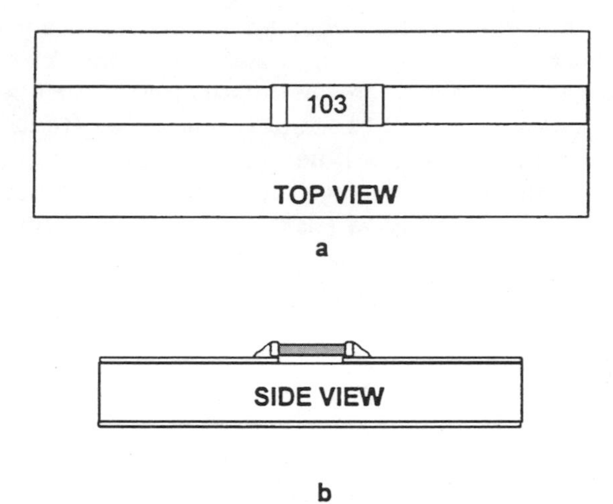

Figure 10.9 Proper component width (a) and component soldering (b) for decreased reflections.

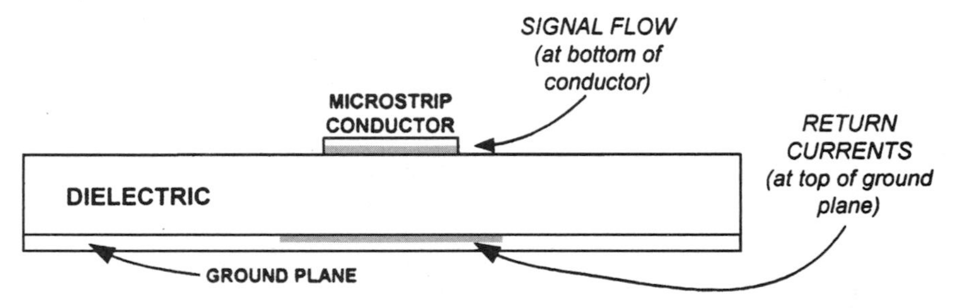

Figure 10.10 Currents in microstrip.

high-speed digital circuit on a wireless board should run only near the PCB's digital ground plane to minimize the current loop area, and thus decrease loop emissions and stray electromagnetic pickup. Tracks should be properly terminated into 50 ohms to minimize reflections and, if these traces are over 2 inches per nanosecond long, then 50-ohm microstrip should be adopted (the maximum track length, in millimeters, should not be more than 46 times the fastest rise or fall time, in nanoseconds, to avoid transmission line effects). If the proper tracks are not employed, or the correct terminations are not used, then ringing and stair stepping of the digital waveform will be created, as well as antenna-like effects (causing EMI). Even when utilizing microstrip, minimize vias along the microstrip to avoid adding capacitance, which lowers the track's impedance, causing delays, a higher VSWR, and reflections. Any *stub* (a short length of copper trace) situated off the main signal track, and which terminates into any high impedance, should be avoided, as this forms an open stub. The open stub would act as an undesired bandstop filter at a certain frequency of $\lambda/4 \times V_P$ (see Sec. 6.2, "Distributed Filters").

Minimize high-impedance nodes, since these areas will add noise caused by induced currents created by EMI propagation, as well as ground loops, which create noise and EMI through induction. Adding extra ground planes between signal tracks of a multilayer board will shield other tracks from electromagnetic coupling, thus reducing cross talk, and permit increased AC coupling to the ground reference plane. It is always wise to check all traces on a PCB design for cross talk at the early layout phase. This can be accomplished by employing field solver or cross talk software. Openings made for connectors, LEDs, switches, and thermal vents within metal enclosures or ground planes can act as slot antennas, especially if the opening is greater than one-tenth wavelength, radiating the fundamental and/or its harmonics.

Keep the power supply separated from the digital and analog sections by shielding and decoupling. Shield the synthesizer and the VCO to prevent EMI from entering the oscillator circuit, causing spurs in their output. The use of *power planes,* which are separate copper layers on a multilayer board that are each attached to the hot end of the power supply, at RF frequencies can cause them to become patch-like antennas, sending spurious signals around the immediate PCB area. Utilizing simple traces for supplying the DC, with adequate decoupling, is all that is recommended for dealing with most high-frequency designs.

Attempt to physically separate high-speed digital circuits from the analog circuits, or allow the digital circuits to function only when the analog circuitry is not being affected adversely, with the RF traces located as far from the digital traces as possible. Modern high clock speeds mean that the propagation velocity is longer than the clock cycle, so propagation delay becomes a large consideration. If this is taken into account, then the digital clock pulses can arrive properly at multiple chips and pins with equal delay.

10.3.4 Board design issues

Even if an MMIC or discrete amplifier is designed to be unconditionally stable, poor PCB layout can cause stability problems. Subband oscillations (below the amplifier's bandwidth), out-of-band oscillations (above the amplifier's bandwidth), and in-band oscillations may all occur unless:

1. The amplifier is connected *directly* to ground.
2. The PCB board itself must be connected at one-eighth wavelength points to the system ground (and preferably closer).
3. Vias must be placed from the PC board's top ground plane to the bottom ground plane to lessen the reactance between these two planes.
4. V_{CC} must be properly decoupled by using both low- and high-value capacitors for both high and low frequencies.
5. All amplifier stages with high gain should be shielded to prevent them from bursting into oscillation.

6. An *even* number of stages in an amplifier strip should be avoided, since this can cause in-phase feedback to the chain's input—either through circuit trace interaction or conductive dirt contamination.
7. Circuit input and output traces should be kept separated from each other to avoid bypassing the circuit itself or creating feedback.
8. Employ only components made for RF service, as the parasitics and low Q that most affect wireless circuit design are due to real-world component inadequacies of inductors and, to a more limited extent, capacitors and resistors. These problems involve:
   a. The capacitance from turn-to-turn in an inductor—an effect lessened by smaller diameter coil turns
   b. The inductance inherent in all leads (a $^1/_4$-inch lead of a through-hole capacitor can reach 10 nH, while even leadless capacitors can still have inductances of 1 nH) can be somewhat alleviated by running capacitors in parallel
   c. The capacitance to ground natural to all components—lessen by employing smaller components
   d. The mutual coupling of inductors—decrease by not running inductors in parallel, only at right angles
   e. Limited inductor Q—helped by using only a manufacturer's specific "high-Q" coils, by winding your own coils in which the length is equal to the diameter, or by utilizing a distributed inductor

Surface mount components are all that should be considered at frequencies above a few 100 MHz. Nonetheless, not all SMDs are created equal when it comes to high-frequency operation. The only resistors, inductors, and capacitors that should be adopted for microwave operation are those that have been specified by the manufacturer to dependably operate above the design frequency—without hitting any series or parallel resonances (except in certain coupling or bypass applications). Since many resistors are not specified for their maximum frequency of operation, they must sometimes be selected only on the basis of the type of high-frequency resistor design employed; usually thin- and thick-film types can be depended on to reach to very high frequencies. Still, as discussed in a previous chapter, as a resistor's resistance values are increased, its ability to operate at microwave frequencies is decreased (Fig. 10.11). In fact, for sensitive or very high frequency circuit operation, candidate RF resistors should be tested for any resonances, as well as a lack of major resistance changes versus frequency, to at least 20 percent above the desired frequency of operation.

As stated, all active and passive components that are subjected to RF must be able to properly operate at the desired frequency, but components that are out of the RF path may be low-cost, low-frequency parts. This will become especially meaningful in DC bias circuits (Fig. 10.8*b*) where the RF choke (RFC) must stop most of the RF from entering the bias supply, with any that gets through being bypassed by the high-frequency capacitor, C_{B1}, to ground,

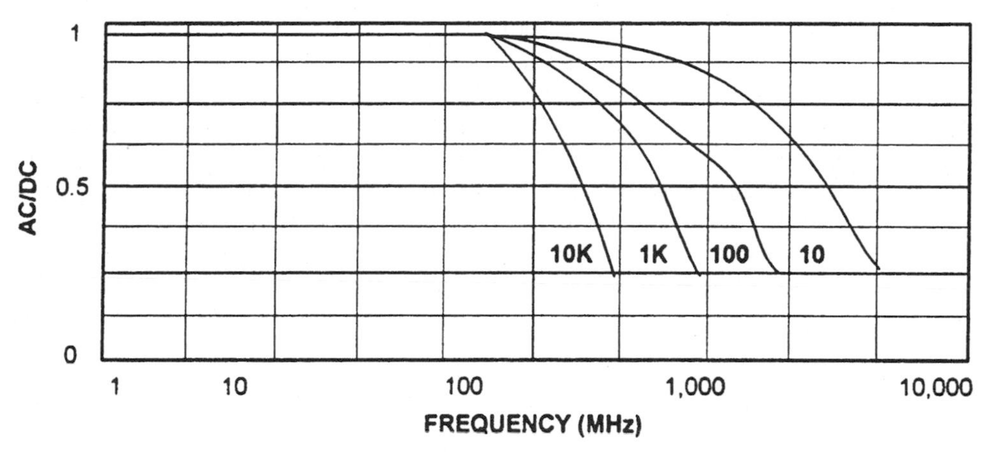

Figure 10.11 Ratio of resistance at DC to resistance at AC at different frequencies for standard SMD resistors.

while C_{B2} and C_{B3} need not, and indeed cannot (because of their high inductance), operate at microwave frequencies. However, all other components beside C_{B2} and C_{B3}, such as C_C, R_B, and RFC, must be capable of operating effectively up to the highest RF of the circuit design.

The place called *ground* on a PCB layout is merely the reference point for a circuit, the common point for all the circuit's voltages. It is the area that is supposed to be at 0 V. But, as mentioned previously, most grounding points are, in actuality, only close to 0 V. This is because any conductor will have a certain amount of resistance, no matter how small. As current flows through the ground conductor with its internal resistance, a voltage is formed across it, and the conductor is now no longer at 0 V. Large circulating currents cause the ground to significantly vary from the perfect 0 V, even at very small resistance values. These unavoidable Ohm's law effects of $V = IR$ are, fortunately, minor in a well-laid-out board. But, with a sensitive circuit, having a ground above 0 V—even a little above—may cause oscillations, as well as a myriad of bizarre circuit problems. This can be ameliorated by maintaining the current in a ground return to a small value by forcing each separate circuit to have its own return line, instead of sharing, which will lower the voltage dropped by each circuit's return. We can also lower the resistance of the return by widening the trace width. However, with currents that vary in a high-frequency AC manner, the impedance of the trace can actually have a far larger effect than its DC resistance. This is why high-frequency circuits will always have an extremely direct return to the board's ground plane.

Even on signal traces that employ microstrip, losses are a fact of life. Depending on the PCB substrate in use, and the frequency of operation, losses of between 0.1 to 1 dB (or more) per inch can be caused by dielectric constant variations, dielectric heating, copper losses, and undesired radiation. In many applications this may not be of much concern, but where every dB matters, such as in the front end of a low-noise receiver, we would want to go to a board material with less loss at our frequency of interest.

To connect the radio's internal PCB board to the outside world, an SMA, N, or BNC RF connector is normally adopted. The actual connector chosen will depend on the frequency of operation and the cost, and can be directly connected to the board's 50-ohm microstrip line in one of two ways. The connector can either be launched directly off the edge of the board (Fig. 10.12), or at a 90 degree angle through the ground plane and substrate (Fig. 10.13).

SAW filters in wireless design are becoming much more common than in the past, so some considerations for the board's layout must be accommodated. Because of the SAW filter's large insertion losses, high isolation between its input and output must be maintained on the circuit board, since a signal must not be permitted to bypass the SAW. The isolation required is calculated by

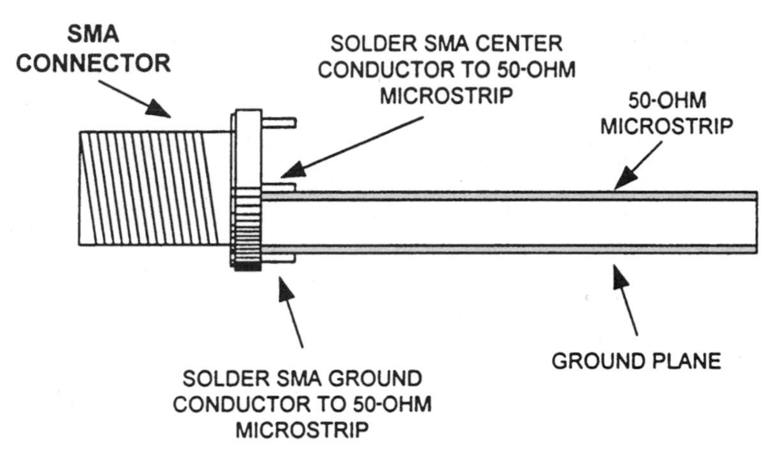

Figure 10.12 SMD connector edge-board-mounted.

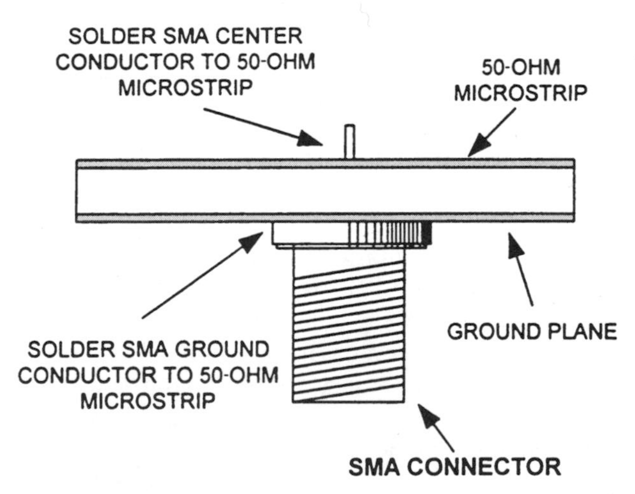

Figure 10.13 SMD connector board-mounted on top of ground plane.

adding the ultimate rejection of the SAW with the SAW's insertion loss. For example, if a SAW has an insertion loss of 20 dB, and its ultimate attenuation is 50 dBc, then the PC board will be required to supply a minimum of 70 dB of isolation to maintain the SAW's expected performance. This can be accomplished by isolating the SAW filter's input and output ports from each other by a *plated-through slot,* in which a piece of metal is placed through the board's dielectric to its ground between the SAW's ports. The plated-through slot will decrease the RF leakage through the substrate material of the PCB. Other methods that can be used in conjunction are:

Placing a metal shield around the SAW, which separates its input/output ports *above* the PCB

Assuring that the SAW's case has multiple direct connections to the ground plane

Positioning the input/output port's matching inductors at right angles to each other

Using shielded inductors

10.4 Software Radio

10.4.1 Introduction

Because of advances in high-speed analog-to-digital converters (ADCs), integrated mixers, and digital signal processors (DSPs), digital radio—at least up to the first IF stage—has become a reality for some high-end communications systems. However, for the foreseeable future, complete software radios will be a design curiosity that will be adopted in a few expensive, and relatively low-frequency, radios. It will take time to perfect this technology to be viable in the lower-cost, higher-frequency radio area.

10.4.2 Software radio designs

The dream of a radio that can economically change from one type of wireless device to another, from one modulation and band to another, and from one bandwidth and frequency to another by simply re-programming is the promise of software radio. A *complete* software radio has yet to be designed for the mass market—but the digital back end is beginning to reach closer and closer to the RF front end of the receiver.

There are basically three types of software radios: *software-defined radio,* which changes a restricted number of distinct hardware functions by software (this is already a reality in some dual-mode handset radios and base stations); *software radio,* in which some of the analog circuitry is replaced by software, while the rest of the analog circuits are *reconfigurable* by software (this is expected to become more common by the year 2004); and *ideal software radio,* in which a limited amount of front-end analog circuitry is fixed and untunable, with the software completely controlling and changing the functions of the radio (expected in about 2006).

As *software-defined radio* (SDR) exists today, albeit in a limited manner, we will concentrate on this technology. A majority of SDRs consist of a transmitter with an analog filter, SSPA, frequency conversion stage, and the digital stages (Fig. 10.14). The receiver is constructed of an analog filter, LNA, frequency conversion stage, and digital IF and baseband functions incorporating signal processing hardware algorithms (Fig. 10.15). SDR architectures must be low-cost, small, and easily configurable to different bands, protocols, bandwidths, and modulation schemes in order to become pervasive.

There are two different kinds of software-defined radios.

Heterodyne SDR. As shown in Fig. 10.15, the antenna receives the RF signal, which is then filtered, amplified, and converted to the IF by these analog stages. A very fast ADC then converts the analog IF into a digital signal, where the receive signal processor (RSP) filters and tunes the signal to the required channel as required to provide baseband I and Q outputs to the DSP (digital signal processor). This allows various channels, frequencies, and standards to be placed within one radio. In fact, the RSP is where the SDR has its tuning and selectivity, data rate, channel bandwidth, and even channel-shaping abil-

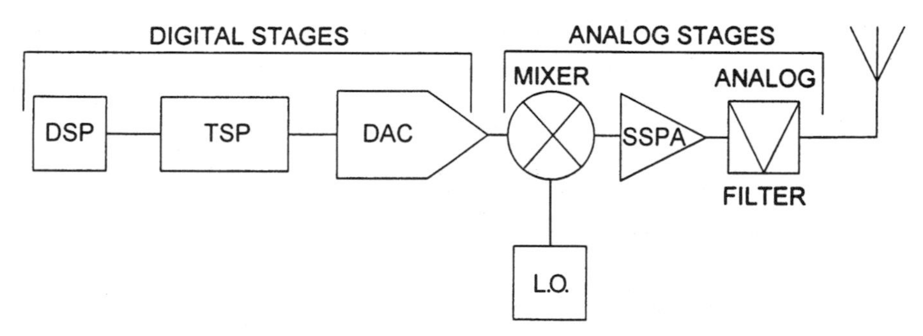

Figure 10.14 A software-defined radio transmitter.

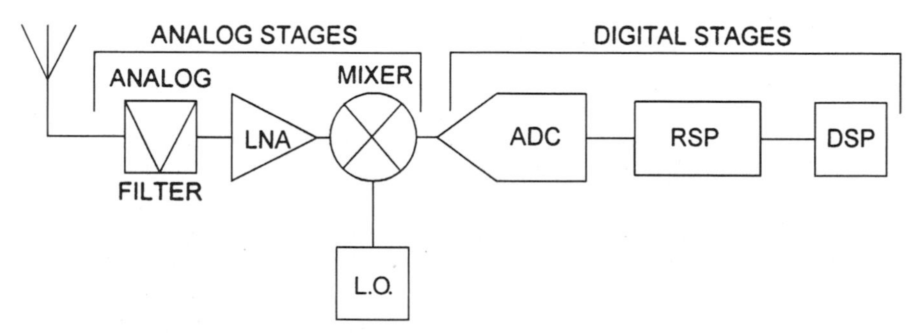

Figure 10.15 A software-defined radio receiver.

ities, and actually supplants the LO, channel select filter, quadrature mixer, and data decimation filter. The RSP is basically a special-purpose DSP, and must be capable of the same speed as the A/D converter. The output of the RSP is then sent to the DSP, which demodulates the baseband signal. The DSP, depending on its programming, can demodulate both digital and analog signals, and can thus receive either/or FM, AM, QPSK, CDMA, etc.

A fast and quiet ADC makes software radio possible. The A/D converter must have a high SNR rating, which embodies both quantization and thermal noise, as well as the A/D sample clock's wideband phase noise. High dynamic range is also a must in order to decrease spurious responses, since in a multicarrier radio the spurious responses of one channel can interfere with the weaker signal of another channel.

The transmitter of a heterodyne SDR, as shown in Fig. 10.14, functions so: The DSP places digital data to be modulated into the transmit signal processor (TSP). The TSP then modulates the carrier with this information. The digital modulated signal is transformed to analog by the D/A converter, sent into the analog stages and frequency up-converted to RF by the mixer/LO, amplified by the power amplifier (SSPA), filtered, and sent out of the antenna.

DC SDR. Another SDR architecture is the *direct conversion* (DC) design. This technique reduces much of the heterodyne system's disadvantages, such as a fixed front-end filter to set the heterodyne radio's frequency and bandwidth, and converts the RF directly into a zero IF with no image frequency. Since front-end RF filters that are not only center frequency tunable, but also maintain their return loss and have a tunable bandwidth, are extremely difficult to design and mass-produce, each wireless standard that the heterodyne SDR would be required to emulate would need, at a minimum, an expensive and large switchable filter bank, along with a fine-tunable LO. The DC SDR removes this problem and lowers system costs.

10.5 Hybrid Circuits

10.5.1 Introduction

Hybrid design and construction predates today's monolithic ICs, but hybrids have most assuredly not been completely supplanted by them in all electronic products and applications. But, hybrids are used only when ICs or common surface mount discrete printed circuit board designs are inferior and weight, space, cost, and performance are vital for specialized or high-reliability situations. Hybrid circuits can also be an excellent choice for low-volume production runs when custom monolithic ICs would have too high an initial design and cost.

RF hybrids will appear to an observer as simply a large IC, or as a small populated PCB if the hybrid is left unpackaged. In most consumer applications, the completed hybrid circuit will be placed in a special plastic package, and silicone or resin is then poured over it, or the circuit itself can be dipped into a liquid plastic and laid out to dry. Both of these techniques, and more, are used to protect the circuit and to displace excess heat from the hybrid assembly.

A completed hybrid will take advantage of different circuit construction techniques, such as thin- and thick-film resistors (which can be printed and trimmed directly on the hybrid's substrate), SMD IC chips, SMD capacitors, flip-chips, trace coils, and microstrip transmission lines. In fact, hybrid manufacture is basically a method of constructing a circuit on a specialized PCB by employing any desired electronics technology that is available so that we may decrease both the size and the cost of a circuit, while increasing reliability and performance.

Most of the actual ICs placed on a hybrid board are in their unpackaged form, even though many hybrids can and do use packaged SMD ICs. But the ability to exploit ICs in their bare die form means that the size and cost of the entire hybrid circuit can be significantly improved. The bare die is first attached to the hybrid PCB with an epoxy that is optimized for thermal conductivity and/or electrical conductivity, or the epoxy may insulate the bare chip both thermally and electrically from the PCB, depending on the application. The chip's bond wires are attached to the hybrid PCB's metallization layer by a wire-bonding process. This process will rely on the type of bonding wire used out of the bare die. Gold wires require thermosonic bonding, while aluminum wire will need an ultrasonic procedure. Both methods demand a special machine that utilizes a chiseled point that contacts the wire, pressing it against the board's surface metallization while vibrating at ultrasonic frequencies. Thermosonic bonding for gold wires, as its name implies, also adds heat to this process.

A superior bare die attachment technique does not use wire bonding at all and obtains much decreased inductive effects. Instead, small solder balls are attached to the IC's connection points. The chip, now called a *flip-chip,* is affixed to the board by turning the IC chip over so that its solder balls are in contact with the pads of the board's metallization layer. The entire assembly is then heated, causing the flip-chip to be permanently melded to the PCB, both mechanically and electrically. There are other methods of attaching a bare die to a PCB, such as tape automatic bonding and adhesive-bonded microbumps. For more detailed information on these processes, consult *Hybrid Microelectronics Handbook* by Sergent and Harper.

All of the other SMT components, such as the transistors, packaged RFICs, capacitors, resistors, and inductors, can be attached to the substrate's metallization layer by the *reflow soldering process,* which employs a solder paste that is "printed" onto the PCB and heated, causing the solder to reflow, electrically and mechanically connecting the components to the board.

10.5.2 Board and conductor materials

The composition of the substrate and conductor materials are an important consideration in any high frequency PCB design, and hybrids are no different.

Metallization is the conductive layer of the interconnecting traces and pads placed onto the substrate. It can be made of copper, gold, or silver deposited on the substrate board material. The entire PCB itself can be called a metallized substrate in these hybrid applications. The board material is usually a type of ceramic, such as *alumina, aluminum nitride,* or *beryllia.* These ceramics are extremely rigid and are quite temperature stable, besides having a very high-strength characteristic, attributes that make ceramics perfect as a substrate material for hybrid applications. Alumina (aluminum oxide) is by far the lowest in cost and most popular. It is a high-frequency (up to 25 GHz), very hard substrate material that does not require a carrier (heavy metal stiffening plate), and has a very high dielectric constant of around 9.8 (for small circuit layout sizes). This material is used in applications that require rigidity, strength, and temperature stability, along with decent thermal conductivity. The next substrate, aluminum nitride, is found only in specialized hybrid applications that require better thermal conductivity, but at a substantially increased cost, over alumina. Beryllia (beryllium oxide), which is even higher in cost than aluminum nitride, is seen in applications where low dielectric constants are needed (around 6), as well as improved thermal conductivity. However, Beryllia dust particles are toxic, and must not be inhaled when this dangerous substrate material is machined.

10.6 Direct-Conversion Receivers

10.6.1 Introduction

Direct-conversion receivers (DCRs, also called *zero-IF* receivers; Fig. 10.16) have seen only limited use because of implementation complexities. A DCR is

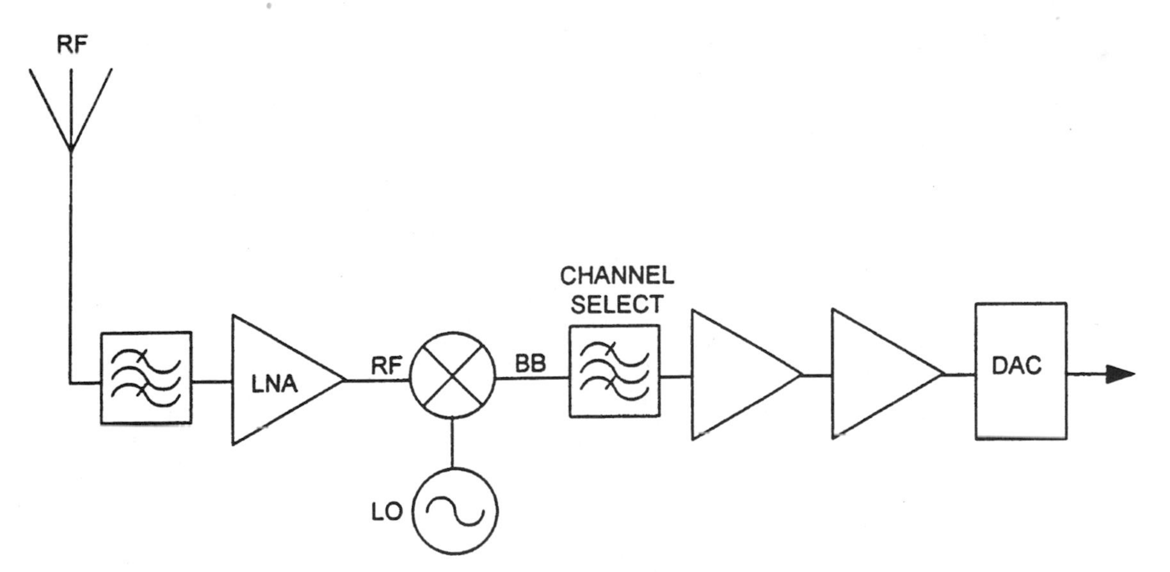

Figure 10.16 A direct conversion receiver block diagram.

a receiver with an "IF" after the first mixer stage, that is fixed at or near zero frequency. The RF input signal has been mixed down *immediately* to baseband, with the middle of the desired band translated to this zero frequency.

DCRs have a much lower parts count, and are thus cheaper to build, than the competing superheterodyne designs. DCRs do not require an image filter or high-frequency IF filters and amplifiers since no image frequency is seen by the DC receiver, and it has no IF. However, DCRs have multiple problems that make a discrete DCR almost impossible, while even RFIC designs are rife with difficulties. Nonetheless, many of the DCR's limitations can, and have been, addressed relatively successfully within the domain of some of the newer RFICs, especially by companies such as *Analog Devices* and *Maxim.*

10.6.2 Direct-conversion issues

Most design choices have tradeoffs, and a DCR is no different. Superheterodyne receivers will have more selectivity and sensitivity than an equivalent direct-conversion receiver, with most DC receivers barely able to function up to 900 MHz while attaining only −95 dBm sensitivity (−105 dBm is a requirement in many systems). And selectivity naturally suffers in most RFIC designs because of the adoption of active low-pass filtering at baseband, so DCRs have less interference rejection than superheterodynes. Many zero-IF receiver RFICs, as well, have the following problems:

1. An incidental offset voltage is caused by the self-mixing in the direct conversion receiver; the LO leakage actually mixes with the original LO signal, creating a DC voltage that can contaminate the signal of interest, lowering the SNR, and can even saturate the baseband amplifier stages.
2. LO leakage through the RF sections is a large consideration in design, since the LO is very close to the frequency of the incoming RF and can be radiated by the receiver's antenna, causing in-band interference.
3. Flicker-effect ($1/f$) noise from the mixer output can be a problem, as the down-converted signal is usually of low amplitude and low frequency (near 0 Hz), decreasing SNR.
4. In very wideband and high frequency DCR receivers, the difficulty of maintaining equal amplitude and phase in both the I and Q legs, called *I and Q mismatch,* has been difficult to solve. This will cause an increase in the BER.
5. Any circuit that employs phase shifts of the incoming signal by 90 degrees, as all modern zero-IF receivers must do, undergoes compromises with noise, linearity, and power.
6. Zero-IF receivers permit distortion, caused by strong signals at the mixer, to reduce sensitivity more quickly than with superheterodynes.

Up until quite recently, the above problems have relegated zero-IF receivers to FSK pagers and a few amateur SSB receivers. But companies have attempted to get around many of these difficulties by some ingenious design techniques. For instance, the above DC offset dilemma can be mitigated by transitioning from a direct conversion, zero-IF architecture to a very low—but not zero—baseband frequency. This allows the operation of AC coupling into the mixer stage, which eliminates the DC offset, while applying an LO that is not at the same frequency as the RF. However, costs rise because of the need for a higher-frequency ADC. (The LO radiation from the antenna can also be attenuated by a high-reverse-isolation LNA.) Many other improvements in DCR design will be forthcoming in the next few years.

Despite the above implementation problems, zero-IF receivers are becoming popular in wireless design simply because they require fewer components than the standard superheterodyne technology, and are therefore much cheaper to build.

10.7 Prototyping

10.7.1 Introduction

The construction of any electronic prototype encompasses many disciplines: engineer, technician, assembler, and mechanic. A properly constructed prototype can sometimes make or break a wireless project, as a prototype that does not function as expected can, frequently, damage the confidence level in an entire wireless design.

10.7.2 Prototyping considerations

When specifying and building a prototype for a wireless project, confirm that all parts and components selected will not be thermally or mechanically stressed; if so, the utilization of a heat sink or a more robust component is indicated. Do not permit components to operate above their maximum voltage, current, or power rating, and allow enough derating to permit long and safe operation. Place the completed circuit or product prototype under thermal, humidity, and vibration stress testing to check for any design or implementation flaws. Do not purchase unknown or suspect parts simply because they are low in cost, and test and specify all parts from unknown manufacturers. In a multiprototype project build only one first, since the redesign or reshuffling of circuits or components may be required after the first unit's test and measurements are complete. Allot enough time not only to construct the prototype, but also to debug the myriad problems that will surface in any complex project.

Presented are some of the more common issues encountered during prototype design and construction, along with their solutions:

Confirm that all digital and analog power and grounds remain separated in order to reduce hash within the RF sections.

Check that all power supplies are fully decoupled, and that all of the appropriate parts have their proper V_{CC} and ground.

Verify that your parts supplier is dependable, since not all components or RFICs that are in the data books will be truly available—either because they are preliminary or because they have been discontinued.

Establish that every trace on the PCB is correctly routed, not only to make sure it reaches the proper board location, but also to avoid unpleasant coupling effects and EMI.

Validate that all active devices are soldered to the correct terminals (especially diodes) and that all electrolytics are inserted with the proper polarity.

Ascertain that components are of the appropriate value, and that they are correctly and neatly soldered.

All of the above should be performed before turning on the prototype for the first time to avoid an expensive, embarrassing, and time-consuming ending to the infamous "smoke test."

10.8 Antennas

10.8.1 Introduction

Antennas are designed to efficiently transform alternating current into electromagnetic waves, and to then send these waves out into free space. The electromagnetic waves are then caught by the receiving antenna, where they are converted into an alternating current. The actual design of the antenna depends on the frequency of its operation, output power, directivity, robustness, cost, and space limitations. However, any resonant antenna will function as a series resonant circuit, and when cut to one-quarter (for a vertical monopole) or one-half (for a dipole) wavelength, maximum current will be allowed to flow through its elements (Fig. 10.17). This will give the maximum signal strength possible for that particular antenna design.

If the antenna length stays the same, but the frequency changes, then the antenna's radiation patterns will vary. These alterations in the radiation pattern will change the antenna's directional characteristics. Now, instead of omnidirectional antennas radiating in all directions with almost equal signal strength, it will have far more power nulls and power lobes than when run on its fundamental. Nonetheless, this makes harmonic operation at multiples of a transmitter's fundamental frequency possible with a single antenna, since the antenna's center-fed feedpoint will remain at the same low input impedance.

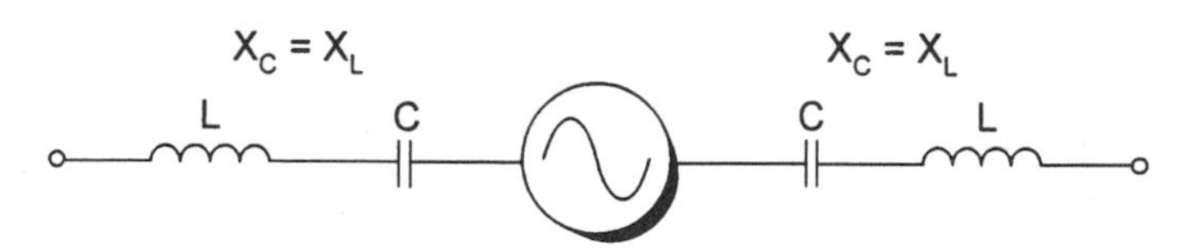

Figure 10.17 A dipole antenna circuit at resonance.

Just as with any RF circuit, a good impedance match between the transmitter, the feed line, and the antenna is important for the maximum transfer of power. A proper match will also stop the energy generated at the transmitter from being reflected back into its output, which can produce severe transmitter damage and transmission line insulation breakdown. This match between the transmitter, its feed line, and the antenna (Fig. 10.18) will occur when the inductive reactances equal the capacitive reactances and cancel, leaving only the resistances (which must equal each other), allowing maximum power to be radiated from the antenna. This prevents standing waves from being created (see Sec. 1.4, "Transmission Lines") on a mismatched transmission line, with much of the reflected power then being given up as heat. So, when the antenna is resonant and the impedances all match between the transmitter, feed line, and antenna, maximum alternating current is sent into the antenna and broadcast as an RF signal.

However, on the antenna itself, maximum output power will be possible only if there are maximum standing waves. This will create maximum voltage at the ends of the antenna—and maximum current in the center—as shown in the diagram of the dipole antenna of Fig. 10.19. Since the power of the RF signal radiated from an antenna is also contingent on the RF currents in the antenna, the more current the higher will be the output power. But for the energy to actually break free of the antenna and radiate far into space, the frequency must reach a point where the field lines (created by the RF currents) cannot fall back into the antenna's elements *before* the RF currents change polarity. The minimum frequency at which this can occur is approximately 30 kHz.

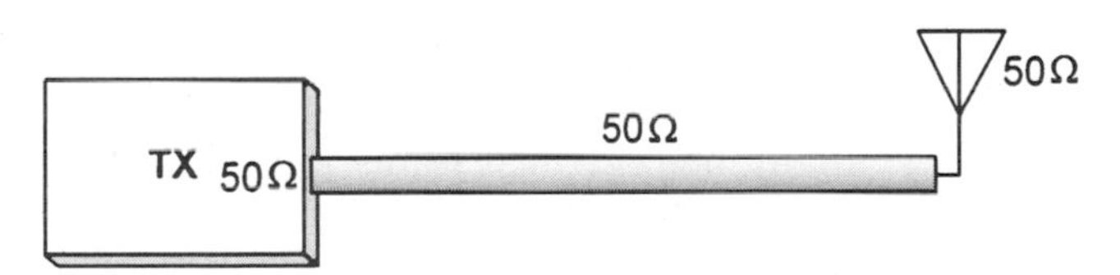

Figure 10.18 Maximum power output results when a perfect impedance match exists between the transmitter, feed line, and antenna.

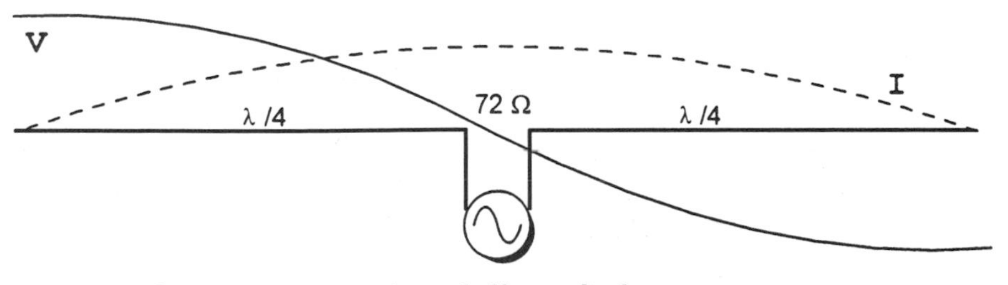

Figure 10.19 Current and voltage along a half-wave dipole.

As the electromagnetic energy travels through space after leaving the transmitting antenna, it will eventually cut the elements of a receiving antenna. This will induce a tiny voltage, which must then be heavily amplified and filtered by the receiver to obtain the desired demodulated signal information at the required amplitude and SNR.

The following is some common antenna terminology:

Beamwidth—The number of degrees, in the horizontal, of the main beam at its 3-dB-down points. Intimately linked with antenna gain, and measured in degrees.

Directivity—The power in the main beam of an antenna compared to an isotropic. Measured as a ratio.

G/T ratio—A figure of merit for microwave satellite receivers, *G* being the gain of the receiver's antenna and *T* being the system's noise temperature, or the ratio of ground (noise) temperature to antenna gain. Minimizing sidelobes maximizes this *G/T* ratio.

Gain—The gain supplied by an antenna over that of a (typically) isotropic antenna, and is measured in dBi. High gain, high directivity, and large electrical size (to wavelength) are interdependent. It is impossible to have a high-gain omnidirectional antenna, or a high-gain antenna that is electrically small. As the gain of an antenna is increased, the beamwidth must decrease; a general rule is that doubling the antenna's physical elements will double the antenna's gain, and halve its beamwidth. (However, losses within the antenna and feed network will soon reach a point where adding more elements to an antenna increases gain very little.)

Isotropic—A theoretical omnidirectional antenna that radiates equally well in all directions. Used as a reference to compare a real antenna's true gain specifications.

Main beam—The dominant lobe of a directional antenna where the majority of the output power radiates.

Polarization—Polarization is the orientation of the *electric* field of the electromagnetic wave as it travels through space. When an antenna's elements are parallel with the ground, it is referred to as a *horizontally* polarized antenna. Such an antenna can receive an electromagnetic wave only from a *vertically* polarized source through the small shift in polarization that takes place over distance. Indeed, if it were not for this slight EM wave change, a perfect horizontal antenna would not be able to induce a voltage into a perfect vertical antenna, and vice versa.

Sidelobes—Antennas will unfortunately emit electromagnetic radiation at other directions than that in the main lobe. These are normally wasted and undesired emission areas, and are referred to as the sidelobes of an antenna. In fact, high-powered outputs from the antenna may contain dangerous levels of EM radiation within these sidelobes. Minimizing sidelobes will increase antenna performance because smaller lobes collect less temperature, and thus less noise, from the earth.

10.8.2 Common antenna types

A very popular antenna for 900 MHz and below for omnidirectional applications is the *vertical antenna,* ordinarily a quarter-wavelength long, with earth or some other ground surface supplying the required additional quarter-wavelength through ground reflection (Fig. 10.20).

The other basic type is the horizontal antenna, which is normally employed when increased directional characteristics are required. However, almost any antenna can be oriented either horizontally or vertically, depending on size, frequency, and radiation pattern constraints. The most popular horizontal antenna is the bidirectional half-wave *dipole,* with the radiation pattern of Fig. 10.21*a*. Another common horizontal antenna is the highly directional parasitic multielement *Yagi* antenna, with a radiation pattern as shown in Fig. 10.21*b*. The Yagi structure is displayed in Fig. 10.22. The parasitic element of this antenna refers to the director and/or reflector elements that are not driven by a physically attached feed line, but instead have the RF voltage induced into them by the single *driven* element. This driven element is a simple dipole fed by the transmission line from the RF transmitter (or receiver). Some of the electromagnetic energy radiated by the driven element will cut the one or more successively longer reflector elements, which bounce the signal back to the dipole in phase, now adding to the driven element's own radiation. For an additional increase in antenna

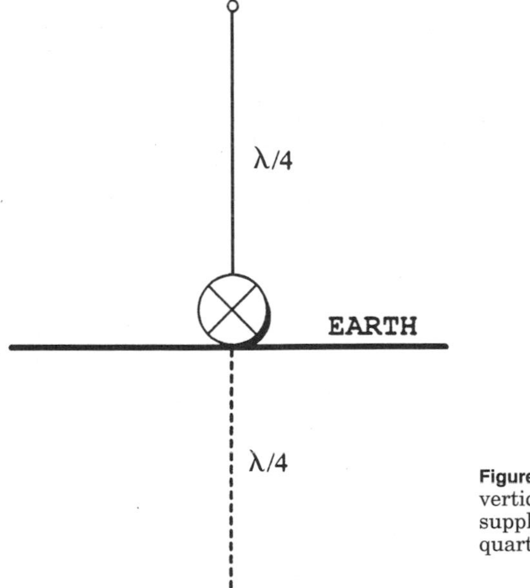

Figure 10.20 A quarter-wave vertical antenna showing earth supplying the other required quarter-wavelength.

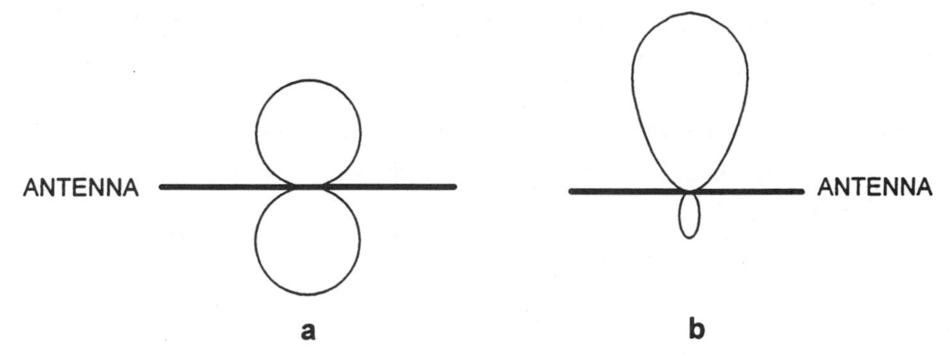

Figure 10.21 The radiation patterns of (*a*) a dipole antenna and (*b*) a Yagi antenna.

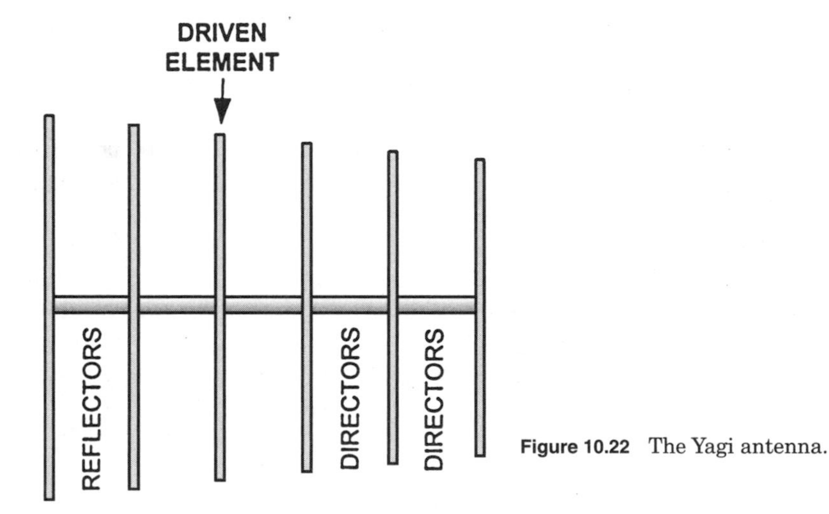

Figure 10.22 The Yagi antenna.

gain and directionality, we can add one or more progressively shorter directors in front of the dipole element.

Microwave antenna structures can be quite different from their low-frequency cousins. There are three prevalent types for the mid- to high-microwave-frequency range: the *patch, dish,* and *helical* antenna.

The patch antenna is very flat and simple to manufacture, and low in cost. However, they have low gain, a narrow bandwidth, and high surface wave losses. Nonetheless, they are a natural for many microwave applications, and are constructed of microstrip placed on a substrate above a ground plane.

The dish antenna (Fig. 10.23) uses a spherical or parabolic focusing surface constructed of solid sheet metal or wire mesh and, located at the antenna's focal point, an integral horn antenna. The horn antenna is fed by a waveguide for transmitting or receiving a signal, and is simply a flared-out section of the waveguide. The horn functions as an impedance match between the waveguide and the surrounding space. Gain of the dish antenna is contingent on

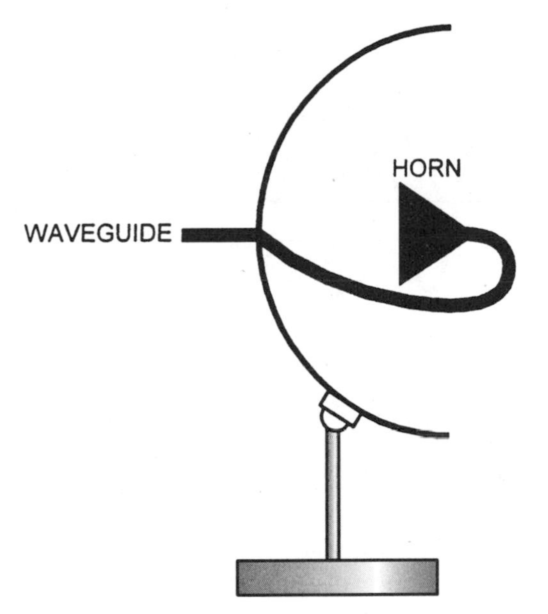

Figure 10.23 A microwave dish antenna.

the actual size of the dish, with its diameter being about 10 times its wavelength of interest for high gain, low loss, and extreme directivity.

Figure 10.24 displays the helical antenna, which is constructed of a nonconducting center strut helically wrapped with wire or tubing. This type of antenna is circularly polarized, so the opposing antenna of the system must be wound with the same sense, or little or no reception is possible. While the helical is a simple and low-cost antenna to design and build, its gain and beam width do not compare favorably to those of the dish antenna.

10.8.3 Antenna issues

Antennas are an important consideration in optimizing wireless system performance, especially in a multipoint, sectorized environment, where one type may act as a sectorized unidirectional antenna, with a different frequency or polarization allotted for each adjacent sector. The following are some considerations in such multipoint applications:

Horizontal beamwidth should be chosen to reduce overlap into adjacent sectors in order to increase BER performance.

Front-to-back ratio and sidelobe suppression must be maximized to reduce cochannel interference, and thus improve the BER (antennas are available with a front-to-back ratio of up to 40 dB, but this figure can be degraded by 10 dB or more by multipath effects).

VSWR, if at or below 2:1, will not normally be a large concern, since the resulting system loss will be only 0.5 dB or less.

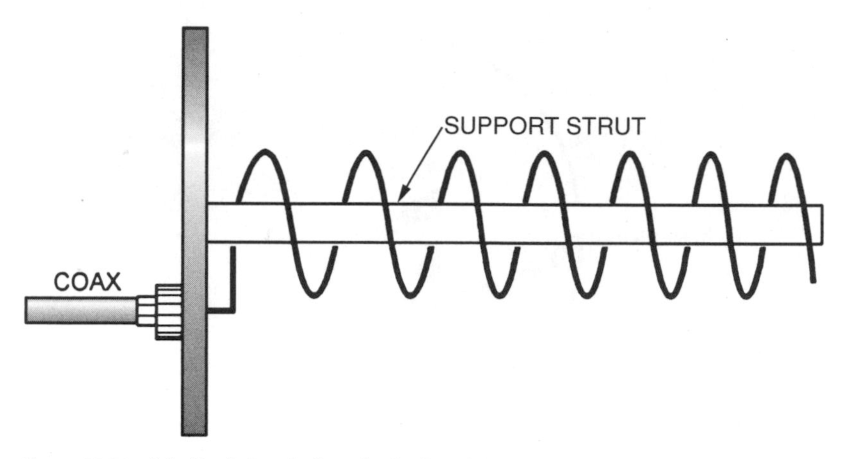

Figure 10.24 A helical circularly polarized antenna.

Special *null-fill* antennas are designed to abate the transmitted signal's nulls that naturally exist within a coverage area. This is accomplished by the selective tweaking of certain antenna parameters. However, wideband antennas for high-bit-rate applications are less successful at obtaining true null-fill performance.

There is a 2.15-dB difference in gain between an antenna that is rated in dB instead of dBi. As mentioned, the dBi rating is comparing the increase in gain over the antenna under consideration to that of an isotropic source, while the dB rating is a comparison to the gain of a dipole antenna, with its superior gain of 2.15 dB. In other words, a dipole *starts out* with an initial gain of 2.15 dBi. Thus, an antenna manufacturer that states that its product's gain is *6 dB* could also state a gain of *8.15 dBi*.

10.9 RF Connectors

10.9.1 Introduction

An RF connector is a part that is adopted to obtain a permanent or temporary connection for the transferring of RF energy from one circuit or cable to the next, preferably at a constant impedance.

All connectors will have a finite lifetime. Indeed, some high-frequency, high-precision units must be discarded after as little as 100 connects/disconnects. This limitation is due to wear between the two mating surfaces causing a change in the connector's geometry, thus increasing the insertion loss and decreasing the return loss.

Any connector that must function reliably out of doors in the wind, rain, ice, and snow should be specifically made for this type of abuse, and must be properly shielded. If not, corrosion will cause damage, sometimes quite rapidly, decreasing the connector's rated specifications.

10.9.2 Types of connectors

There have been many connectors invented and manufactured over the years for different frequencies and applications, as well as to improve ease of use and performance. The following connectors are the most common found today for coaxial connections from HF to SHF:

UHF—Developed by Amphenol in the 1930s, the UHF connector is utilized in RF applications up to only 300 MHz. This frequency limitation is a result of their undesirable nonconstant impedance characteristics. Also called a PL-259 connector, it is operated in low-cost, undemanding service.

N—Developed at Bell Labs, the N connector is the foremost connector for test equipment and antenna interfaces up to 4 GHz. They are 50-ohm threaded units that are capable of operating up to 11 GHz.

BNC—Can function up to 3 GHz, but normally found on lower-frequency, lower-cost test equipment, as well as antenna connections. Both 50- and 75-ohm versions are available.

TNC—A threaded version of the BNC connector, but designed for high-vibration environments up to 10 GHz.

SMA—The dominant 50-ohm microwave connector, capable of operation at up to 20 GHz. They are quite small and have threaded connections.

SMB—Nonthreaded, push-on connector operational up to 4 GHz, and available in both 50- and 75-ohm versions.

APC-7—A sexless screw-in, very high performance microwave connector qualified up to 18 GHz. Because of their very high cost, they are almost exclusively found only on high-end (Agilent) test equipment.

3.5MM—A precision version of the SMA connector, it can function up to 34 GHz. It has an air dielectric for increased performance over that of the SMA type. 3.5MMs will maintain their operational characteristics for thousands of connects/disconnects.

Wiltron-K—A sturdy, high-frequency (40-GHz) connector that will also mate with the SMA type if so required.

10.10 Wireless Design Software

10.10.1 Introduction

The computer and specialized simulation software have become invaluable tools in wireless circuit design during the last 20 years. Many software programs are available that can assist the engineer in optimizing a circuit for increased gain, frequency, output power, stability, etc., that would just not be possible without computer simulation capabilities in the shortened time frame permitted in modern industry. This is true not only on the circuit design level, but also in wireless system design.

Computer wireless circuit simulation programs can involve one or more methodologies. The most common is *Spice,* developed at Berkley University some 30 years ago. Spice works quite well at frequencies below 20 MHz (depending on the Spice model, some of which are optimized for much higher frequencies), and permits simulation of linear and nonlinear circuit behavior in the time, frequency, and transient domains. It also allows the modeling of most real-life effects that occur when a transistor or diode is biased to almost any desired value. However, Spice is *very* slow (a relatively simple circuit may take up to an hour to simulate), high-frequency component models are quite rare, and convergence can be a problem (convergence is the ability for Spice to come up with a correct answer during a simulation run). Spice is still invaluable in the simulation of many circuit designs even when another simulation methodology is used, especially for confirming proper active device bias and viewing the output waveform of a circuit in the time domain.

Linear simulators, such as Eagleware's *Genesys* and the included Caltech *Puff,* are the dominant program types employed in the RF and microwave world. These operate with *S-parameter models,* which are the most accurate way of representing a device in the RF and microwave regions (see Sec. 1.5, "S Parameters"). However, S parameters of active devices are already biased with a certain I_C and V_{CE} when these models are taken, so a linear simulator's main limitation is that a circuit cannot be biased to any chosen current and voltage level. In fact, the model's input and output parameters will be quite different in a real circuit if the designer later chooses to bias the transistor to any other value. Time domain views are normally not possible (unless a reverse FFT is taken), and the linear simulator's display, using the program's own Bode analyzer tool, will always be in the frequency domain. Nonetheless, linear modeling is very fast, allows rapid circuit tuning and optimization, and is quite accurate, and the models are prevalent.

Harmonic Balance (HB) simulation methods are practiced in very expensive, high-end simulators to model both linear and nonlinear circuit effects. Unfortunately, HB component models are not only difficult to obtain, but it may take a very long time to simulate even a simple circuit. Harmonic Balance will also not measure transient effects and is very poor at measuring the IMD products and higher-order harmonics of mixers and saturated amplifiers. If this type of simulator can be afforded ($30,000 and up), it is still an incredible and enlightening look into the functionality of a particular RF circuit design.

A cousin to Harmonic Balance is the *Volterra Series,* which accurately models the slightly nonlinear effects present in all linear circuits. This method is not meant to model grossly nonlinear circuits such as mixers, Class C amplifiers, oscillators, or frequency multipliers.

Three-dimensional planar electromagnetic analysis software employs the *method of moments* or the *method of lines* technique to simulate planar (microstrip, stripline, etc.) microwave structures. This type of simulator is able to display the gain and return loss of distributed microwave filters, transmission lines, waveguides, spiral inductors, planar antennas, and more, as well as exhibit the actual RF current flow and density running through these structures. The common *circuit theory*–based algorithms of other simulation programs are typically inadequate to accurately model these types of microwave structures and their stray coupling interactions, box modes, and discontinuities. In fact, the EM simulator supplied free with this book, *Sonnet Lite,* is the light version of one of the premier EM simulator packages on the market today, *Sonnet em® Suite,* and is perfect for this type of wireless simulation.

If some type of valid computer simulation is not performed—even after careful design and layout—a lot of rework will normally have to be performed in order to get a microwave or RF circuit to function properly. This is because of the abundance of undesired real-world internal component reactances and resistances, along with the various component tolerances and temperature effects, all conspiring to lower the expected performance of the circuit. And unless an EM simulation is run, complex interactions of electromagnetic coupling and certain stray parasitic reactances will decrease the resonant frequency of a design below what was expected. Eagleware's *Genesys* package is one popular program that will simulate the unpredictable effect of lumped components at high frequencies by permitting the designer to build a virtual RF circuit within the computer itself, and then tuning it to function as desired. With the inclusion of Eagleware's EM module *Empower,* both circuit theory and electromagnetic effects can be combined to allow the design of a more predictable circuit. However, even with less expensive linear computer simulation packages—such as Caltech's Puff—the development cycle can be tremendously reduced by viewing how various circuit parameters are affected by component tolerances, temperature variations, internal component reactances, and other annoying contributors to disappointing circuit operation.

10.10.2 RF programs

There are many free and low-cost software programs that will assist the RF designer in producing an electronic circuit—or even a complete wireless system—in one-hundredth the time it would take to design with a hand-held calculator, and will permit the engineer to tweak or change the design until it is optimized. This is just not possible with hand calculation techniques in any reasonable time. All of these programs as presented below run quickly and reliably on everyday PCs, some with as old a CPU as a 486.

The first is the Windows™ version of Agilent's *AppCad.* (The older DOS version was indispensable in RF design, but is difficult to get to run on all Pentium™-based Windows systems.) AppCad is a completely free program from Agilent, and is included for your convenience with this book's CD ROM. AppCad helps the engineer to instantly design bias networks for BJTs, FETs, and MMICs, as well as detector circuits and microstrip and stripline transmission lines. It also has a reflection calculator to compute VSWR, return loss, and mismatch loss for any desired input and output impedance of a circuit, a noise calculator to compute a receiver's NF, a standard-value calculator for

radiators in RF communications would comprise receivers that create high levels of RF radiation internally for frequency conversions, but do not purposely propagate this energy beyond their own cabinet. Both intentional and unintentional radiators must obtain FCC authorization to be marketed within the United States. (There are *some* exceptions to this rule; see “Keeping the FCC Happy—A Reference Article,” Mitchell Lazarus, *Wireless Design and Development Magazine,* November 1999.) The third category, incidental radiators, is not of interest to the average wireless designer. This category comprises devices that are not intended to generate RF internally or externally, but do anyway, such as light dimmers, neon signs, and electric motors. No FCC authorizations are needed for any of these devices under the incidental radiator category, but they must not produce RF radiation above similar devices on the market.

Both intentional and unintentional radiators must have one of three new FCC authorizations, depending on the product and its use (or possible misuse). The old *Type Approval, Notification,* and *Type Acceptance* authorizations have been replaced with the *Verification* (similar to the old Notification), *Declaration of Conformity* (DoC), (similar to the new Verification), and *Certification,* (similar to the old Type Acceptance). There is no old Type Approval replacement.

Because of the legal complexity, a lawyer that specializes in FCC technical issues should be consulted before marketing any wireless device, but the following is a general guide to the type of equipment and the type of authorization that typically must be obtained. A major communications law firm such as Fletcher, Heald & Hildreth of Washington, D.C., is a full-service telecommunications firm founded in 1936, and is quite capable of handling most FCC related technical issues. Other practices, such as Latham & Watkins and Miller & Van Eaton, are also very competent in this area.

Fixed microwave point-to-point and licensed broadcast transmitters (intentional radiators), as well as television and FM stereo and mono receivers (unintentional radiators), will usually obtain a Verification Authorization. This type of authorization simply requires the manufacturer of the device to test for FCC technical compliance. A Verification Authorization is by far the least complicated to acquire, as your company need not even file FCC compliance documents and, as soon as the device successfully passes, sales can actively begin. All test and design paperwork should be preserved over the manufacturing life of the product, plus 2 years. All that is demanded is that certain devices must be labeled as directed by the FCC rules.

Virtually all other consumer unintentional radiators are commonly authorized under the Declaration of Conformity. The DoC also allows your company to obtain its own equipment authorization and, as with the Verification Authorizations, the FCC is not even notified of the product’s existence. However, the DoC is far more expensive and complex to obtain than the simple verification, since all RF equipment tests must be performed by an accredited test facility. If passed, the equipment must be sold with the FCC logo attached, as well as with copies of paperwork called the *Compliance Information Statement* (which contains information on the product and its manufacturer). If the manufacturer so desires, it may also opt to obtain a Certification.

Mobile radios, licensed or unlicensed, as well as virtually all high-volume unlicensed transmitters of any kind, must obtain an FCC Certification. This covers the great majority of wireless devices, such as cordless and cellular phones, wireless local-area networks (WLANs), 802.11 devices, handi-talkies, and citizen band radios. Certification is quite intensive and expensive. It involves testing, substantial paperwork, and a large filing fee. Even then the device cannot be sold until the FCC approves the application, which can take 3 months or longer (private testing facilities may soon be approved to issue such Certifications, and in much shorter time periods).

After any engineering changes to an already authorized (either by Verification, DoC, or Certification) device that may have altered its RF compliance—such as a change in power, frequency, shielding, or isolation—a retest should be conducted to confirm that these changes have not dropped the product from the compliance standards. If the compliance tests on the modified device fail, all sales must be halted, and any problems fixed. However, if the engineering changes either have not affected the device, or have actually *improved* its compliance, then nothing need be done in the Verification or DoC cases (except maintaining all new test and design paperwork). In the case of Certification, things become more complicated. Almost any changes to the radio's (or other authorized device) frequency-generating stages, or any increase in output power, must typically obtain a completely new Certification. Still, if the modifications were to other circuits that are not involved in the normal compliance tests, and no degradation in test results occurred, then the device may be allowed to continue to operate as before, with no new FCC filing necessary. Even if the device now has test results that are slightly inferior to the original device, then the manufacturer may still file a "Class 2" application, along with the current test data, and hope for FCC approval.

If the original or modified product is denied authorization by the FCC because of a specific, and *correctly* interpreted, rule or regulation, then it may be possible to obtain an FCC waiver that will permit your product to be sold. Or you may actually attempt to change the rules that are impeding your FCC authorization. This, however, is a highly complex, time-consuming, and expensive alternative—with no guarantee of success.

10.12 Support Circuits

10.12.1 Introduction

The following introduces assorted support circuits that are used frequently in wireless design. This book does not go into any design detail because many of them are usually placed in ICs (and may be but a small part of an overall RFIC) and rarely need to be designed by the wireless engineer today.

10.12.2 Circuits

Speech processing. Speech processing is a general term for a circuit that adjusts an input audio signal in amplitude, frequency, or both, before it is placed into a transmitter's modulator. *Speech compression* and *companding* are the most common of these processing techniques. A special form of speech processing, called *automatic level control* (ALC), affects the RF, instead of the audio, of a transmitter system.

Considering that the modulation frequency of FM, SSB, and AM transmitters will affect the transmitted bandwidth, a method to limit the maximum baseband frequency must be utilized. This can be accomplished by an active low-pass filter placed within the audio sections.

Speech compression prevents a much wider bandwidth from forming outside the desired AM passband because of the deleterious effects of overmodulation, which produces spectral *splatter.* Splatter is constructed of the additional harmonics created in the baseband signal by overmodulation, which further modulates the carrier, causing extra sideband components and a widening of the bandwidth. An overdriving of the IF and/or RF amplifiers of the transmitter may also occur, creating IMD. Splatter and IMD generate *adjacent channel interference* (ACI) and a less intelligible baseband signal. A speech compression circuit decreases these negative effects by amplifying a signal normally up to only a predetermined level, but then will begin to reduce gain by 1 dB for every 2 dB of audio input signal. These basic speech compression circuits will simply confine the maximum AM or SSB audio amplitude to some maximal value, while *dynamic compression* helps intelligibility by increasing the smaller baseband amplitude levels as well. Compression schemes in general are quite capable of raising the average output power of an SSB transmitter, while decreasing distortion and splatter and limiting modulation to 100 percent or below. Compression is so effective at increasing the average transmit power because human speech has voice amplitudes that are highly complex and irregular, yet the transmitter must be prepared to send out the *highest* voice peak with low distortion and splatter—a peak that may be 10 to 12 dB higher than the average energy contained within the entire waveform. Thus, compression smooths over much of these amplitude variations in order to allow a much higher average output power, and thus an increase in the range of the wireless link. Simply, basic speech compression acts in the same way as standard AGC, but is located in the audio stages (Fig. 10.25).

The technique called *companding* can take compression to outrageous levels by almost completely compressing all of the peaks to be at close to the same level as the speech's valleys, and then expanding them back to normal amplitudes at the receiver (Fig. 10.26). This permits a high dynamic range, better signal-to-noise ratio, and much higher average output powers.

Many FM, and some AM and SSB, voice transmitters may even employ a form of processing called *speech clipping.* This clipper circuit actually hard-limits the voice signal if it reaches a certain high amplitude. A low-pass filter at the output of the audio section not only removes the harmonics produced by this clipping action, but also limits the maximum frequency possible of the baseband signal. A similar concept is an *audio clipper circuit* (Fig. 10.27), which can provide a degenerative out-of-phase feedback signal for any audio

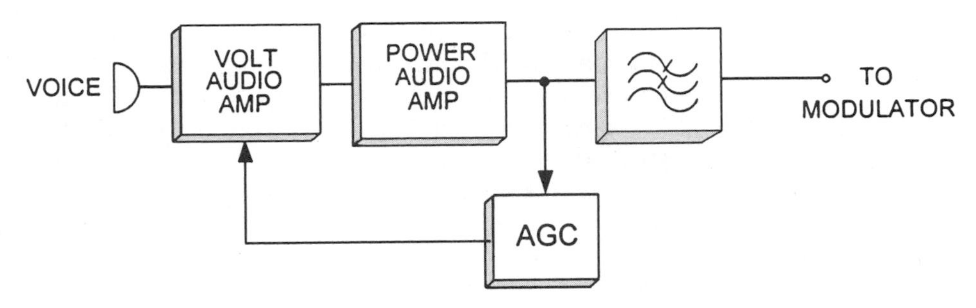

Figure 10.25 Transmitter speech compression in the audio section.

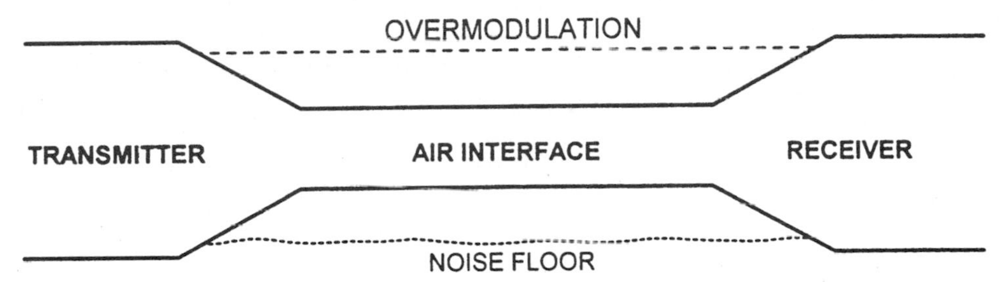

Figure 10.26 Companding action between a single-sideband transmitter and receiver.

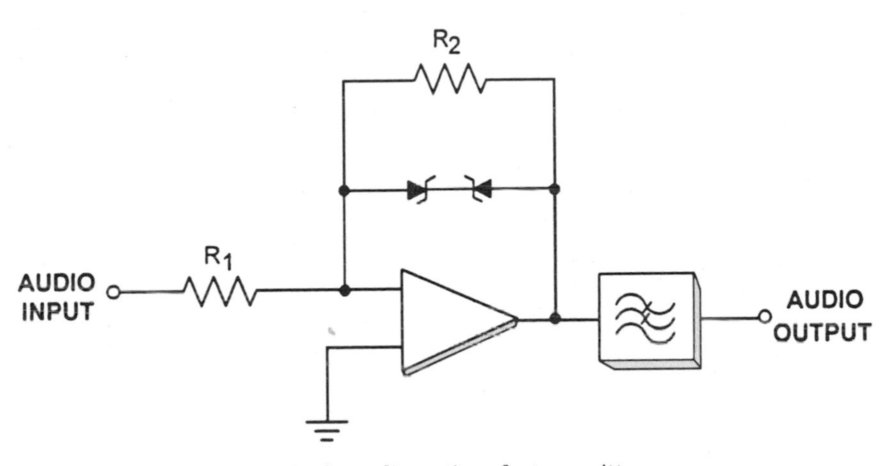

Figure 10.27 A voice clipper in the audio section of a transmitter.

peak voltage over a certain preset level. This circuit consists of an operational amplifier with back-to-back zeners, and a resistor that sets the audio gain, with the peak value of the baseband waveform being governed by the value of the diodes.

The form of compression called *automatic level control,* or *RF compression,* is found in SSB, AM, and even some FM transmitters. ALC essentially uses a standard AGC circuit, but it is designed to operate on the transmitter's IF (Fig. 10.28). This automatic level control decreases high-amplitude transmitted signals, without affecting nominal or lower signal levels. ALC controls the gain of the IF stages by sending a voltage to the gate or base of a variable-gain amplifier (VGA), which then decreases or increases the input power into the final amplifier. ALC does this by tapping the output of the power amplifier, rectifying and filtering this RF to DC, and then placing this voltage at the gain control input of the IF amplifiers. Thus, no matter what the amplitude of the original baseband signal, the linear power amplifier cannot be overdriven and create excessive adjacent channel interference.

Automatic frequency control. In low-cost non-PLL or non-crystal-controlled transmitters or receivers, an automatic frequency control (AFC) can be used to steady the system's *LC* local oscillators. General frequency stability is obviously required for any superheterodyne receiver, especially when internal and external temperature changes occur, or the resultant frequency drift would convert any incoming RF signal to an improper IF, causing attenuation and distortion of the baseband signal. A transmitter also has the same reasons to maintain its stability for the receiver, and because it is capable of interfering with other wireless devices within adjacent channels, as well as the negative legal ramifications of a wandering transmitter.

A basic example of one type of AFC is shown in Fig. 10.29 for an FM receiver. The IF signal is tapped from the last stage of the IF strip and sent into an FM demodulator, which changes its own voltage output in step with the IF's frequency drift. This drift is due to the unstable *LC* local oscillators. The demodulated signal is then inserted into the low-pass filter to obtain a DC control voltage

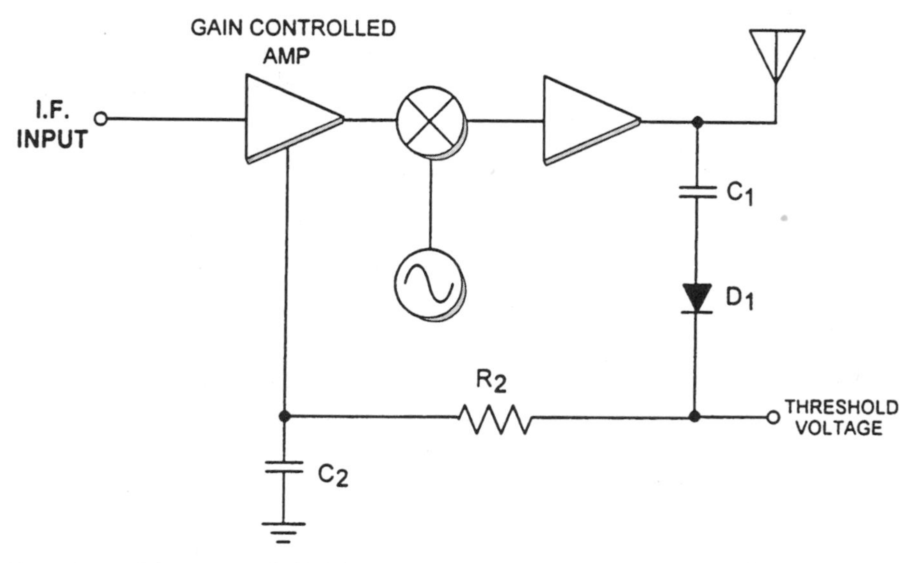

Figure 10.28 A transmitter's ALC circuit.

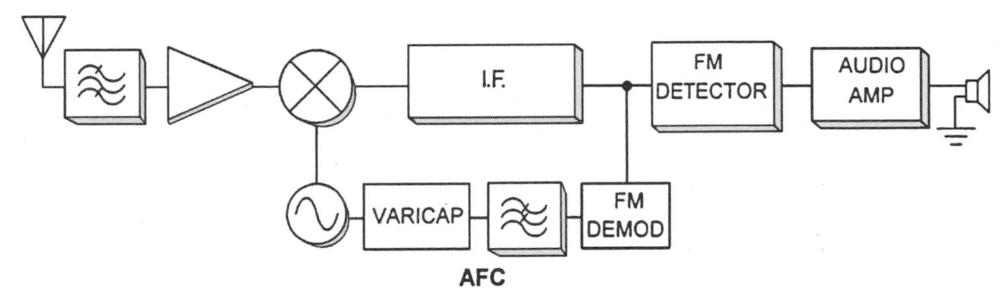

Figure 10.29 An FM receiver's AFC circuit for frequency stability.

for the varactor. The control voltage adds to or subtracts from the diode's center frequency bias. Since the varactor is placed across the tuned circuit of the *LC* oscillator, it has some control over the output frequency of the LO. Thus, if the LO begins to move off frequency, the FM demodulator/filter will change from its center frequency output voltage level and place the appropriate correction voltage into the varactor, altering its capacitance, and coercing the LO back to its proper frequency.

Squelch. Squelch circuits mute the annoying static that occurs when an AM or FM receiver has no RF input signal since, because of normal AGC action, this will be when the IF stage gain will be at its peak. Another positive attribute to squelch circuits is that they save a significant amount of battery power in a portable device. An unsquelched receiver may use 100 mA or more, while a squelched receiver may run at only 15 mA or less.

A squelch circuit functions by stopping static from reaching the radio's output speaker. This can be realized in one of three ways: The supply voltage to the audio amplifiers can be switched off; the audio amplifiers can be disabled by furnishing a reverse bias to their base; or the noise energy is actually prevented from arriving at the audio amplifiers by either blocking the energy with a series pass transistor, or by shorting it to ground. Nevertheless, a majority of all squelch circuits detect the presence of an RF input signal by simply looking at the DC output of the AGC loop.

One common squelch circuit is as shown in Fig. 10.30. When the receiver is not receiving an RF signal of the proper input level—or perhaps no signal is present at all—then the IF stages of the receiver will be biased by the AGC loop for maximum amplification. The static level would be very high into the radio's speakers if this squelch circuit were not present, and shunting the noise to ground through the squelch gate (an on transistor switch). But if we couple this IF AGC bias voltage into the AGC IN of the SQUELCH AMP and on to the squelch gate, then when an RF signal of the proper strength is finally received, it will generate an AGC voltage of a sufficient amplitude capable of cutting off the squelch gate, switching it into a nonconducting state. This will permit the detected baseband audio to now proceed into the AUDIO AMP to be amplified and sent onto the speaker.

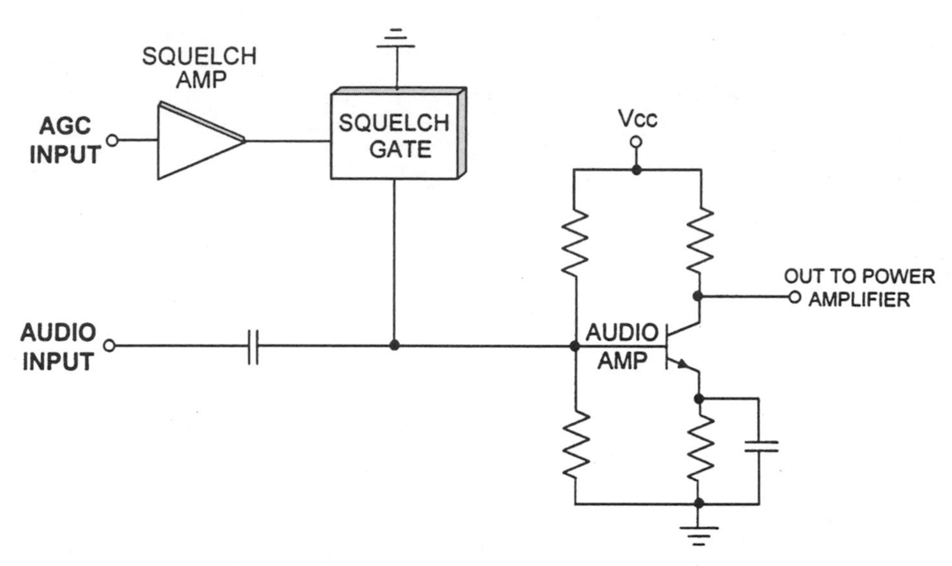

Figure 10.30 An AGC-based squelch circuit.

Another squelch system is shown in Fig. 10.31, and filters the noise frequencies, which are at about 5 kHz to 6 kHz, with a bandpass filter that taps a small amount of the signal from the IF. It then passes only these noise components on to a STATIC AMP, and into the RECTIFIER/LPF to be changed into a DC control voltage for the squelch gate. For instance, if no RF signal is being received, the noise amplitude will be quite high. The noise will be further amplified by the STATIC AMP, changed to DC by the RECTIFIER/LPF, and placed at the squelch gate input to bias it on. This short-circuits the output of the detector to ground, preventing the AUDIO AMP from receiving this noise, and thus quieting the audio output. However, if a signal of sufficient strength were received, the noise levels would naturally decrease, biasing off the squelch gate, and thus permitting the baseband to reach the AUDIO AMP and to be sent on to the radio's speaker.

We can employ integrated circuits to assist in performing squelch functions (Fig. 10.32). As most wireless devices, at a minimum, utilize an IF IC strip, we may simply tap the IC's *received signal strength indicator* (RSSI) pin to obtain a DC voltage level that corresponds to the actual RF received signal strength. We can then feed this into a comparator that is adjustable for the selected squelch threshold by the control R_1. As an example, if IN1 from the chip's RSSI output is lower than the reference level at IN2 (as set by the voltage divider of R_1), then the comparator will swing near to its positive rail ($+V_{CC}$), turning on the squelch gate. This will short the AUDIO AMP's collector to ground, preventing the static signal from reaching the power amplifier and speaker of the next stages.

There are highly integrated RFICs that employ their own internal squelch circuits, with an internal voltage audio amplifier, and that demand only a discrete potentiometer at the appropriate pin to fully adjust the squelch to any desired level.

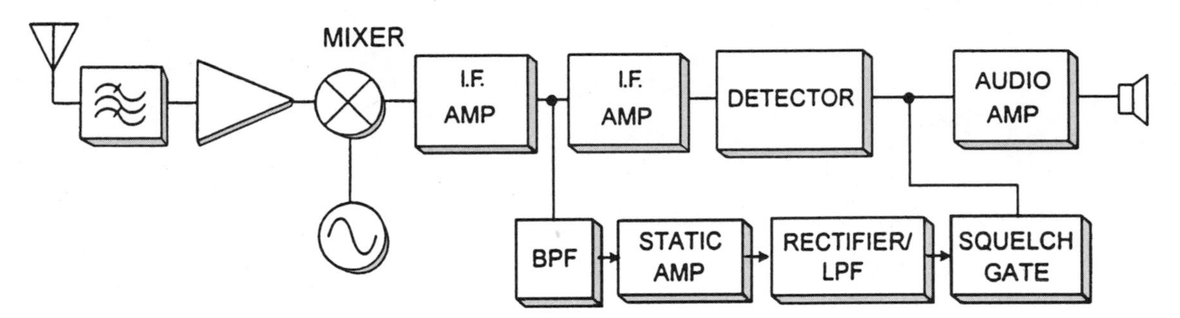

Figure 10.31 A noise-based squelch circuit.

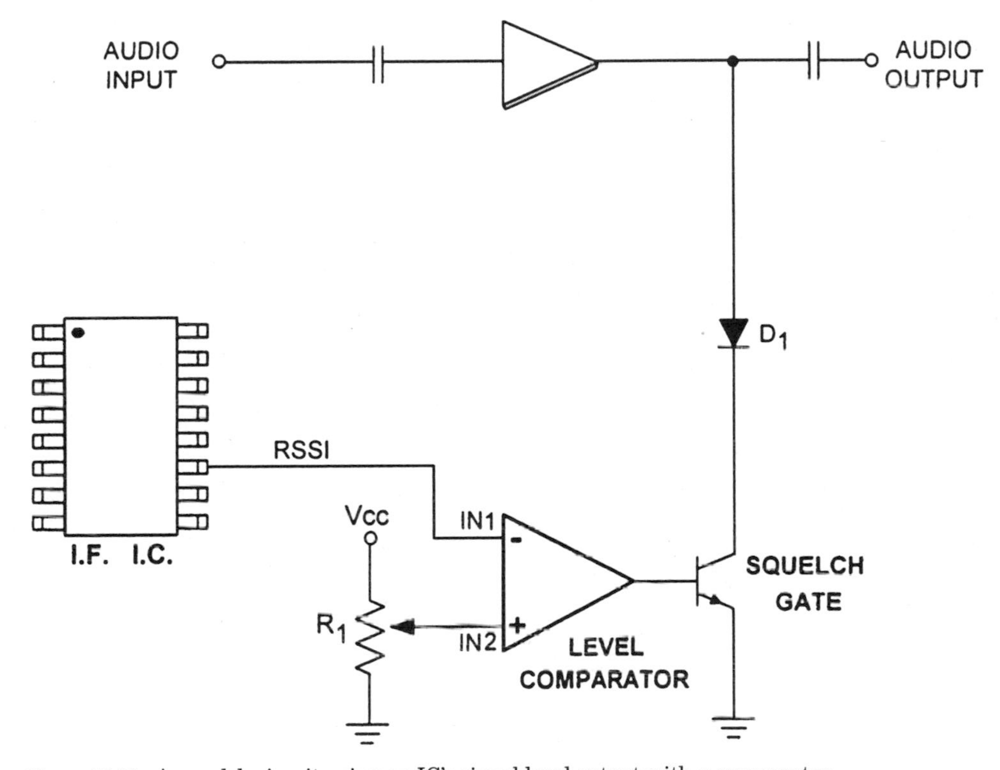

Figure 10.32 A squelch circuit using an IC's signal level output with a comparator.

Appendix

A

Puff Manual

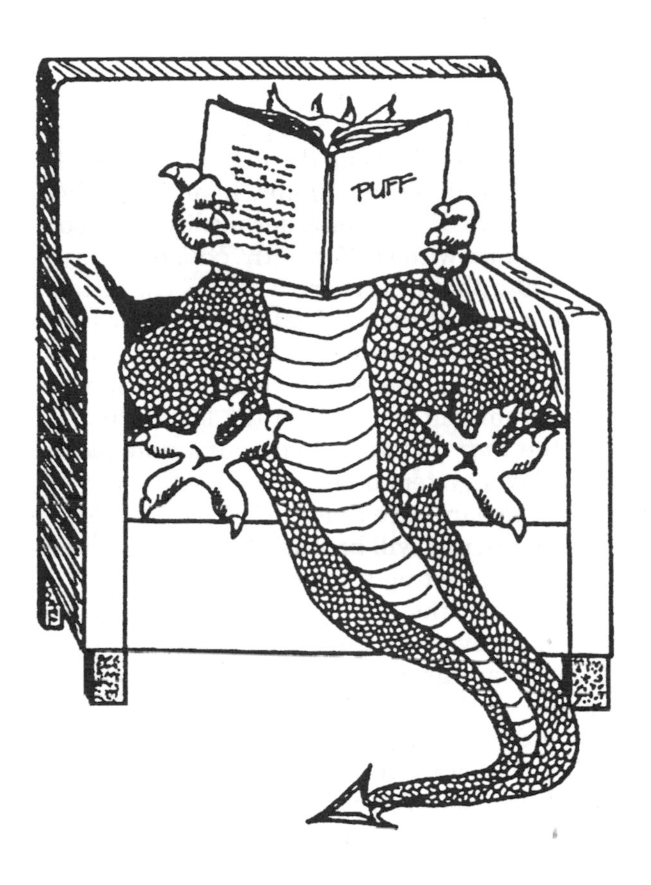

1 Getting Started

TO RUN *Puff,* you need an IBM PC, AT, PS/2 or compatible, a graphics card and display, and DOS 3.0 or later. For best performance we recommend a computer with 640 kilobytes of memory, a math-coprocessor chip, and a hard disk. Limited operation is possible with less memory, and floating point emulation is used if no coprocessor is present. The program works best with a *Video Graphics Array* (*VGA*) display and adapter. All the screens shown in this manual use the *VGA*. The smaller text in the *VGA* mode makes room for more characters, allowing an extra decimal point of numerical precision to be displayed. *Puff* will also work with the *Color Graphics Adapter* (*CGA*) and the *Enhanced Graphics Adapter* (*EGA*), but the screens will appear different than shown here and some of the numbers displayed will lack the extra decimal. In order to make a hard copy, you will need a screen-dump routine for your display and printer. The following programs have been included on your *Puff* diskette:

```
ega2eps.com     vga2eps.com
ega2pro.com     vga2pro.com
ega2lasr.com    vga2lasr.com
```

These are stand-alone programs for doing *EGA* and *VGA* screen dumps on *Epson, IBM Proprinters,* and *HP LaserJet* printers, respectively. Use them in the same way as the `graphics.com` program that comes with DOS. Just run the

```
\b{oard} {setup.puf file for PUFF, version 2.0}
D      0        {display: 0 VGA or PUFF chooses, 1 EGA, 2 CGA, 3 Mono}
o      1        {artwork output: 0 dot-matrix, 1 LaserJet, 2 HPGL file}
t      0        {type: 0 microstrip, 1 stripline, 2 Manhattan}
zd     50.000 Ohms     {normalizing impedance. zd>0}
fd     5.000 GHz       {design frequency. fd>0}
er     10.200 {dielectric constant. er>0}
h      1.270 mm        {dielectric thickness. h>0}
s      25.400 mm       {circuit-board side length. s>0}
c      19.000 mm       {connector separation. c>=0}
r      0.200 mm        {circuit resolution, r>0, Um=micrometers}
a      0.000 mm        {artwork width correction.}
mt     0.010 mm        {metal thickness, Um=micrometers.}
sr     0.000 Um        {metal surface roughness.}
lt     0.0E+00{dielectric loss tangent.}
cd     5.8E+07{conductivity of metal in mhos/meter.}
p      5.000   {photo reduction ratio. p<=203.2mm/s}
m      0.600   {mitering fraction. 0<=m<1}
\k{ey for plot window}
du     0        {upper dB-axis limit}
dl     -20      {lower dB-axis limit}
fl     0        {lower frequency limit. fl>=0}
fu     10       {upper frequency limit. fu>fl}
pts    51       {number of points, positive integer}
sr     1        {Smith-chart radius. sr>0}
S      11       {subscripts must be 1, 2, 3, or 4}
\p{arts window} {0=Ohms, D=degrees, U=micro, |=parallel}
lumped 1500
tline 500 90D
qlinc 500 130D
xformer 1.73:1
atten 4dB
device fhx04
clines 600 400 90D
```

Figure 1.2 Listing of `setup.puf`. It is safest to edit a copy of this file, rather than the original. Be very careful when you edit; if some of the board parameters are missing, the program will abort. Comments may be added at the end of the lines in braces, but they should not extend into the next line.

Exercise 1.2 The spacing of the connectors in the *Layout* window is controlled by parameter c in the *Board* window. Hit *F4* to move to the *Board* window. Use the cursor keys and the number keys at the top of your keyboard to set c to zero. Type *F1* and see what happens to the connectors. When done, return c to its original value.

A step-by-step *Puff* example is now given. We begin with the screen in Fig. 1.1. Hit the *F3* key to make the *Parts* window active. This is indicated by the flashing cursor for part a, and the highlighted F3 at the top of the window. Now hit *F1* to move to the *Layout* window. A large white × will appear in the center of the circuit board, and the first line in the *Parts* window will become highlighted. Part a is a lumped 150-Ω resistor. Push ↓, and *Puff* will draw a hollow blue rectangle, labeled a, down the screen, and the × will move to the other end of the resistor. In the *Message* box, Δy -2.540mm will appear. This shows you the change in the y coordinate. When no length is given for a lumped part, *Puff* assumes it to be one-tenth the size of the layout. Now hit the = key to ground the end of the resistor and type ↑ to move to the other end. Without the ground, *Puff* thinks that the end of the resistor is open-circuited. Now a 90° section of 50-Ω transmission-line shall be added. This is the `tline` part specified on line `b` of the *Parts* window. Type `b` to select the part, and ↑ to draw it. A copper rectangle is drawn representing a quarter-wavelength transmission-line section at the 5 GHz design frequency (`fd`). The width and length of the line are drawn to scale. In the *Message* box, Δ`y 5.690mm` appears, telling you the line length. Next type *F3* to move back to the *Parts* window, and move the cursor down into the description for part `b`. Hit the = key again. In the *Parts* window the = key is used to display the length and width of the `tline` in the *Message* box. Next, return to the *Layout* window by typing *F1* and then type `1`. You should use the number key at the top of the keyboard rather than the numeric keypad. In fact, the *NumLock* key is disabled when *Puff* runs, to avoid mix ups between arrows and digits. *Puff* will make a gray outlined path up and to the left to the first connector. The width of the gray path is the same as that for a 50-Ω line. The point at which it connects to the circuit will be the reference point for calculating phase, and so the length of the gray path has no effect on the analysis. The × will not have moved. This completes our circuit, which consists of a section of 50-Ω line with a 150-Ω load. The screen is shown in Fig. 1.3.

When you type a key, *Puff* first checks for a valid keystroke. If it is not, *Puff* will beep, and do nothing else. You can press `z` to hear the beep. Next, *Puff* checks to see if it can carry out the command. If it cannot, it will give an error message in the red *Message* box. For example, if you retype `1` in the present circuit, *Puff* will say, `Port 1 is already joined`, because multiple connections to the same port are not allowed.

This circuit is ready to be analyzed. Push *F2* to go to the *Plot* window and `p` to plot. The calculated s_{11} values will appear on the Smith chart as small dots joined by a cubic spline curve. A square marker indicates the reflection coefficient at the design frequency, and the numerical values are given in the *Plot* window. Any magnitude greater than 100 dB will be reported as ∞, and any magnitude as small as −100 dB will be listed as zero. When *Puff* finishes, the *Message* box indicates the time required for the calculation. The plot is a circle of radius one-half on the Smith chart. The rectangular graph shows the magnitude. It is not very exciting because it is a constant, −6.02 dB. To look at the impulse response, push `i`, and give a carriage return when *Puff* prompts for an integer. *Puff* will first calculate the frequency response and then use an inverse fast Fourier transform to calculate the time response. The result is a reflected pulse of height 0.5, delayed 100 ps. The delay comes from the transmission line. The resulting *Puff* screen is shown in Fig. 1.3.

The 150-Ω load may be matched with a quarter-wavelength section. Push the space bar to leave the time-domain plot. A section of 87-Ω transmission

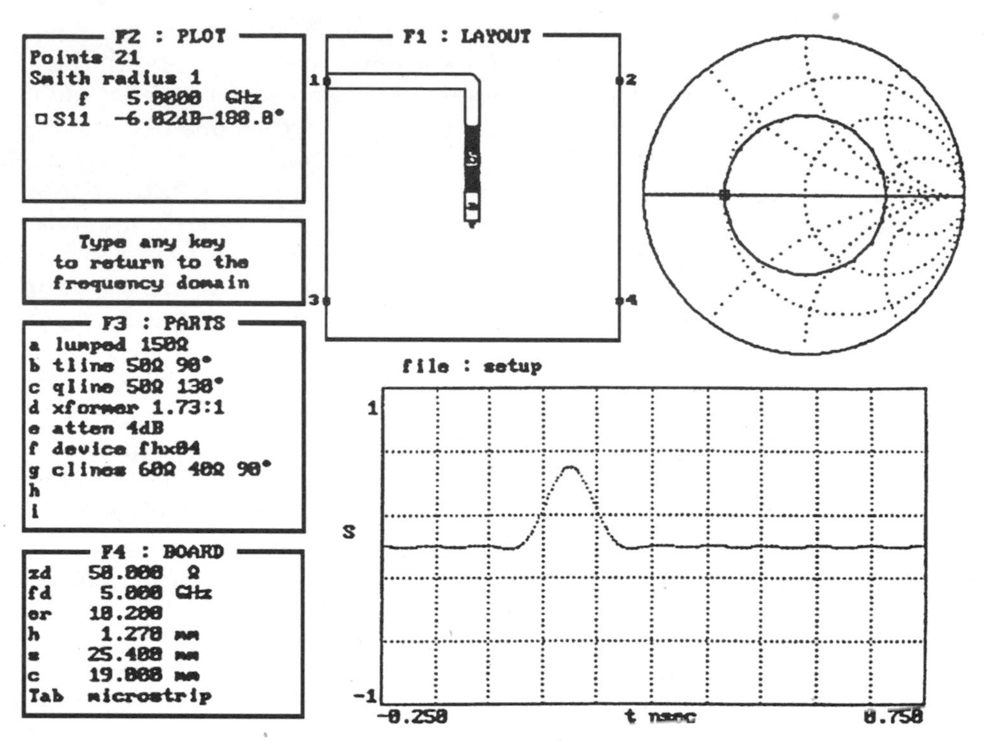

Figure 1.3 Circuit consisting of a quarter-wavelength section of 50Ω line with a 150-Ω load showing the impulse response.

line will match a 50-Ω line to the load. The impedance of part b is changed by first pushing *F3* to move to the *Parts* window. Use the cursor keys to move to the 5 on line b, and type 87 to change the impedance. Typing the = key displays the new dimensions of the tline in the *Message* box. Now push *F2* to return to the *Plot* window. *Puff* will update the circuit, and you will notice that the transmission-line section has become narrower, because of the impedance increase. Push p to plot again. The screen will appear as in Fig. 1.4. The circle on the Smith chart has shrunk so that it passes through the origin at the design frequency. The curve on the magnitude plot is off the screen at fd, but you can check the numerical value in the *Plot* window, which shows s_{11} equal to −46.78 dB.

We conclude by changing to a stub-matching circuit. Type *F3* to return to the *Parts* window and change both impedance and length specifications for part b to read tline 50Ω 60°. If you need to replace an Ω or ° symbol, use the *Alt*-o or *Alt*-d keys, respectively. Type *F1* to go to the *Layout* window, and check that the × is at the top of the b transmission-line. Type c and → to add the open circuit stub-matching section shown in Fig. 1.5. As specified here, the qline part used for c is identical to a tline. Push *F2* and p to analyze. This circuit matches over a narrower frequency band than the quarter-wave match.

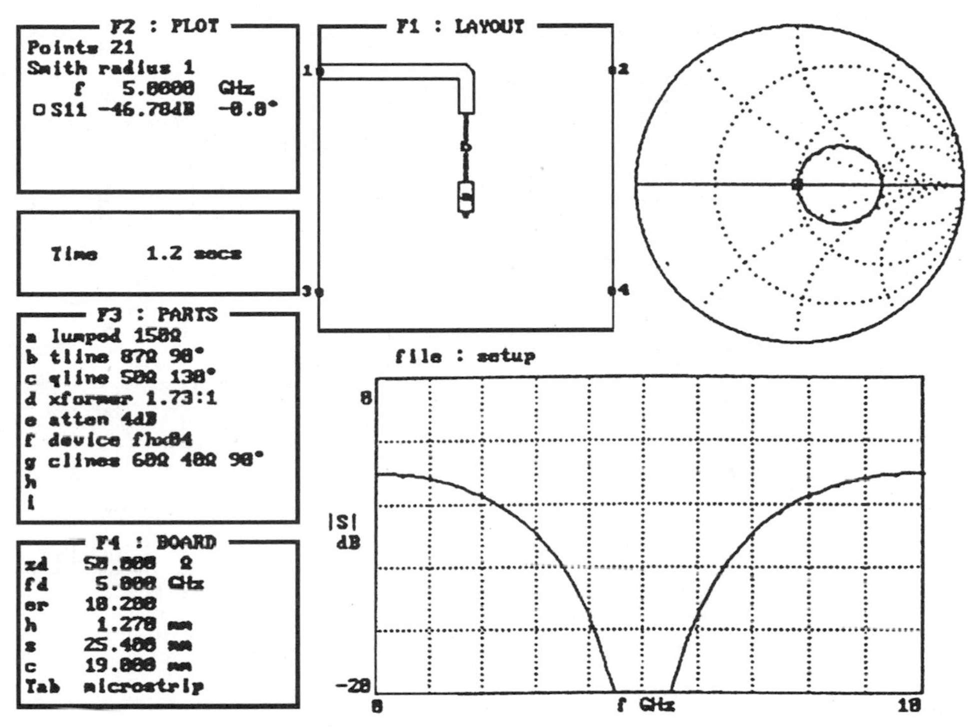

Figure 1.4 Quarter-wave matching.

Exercise 1.3 The qline is equivalent to the tline, but with loss effects included. To change the qline from a lossless transmission line to that with $Q = 10$ at fd, append the specification for part c to read qline 50Ω 130° 10Q. What effect does this have on the analysis? What are the differences between substituting 10Qc and then 10Qd?

Exercise 1.4 Go into the *Layout* window and make sure that the × is at the right of part c. Hold down the shift key and type ←. Such a shift-cursor operation in the direction of a part will cause it to be erased. Hold down the shift key and type 1 to erase the path to the connector. What happens if the shift-cursor operation is not in the direction of a part?

Exercise 1.5 Erase the entire layout by typing *Ctrl*-e from either the *Parts* or *Layout* window. Reattempt the 150-Ω matching problem using the xformer part. How does the bandwidth compare?

Exercise 1.6 *Puff* will automatically analyze a tline or cline for losses and dispersion when their specification includes an exclamation point. Repeat the quarter-wave matching procedure, but this time substitute tline! for tline. Compare the plots with and without the exclamation point.

Exercise 1.7 The file fhx04.dev that was included on your *Puff* diskette contains the two-port scattering parameters for the Fujitsu FHX04 high-electron mobility

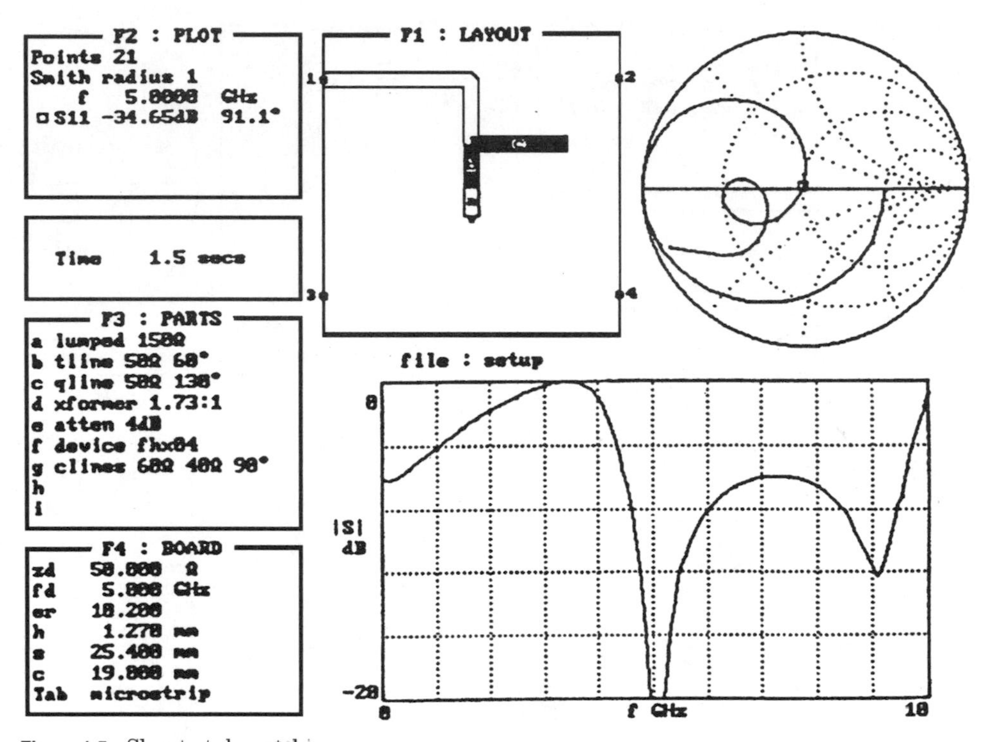

Figure 1.5 Shunt-stub matching.

transistor. Erase the current layout by typing *Ctrl*-e, select and draw this `device` part, and connect it between ports 1 and 2. Plot its scattering parameters. What happens when you repeat this procedure after substituting `indef` for `device` in the part's specification. How can you get the same scattering parameters using `indef` as you did with `device`?

Exercise 1.8 Erase the layout, go to the *Board* window, and use the `Tab` key to select a *Manhattan* circuit. See what happens when you then draw the `tline` and `clines`.

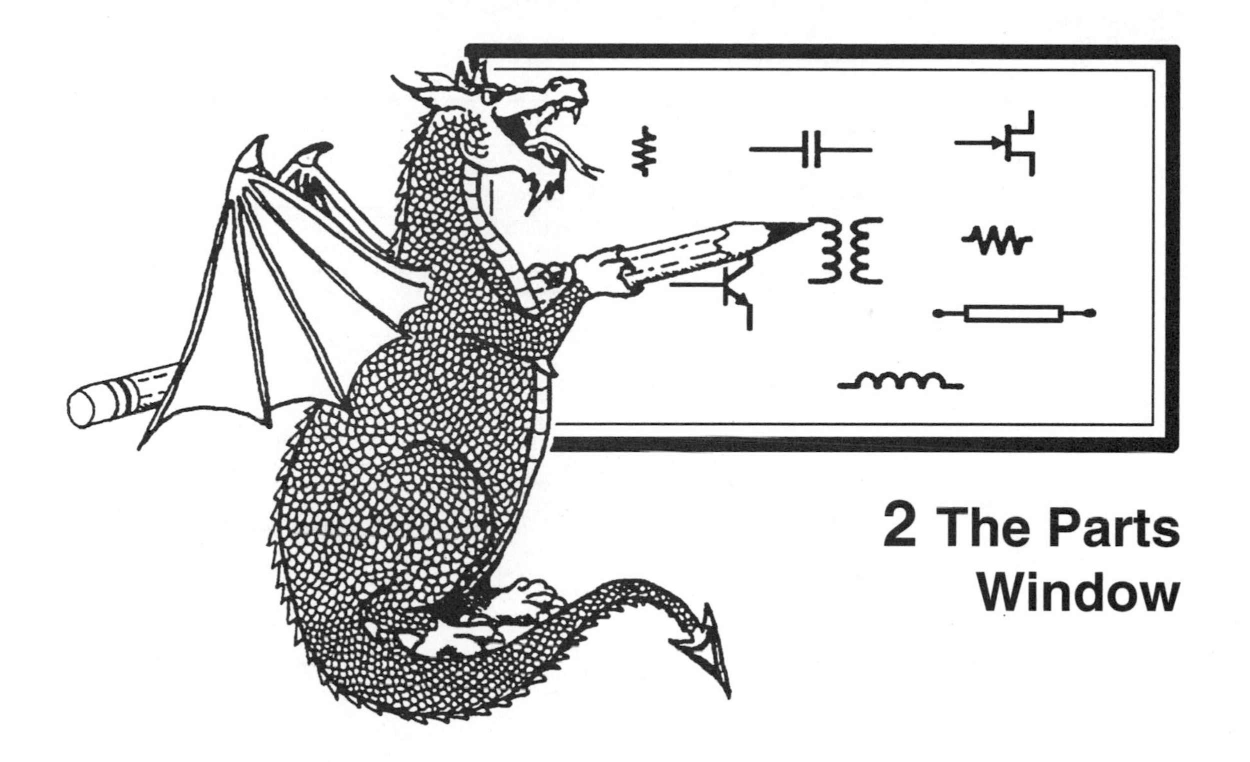

2 The Parts Window

THE PROGRAM BEGINS in the *Parts* window. The initial parts list is taken from the setup file. A previously saved circuit file can be read in by typing *Ctrl*-`r`. *Puff* will prompt for a file name, and will add `.puf` if no extension is given. You can edit the parts list with the arrow keys, the backspace key, the carriage return (or enter) key, *Ins,* and *Del.* The first letter on each line, from `a` to `i`, identifies each part for use in the layout. By hitting the *Tab* key, the parts list may be doubled in size, allowing extra parts `j` through `r` to be defined. Hitting the *Tab* key a second time will shrink the parts list back to its normal size, unless an extra part was defined and used in the layout. The types of parts available in *Puff* are listed in Fig. 2.1, including examples of their use. Only the first letter of a part name is needed for it to be recognized: `l` is equivalent to `lumped`, `t` is equivalent to `tline`, `q` is equivalent to `qline`, etc. Free up space in the parts list by using single character part descriptions. Commas and periods are both treated as decimal points. Spaces are ignored, except around `device` and `indef` file names. When you leave the *Parts* window, the list will be checked. There will be an error message if a part is longer than the board or shorter than the circuit resolution `r` given in the circuit file. *Puff* will redraw the circuit if it has changed, and report an error if changes result in a part being drawn off the board.

The simplest part available in *Puff* is the attenuator, written as `atten`. It is drawn as an open blue square with red dots denoting its two ports. It is ideal; *Puff* will always consider it as matched to the normalizing impedance `zd`, and its attenuation will be frequency independent. The desired amount of

Part	*Description and Examples*
`atten`	Ideal attenuator. Enter attenuation in dB: `atten 3dB`
`xformer`	Ideal transformer. Enter turns ratio: `xformer 2:1`
`lumped`	Resistor, capacitor, and/or inductor: `lumped 50Ω││1nH││1pF` {parallel RLC} `lumped 10 + j10 - j10Ω` {series RLC resonant at `fd`}
`tline`	Ideal transmission line. Enter impedance and length: `tline 50Ω 90°` {γ/4 length at `fd`} `tline 1.0z 5mm +1.0h` {artwork length correction added} `tline! 50Ω 90°` {analyze with advanced models}
`qline`	Finite Q `tline`: `qline 50Ω 90° 75Qd` {$Q = 75$ due to dielectric losses} `qline 20mS 90° 50Qc` {$Q = 50$ due to conductor losses}
`clines`	Ideal coupled transmission lines. Enter Z_e, Z_o, and length: `clines 60Ω 90°` {Z_e specified} `clines 60Ω 40Ω 6mm` {Z_e and Z_o specified} `clines! 60Ω 90°` {analyze with advanced models}
`device`	Read file with s-parameter data. Specify filename: `device fhx04` {read filename `fhx04.dev`} `device atf131.s2p` {read file in EEsof format}
`indef`	Indefinite s-parameter generator. Converts n port s-parameter file to $n + 1$ port device. Specify filename as in device.

Figure 2.1 Descriptions and examples of the parts available in *Puff*.

attenuation is entered in `dB`. If no units are entered, *Puff* will assume dB. The `atten` part is reciprocal and symmetrical.

Exercise 2.1 Plot the scattering parameters for an `atten` part that has a negative value of attenuation. Why is this different than an ideal amplifier?

The `xformer` part denotes a lossless and frequency independent transformer. The only numerical parameter needed is the (unitless) turns ratio. *Puff* accepts a single real number for the turns ratio, not a ratio of two numbers. However, a colon may be used to write the number as a ratio to one. Therefore, `1.5:1` is a valid entry, whereas `3:2` is not. The transformer is antisymmetric, so it is drawn in the layout as a trapezoid. Red dots are again used to denote the ports. For an $n:1$ transformer, the wide end of the trapezoid represents the n side, while the narrow end represents the 1 side. If a negative turns ratio is entered, *Puff* will add a 180° phase shift to the transmission parameters.

Exercise 2.2 Layout and plot the scattering parameters for an ideal 180° phase shifter using an `xformer` with a −1:1 turns ratio. How would the schematic diagrams differ for a positive and negative turns ratio? Create a three-port 0°/180° power splitter using two transformers with turns ratios of $\sqrt{2}$ and $-\sqrt{2}$.

The lumped part is used to specify series and parallel combinations of resistance, capacitance, and/or inductance. It is drawn as an open blue rectangle. *Puff* understands four units for impedance and admittance: Ω (ohms, typed as *Alt*-o), S (siemens), z (normalized impedance), and y (normalized admittance). For example, a 100-Ω resistor may be specified as 100Ω, 0.01S, 2z, or 0.5y, assuming that zd is 50Ω. Capacitance and inductance values are entered in units of Farads (F) and Henries (H), respectively. Note that these must be upper case. Resistance, capacitance, and inductance values may be positive, zero, or negative. Reactance values are specified by placing a j before or after a number, and using the impedance units Ω or z. For example, 25jΩ and 0.5jz specify a 25-Ω reactance at fd. *Puff* scales positive reactances proportionally and negative reactances inversely with frequency, interpreting them as inductive and capacitive reactances, respectively. This means that at 2fd, $50j\Omega$ becomes $100j\Omega$, and $-50j\Omega$ becomes $-25j\Omega$. Specifications with the admittance units S and y are treated in a dual way: positive susceptances scale proportionally with frequency, and the negative susceptances scale inversely.

Series circuits with two or three different lumped elements are specified by combining a real number, a positive imaginary number, and a negative imaginary number, all in the same lumped part. For example, 1+j10-j10z specifies a circuit that is resonant at the design frequency with a Q of 10. The resistance is equal to zd and the inductive and capacitive reactance are 10 zd at the design frequency. Note that the unit appears only once, after all the numbers. This type of lumped element entry is very convenient, for example, when taking values from filter tables. Parallel circuits are specified in a dual way using admittance units S and y. Series and parallel lumped elements that combine resistance and values of inductance and capacitance may also be formed. *Puff* has a special character for forcing a lumped part to be a parallel circuit when it does not contain admittance units. This character is ||, typed as *Alt*-p, and it is referred to as the parallel sign. When used with a lumped part, the parallel sign forces the part's description to be interpreted as a parallel circuit, regardless of the unit. This allows part entries such as 50Ω||1nH||1pF for a parallel RLC circuit. Note that you can use symbols for the engineering prefixes *nano* and *pico*. These symbols may appear in front of units in the parts list, in .puf files, and in some *Board* window parameters. *Puff* recognizes the symbols for the following engineering prefixes:

Multiplier	Prefix	Symbol
10^{18}	exa	E
10^{15}	peta	P
10^{12}	tera	T
10^{9}	giga	G
10^{6}	mega	M
10^{3}	kilo	k
10^{-3}	milli	m

(*Continued*)

Multiplier	Prefix	Symbol
10^{-6}	micro	μ
10^{-9}	nano	n
10^{-12}	pico	p
10^{-15}	femto	f
10^{-18}	atto	a

Note that each symbol is case sensitive, and there is a big difference, for example, between `10mΩ` and `10MΩ`. The μ symbol is obtained by typing *Alt*-`m`.

Exercise 2.3 Ground one terminal and compare the one-port scattering parameters for the following lumped parts:

lum 50Ω 1nH 1pF
lum 50Ω||1nH||1pF
lum 50Ω +1nH +1pF
lum 50Ω -1nH -1pF
lum 50Ω||-1nH||-1pF

Use *Alt*-`p` to obtain the || sign and *Alt*-`o` to obtain the Ω sign. Use the Tab key, while in the *Plot* window, to align your plots with the Smith chart circles.

The `atten`, `xformer`, and `lumped` parts do not have electrical length. When included in a circuit they are drawn in a default length, referred to here as the *Manhattan* length. An alternate length may be entered for the `lumped` part, although it is optional. This comes in handy when trying to align parts in the layout. The `lumped` length specification should come last, and the units must be in meters (`m` with an appropriate prefix). The *Manhattan* length is one-tenth the size of the layout. All parts using *Manhattan* dimensions will be drawn with the same spacing between terminals.

The `tline` part is an ideal transmission line section. It is lossless and without dispersion. In the layout, it is drawn as a copper rectangle. The characteristic impedance or admittance of a `tline` may be specified in the same way as the resistance of a `lumped` part, except that it must be positive. Length units are required. Valid units include: meters (`m` with an appropriate prefix), `h` (the substrate thickness), and ° (degrees, typed as *Alt*-`d`). A transmission line specified with a 360° length will be one wavelength long at the design frequency `fd`. It is usually convenient to specify degrees other than, say millimeters (`mm`), but sometimes the physical units are useful for aligning a `tline` with other parts. You may also specify an artwork length correction to compensate for discontinuities. This is done by adding a plus or a minus sign and a second length. These corrections are made on the screen and in the artwork, but do not affect the electrical length used in the analysis. For example, the length `90°-0.5h` will be treated in the analysis as a quarter-wavelength section, but it will be drawn shorter on the screen and in the artwork. The `h` units are particularly useful here because open-circuit end corrections and phase shifts in tee junctions are proportional to the substrate thickness (see Figs. 7.2 and 7.3).

Often it happens that a desired electrical length results in a physical length larger than the size of the board, or is perhaps negative for a deimbedding problem. *Puff* provides for this eventuality by allowing you to force *Manhattan* dimensions upon the `tline`. This is done by placing an `M` (must be uppercase) as the last character in the part's description. Then, regardless of the parameters specified, *Puff* will make the `tline` length one-tenth the size of the layout. To find out the physical dimensions of a `tline`, place the cursor on the part, and hit the = key. *Puff* will calculate the dimensions for the part, and list them in the *Message* box. The dimensions listed will include any artwork corrections, or will be Manhattan dimensions if these were called for. Manhattan dimensions may be forced on all parts using the `Tab` key in the *Board* window.

The `qline` is similar to the `tline`, but is lossy. In addition to impedance and length, you can add a value for the quality factor, or Q. The attenuation in the line is calculated by enforcing the given Q at the design frequency `fd`. Outside the design frequency, the attenuation is made to follow one of two models. Specify `Qd` or `Q` if you wish a dielectric loss model, and `Qc` if you prefer a conductor loss model. Manhattan dimensions are requested by placing an `M` (must be uppercase) as the last character in the part's description. Artwork corrections are not allowed for the `qline`.

Exercise 2.4 Plot and compare the two-port scattering parameters s_{11} and s_{21} for the following `qline` examples:

qline 70Ω 90°
qline 70Ω 90° 10Qd
qline 70Ω 90° 10Qc
qline 70Ω −90° 10QM

Use a 5 GHz design frequency and plot from 0 to 50 GHz. If needed, enlarge the log and Smith chart plots for the last example.

The `clines` part is a pair of coupled transmission lines. Its specification is similar to a `tline`, except that either one or two impedances or admittances may be given. If only one appears, then the specification looks the same as a `tline`. If this impedance is larger than `zd`, *Puff* interprets it as the even-mode impedance, and if smaller, as the odd-mode impedance. The other mode impedance is chosen to match the lines by forcing the product of even and odd mode impedances to be `zd`^{2}. If two impedances are given, the larger is the even-mode, and the smaller the odd-mode impedance. As with the `tline`, use an upper case `M` at the end of a `clines` specification to force Manhattan dimensions.

The `device` part is used to read in a file containing multiport scattering parameters. Files may contain transistor data, measured data to be plotted, or parameters that define idealized parts, meters, or sources. A filename specifying the s-parameters must be given, preceded and followed by a space. *Puff* will assume a `.dev` extension if none is given. An optional length may be specified after the filename, as with the `lumped` part. The `device` is drawn in the *Layout* window as a blue arrowhead with red dots denoting the ports. The dot at the wide end of the arrowhead represents port 1 in the file, and the remaining ports follow at equal intervals along the arrow's axis.

If you wish to make your own `device` files, you should study the format of the file `fhx04.dev` given in Fig. 2.2. There is an optional comment line in braces, followed by a template line. The `f` at the beginning of the template stands for frequency. If left out, *Puff* assumes that the scattering coefficients are independent of frequency. The scattering parameters that follow the frequency in the template may appear in any order, and *Puff* will assume that any parameters that are not given on the template line are zero. The program will figure out how many ports the device has from the highest port number that appears in the template line. *Puff* can handle up to four-port device files. The numbers that follow the template are separated by one or more spaces or carriage returns. The frequency is first (if it appears) followed by the magnitude (linear, not dB) and phase (in degrees) of each of the scattering parameters in turn. When *Puff* is calculating scattering parameters in the *Plot* window, it will interpolate linearly between points in the device file, as necessary. *Puff* will not extrapolate beyond a `device` file's frequency range, and will give an error when this is attempted. A previously saved `.puf` file (that includes saved scattering parameters) can also be recalled as a `device` file. Complex networks can then be formed by combining many smaller circuits.

The `indef` part is similar to the `device`, but it is used to generate *indefinite* scattering parameters from a file containing definite scattering parame-

```
{FHX04 Fujitsu HEMT (89/90), f=0 extrapolated; Vds=2V, Ids=10mA}
  f         s11                s21                  s12                 s22
  0.0   1.000     0.0    4.375   180.0   0.000     0.0   0.625     0.0
  1.0   0.982   -20.0    4.257   160.4   0.018    74.8   0.620   -15.2
  2.0   0.952   -39.0    4.113   142.0   0.033    62.9   0.604   -28.9
  3.0   0.910   -57.3    3.934   124.3   0.046    51.5   0.585   -42.4
  4.0   0.863   -75.2    3.735   107.0   0.057    40.3   0.564   -55.8
  5.0   0.809   -92.3    3.487    90.4   0.065    30.3   0.541   -69.2
  6.0   0.760  -108.1    3.231    75.0   0.069    21.0   0.524   -82.0
  7.0   0.727  -122.4    3.018    60.9   0.072    14.1   0.521   -93.6
  8.0   0.701  -135.5    2.817    47.3   0.073     7.9   0.524  -104.7
  9.0   0.678  -147.9    2.656    33.8   0.074     1.6   0.538  -115.4
 10.0   0.653  -159.8    2.512    20.2   0.076    -4.0   0.552  -125.7
 11.0   0.623  -171.1    2.367     7.1   0.076   -10.1   0.568  -136.4
 12.0   0.601   178.5    2.245    -5.7   0.076   -15.9   0.587  -146.4
 13.0   0.582   168.8    2.153   -18.4   0.076   -21.9   0.611  -156.2
 14.0   0.564   160.2    2.065   -31.2   0.077   -28.6   0.644  -165.4
 15.0   0.533   151.6    2.001   -44.5   0.079   -36.8   0.676  -174.8
 16.0   0.500   142.8    1.938   -58.8   0.082   -48.5   0.707   174.2
 17.0   0.461   134.3    1.884   -73.7   0.083   -61.7   0.733   163.6
 18.0   0.424   126.6    1.817   -89.7   0.085   -77.9   0.758   150.9
 19.0   0.385   121.7    1.708  -106.5   0.087   -97.2   0.783   139.1
 20.0   0.347   119.9    1.613  -123.7   0.098  -119.9   0.793   126.6
```

Figure 2.2 The file for the Fujitsu FHX04 HEMT, fhx04.dev.

ter data. Indefinite parameters are those with an undefined ground terminal. If a one-port file is specified, `indef` will convert it to a two-port. If a two-port file is specified, `indef` will convert it to a three-port; and so on up to a four-port to five-port conversion. The `indef` part is most often used to model a transistor as a three-port. The n port to $n + 1$ port conversion is made possible by assuming that Kirchhoff's current law is valid, allowing what was the ground terminal to be converted to a port. The `indef` part is drawn on the screen in the same way as the `device` part, except the port created from ground is drawn as a yellow dot. The extra port generally appears as the last port, except for the two-port to three-port `indef` where the extra port appears in the center of the part. Note that to turn an `indef` part into its `device` equivalent, the extra port should be shorted.

In addition to the `device` file format given in Fig. 2.2, *Puff* can also read s-parameter data files in the EEsof format. This is done when the appropriate file extension is given for `device` or `indef` parts: a one-port file requires an `.S1P` extension, the two-port uses `.S2P`, and so on up to a four-port `.S4P`. *Puff*, however, cannot read every possible data format for these files. It is assumed that they possess the following format:

*x*HZ S MA R *yy*

where x is the same engineering prefix used for `fd` in the *Board* window, and *yy* is the value for `zd`. *Puff* reads only scattering parameters given in the magnitude/angle format. These restrictions require some caution. If a device file contains frequencies given in GHz, *Puff* will give erroneous results if you try to plot data in MHz. The prefix for `fd` and the impedance of `zd` in the *Board* window *must* coincide with those used in all `device` and `indef` files. This goes for files in both EEsof and `.dev` formats. Noise parameters present in files are ignored.

Exercise 2.5 Create a two-port circuit and plot all of its scattering parameters (s_{11}, s_{21}, s_{12}, and s_{22}). Use *Ctrl*-s from the *Plot* window to save it to a file. Erase the circuit, and then make a `device` part that will recall the scattering parameters from the previously saved file. Be sure to use the `.puf` extension. What happens if the number of points is different?

Exercise 2.6 Generate `device` files for an ideal microwave isolator and circulator. Make them frequency independent by leaving out the `f` frequency designation in the template. Each should be a simple two-line file.

3 The Layout Window

IN THE UPPER PORTION of the screen is the *Layout* window. The square represents the substrate, and the numbers on the sides represent connectors. Typing an arrow key will draw the selected part in the *Parts* window in the direction of the arrow. The *Message* box will show the change in the x and y coordinates. *Puff* starts out drawing part a, but you can select another part in the list by typing the letter for the part desired. The circuit can be grounded at any point by pushing the = key. If there is already a part in the direction of the arrow key, *Puff* will move to the other end rather than draw over it. If the ends of two parts are closer together than the circuit resolution r, *Puff* will connect them together. *Puff* will stop you from drawing a part off the edge, but it will not stop you from crossing over a previous part. You can make a path to a connector by pushing one of the number keys 1, 2, 3, or 4 on the top row of the keyboard. Notice that *Puff* does this by first moving up or down, then right or left. The electrical length of a connector path is not taken into account in the analysis, and it is drawn in a different color to indicate this. You can erase the entire circuit and start over by pushing *Ctrl*-e. *Ctrl*-n moves the × to the nearest node. This can be useful if you are off the network and want to get back, or if you want to see if two nodes are connected.

The shift key is used for erasing and moving around the layout. *Puff* will erase a part rather than move over it if you hold the shift key down while pushing an arrow key. A ground can be removed with shift =. The path to a connector is erased by holding the shift key down while typing the connector number. If you are not at a connection to a port, this shift-number operation will move you to that port number without drawing the path. This is useful if you would rather start a circuit at a connector than in the center. The shift-arrow operation moves the × where there is no part to erase. The × moves half the length of the currently selected part. Half steps are used rather than full steps to allow centered and symmetrically laid out circuits. To move the full length of the part, step twice. It is important to note that this shift-arrow operation is actually drawing an invisible part. If you later change the size of an invisible part, it will have an impact on the layout, possibly resulting in a part being drawn off the board. If this happens, *Puff* will give you an error message that you may not believe. As a precaution, keep the number of invisible parts to a minimum.

There are special rules for drawing `clines`. Use the arrows to move along the lines and *Ctrl*-`n` to jump from one line to another. When connecting `clines` together, if you draw the second `clines` in the same direction as the first, the new lines will join the previous pair. This is the usual arrangement in a directional coupler. If you change directions, the `clines` will be staggered so that only one of the lines in each pair is connected. This is appropriate for a band-pass filter.

Typing *Ctrl*-a from the *Plot* window will activate *Puff*'s photographic artwork routines. The layout produced will be magnified by the photographic reduction ratio (`p`) in the circuit file. The artwork output parameter (`o`) in the circuit file allows you to specify dot-matrix or HP LaserJet printouts, or the production of a Hewlett-Packard Graphics Language (HP-GL) file. *Puff* will prompt for titles to be placed atop the printout, or for an HP-GL file name. Only the `tlines`, `qlines`, and `clines` will appear in the artwork, and the corners will be mitered. You may adjust `tline` and `clines` lengths using discontinuity corrections in the parts list. Widths may be adjusted using the artwork width correction factor `a` in the circuit file.

The rest of this chapter is a summary of the synthesis formulas used for determining the layout dimensions on the screen and in the artwork. We start with the `tline` and `qline` transmission line sections. The microstrip width W (Fig. 3.1a) is given by Wheeler [1,2]. For high-impedance lines,

$$\frac{W}{h} = \frac{8 \exp H'}{\exp (2H') - 2} \tag{3.1}$$

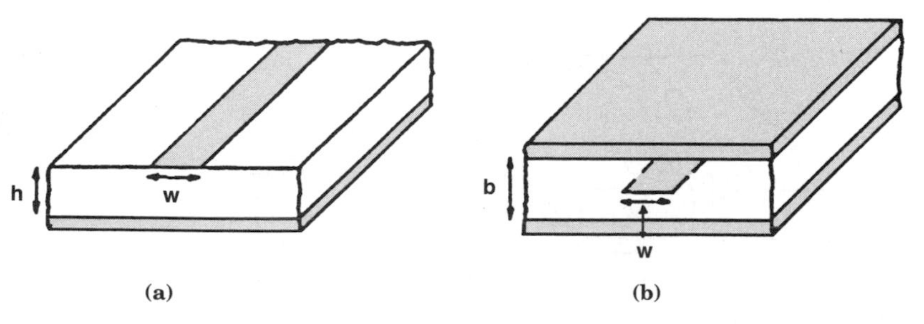

Figure 3.1 The `tline` dimensions for microstrip (a), and stripline (b).

where

$$H' = \frac{Z_0 \sqrt{2(\varepsilon_r + 1)}}{120} + \frac{1}{2}\left(\frac{\varepsilon_r - 1}{\varepsilon_r + 1}\right)\left(\ln\frac{\pi}{2} + \frac{1}{\varepsilon_r}\ln\frac{4}{\pi}\right), \tag{3.2}$$

where ε_r is the dielectric constant and Z_0 is the given characteristic impedance. For lines with characteristic impedances smaller than $(44 - 2\varepsilon_r)\ \Omega$ [3], the microstrip formulas are

$$\frac{W}{h} = \frac{2}{\pi}\left[d_\varepsilon - 1) - \ln(2d_\varepsilon - 1)\right] + \frac{\varepsilon_r - 1}{\pi\varepsilon_r}\left[\ln(d_\varepsilon - 1) + 0.293 - \frac{0.517}{\varepsilon_r}\right] \tag{3.3}$$

where $d_\varepsilon = 60\pi^2/(Z_0\sqrt{\varepsilon_r})$. Microstrip lengths in degrees are related to lengths in millimeters by the effective dielectric constant ε_{re} [4],

$$\varepsilon_{re} = \frac{\varepsilon_r + 1}{2} + \frac{\varepsilon_r - 1}{2}\left(1 + 10\frac{h}{W}\right)^{-1/2} \tag{3.4}$$

The stripline formulas are simpler because the effective dielectric constant is just ε_r. The width W (Fig. 3.1*b*) is given by Cohn's equations [5]

$$\frac{W}{b} = \frac{2}{\pi}\tanh^{-1} k \tag{3.5}$$

where

$$k = \begin{cases} \sqrt{1 - \left[\dfrac{e^{\pi x} - 2}{e^{\pi x} + 2}\right]^4}, & 1 < x; \\ \left[\dfrac{e^{\pi/x} - 2}{e^{\pi/x} + 2}\right]^2, & 0 \le x \le 1; \end{cases} \tag{3.6}$$

with $x = Z_0\sqrt{\varepsilon_r}/(30\pi)$.

For `clines`, we start with the even and odd mode impedances Z^e and Z^o. In microstrip, the dimensions W and S (Fig. 3.2*a*) are obtained by solving the following equations [6],

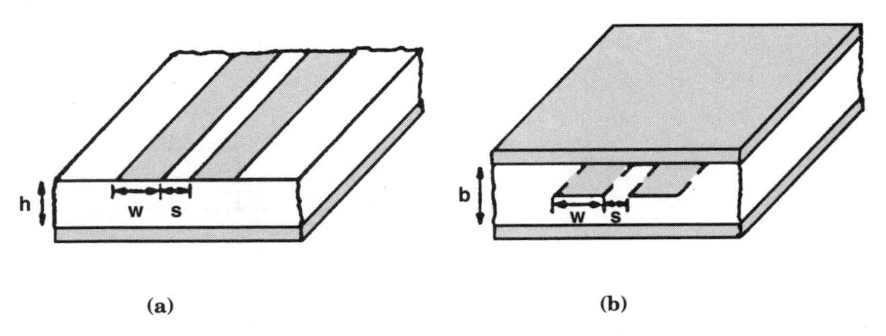

Figure 3.2 The `clines` dimensions for microstrip (*a*), and stripline (*b*).

$$\frac{W_e}{h} = \frac{2}{\pi}\cosh^{-1}\left(\frac{2d - g + 1}{g + 1}\right)$$

$$\frac{W_o}{h} = \begin{cases} \frac{2}{\pi}\cosh^{-1}\left(\frac{2d - g - 1}{g - 1}\right) + \frac{1}{\pi}\cosh^{-1}\left(1 + \frac{2W}{S}\right), & if\ \varepsilon_r \geq 6 \\ \frac{2}{\pi}\cosh^{-1}\left(\frac{2d - g - 1}{g - 1}\right) + \\ \qquad \frac{4}{\pi(1 + \varepsilon_r/2)}\cosh^{-1}\left(1 + \frac{2W}{S}\right), & if\ \varepsilon_r < 6 \end{cases} \tag{3.7}$$

where $g = \cosh(\pi S/2h)$ and $d = \cosh(\pi W/h + \pi S/2h)$, and W_e and W_o are the widths of a single microstrip transmission line with characteristic impedance $Z^e/2$ and $Z^o/2$. For ε^e_{re} and ε^o_{re}, we follow Garg and Bahl [7]. These formulas are long, and are not given here. The length that appears on the screen and in the artwork is the average of the lengths for the even and odd modes.

The stripline formulas for `clines` are simpler. The effective dielectric constant is ε_r for both modes. The following formulas give W and S (Fig. 3.2*b*) [8],

$$\frac{W}{b} = \frac{2}{\pi}\tanh^{-1}\sqrt{k_e k_o} \tag{3.8}$$

$$\frac{S}{b} = \frac{2}{\pi}\tanh^{-1}\left[\frac{1 - k_o}{1 - k_e}\sqrt{\frac{k_e}{k_o}}\right]$$

where k_e and k_o are obtained by substituting Z^e and Z^o into Eq. 3.6.

You can inspect *Puff's* layout calculations from the *Parts* window. Place the cursor in any `tline`, `qline`, or `clines` description and hit the = key. *Puff* will tell you the length and width for these parts, as well as the spacing for `clines`.

Exercise 3.1 Lay out a simple circuit consisting of `tlines` and `clines`. Go to the *Board* window and use the `Tab` key to change the circuit type. Return to the *Layout* window and see how the circuit is affected. Repeat for microstrip, stripline, and Manhattan layouts. Can you make a microstrip circuit that is difficult to realize in stripline? What stripline circuits cannot be realized in microstrip?

Exercise 3.2 Lay out a simple circuit. Go to the *Plot* window and save the circuit using *Ctrl*-`s`. Exit *Puff*, then use an ASCII editor to open the saved *Puff* file. Go to the section of the file that begins `\c{ircuit}`. *Puff* has saved your keystrokes in the *Layout* window in what we call a *keylist*. When reading a new file, *Puff* uses the keylist to redraw the circuit. What keystrokes are not saved? Is it better to erase all the parts using *Ctrl*-`e`, or with repeated shift-arrow operations? Can you think of some advantages in keeping the keylist as short as possible?

Exercise 3.3 The most common *Puff* layout errors involve invisible parts. Starting from a blank layout, select a `tline` part and make repeated shift-arrow operations to move the × cursor about the *Layout* window. Go to the *Parts* window, increase the length of the `tline` used, then return to the *Layout* window. See how long you can make the `tline` before causing a layout error. What happens if you try to delete the `tline` from the parts list?

References

1. H. A. Wheeler, "Transmission line properties of parallel wide strips by a conformal mapping approximation," *IEEE Trans. on Microwave Theory and Tech.,* vol. MTT-12, pp. 280–289, May 1964.
2. H. A. Wheeler, "Transmission line properties of parallel strips separated by a dielectric sheet," *IEEE Trans. on Microwave Theory and Tech.,* vol. MTT-13, pp. 172–185, March 1965.
3. R. P. Owens, "Accurate analytical determination of quasi-static microstrip line parameters," *The Radio and Electronic Eng.,* vol. 46, pp. 360–364, July 1976.
4. M. V. Schneider, "Microstrip lines for microwave integrated circuits," *Bell System Technical Journal,* vol. 48, pp. 1421–1443, May/June 1969.
5. S. B. Cohn, "Problems in strip transmission lines," *IRE Trans. on Microwave Theory and Tech.,* vol. MTT-3, pp. 119–126, March 1955.
6. S. Akhtarzad, T. R. Rowbotham, and P. B. Johns, "The design of coupled microstrip lines," *IEEE Trans. on Microwave Theory and Tech.,* vol. MTT-23, pp. 486–492, June 1975.
7. R. Garg and I. J. Bahl, "Characteristics of coupled microstriplines," *IEEE Trans. on Microwave Theory and Tech.,* vol. MTT-27, pp. 700–705, July 1979, and corrections *MTT-28,* pp. 272, Mar. 1980.
8. S. B. Cohn, "Shielded coupled-strip transmission line," *IRE Trans. on Microwave Theory and Tech.,* vol. MTT-3, pp. 29–38, Oct. 1955.

4 The Board Window

THE RELATIVE DIMENSIONS of the circuit board in the *Layout* window are specified from the *Board* window. These dimensions set the scale used to draw the distributed components on the screen. Figure 4.1 gives a brief description of each of the parameters available. Access the *Board* window by pressing function key *F4*. Edit the parameters using the same keys as in the *Parts* window. *F10* brings up a help window explaining the board parameters.

The normalizing impedance `zd` is used to calculate the scattering parameters. It also defines the normalized impedance and admittance values (`z` and `y`) that may appear in the *Parts* window. Paths to connectors are drawn with transmission lines with impedance `zd`. The design frequency `fd` is used to determine the physical lengths of distributed components entered in degrees. The Hz units for `fd` may use any prefix that appears in the chapter 2 table, although MHz and GHz are generally the most practical. The *Plot* window will inherit the frequency prefix. Change the prefix when you find yourself entering lots of zeros below the log-magnitude plot. Use caution. The prefixes are case sensitive. Also beware of frequency data in `device` and `indef` files that does not coincide with the `fd` prefix, e.g., *Puff* will give meaningless results if you use a device file with GHz frequency data when MHz is used for `fd`. Device scattering parameter data must also match the *Board* window's definition of `zd`.

.*Parameter*	*Description*
`zd`	Normalizing or characteristic impedance. Used in the calculation of scattering parameters. Units are Ω's with optional prefix.
`fd`	Design frequency. Used to compute electrical length of parts entered in degrees. Also the frequency used for the component sweep. Prefix given is carried over into the *Plot* window.
`er`	Relative dielectric constant of substrate. Used to calculate dimensions for microstrip and stripline components. Unitless.
`h`	Substrate thickness. One of three parameters that specify the equivalent dimensions for the *Layout* window. Significant in transmission line calculations.
`s`	Board size. Specifies the equivalent length of each side of the square circuit board that makes up the *Layout* window.
`c`	Connector separation. Sets the spacing between ports 1 and 3, and ports 2 and 4 in the layout. Set to zero to create a centered two-port.
`Tab`	Circuit type. Use the `Tab` key to select a microstrip, stripline, or Manhattan layout.

Figure 4.1 Description of parameters that may be modified from the *Board* window.

The `Tab` key toggles between microstrip, Manhattan, and stripline layouts. Resort to the Manhattan mode when `tlines` or `clines` become too long or short. All distributed parts are then drawn with widths 1/20 the board size, and lengths 1/10 the board size. This mode also permits components with unrealizable values, such as `tline`'s with negative electrical lengths. However, if *Puff* requires an electrical length calculation, stripline models will be used, and physically unrealizable parameters may not be allowed. If enabled, Manhattan dimensions will appear in the artwork, and artwork corrections will be ignored.

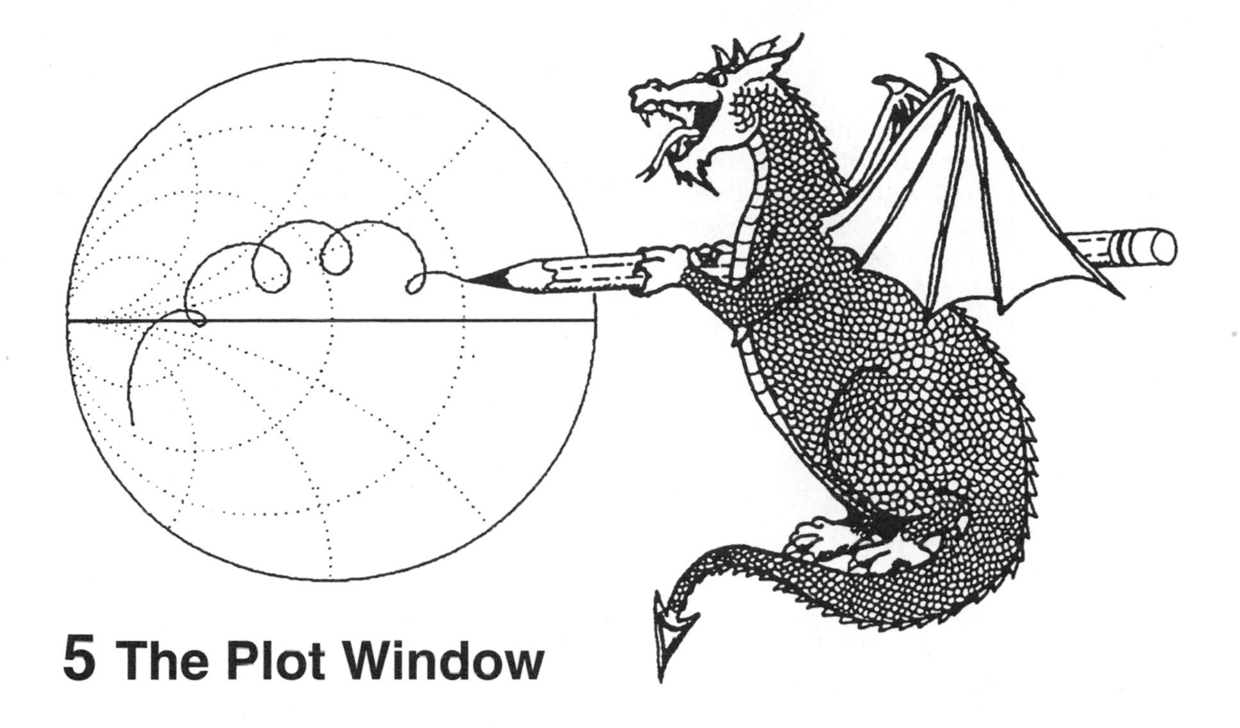

5 The Plot Window

TO REACH THE *Plot* window, push function key *F2*. The circuit in the *Layout* window is analyzed by typing p. If you type *Ctrl*-p, the previous plot will be drawn before the new plot, allowing a comparison of results. After an analysis has been completed, the *Plot* window lists the values of the scattering coefficients at design frequency fd. Use the *PgUp* and *PgDn* keys to move the markers and show the scattering coefficients at the other frequencies. The ↑ and ↓ keys can be used to move the cursor to various parameters. They cycle through a loop that includes the *Plot* window and the *x* and *y* axes on the rectangular plot. You can type over any parameter to change it. *Puff* can plot up to four different scattering parameters simultaneously. To select an *s*-parameter, move the cursor down toward the bottom of the *Plot* window, and a marker will appear, together with the letter S. Then type in the port numbers for the desired *s*-parameters. If you leave a line blank, it will be erased when you move the cursor. Those with *EGA* and *VGA* graphics can use the Tab key to change the Smith chart from an impedance chart to an admittance chart. Typing *Alt*-s with *VGA* graphics toggles an enlarged Smith chart.

The *Plot* window allows you to select the number of frequency Points to analyze. This must be a positive integer no greater than 500, assuming you have a full complement of memory. *Puff* interpolates between calculated points with a cubic spline. The interpolation is performed by splining the real and imaginary parts of the scattering coefficients separately. The independent variable for calculating the spline curve is the spacing on the Smith chart. This gives better results than using frequency as the splining parameter. If a curve kinks on the Smith chart, or gives erroneous ripples on the rectangular plot, it is an indication that the number of `Points` is too small.

You can plot an impulse response by typing `i`, or a step response by typing `s`. *Puff* will request a frequency interval specified by the ratio `fd/df`. It will then do a 256-point inverse fast Fourier transform of the scattering coefficients, and plot the results on a linear scale. The amplitude of the time-domain plot is the same as the Smith chart radius. The ratio `fd/df` will determine the time axis for the plot, which goes from −1/8`df` to 3/8`df`. The upper and lower limits on the frequency axis of the magnitude plot are used for windowing the Fourier transform. The window is a raised cosine that goes to zero at the upper frequency limit. The scattering coefficients are set to zero outside the window.

A convenient way to see the impulse or step inputs is to draw an open-circuited connector path at an unused port. The reflection coefficient for this port is the input waveform. *Puff* normalizes the input waveforms so that the peak value is 1. The high-frequency limit controls the rise and fall times, and the low-frequency limit affects the ringing. Be aware that the time-domain waveforms are actually periodic, with period 1/`df`, and aliasing from the previous pulse may affect the response. The step input is actually a square wave, and the response to the previous falling edge will affect the rising step that follows.

To save a network in a circuit file, type *Ctrl*-`s` and give a file name. The *Parts* window and the data in the *Plot* window will be saved along with the circuit. The `.puf` extension will be added to the filename if one is not specified. Typing *Ctrl*-`a` from the *Plot* window will activate *Puff*'s photographic artwork routines. The layout produced will be magnified by the photographic reduction ratio (`p`) in the circuit file. Only the `tlines`, `qlines`, and `clines` will appear in the artwork, and the corners will be mitered. The artwork output parameter (`o`) in the circuit file allows you to specify dot-matrix or HP LaserJet printouts, or the production of a Hewlett-Packard Graphics Language (HP-GL) file. *Puff* will prompt for titles to be placed atop the printout, or for an HP-GL file name. The HP-GL file will be created with the `.hpg` extension. The *Puff* diskette includes the program `hpg2com`. Use this to dump `.hpg` files to a serial plotter connected at port COM1.

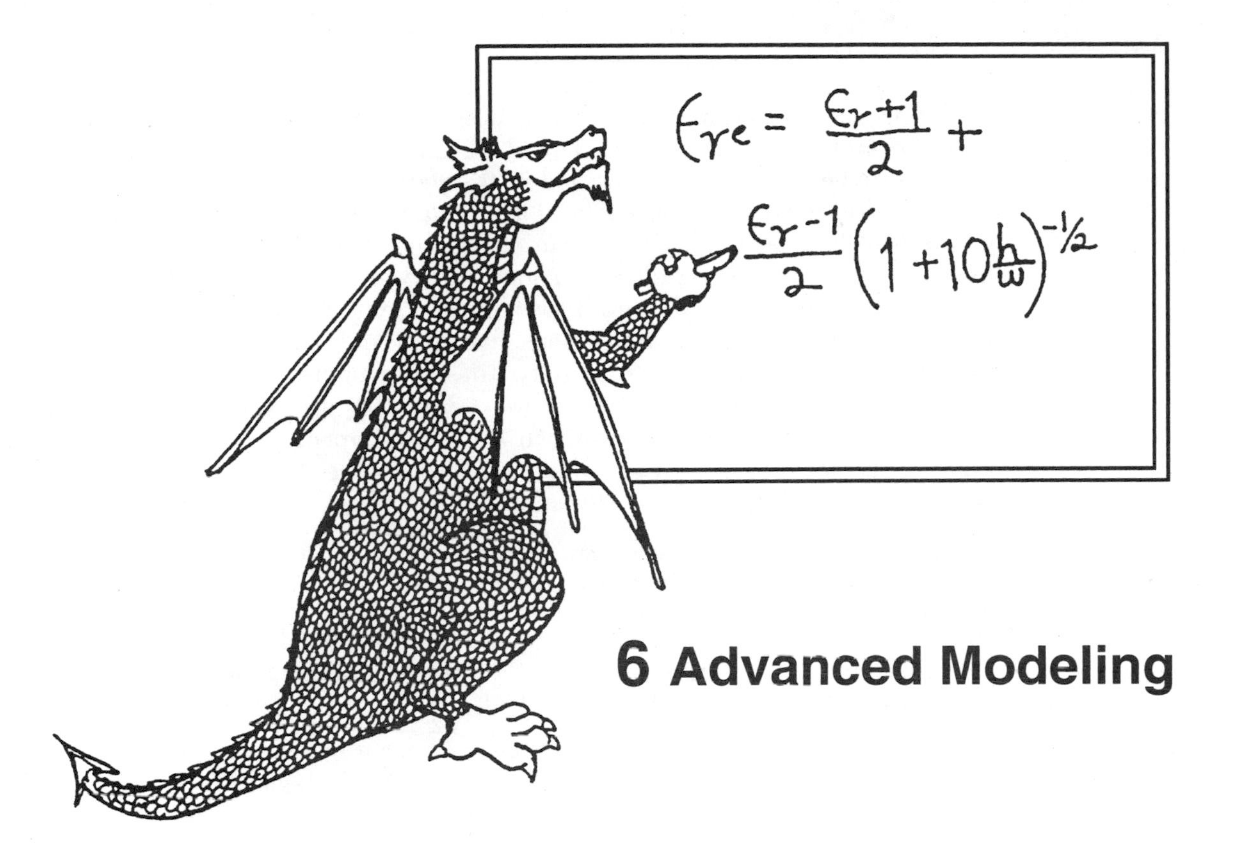

6 Advanced Modeling

THE SINGLE AND COUPLED TRANSMISSION LINES represented by the `tline` and `clines` parts are modeled as *ideal* distributed components: their effective dielectric constants and impedances are assumed constant over frequency, and they are lossless. These models neglect dispersion and the many loss effects common in microwave circuits. When these effects must be included in the analysis, more accurate models may be invoked by the addition of an exclamation point (`!`) to the part's description. The `tline!` and `clines!` are therefore referred to as *advanced* models. An analysis with advanced models may be made for a microstrip or stripline circuit and will include effects due to finite strip thickness, conductor and dielectric losses, surface roughness, and dispersion. When using these models, be sure to first specify accurate values for metal thickness, surface roughness, loss tangent, and conductivity in your `.puf` file.

The ideal and advanced parts are not entirely distinct. When either type is entered into the *Parts* window, *Puff* will calculate their effective dielectric constants and physical layout dimensions using the quasi-static *synthesis* formulas of Chapter 3. These calculations permit a scale drawing of the part in the *Layout* window, and artwork generation. A transmission line, for example, will therefore be the same size, be it specified a `tline!` or a `tline`.

The differences between ideal and advanced models arise when *Puff* is analyzing for scattering parameters. The ideal `tline` and `clines` are analyzed using the impedance values specified in the parts list and quasi-static effective dielectric constant values. These are held constant during the frequency analysis. Analyses using the advanced `tline!` and `clines!` models are based solely on the physical dimensions that were derived for the layout. Impedance and electrical length parameters entered in the parts list, although used to derive the dimensions, are ignored during the advanced frequency analysis. This is done to allow these parameters to vary over frequency due to dispersion. Often an advanced analysis based on a component's dimensions will result in different impedances and electrical lengths than were originally specified. This is because the *analysis* is generally more accurate than the *synthesis*.

Exercise 6.1 Go to the *Parts* window, move the cursor into a `tline` description, and hit the = key. If a valid part, the message box will reveal its layout dimensions. Change the part's specification to a `tline!` and again hit the = key. The message box now lists low frequency values for impedance and electrical length based on an accurate analysis of the dimensions. Note how the values differ from those in the part's description. Repeat this procedure with `clines`.

Exercise 6.2 Plot s_{11} and s_{21} for a single non 50Ω microstrip `tline` over several wavelengths. Change the `tline` to a `tline!` and re-plot by typing *Ctrl*-p from the *Plot* window. Note the differences caused by the addition of losses and dispersion.

Puff calculates the scattering parameters for distributed components using the formulas given in Fig. 6.1. The calculations require the physical length l of the part, and its impedance(s) and propagation constant(s) (even and odd mode for coupled lines). The complex propagation constant γ is given by

$$\gamma = \alpha + j\beta \tag{6.1}$$

where α is the attenuation factor and β is the (real) propagation constant. The latter is given by

$$\beta = \sqrt{\varepsilon_{re}(\omega)}\, k_0 \tag{6.2}$$

where $\varepsilon_{re}(\omega)$ is the effective dielectric constant and k_0 is the TEM propagation constant $k_0 = \omega\sqrt{\varepsilon_0\mu_0}$ for the equivalent transmission line with an air dielectric. For the ideal `tline`, the propagation constants simplify to

$$\alpha = 0 \qquad \beta = \sqrt{\varepsilon_{re}}\, k_0 \tag{6.3}$$

where, for microstrip, ε_{re} is the frequency independent quasi-static value given by (3.4). For stripline, ε_{re} is simply the relative dielectric constant ε_r of the substrate. If a `tline` specification is made using an electrical length in degrees at `fd`, then *Puff* will simply frequency scale the product $\theta = \beta l$ when computing the scattering parameters. For the ideal `clines`, (6.3) still applies, and is used for both even and odd mode propagation constants.

Part	s_{ij}

`tline`
`tline!`
`qline`
	$\dfrac{(z-y)\sinh\gamma\ell}{2\cosh\gamma\ell+(z+y)\sinh\gamma\ell}$, for $i=j$;
	$\dfrac{2}{2\cosh\gamma\ell+(z+y)\sinh\gamma\ell}$, for $i\neq j$.

`clines` `clines!`	$\dfrac{(z_e-y_e)\sinh\gamma_e\ell}{4\cosh\gamma_e\ell+2(z_e+y_e)\sinh\gamma_e\ell}+\dfrac{(z_o-y_o)\sinh\gamma_o\ell}{4\cosh\gamma_o\ell+2(z_o+y_o)\sinh\gamma_o\ell}$, for $i=j$; $\dfrac{(z_e-y_e)\sinh\gamma_e\ell}{4\cosh\gamma_e\ell+2(z_e+y_e)\sinh\gamma_e\ell}-\dfrac{(z_o-y_o)\sinh\gamma_o\ell}{4\cosh\gamma_o\ell+2(z_o+y_o)\sinh\gamma_o\ell}$, coupled port; $\dfrac{1}{2\cosh\gamma_e\ell+(z_e+y_e)\sinh\gamma_e\ell}+\dfrac{1}{2\cosh\gamma_o\ell+(z_o+y_o)\sinh\gamma_o\ell}$, through port; $\dfrac{1}{2\cosh\gamma_e\ell+(z_e+y_e)\sinh\gamma_e\ell}-\dfrac{1}{2\cosh\gamma_o\ell+(z_o+y_o)\sinh\gamma_o\ell}$, isolated port.

Figure 6.1 Expressions used to calculate the scattering parameters for distributed components: the single transmission line (`tline`), transmission line with finite Q (`qline`), and coupled transmission lines (`clines`). In the table, z is the normalized impedance, y is the normalized admittance, γ is the complex propagation constant $\gamma = \alpha + j\beta$, and l is the physical length of the component. The subscripts e and o refer to even and odd modes of the clines. For the `clines`, the ports are labeled in a way that is appropriate for a directional coupler: the through port is on the same line as the input port, but at the opposite end. The isolated port is the port diagonally across from the input, and the coupled port is at the same end as the input, but on the other line.

When a `tline!` or `clines!` model is invoked, *Puff* will calculate α at every analysis frequency for both microstrip and stripline circuits. It will also calculate β at each frequency for microstrip circuits. Dispersion can be a strong effect in microstrip transmission lines due to their inhomogeneity. Typically, as frequency is increased, ε_{re} increases in a non-linear manner, approaching an asymptotic value. For single and coupled microstrip transmission lines, ε_{re} is modeled using Getsinger's expression [1,2]

$$\varepsilon_{re}i\,(\omega) = \varepsilon_r - \frac{\varepsilon_r - \varepsilon_{re}i\,(0)}{1 + F_i\,(\omega)} \tag{6.4}$$

where ε_r is the relative dielectric constant of the substrate material, $\varepsilon_{re}i(0)$ is the low frequency quasi-static value, and $F_i(\omega)$ is an increasing function of frequency. The subscript i is used to distinguish between the functions and values used for single microstrip, and for even and odd modes in coupled microstrip. Accurate closed form expressions for $F_i(\omega)$ are complicated. *Puff* uses the expressions derived for single and coupled microstrip transmission lines by Kirschning and Jansen [3,4]. Dispersion affects characteristic impedances in a similar way. It is modeled using Bianco's expression [5]

$$Z_{0_i}(\omega) = Z_{0_i}^{\ s} - \frac{Z_{0_i}^{\ s} - Z_{0_i}(0)}{1 + F_i(\omega)} \tag{6.5}$$

where $Z_{0_i}^{\ s}$ is twice the characteristic impedance of an equivalent (single or coupled) stripline with twice the thickness of the microstrip, and $Z_{0_i}(0)$ is the quasi-static impedance. *Puff* uses the same $F_i(\omega)$ functions for both (6.5) and (6.4).

Calculation of attenuation factors for the `tline!` and `clines!` includes contributions from a dielectric attenuation factor α_d and a conductor attenuation factor α_c where

$$\alpha = \alpha_c + \alpha_d \tag{6.6}$$

The calculations for conductor attenuation have the form given by

$$\alpha_c^{\ i} = \frac{R_s}{Z_i W_i} K_i F_{sr} \tag{6.7}$$

where Z_i is the line impedance (even or odd mode for couplers), W_t is the effective width of the line (taking into account finite strip thickness), and K_i is the current distribution factor. The subscript and superscript i is again used to distinguish between the single line, and even and odd mode values for the coupled line. The surface resistance R_s is given by

$$R_s = \sqrt{\frac{\pi f \mu_0}{\sigma}} \tag{6.8}$$

at frequency f, free-space permeability μ_0, and conductivity σ. The substrate surface roughness factor F_{sr} has been evaluated by Hammerstad and Bekkadal [6]. They find it well approximated by the expression

$$F_{sr} = 1 + \frac{2}{\pi} \arctan\left\{ 1.4 \left(\frac{\Delta}{\delta_s} \right)^2 \right\} \tag{6.9}$$

where Δ is the rms surface roughness and δ_s is the skin depth at the operating frequency. This factor is necessary to account for an asymptotic increase seen in the apparent surface resistance with decreasing skin depth. The expressions used for the current distribution factor K_i and effective width W_t have been given by Pucel, et al. [7] and Gupta, et al. [8] for single microstrip. Expressions for coupled microstrip have been given by Garg and Bahl [9] and Hammerstad and Jensen [10], while expressions for single and coupled stripline have been given by Gupta, et al. [11]. In most cases, K_i is derived from an application of Wheeler's incremental inductance rule [12].

Dielectric loss is due to the effects of finite loss tangent, tan δ. For the inhomogeneous line, an effective dielectric filling fraction gives that proportion of the transmission line's cross section not filled by air. For microstrip lines, the result is [13,14]

$$\alpha_d^i = \frac{\pi\varepsilon_r}{\varepsilon_r - 1}\frac{\varepsilon_{re_i} - 1}{\sqrt{\varepsilon_{re_i}}}\frac{\tan\delta}{\gamma_0} \tag{6.10}$$

where again i has been used to distinguish between values for single and coupled lines. For homogeneous stripline transmission lines, (6.10) simplifies to

$$\alpha_d = \frac{\pi\sqrt{\varepsilon_r}\tan\delta}{\gamma_0}. \tag{6.11}$$

The `tline!` is very useful when an accurate analysis of a microstrip or stripline transmission line is required. Often, however, one may only be interested in transmission line losses. For this occasion, *Puff* has a special transmission line model that is identical to the `tline`, but allows one to specify its loss in terms of quality factor, or Q. Called the `qline`, it is specified in the same way as the `tline`, but a value for Q may be included in addition to impedance and electrical length. The definition of the quality factor, or Q, for a transmission line is the same as with other components. It is the well-known figure of merit for the ratio of stored energy to dissipated energy. In terms of group velocity, it is written [15]

$$Q = \omega\frac{\text{energy per unit length}}{\text{power loss per unit length}} = \frac{\omega P/v_g}{P_L} = \frac{\omega}{2v_g\alpha} \tag{6.12}$$

where P_L is the power loss per meter, and power flow P equals the product of the energy density and the energy transport (group) velocity v_g. In the absence of dispersion, Q becomes the widely used expression

$$Q = \frac{\beta}{2\alpha} \tag{6.13}$$

In a line dominated by dielectric loss, both β and α will increase proportionately over frequency, and the Q is therefore constant. For the line dominated with conductor loss, both α and Q have $\sqrt{f}$ behavior. When using the `qline`, a conductor or dielectric loss model may be specified. A transmission line dominated by conductor loss with $Q = 100$, is specified as

qline 50Ω 90° 100Qc

The same line dominated by dielectric loss is specified as

qline 50Ω 90° 100Qd

Puff computes ß for the `qline` using the same quasi-static formulas used for the `tline`. It then computes α at design frequency `fd` using the specified Q value and (6.13). At frequencies other than `fd`, *Puff* scales the α according to either the dielectric or conductor loss model.

Exercise 6.3 Edit a `.puf` file to include values for metal thickness, surface roughness, loss tangent, and conductivity for your favorite microwave substrate. In the *Layout* window, connect a `tline!` between ports 1 and 2, and a comparable `qline` between ports 3 and 4. While comparing dB plots for s_{21} and s_{43} over many wavelengths, find the best Q value to model the losses seen in the `tline!` Is it better to use the `Qc` or `Qd` model?

Exercise 6.4 Design a three section `clines` bandpass filter and plot its frequency response. For a simple dry-lab, change the `clines` to `clines!` and re-plot using *Ctrl*-p. Can you account for the differences?

References

1. W. J. Getsinger, "Microstrip dispersion model," *IEEE Trans. Microwave Theory Tech.,* vol. MTT-21, pp. 34–39, Jan. 1973.
2. W. J. Getsinger, "Dispersion of parallel-coupled microstrip," *IEEE Trans. Microwave Theory Tech.,* vol. MTT-21, pp. 144–145, Mar. 1973.
3. M. Kirschning and R. H. Jansen, "Accurate model for effective dielectric constant of microstrip with validity up to millimeter wave frequencies," *Electron. Lett.,* vol. 18, pp. 272–273, Mar. 1982.
4. M. Kirschning and R. H. Jansen, "Accurate wide-range design equations for the frequency dependent characteristics of parallel coupled microstrip lines," *IEEE Trans. Microwave Theory Tech.,* vol. MTT-32, pp. 83–90, Jan. 1984. Corrections: *IEEE Trans. Microwave Theory Tech.,* vol. MTT-33, p. 288, Mar. 1985.
5. B. Bianco, A. Chiabrera, M. Granara, and S. Ridella, "Frequency dependence of microstrip parameters," *Alta Freq.,* vol. 43, pp. 413–416, July 1974.
6. E. O. Hammerstad and F. Bekkadal, *A Microstrip Handbook,* ELAB Report, STF 44 A74169, N7034, University of Trondheim, Norway, 1975.
7. R. A. Pucel, D. J. Masse, and C. P. Hartwig, "Losses in microstrip," *IEEE Trans. Microwave Theory Tech.,* vol. MTT-16, pp. 342–350, June 1968.
8. K. C. Gupta, R. Garg, and I. J. Bahl, *Microstrip Lines and Slotlines,* Artech House, Dedham, Mass., 1979.
9. R. Garg and I. J. Bahl, "Characteristics of coupled microstriplines," *IEEE Trans. Microwave Theory Tech.,* vol. MTT-27, pp. 700–705, July 1979.

10. E. Hammerstad and O. Jensen, "Accurate models for microstrip computer-aided design," *IEEE MTT-S Int. Microwave Symp. Dig.* (Washington, D.C.), pp. 407–409, 1980.
11. K. C. Gupta, R. Garg, and R. Chadha, *Computer-Aided Design of Microwave Circuits,* Artech House, Dedham, Mass., 1981.
12. H. A. Wheeler, "Formulas for the skin effect," *Proc. IRE,* vol. 30, pp. 412–424, Sept. 1942.
13. M. V. Schneider, "Dielectric loss in integrated microwave circuits," *Bell System Technical Journal,* vol. 48, pp. 2325–2332, Sept. 1969.
14. B. Ramo Rao, "Effect of loss and frequency dispersion on the performance of microstrip directional couplers and coupled line filters," *IEEE Trans. Microwave Theory Tech.,* vol. MTT-22, pp. 747–750, July 1974.
15. R. E. Collin, *Field Theory of Guided Waves,* IEEE Press, New York, 1991.

7 Discontinuities

IN THIS CHAPTER the four dominant discontinuity effects in microstrip shall be considered: the excess capacitance of a corner, the capacitive end effect for an open circuit, the step change in width, and the length correction for the shunt arm of a tee. *Puff* will automatically miter corners in the artwork to reduce their effect. However, you must compensate for the other discontinuities yourself; they are neglected in the *Puff* analysis.

When a sharp right-angle bend occurs in a circuit (Fig. 7.1*a*), there will be a large reflection from the corner capacitance. *Puff* miters corners to reduce the capacitance and minimize this reflection, as shown in Fig. 7.1*b*. You can change the value of the miter fraction `m` defined to be

$$m = 1 - b/\sqrt{W_1^2 + W_2^2}\,. \tag{7.1}$$

When the two line widths are equal, this formula reduces to the conventional definition [1]. In `setup.puf`, `m` is set to 0.6.

In an open-circuit, the electric fields extend beyond the end of the line. This excess capacitance makes the electrical length longer than the nominal length, typically by a third to a half of the substrate thickness. This will cause the design frequencies of patch antennas and filters to be shifted. To compensate for this effect in the artwork, a negative length correction can be added to the

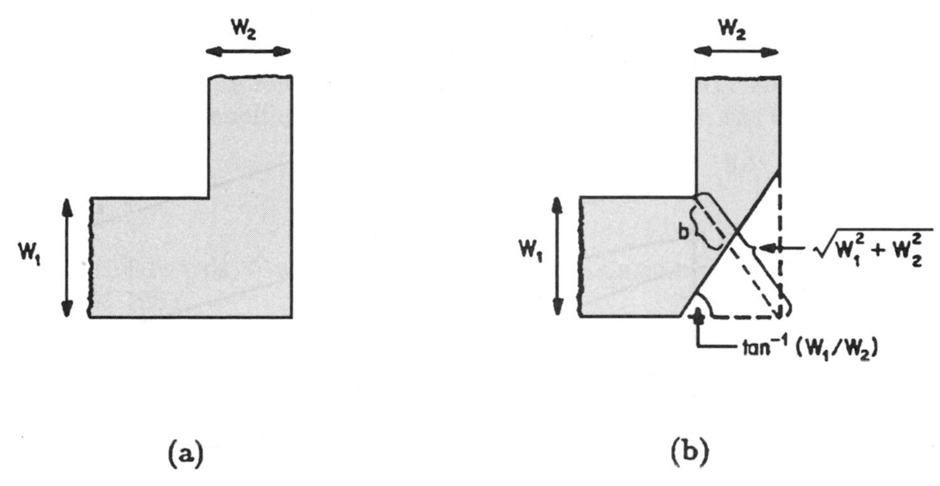

Figure 7.1 Mitering a right-angle bend.

parts list. Hammerstad and Bekkadal give an empirical formula for the length extension l in microstrip [2],

$$\frac{l}{h} = 0.412\left(\frac{\varepsilon_{re} + 0.3}{\varepsilon_{re} - 0.258}\right)\left(\frac{W/h + 0.262}{W/h + 0.813}\right), \tag{7.2}$$

where h is the thickness of the substrate. This formula is plotted in Fig. 7.2 for several different dielectric constants.

A similar method may be used to compensate for a step change in width between high and low impedance lines. The discontinuity capacitance at the end of the low impedance line will have the effect of increasing its electrical length. Assuming the wider low impedance line has width W_2, and the narrower high impedance line has width W_1, compensate using the expression [1]

$$\frac{l_s}{h} \approx \frac{l}{h}\left(1 - \frac{W_1}{W_2}\right) \tag{7.3}$$

where l_s is the step length correction for line W_2, and l/h is the value obtained from (7.2) and Fig. 7.2.

In the tee, shown in Fig. 7.3, the electrical length of the shunt arm is shortened by distance d_2. The currents effectively take a short cut, passing close to the corner. It is particularly noticeable in the branch-line coupler because there are four tees. Hammerstad and Bekkadal give an empirical formula for d_2 in microstrip [2],

$$\frac{d_2}{h} = \frac{120\pi}{Z_1\sqrt{\varepsilon_{re}^1}}\left(0.5 - 0.16\,\frac{Z_1}{Z_2}\,[1 - 2\ln(Z_1/Z_2)]\right). \tag{7.4}$$

8 The Component Sweep

THE SIMPLE OPTIMIZER included in *Puff* is called the *component sweep*. Instead of sweeping with respect to frequency, a circuit's scattering coefficients may be swept with respect to a changing component parameter. This feature is invoked by placing a question mark (`?`) in front of the parameter to be swept in the appropriate position of a part's description. For example, to find the optimum value for a tuning capacitor, one could specify a part as `lumped ?5pF`. When plotting in the *Plot* window, the frequency will then be held constant at the design frequency `fd`, and the values specified in the `x`-axis of the rectangular plot will be substituted for (in the above example) capacitance values in picofarads. In this manner, any parameter used in the parts list may be designated as a sweep parameter, but only one parameter may be swept at a time. Swept lumped elements are restricted to single resistors, capacitors, or inductors. A description such as `lumped ?1+5j-5j`Ω is not allowed since it is a series RLC circuit. In addition, the parallel sign `||` may not be present in the lumped specification. The unit and prefix given in the part's description (following the `?`) is inherited by the component sweep.

As an example, we will use the component sweep to make a stub matching circuit for the Fujitsu FHX04 HEMT. We begin with the circuit of Fig. 8.1. The electrical length (in degrees) of the `tline` at part `b` has been swept from 0° to 100°. At 62°, the input impedance of the circuit is seen to cross the upper half of the $g = 1$ normalized conductance circle (use the `Tab` key to toggle between impedance and admittance circles). This means we can then match with an open circuit stub. A stub has been added in Fig. 8.2. The length of the `tline` at part `b`

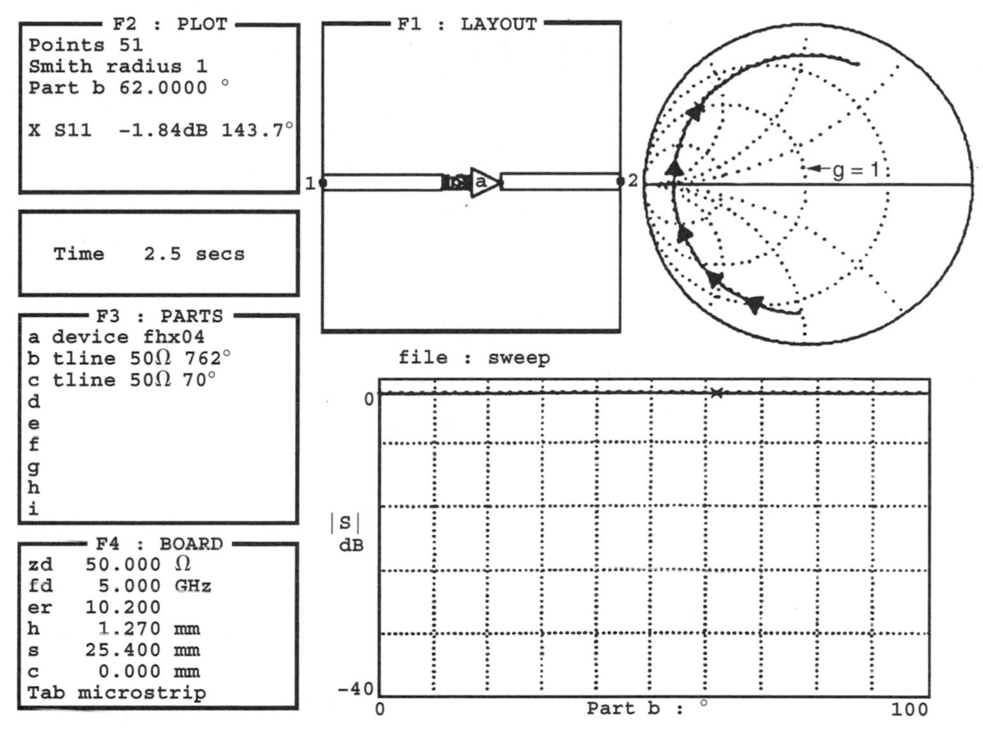

Figure 8.1 The `tline` at part `b` is swept from 0° to 100° while monitoring the input match of the Fujitsu FHX04 HEMT. At 62°, the locus intercepts the $g = 1$ normalized conductance circle.

is now held constant, and we sweep the electrical length of the stub at part `c`. With increasing stub length, the impedance slides down the $g = 1$ circle and becomes well matched at 70°. We thus have the matching circuit for operation at `fd`. Granted, this is a narrow bandwidth match, but this process can be repeated with more components and frequency domain checking to create a broader bandwidth circuit.

Be aware that values are easily encountered in the component sweep that cannot be physically realized. *Puff* must change the swept part to Manhattan dimensions to avoid problems. When specifying `tline` lengths in other than degrees, or when sweeping `clines`, this may cause the alternate parameter sweep to yield *s*-parameter values different from the standard *Puff* analysis. This occurs because electrical lengths are also a function of impedance via effective dielectric constant calculations. To reduce this disparity, specify electrical lengths in degrees whenever possible. When sweeping a `tline` or `clines` impedance, include a best guess value. For example, the specification `tline ?Ω 5mm` for a microstrip circuit is a poor one, because for *Puff* to compute the electrical length of a 5 mm length of line, it needs a W/h value from the impedance to compute the effective dielectric constant. When *Puff* encounters such a specification, it will compute an effective dielectric constant based

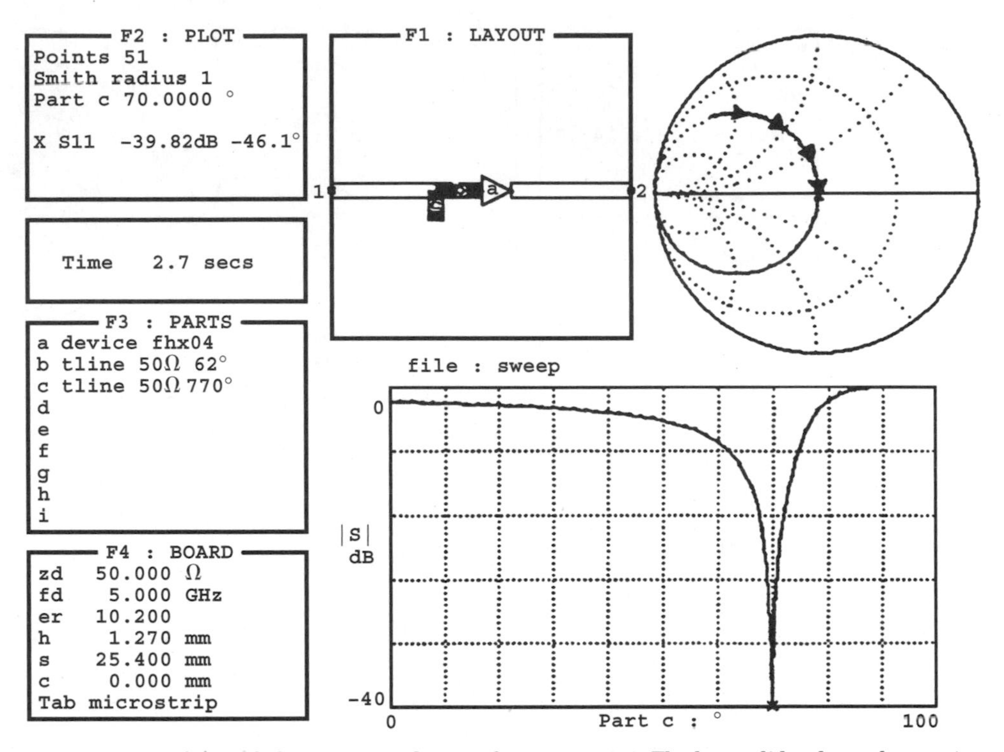

Figure 8.2 A stub is added at part c and swept from 0° to 100°. The locus slides down the $g = 1$ circle, and becomes well matched at 70°.

on a characteristic impedance of `zd`. If it was known that the impedance would be near 70Ω, a better specification would be `tline ?70Ω 5mm`. Then *Puff* would use an effective dielectric constant calculation based on a 70Ω impedance and this value would be used for the impedance sweep.

Similar problems arise for `clines`. The description `clines ?Ω 90°` is ambiguous, because *Puff* cannot determine if the value to be swept is an even or odd mode impedance. In addition, `clines` require the calculation of both even and odd mode effective dielectric constants; the electrical length specified is the average of the two. For this reason it is always better to first enter a best guess component designation, and then add the question mark. If, for example, one wanted to sweep the even mode impedance with a 40Ω odd mode impedance, then by entering

clines ?60Ω 40Ω 90°

Puff has enough information to calculate the propagation constants, and can recognize that the even mode impedance is to be swept. Whenever possible, enter a valid best guess component specification, then add the question mark.

9 Using Puff

THE CALTECH MICROWAVE course has a weekly laboratory in which students use *Puff* to design, fabricate, and measure microwave integrated circuits [1]. Over the duration of the course, each student generates a minimum of eight different circuits. Six examples are given in Fig. 9.1. Quarter-wave low-pass filters (as in Fig. 9.1*a*) are used to study open-circuit end effects in low impedance transmission lines, periodicity in the frequency domain, and control of filter passband ripple. Students build either a branch-line (Fig. 9.1*b*) or rat-race 3 dB directional coupler to examine the properties of symmetrical four-port structures and to investigate the effects of tee discontinuities. A bandpass filter (Fig. 9.1*c*) is employed to study coupled transmission lines. Next, the optimum input and output matching circuits are generated for a GaAs FET low-noise amplifier (Fig. 9.1*d*), followed by the design of a broadband amplifier using resistive feedback. A GaAs FET is also used to build a microwave oscillator using either coupled-line feedback (Fig. 9.1*e*) or a common gate circuit. Students also use *Puff* to build patch antennas (Fig. 9.1*f*) for class projects [2].

The microwave lab procedure is streamlined so that it can be completed in a single two-hour period. Students begin by completing their circuit designs and then use *Puff* to produce the photographic artwork on an HP LaserJet printer. The artwork is then photographically reduced onto 2.5″ square glass emulsion plates. These are developed and used as masks to expose photoresist covered, copper clad *RT/duroid* substrates. The etching process leaves dimensions typically accurate to 25 μm. An alternative lower-cost fabrication

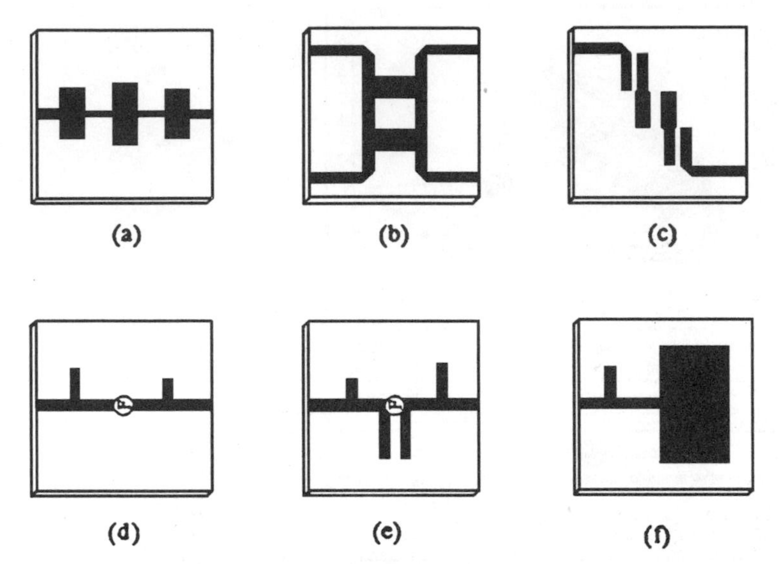

Figure 9.1 Example microstrip circuits designed with *Puff*: (*a*) a five-section low-pass filter; (*b*) a branch-line coupler; (*c*) a coupled-line bandpass filter; (*d*) a reflective matched GaAs FET low-noise amplifier; (*e*) an oscillator with coupled-line feedback; (*f*) a patch antenna with matching circuit. The circuits are typically fabricated on 0.635-mm (25-mil) and 1.27-mm (50-mil) *RT/duroid* substrates with $\varepsilon_r = 10.2$ and one-ounce copper cladding.

procedure is also possible [3]. When fabrication is complete, the circuits are placed in a brass test fixture, as shown in Fig. 9.2. Scattering parameter measurements are made with an HP-8720 microwave network analyzer. A personal computer controls the process, placing the measured *s*-parameters into a device file readable by *Puff*. The students then compare *Puff*'s predictions alongside the measured data. Correlation between the two is usually not perfect, so the students must determine where the analysis goes wrong. *Puff* does not automatically compensate for discontinuities, and it is a good puzzle for the students to adjust their circuit models to include their effects.

We shall now look at several designs and compare *Puff* predictions with measurements. Discontinuities have been taken into account using the length corrections described in Chapter 7. Figure 9.3*a* shows a low-pass filter with alternating sections of high and low impedance lines. This *Puff* analysis was performed using `tline` sections with length corrections included for the end-capacitance of the low impedance lines. Figure 9.3*b* is a bandpass filter with three quarter-wave coupled line sections. This *Puff* analysis was performed using `clines!` to model the losses and dispersion. Each open circuit of the coupled lines was modeled using `tlines` with widths equal to the coupled line branches, and lengths equal to half the open-circuit end correction.

Figure 9.4 shows the design of a single-stage low-noise amplifier using a Fujitsu FSC10 field effect transistor. The gain of the amplifier was measured using an HP-8970A noise-figure meter, and the transistor bias was applied

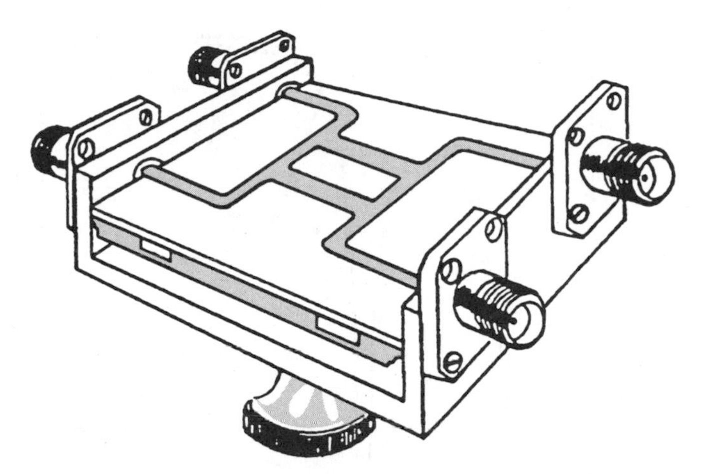

Figure 9.2 Brass test fixture for microstrip circuits. The plate underneath the circuit flexes and pushes against the sides to make a ground connection. The thumbscrew on the bottom locks the plate into place. The VSWR of the coax-to-microstrip transition is typically 1.1 or less for frequencies less than 6 GHz.

using two external bias tees. A maximum gain of 13.2 dB was obtained at 3.7 GHz with a noise figure of 1.9 dB.

We use an approach to oscillator design that differs from the traditional. With *Puff,* it is natural to analyze the circuit looking in from the external ports, whereas a conventional design looks at conditions inside the circuit. The oscillation condition is $s_{11} = 1/s_L$, where s_{11} is the reflection coefficient at the input of the transistor circuit and s_L is the reflection coefficient of the load. Figure 9.5*a* shows how this condition can be achieved. The s_{11} curve loops around the point $1/s_L$ (Fig. 9.5*a*). It is assumed that the s_{11} curve circles clockwise as the frequency increases and that as the transistor saturates, the loop contracts. Oscillations build up in the circuit, the transistor saturates, and the s_{11} loop shrinks until the oscillation condition is met.

However, an interesting problem arises when we try to follow this procedure with *Puff.* The load is the matched port, so that $1/s_L = \infty$. It is not clear how we go about drawing a clockwise loop about the point at infinity. The solution is to add circuit elements that produce a *counterclockwise* loop. Because of the properties of bilinear transforms, a region *inside* the previous clockwise loop is mapped to the *outside* of the counterclockwise loop. This means that when the transistor saturates the loop expands until it intersects the point at infinity, and the oscillation condition is satisfied (Fig. 9.5*b*). It is easy to demonstrate this effect by reducing the transistor gain to simulate saturation. This design procedure has been used to build several oscillators.

The final example shows how to calculate the voltage waveforms in circuits that are not terminated in the normalizing impedance. On your disk are a voltmeter, `vmeter.dev`, and voltage source, `vsource.dev`. The voltmeter is connected with the wide end, port 1, at some point in the network. The other end

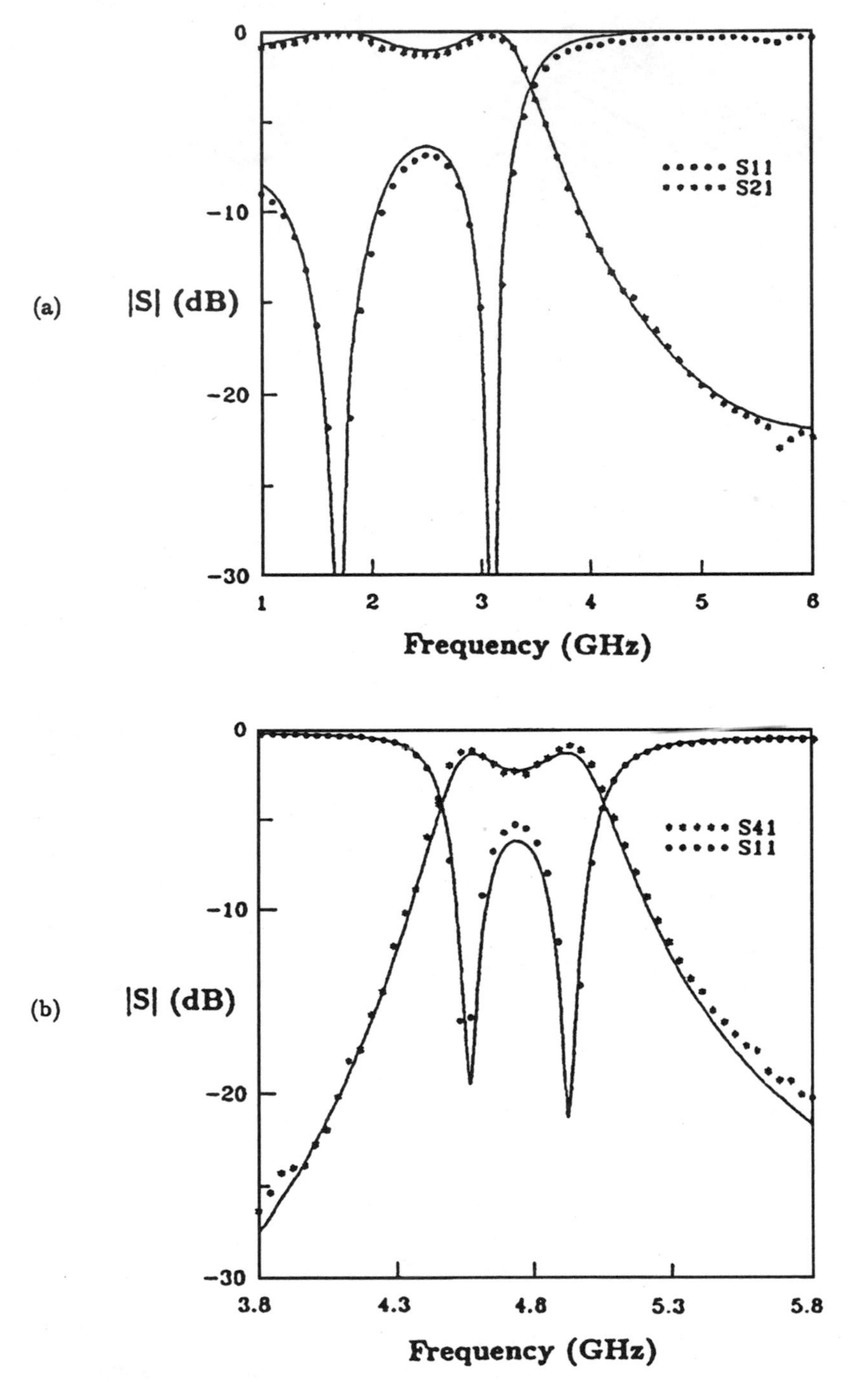

Figure 9.3 Comparison between *Puff*'s predictions (solid lines) and HP-8720 measurements (• and *) for a low-pass filter (*a*), and a bandpass filter (*b*).

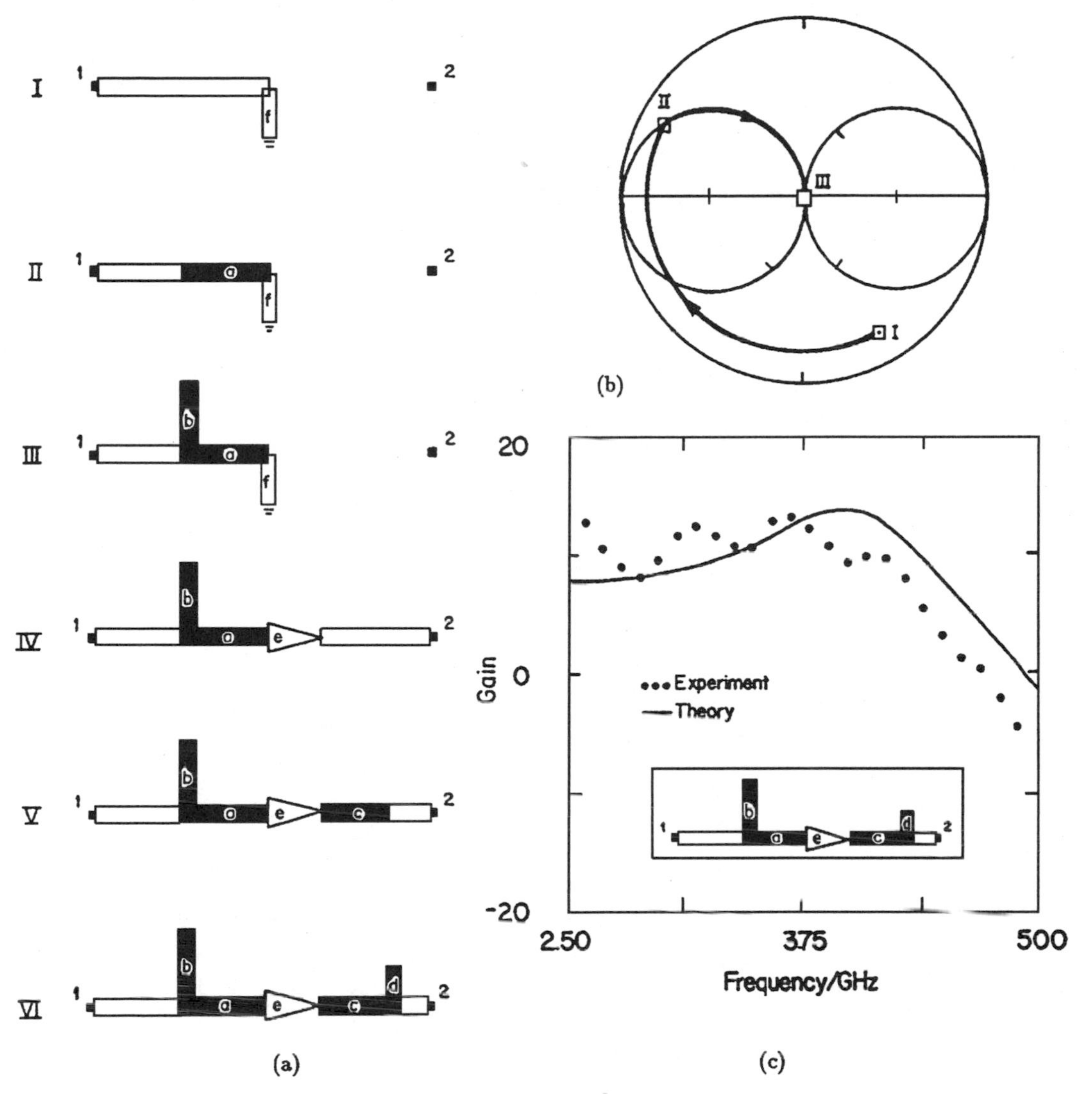

Figure 9.4 Designing a low-noise FET amplifier. Successive views of the *Layout* window are on the left (*a*), and the Smith chart (*b*) and the *Plot* window (*c*) are on the right. We start with a noise match of the input circuit in (*b*). In (I), `lumped f` is the conjugate of the optimum source impedance, given in the manufacturer's data sheet. In (II), `tline a` moves s_{11} around to the $g = 1$ circle and in (III) a stub, `tline b`, brings it into the center of the Smith chart. In a similar way, we power match the output circuit. In (IV–VI), `tline c` and `tline d` bring s_{22} into the center. The rectangular plot, (*c*), compares *Puff*'s predictions with gain measurements.

is joined to a connector port. The voltmeter s_{11} is 1, an open, so that it does not load the network, and the s_{21} is 2, because the total voltage at an open circuit is twice the incident voltage. The voltage source is also connected with the wide end at some point on the circuit and the narrow end to another connector port. The source s_{11} is -1, a short, and the source s_{12} is 1. The source s_{22} is

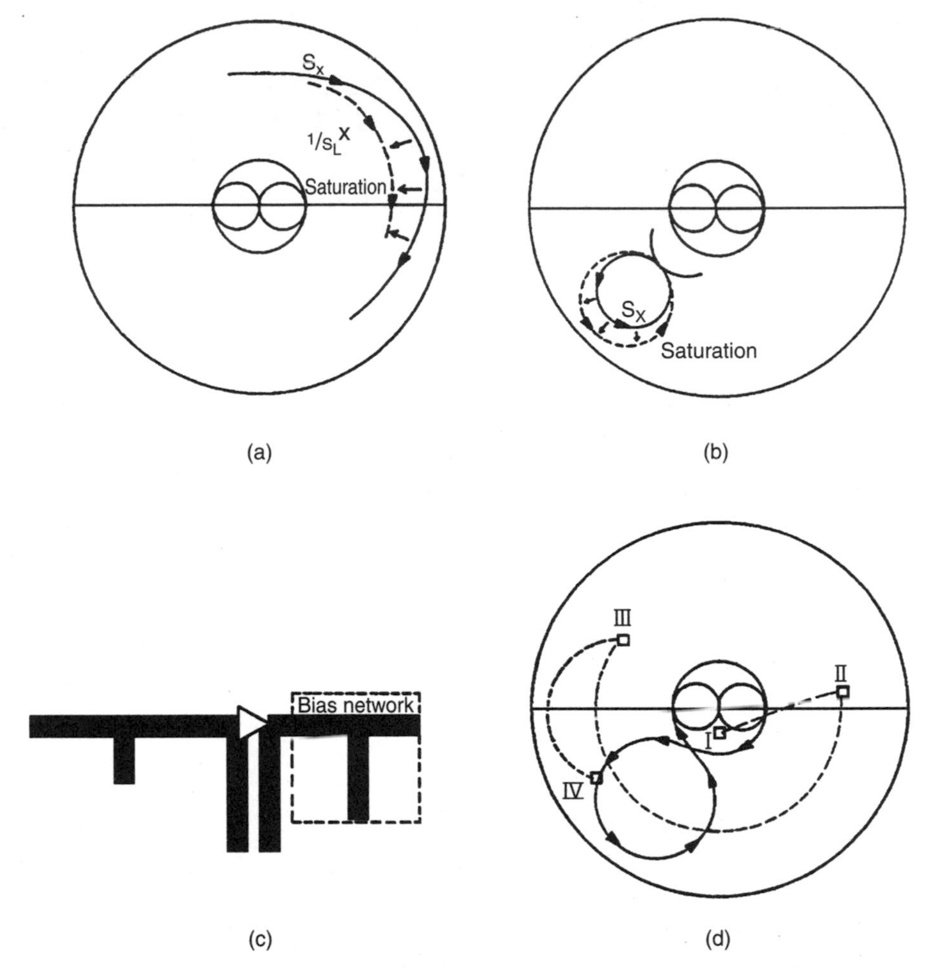

Figure 9.5 Expanded Smith chart of radius 4 for the FET oscillator design. A clockwise loop enclosing the point $1/s_L$ results in oscillations (*a*). A counterclockwise loop gives oscillations for a matched load, where $1/s_L = \infty$ (*b*). Layout for a microstrip FET oscillator (*c*). The coupled-line feedback loop increases $|s_{11}|$ to 3 (I–II) and a stub is positioned to produce a counterclockwise loop (III–IV) (*d*). This circuit generated 10 mW at 5.6 GHz.

1; this allows us to see the input waveform. With these devices in the circuit, we can interpret scattering coefficients as voltage-transfer parameters. Figure 9.6 shows the step response at 50-Ω line section that is terminated with a 150-Ω resistor. This approach can be extended to make a wide variety of probes and sources, and to many different kinds of plots.

References

1. S. W. Wedge and D. B. Rutledge, "Computer-aided design for microwave education," *Electrosoft,* vol. 2, pp. 13–20, Mar. 1991.

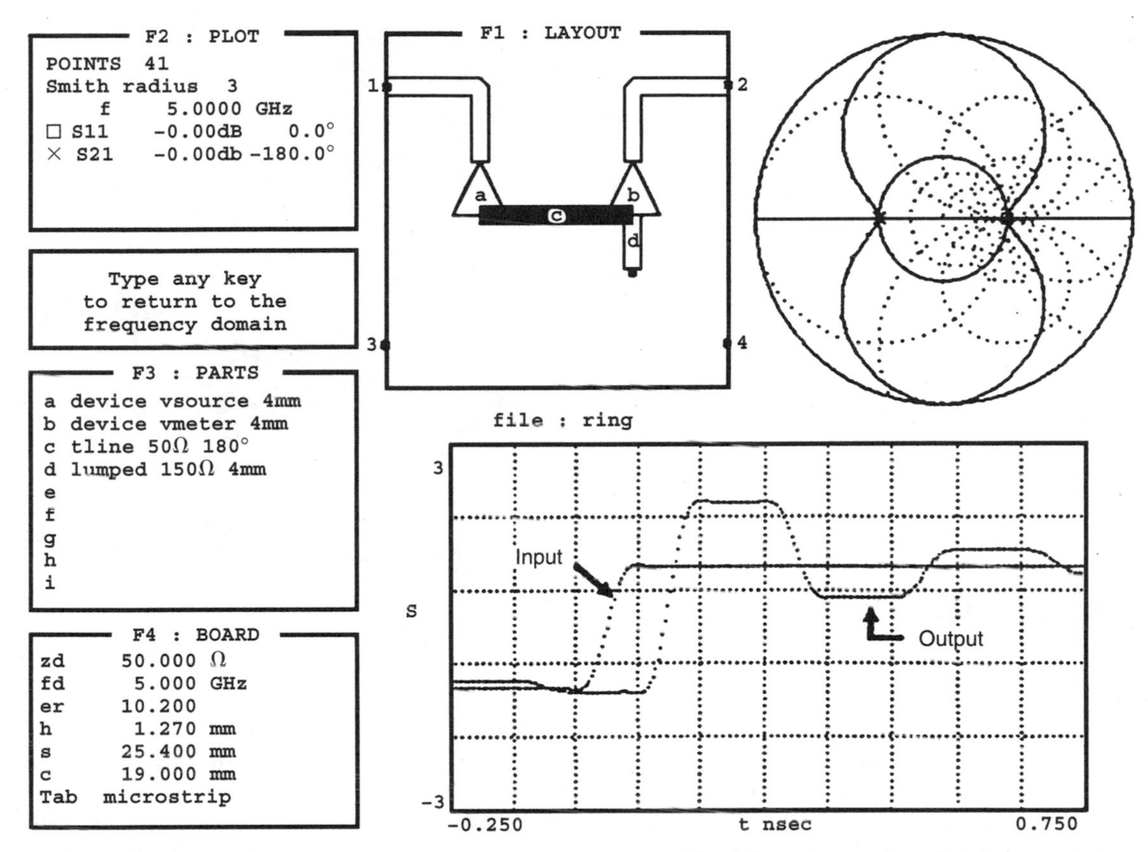

Figure 9.6 Step response for an ideal voltage source driving a 150-Ω load through a section of 50-Ω transmission line. The lower frequency limit is 0, and the upper limit is 20 GHz. Part a is the source, part b is the meter, s_{11} is the input, and s_{21} is the output. The ringing occurs because neither the source nor the load is matched.

2. R. A. York, R. C. Compton, M. Kim, S. Wedge, and D. B. Rutledge, "An interactive approach for designing microwave circuits using a personal computer," *IEEE AP-S Int. Symp. Dig.*, pp. 6–9, 1988.

3. R. C. Compton and R. A. York, "A hands-on microwave laboratory course using microstrip circuits," *IEEE Trans. Educ.*, vol. E-33, pp. 161–163, Feb. 1990.

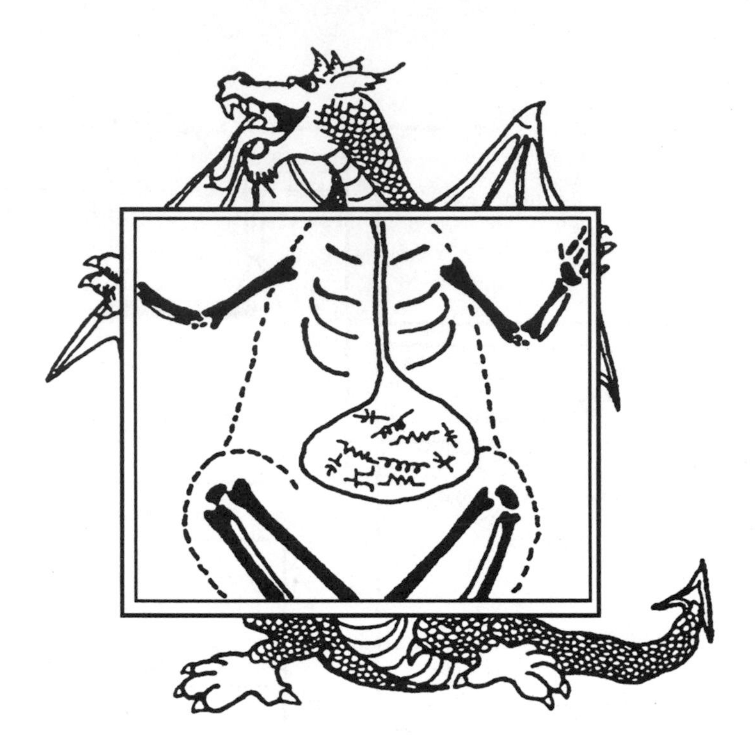

10 Inside Puff

YOU MAY HAVE STUDIED how to analyze a circuit with an admittance or impedance matrix, and solved the matrix by Gaussian elimination. This approach has the advantage of being easy to program, but it is relatively slow for microwave circuit analysis. One problem is that microwave networks are characterized by scattering parameters. If we want to analyze a microwave circuit in terms of an admittance matrix, we must first convert the s-parameters of the components to admittance matrices, then find the admittance matrix of the entire circuit, and finally convert back to s-parameters. Usually these conversions take so much time that it is better to work with scattering coefficients throughout. We can also improve on the simple Gaussian elimination. A matrix that describes a complete microwave circuit is usually large. For example, a branch-line coupler requires the solution of a system of 20 linear equations with complex coefficients. The important thing to realize is that the matrix is sparse, that is to say, it has a large number of zeros. In the branch-line coupler, for example, about two-thirds of the 400 coefficients in the matrix are zero. In an ordinary Gaussian elimination, most of the time would be spent multiplying, adding, or storing zeros.

Puff computes scattering matrices using a fast and memory efficient algorithm called *subnetwork growth.* It is applicable to arbitrarily interconnected components, and involves dividing the network to be solved into subcircuits to simplify the calculations. The approach was first described by Murray-Lasso [1] and applied to scattering matrices by Monaco and Tiberio [2]. The calculations use small and dense scattering matrices, and conversions to impedance or admittance matrices are unnecessary.

Puff must solve the network analysis problem illustrated in Fig. 10.1. The components, represented by scattering matrices $\mathbf{S}_1$, $\mathbf{S}_2$, ... $\mathbf{S}_m$, have been combined to form an aggregate network with unknown scattering matrix $\mathbf{S}_{net}$. The analysis begins with a calculation of the scattering matrices for each component that appears in the circuit. Chapter 6 describes the calculations for the transmission lines and coupled lines. The remaining components are calculated using the expressions given in Fig. 10.2. *Puff* keeps track of the connections present in the circuit and inserts open circuits, grounds, tees, and crosses, as needed. The open circuit is a one-port that occurs when the end of a part is left hanging, unconnected to anything. The one-port ground connection appears when = is typed. Tees occur where three parts join at an ungrounded node, and crosses, where four parts meet.

Associated with each component scattering matrix $\mathbf{S}_k$ are input and scattered wave variable vectors $\mathbf{a}_k$ and $\mathbf{b}_k$, respectively. Organizing the scattering matrices for all components into block diagonal form results in the matrix equation

$$\begin{pmatrix} \mathbf{b}_1 \\ \mathbf{b}_2 \\ \vdots \\ \mathbf{b}_m \end{pmatrix} = \begin{pmatrix} \mathbf{S}_1 & \mathbf{0} & \cdots & 0 \\ 0 & \mathbf{S}_2 & \cdots & 0 \\ \vdots & \vdots & \ddots & \vdots \\ 0 & 0 & \cdots & \mathbf{S}_m \end{pmatrix} \begin{pmatrix} \mathbf{a}_1 \\ \mathbf{a}_2 \\ \vdots \\ \mathbf{a}_m \end{pmatrix} \tag{10.1}$$

Apparent in Fig. 10.1, the aggregate network will have internal connections, as well as external terminals. By separating the internal and external wave variables, a partitioned version of (10.1) is formed:

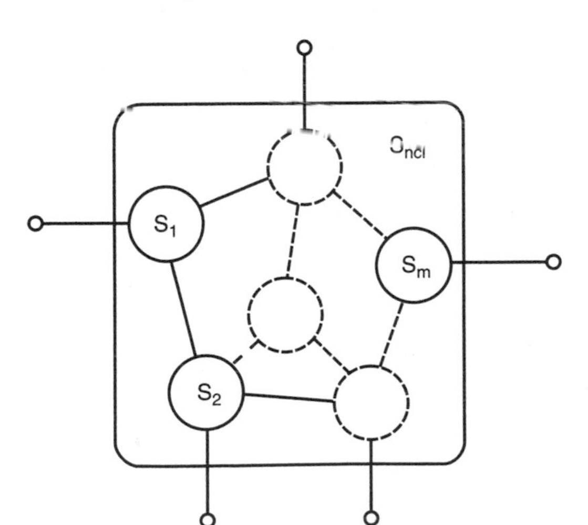

Figure 10.1 Schematic diagram of an aggregate network, $\mathbf{S}_{net}$, consisting of many interconnected components, each characterized by a scattering matrix. Solid lines depict external terminals; dashed lines depict internal connections.

part	s_{ij}
open	1
ground	−1
tee	$-1/3$, for $i = j$; $2/3$, for $i \neq j$.
cross	$-1/2$, for $i = j$; $1/2$, for $i \neq j$.
`atten`	0, for $i = j$; $10^{-\alpha/20}$, for $i \neq j$.
`lumped`	$\frac{z}{z+2} = \frac{1}{1+2y}$, for $i = j$; $\frac{2}{z+2} = \frac{2y}{1+2y}$, for $i \neq j$.
`xformer`	$\pm\frac{n^2-1}{n^2+1}$, for $i = j$; $\frac{2n}{n^2+1}$, for $i \neq j$.
`device`	Values specified in the file; 0 if not specified.
`indef`	Values specified in the file; $\mathbf{S}_n \Rightarrow \mathbf{S}_{n+1}$.

Figure 10.2 Expressions used to calculate the scattering parameters for most *Puff* components. In the table, α is the attenuation in dB, z is the normalized impedance, y is the normalized admittance, and n is the transformer turns ratio. $\mathbf{S_n} \rightarrow \mathbf{S_{n+1}}$ refers to the `indef`'s conversion of an n port scattering matrix into an $n + 1$ port scattering matrix. Scattering parameters for the `tline`, `qline`, and the `clines` are calculated using the equations given in Fig. 6.1.

$$\begin{pmatrix} \mathbf{b}_e \\ \mathbf{b}_i \end{pmatrix} = \begin{pmatrix} \mathbf{S}_{ee} & \mathbf{S}_{ei} \\ \mathbf{S}_{ie} & \mathbf{S}_{ii} \end{pmatrix} \begin{pmatrix} \mathbf{a}_e \\ \mathbf{a}_i \end{pmatrix} \tag{10.2}$$

where the external waves are denoted by subscript **e,** and those internal by subscript **i.** The internal connections of the network impose constraints on the wave variables. Namely, where a connection exists, there will be an equality established between incident and scattered waves. This is apparent upon examination of the single connection shown in Fig. 10.3. The waves in the figure must satisfy

$$a_k = b_l \qquad a_l = b_k. \tag{10.3}$$

These relations exist at every internal connection allowing the construction of a *connection matrix* Γ that satisfies

$$\mathbf{b_i} = \Gamma \mathbf{a_i} \tag{10.4}$$

Since Γ designates equality between internal incident and scattered waves, its elements are either 1 or 0 based on the circuit topology. This is sufficient information to calculate the scattering matrix for the connected network. With $\mathbf{S_{net}}$ defined as the ratio of external incident and scattered waves:

$$\mathbf{b_e} = \mathbf{S_{net}}\,\mathbf{a_e} \tag{10.5}$$

solving (10.2–10.5) yields the connection equation:

$$\mathbf{S_{net}} = \mathbf{S_{ee}} + \mathbf{S_{ei}}\,(\Gamma - \mathbf{S_{ii}})^{-1}\mathbf{S_{ie}}. \tag{10.6}$$

A single application of connection equation (10.6) is all that is needed to carry out the scattering parameter analysis. Unfortunately, for a circuit with many components, the matrices involved become very large, and the computer time and memory required for the $(\Gamma - \mathbf{S_{ii}})$ matrix inversion are substantial. These computational requirements are reduced with the *subnetwork growth* process. In subnetwork growth, a large circuit is analyzed by first calculating the *s*-parameters of smaller intermediate subnetworks [3]. The subnetworks are then connected to form the overall network. Many applications of connection equation (10.6) are required, but the smaller subnetworks require simpler matrix inversions. *Puff* builds up networks by making one connection at a time, effectively reducing the size of the $(\Gamma - \mathbf{S_{ii}})$ matrix to a 2×2. The low order connection calculations continue until the complete network is formed.

As an example of the subnetwork growth process, consider how *Puff* analyzes the four-port branch-line coupler shown in Figure 10.4*a*. *Puff* interprets the coupler to be a network of eight parts: two quarter-wave 50-Ω transmis-

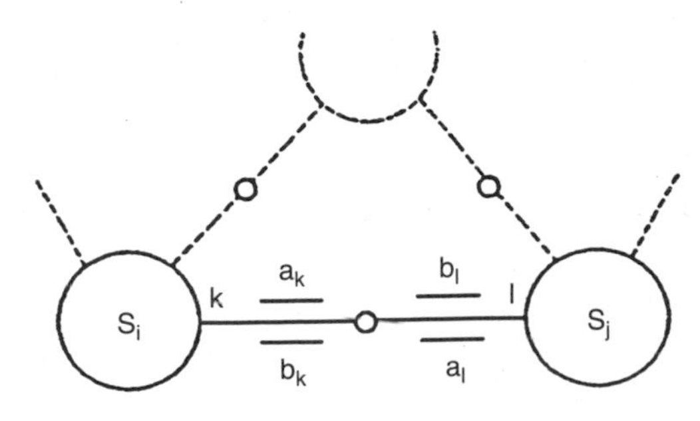

Figure 10.3 Detail showing an incident and scattered waves for connected components $\mathbf{S_i}$ and $\mathbf{S_j}$. The connection requires equalities $a_k = b_l$ and $a_l = b_k$.

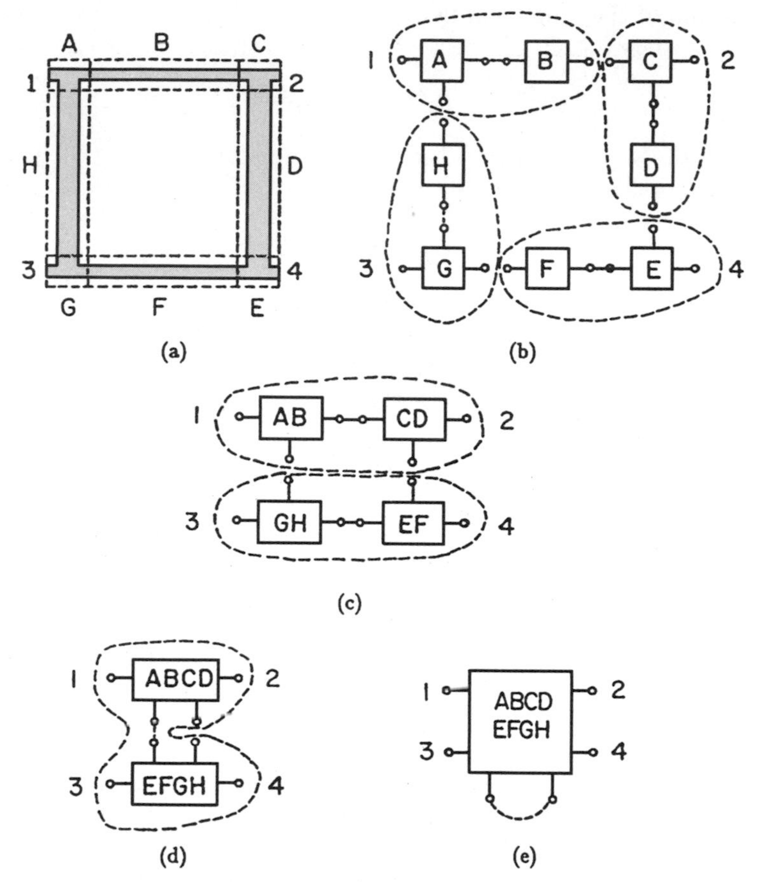

Figure 10.4 Analysis of a branch-line coupler. Sketch of the coupler, showing the different parts (*a*). Representing the eight parts by unconnected boxes (*b*). Joining the eight parts by pairs to make four three-ports (*c*), and joining these to make two four-ports (*d*). Then one connection is made to a single six-port. Finally the last two internal ports are connected (*e*).

sion lines, two quarter-wave 35-Ω lines, and four tees. These components are labeled *A* through *H* in the figure, and the external ports are numbered from 1 to 4. In the first step, the parts are joined in pairs (Fig. 10.4*b*) to make four new three-ports shown in Fig. 10.4*c*. These are combined by pairs to form two four-ports, *ABCD* and *EFGH,* as shown in Fig. 10.4*d*. *Puff* makes a connection between the four-ports to create the six-port shown in Fig. 10.4*e*. Two remaining terminals of the six-port are connected to complete the analysis.

When making one connection at a time, matrix equation (10.6) is equivalent to applying two distinct connection operations. These depend on whether the ports to be joined are on different networks or on the same network (Fig. 10.5). We can apply Mason's theory of signal flow-graphs [4] to calculate the s-parameters for each case. Figure 10.6a shows the signal flow graphs when the two

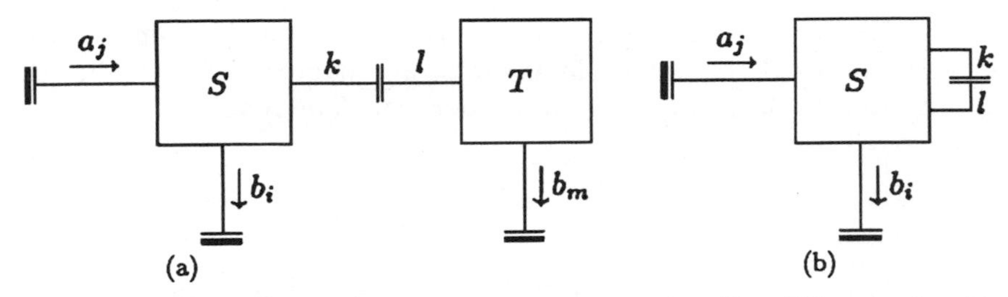

Figure 10.5 The two kinds of joints. A joint between ports k and l on different networks (a), and on the same network (b).

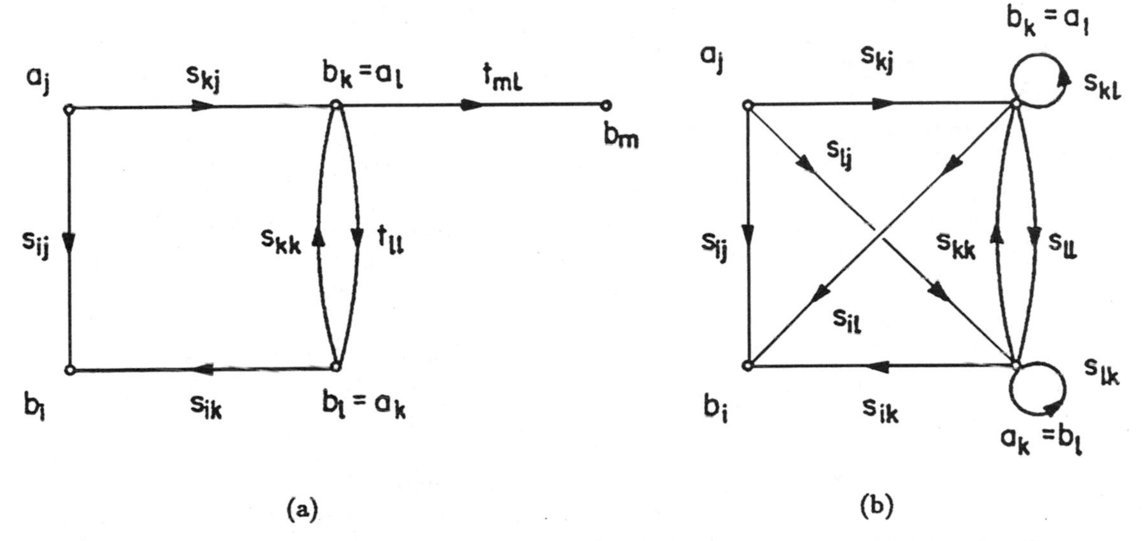

Figure 10.6 The signal flow graphs for joining ports k and l on different networks S and T (a), and on the same network (b).

ports are on different networks. We call the original pair of networks S and T, and the resulting single network S'. The ports to be joined are k and l. We let the input port be j, and consider an output port i, in S, and another output port m, in T. Using Mason's rule, we can write down two formulas for the scattering coefficients:

$$s_{ij}' = s_{ij} + \frac{s_{kj} t_{ll} s_{ik}}{1 - s_{kk} t_{ll}} \tag{10.7}$$

and

$$s_{mj}' = \frac{s_{kj} t_{ml}}{1 - s_{kk} t_{ll}} \tag{10.8}$$

The signal-flow graph for the more complicated internal connection is shown in Fig. 10.6*b*. Mason's rule has to be applied carefully, because there are three first-order loops and a second-order loop. The result is

$$s_{ij}{}' = s_{ij} + \frac{s_{kj}s_{il}(1 - s_{lk}) + s_{lj}s_{ik}(1 - s_{kl}) + s_{kj}s_{ll}s_{ik} + s_{lj}s_{kk}s_{il}}{(1 - s_{kl})(1 - s_{lk}) - s_{kk}s_{ll}} \tag{10.9}$$

The execution time of this algorithm is sensitive to the order in which the joins are made. How do we choose the next pair of ports to combine? We do not know of any proofs that answer this question, but it is easy to see by experimenting that it does make a difference. *Puff* uses a simple approach that searches for a connection that will result in the new subnetwork with the smallest number of *s*-parameters to calculate [5].

There are two classes of singularities that must be avoided when calculating the *s*-parameters. One type occurs when the *s*-parameters become infinite. For example, the reflection coefficient of a resistor connected to ground with a resistance *exactly* equal to `-zd` is infinite. To avoid this problem, *Puff* multiplies all negative resistances by a factor exceedingly close to one: $1 - (1.235 \times 10^{-12})$. This is equivalent to adding a tiny resistor, which does not affect ordinary calculations. This permits the scattering parameters for the singular resistor, although large, to be carried through further calculations without overflows.

The other singularity occurs when a denominator in (10.7–10.9) goes to zero, but the numerator also goes to zero at the same time, in such a way that the scattering parameter remains finite. This is associated with resonant loops in specific circuits involving tees, crosses, opens, and shorts. *Puff* avoids this singularity by multiplying the scattering parameters for these parts by the same factor in the previous paragraph. Physically, this is equivalent to adding tiny attenuators at each port, and it is numerically equivalent to applying l'Hôpital's rule.

Puff is written in *Turbo Pascal* and stores circuits in dynamic variables that are addressed with pointers. The process of laying-out a network consists of building up linked lists of subnetworks and connections. In the analysis, *Puff* collapses these linked lists by applying the reduction formulas (10.7–10.9) until every connection is made. Computation time is reduced by using routines that exploit the power of the math coprocessor. If a coprocessor is not present, it is emulated.

References

1. M. A. Murray-Lasso, "Black-box models for linear integrated circuits," *IEEE Trans. Educ.*, vol. E-12, pp. 170–180, Sept. 1969.
2. V. A. Monaco and P. Tiberio, "Automatic scattering matrix computation of microwave circuits," *Alta Freq.*, vol. 39, pp. 59–64, Feb. 1970.

3. G. Filipsson, "A new general computer algorithm for S-matrix calculation of interconnected multiports," *Proc. 11th Euro. Microwave Conf.,* pp. 700–704, 1981.
4. S. J. Mason, "Feedback theory-further properties of signal flow graphs," *Proc. IRE,* vol. 44, pp. 920–926, July 1956.
5. V. A. Monaco and P. Tiberio, "Computer-aided analysis of microwave circuits," *IEEE Trans. Microwave Theory Tech.,* vol. MTT-22, pp. 249–263, Mar. 1974.

Puff Index

Puff is distributed at-cost by Caltech as a public service. It is provided "as is" without warranty of any kind, express or implied, including but not limited to implied warranties of merchantability and fitness for a particular purpose, with respect to the software and the accompanying written materials. The entire risk as to the quality and performance of *Puff* will lie with the user. In no event shall Caltech, the authors, or any other person, institution, or establishment be liable for any damages whatsoever (including, without limitation, damages for loss of business profits, business interruption, loss of business information, or other pecuniary loss) arising out of the use of or inability to use *Puff*, regardless of the possibility of such damages. There is no guarantee that the operation of the program will be uninterrupted or error free.

Appendix B

Useful Tables

TABLE B.1 Conversion Chart for dBm to Milliwatts to Millivolts

dBm	mW	dBmV	mV_{RMS}	mVp	mVpp
−50	0.000	−3.0	0.7	1.0	2.0
−45	0.000	2.0	1.3	1.8	3.6
−40	0.000	7.0	2.2	3.2	6.3
−35	0.000	12.0	4.0	5.6	11.2
−30	0.001	17.0	7.1	10.0	20.0
−25	0.003	22.0	12.6	17.8	35.6
−20	0.010	27.0	22.4	31.6	63.2
−15	0.032	32.0	39.8	56.2	112.5
−10	0.100	37.0	70.7	100.0	200.0
−5	0.316	42.0	125.7	177.8	355.7
0	1.000	47.0	223.6	316.2	632.5
1	1.259	48.0	250.9	354.8	709.6
2	1.585	49.0	281.5	398.1	796.2
3	1.995	50.0	315.9	446.7	893.4
4	2.512	51.0	354.4	501.2	1002.4
5	3.162	52.0	397.6	562.3	1124.7
6	3.981	53.0	446.2	631.0	1261.9
7	5.012	54.0	500.6	707.9	1415.9
8	6.310	55.0	561.7	794.3	1588.7
9	7.943	56.0	630.2	891.3	1782.5
10	10.000	57.0	707.1	1000.0	2000.0
11	12.589	58.0	793.4	1122.0	2244.0
12	15.849	59.0	890.2	1258.9	2517.9
13	19.953	60.0	998.8	1412.5	2825.1
14	25.119	61.0	1120.7	1584.9	3169.8
15	31.623	62.0	1257.4	1778.3	3556.6
16	39.811	63.0	1410.9	1995.3	3990.5
17	50.119	64.0	1583.0	2238.7	4477.4
18	63.096	65.0	1776.2	2511.9	5023.8
19	79.433	66.0	1992.9	2818.4	5636.8
20	100.000	67.0	2236.1	3162.3	6324.6
21	125.893	68.0	2508.9	3548.1	7096.3
22	158.489	69.0	2815.0	3981.1	7962.1
23	199.526	70.0	3158.5	4466.8	8933.7
24	251.189	71.0	3543.9	5011.9	10023.7
25	316.228	72.0	3976.4	5623.4	11246.8

TABLE B.1 (*Continued*)

dBm	mW	dBmV	mVRMS	mVp	mVpp
26	398.107	73.0	4461.5	6309.6	12619.1
27	501.187	74.0	5005.9	7079.5	14158.9
28	630.957	75.0	5616.7	7943.3	15886.6
29	794.328	76.0	6302.1	8912.5	17825.0
30	1000.000	77.0	7071.1	10000.0	20000.0

TABLE B.2 Conversion Chart for Loss, VSWR, and Transmission/Reflection

Return loss, dB	$\lvert\Gamma_0\rvert$	VSWR	Insertion loss, dB	Power transmitted, %	Power reflected, %
−1.0	0.891	17.391	−6.87	20.57	79.43
−1.5	0.841	11.610	−5.35	29.21	70.79
−2.0	0.794	8.724	−4.33	36.90	63.10
−2.5	0.750	6.997	−3.59	43.77	56.23
−3.0	0.708	5.848	−3.02	49.88	50.12
−3.5	0.668	5.030	−2.57	55.33	44.67
−4.0	0.631	4.419	−2.20	60.19	39.81
−4.5	0.596	3.946	−1.90	64.52	35.48
−5.0	0.562	3.570	−1.65	68.38	31.62
−6.0	0.501	3.010	−1.26	74.88	25.12
−7.0	0.447	2.615	−0.97	80.05	19.95
−8.0	0.398	2.323	−0.75	84.15	15.85
−9.0	0.355	2.100	−0.58	87.41	12.59
−10.0	0.136	1.925	−0.46	90.00	10.00
−15.0	0.178	1.433	−0.14	96.84	3.16
−20.0	0.100	1.222	−0.04	99.00	1.00
−25.0	0.056	1.119	−0.01	99.68	0.32
−30.0	0.032	1.065	0.00	99.90	0.10
−35.5	0.017	1.034	0.00	99.97	0.03
−40.0	0.010	1.020	0.00	99.99	0.01

Glossary

Active device Any component that can amplify a DC and/or AC signal, such as a JFET, BJT, MOSFET, etc.

Active region The 0.2-V range of V_{BE} in which a BJT is capable of amplifying an incoming signal. It is the region between saturation (V_{BE} = 0.8 V) and cutoff (V_{BE} = 0.6 V); within these two V_{BE} values the I_B, and thus the I_C, is controlled.

AGC saturation point An area in which any further increase in AGC voltage becomes nonlinear, or saturates, at the AGC knee.

Application specific integrated circuit (ASIC) A custom-designed and -built integrated circuit.

Automatic noise limiter (ANL) A diode circuit that is located after the baseband detector to cut off any noise spikes that would reach the audio section of an AM or SSB receiver.

Average power Since peak power is the power at the highest amplitude of a digital pulse, most measuring devices would have difficulty measuring a rapid and changing peak amplitude. They instead will normally measure only the average power of a signal over time (about 1 second). The signal itself may have peaks that are 10 times or more higher in amplitude than this average power over time, but these are not uniform, nor are they predictable.

Backplane A common connection point for multiple components, circuits, or systems.

Balanced amplifier A push-pull power amplifier.

Ballast resistor Certain bipolar junction power transistors incorporate a very low value of emitter resistor within their internal structure to make them less open to thermal runaway.

Balun A wideband transformer that is well suited to matching balanced line or stages to unbalanced line or stages, as well as modifying impedances.

Bandgap reference A temperature-stable voltage reference that is similar to a low-voltage zener.

Baseband signal The low-frequency modulation signal, such as music, voice, or digital data.

Bilateral amplifier A single amplifier that amplifies in both directions, and is normally utilized in circuit-sharing transceivers to function in transmit and receive systems.

Bit error rate (BER) The number of bit errors in a digital data communications system. For example, if a communications system has a BER of 10^{-6}, then it is stating that there is 1 error for every 1 million bits sent.

Blanketing A powerful RF signal that damages or destroys reception for all wireless receivers within a certain frequency band. This undesired signal can be received throughout, or only at certain points within, the band.

Blocking (desensing) A strong undesired signal causes a receiver's front-end amplifier to saturate, lowering gain and, in some cases, completely blocking the signal of interest.

Blocking dynamic range A measurement, in dB, between the noise floor and the desired signal that will cause 1 dB of gain compression. This represents the signal level that begins desensitization, producing decreased receiver sensitivity.

Blocking interference A powerful undesired signal can overload a receiver, producing a reduction of the desired signal's amplitude that is either partial or complete.

Bluetooth (BT) A new wireless standard that will allow multiple radios to communicate on their own wireless LAN. At the moment, the radio and the MAC are on two separate chips, but soon these will be combined into a low-cost (around $6) single chip. BT works in the 2.4-GHz Industrial Scientific Medical) ISM band, uses Frequency Hopping Spread Spectrum (FHSS), has a total bandwidth of 83.5 MHz within which it can perform 79 hops with 1-MHz bandwidths (nominal hopping rate is 1600 hops/s) with an effective isotropic radiated power (EIRP) of 0 dBm (a high-power version is 20 dBm). The passband in the United States is 2.402 to 2.480 GHz or, including the guard bands, 2.400 to 2.4835 GHz. Maximum transmit distance in open space for 0 dBm is 10 meters, while 20 dBm is 100 meters. The 20-dBm high-powered BT device must have power control built in, with a monotonic power control step size of not more than 8 dB and not less than 2 dB, all the way down to an EIRP of +4 dBm or less. The 0 dBm low-powered version does not need to implement any power control. Modulation is Gausian Frequency Shift Keying (GFSK), with a binary 1 equaling a positive frequency deviation and a binary 0 equaling a negative frequency deviation. The minimum deviation must always be greater than 115 kHz, but less than 175 kHz. Symbol timing must be ±20 ppm. Bluetooth can work point to point, or point to multipoint. Point-to-multipoint called a *piconet*; one BT unit acts as a master unit, and the other BT units act as slaves. Multiple piconets can form a larger *scatternet* if any of their coverages overlap. Both master and slave units of one piconet can even operate in another piconet with a master in one piconet functioning as a slave in another piconet. BT units employ time division multiplexing (TDM) to be able to work in more than one piconet at a time, with each piconet having its own hopping channel. This channel is one pseudo-random hopping sequence that hops through 79 RF channels, with the psuedo-random noise (PN) code unique to a particular piconet. The PN code is governed by the master unit. With the nominal rate of 1600 hops/s, the channel is divided into time slots of 625 μs, with each time slot also numbered by the master. In this way, the master begins a transmission on even numbered slots, while the slave is allowed to send only during odd-numbered time slots. This is a duplex TDD method. However, the transmitted packets may last to five time slots, with the RF hop frequency frozen until completion of the packet.

Butterworth characteristic A classification of a Butterworth filter's response, which has a very flat bandpass with no peak or ripple.

Bypass capacitor A capacitor used to shunt AC around a component or circuit.

***C/N* ratio** The ratio between a signal's carrier and the noise amplitude, in dB.

Carrier The frequency utilized to convey data or voice modulation from the transmitter to the receiver through the air interface.

Cascaded amplifier One or more amplifiers connected in series to increase the gain over a single amplifier. This multiplies the voltage gain of each amplifier by the next amplifier, or adds the gain in decibels.

Channel A specific section of the frequency spectrum that contains a carrier frequency, with its sidebands; or a passband standardized by common agreement.

Chebyshev characteristic A classification of a filter's response, which has a certain amount of passband ripple, but a sharp frequency cutoff.

Circuit Q (loaded Q) Figure of merit equal to $f_c/(f_2 - f_1)$ and f_2 and f_1 are the upper and lower limits of the bandwidth. The narrower the bandwidth, the higher the Q of a circuit.

Circuit sharing A circuit design methodology that allows the joint use of some circuits in a system, such as filters, oscillators, power supplies, frequency synthesizers, and amplifiers, for both the transmit and receive sections of a transceiver.

Circulators A device that can be employed as a duplexer or as an isolator. Circulators are important because of their unilateral nature, since they will allow the RF's signal power to flow in one direction only. In the duplexer role, one port is attached to an antenna, while the other ports are attached to the transmitter and receiver, respectively. This will permit almost all of the power to flow from the transmitter to the antenna port, while any signal at the antenna will only flow from the antenna to the receiver. Circulators can be mounted on waveguides or on microstrip.

Code division multiple access (CDMA) This multipoint technology allows all client transceivers to occupy the same physical communications channel during the same time period. The modulated signal of each user is spread by an uncorrelated code sequence, which expands the data or voice signal to a wider bandwidth than that required by the information actually being carried. The code itself is unique to each client, which permits any receiver with the appropriate code to accept the transmitted signal, while rejecting all others. The receiver, which possesses the same code as the transmitter, amplifies the received signal, then multiplies it by a duplicate code. This action cancels out the original code sequence as sent by the CDMA transmitter. The device that removes the code is referred to as a *correlator*. After correlation, the now narrowband signal (a duplicate of the originally transmitted information before the code was added at the transmitter) is sent through a bandpass filter, and demodulated in the normal manner.

Common-mode conducted signals Wires will carry common-mode conducted signals if they function as one conductor, with all of their signals in phase, performing as a receiver or radiator for undesired signals, and ordinarily having a return through ground. These conducted signal types can be suppressed by using chokes and ferrite beads (see *differential-mode conducted signals*).

Conductive pad Transistors may employ an electrically and thermally conductive pad placed between their package and their heat sink to improve heat removal. These units do not require thermal grease.

Converter (DC-to-DC converter) A DC power supply that changes one DC voltage to another higher or lower DC voltage.

Copper losses The losses in conductors produced by I^2R losses, skin effects, and crystallization.

Coupling capacitor A capacitor that freely sends AC on to other stages with low impedance, while blocking DC.

Cross modulation The undesired shift of baseband intelligence from an adjacent, strong channel to a desired, but weak, channel. The undesired signal modulates the desired signal in any circuit's nonlinearities in the receiver's front end, creating internal amplitude-modulated interference.

Cross talk Electromagnetic or electrostatic coupling of signals between PCB traces or components.

CSMA/CA Carrier sense multiple access with collision avoidance, a multipoint protocol that controls when a wireless station may transmit, and when it may not. It is used in wireless local-area networks (WLANs) and other multiuser environments. It is quite effective in low-data-rate, bursty, wireless applications. CSMA/CA attempts to avoid data collisions by checking the receiver signal strength indicator, (RSSI) level of its receiver section. If it senses that no other clients are transmitting within the LAN, it will then transmit its data packets. However, it will not attempt to transmit if a sufficiently strong signal is sensed, and will try another transmission only after a random time period has elapsed. Optional additions to the CSMA/CA protocol permits the radio to not only sense spectral energy from the RSSI, but also to determine whether the signal is actually of the same type.

Cutoff The off condition, in which a transistor conducts zero current, with $V_{CE} = V_{CC}$.

Desensitization (desensing) Effect of a powerful undesired signal within the bandpass of a receiver that creates an overload in the receiver's first RF amplifier or mixer, producing decreased sensitivity to the desired signal, as well as distortion due to overdrive, since the LNA or DBM may be placed into a nonlinear part of its operation.

Die form An integrated circuit utilized without a package to conserve space.

Dielectric resonant oscillator (DRO) A microwave oscillator allowing very high frequency operation. The heart of the DRO is a dielectric "puck," which functions as a resonant cavity.

Differential-mode conducted signals Signals that are out of phase with each other by 180 degrees on two-conductor wire. The bulk of desired signals are in differential-mode (see *common-mode conducted signals*).

Digital signal processing (DSP) A method of altering an incoming signal digitally by converting it to digital pulses, manipulating it, then converting the signal back to an analog waveform. DSPs can dramatically improve the signal-to-noise ratio, as well as the fidelity, of a signal.

Dispersion An effect, especially important at high microwave frequencies, that is present in transmission lines in which a wideband (multifrequency) signal becomes distorted because of the different phase velocities of each of the frequencies that make up the signal. This disrupts the phase relationships of the signal before it can reach the other end of the line. In a digital, wideband signal this can cause the eye diagram to begin to close, which increases the BER.

DMOS A type of high-power, high-gain, high-frequency RF power metal oxide semiconductor field-effect transistor (MOSFET).

Driver stage An amplifier that supplies enough power to drive a power amplifier. A *predriver* drives the driver.

Dummy load A 50-ohm resistive load for transmitters; used for testing without radiating.

Dynamic range The measurement between the lowest and the highest signals reproduced without significant noise or distortion in a system.

E_b/N_0 The ratio of the average signal power to the average noise spectral density. It is equivalent to standard signal-to-noise ratio (SNR) normalized over every transmission frequency. Measured in dB.

Effective radiated power (ERP) The true power delivered to the antenna multiplied by the power gain of the major lobe of the antenna. *Effective isotropic radiated power* (EIRP) is the product of the transmitted power multiplied by the gain over an *isotropic* antenna, while ERP employs a dipole instead of an isotropic source as its reference. The ERP of a transmitter system can be calculated by adding the output power, in dB, with the transmitter's antenna gain, in dB, and subtracting the feedline losses, or:

$$P\,(\text{dBm}) = P_{TX} + G_{TX} - L_{TX}$$

where P (dBm) = total amount of power transmitted from the transmitter's antenna, in dBm; called ERP (or EIRP)

P_{TX} = output power of the transmitter, in dBm, before the antenna

G_{TX} = transmitter's antenna gain, in dB (dBi for EIRP)

L_{TX} = feed line losses, in dB, between the transmitter and the antenna

Electromagnetic compatibility (EMC) Tests of electronic devices are conducted to guarantee that they do not radiate excessive interfering electromagnetic signals. These devices should also not be unduly susceptible to another device's signals, or to small levels of EMI. The EMC tests may span the spectrum from 60 Hz all the way up to 3 GHz, and above.

Electromagnetic interference (EMI) The interference caused to any electronic device, such as radios, televisions, and telephones, due to a creator of electromagnetic energy, especially that of electric motors, transmitters, relays, and computers.

Engineering change order (ECO) An explicit instruction to modify components, circuits, or specifications on a manufactured device.

Error-vector-magnitude (EVM) analyzers Agilent is dominant in the field of EVM measurements with their Vector Signal Analyzers (VSA). These devices are utilized mainly in digital radio transmitters, and in some receivers. Vector Signal Analyzers are a relatively new piece of RF test equipment, and are capable of measuring any phase and amplitude impairments present in digital radio systems. Since advanced digital modulation schemes use both amplitude and phase to send information, and these phase and amplitude states can become very close together—as in QAM-16 and above—then noise and other impairments can severely degrade their BER. The VSA is perfectly suited to quantify these types of errors, and is important for checking all digital signal impairments, especially those caused by overdriven amplifiers, excessive group delay variations, amplitude ripple in the passband, LO feedthrough, phase noise, symbol timing, intersymbol interference (ISI), white noise, spurs, frequency offsets, etc.

Faraday shield An electrostatic metallic shield located between the primary and secondary of an RF air-core transformer. This shield significantly decreases the capacitance that occurs between any two conductors that are separated by a dielectric, such as air. Thus, harmonics, noise, and spurious signals cannot as easily pass through the now-increased capacitive reactance of the two coils of the transformer, but the shield still permits the resonant frequency itself to be coupled to the secondary coil through normal transformer action.

Fast Fourier transform (FFT) A technique employing algorithms to convert a time domain signal to the frequency domain. Frequency domain signals are far superior for displaying harmonics than those in the time domain.

Feedthrough capacitor An *EMI filter* device that permits DC and low-frequency AC to penetrate an enclosure, while grounding noise and higher-frequency AC. It functions as a pass-through wire with a shunt capacitor to ground. Commonly used at the input to a metal cabinet or shield that contains a transmitter, receiver, or power supply.

Fencing Internal copper shielding placed within a transmitter or receiver.

Field strength A measurement of the electromagnetic energy radiating from an antenna, or other structure, in microvolts per meter.

Final power amplifier (FPA) The last amplifier that is located before the antenna of a transmitter.

Flat fading A phenomenon that causes the entire transmitted signal to decrease in amplitude, significantly lowering signal-to-noise (SNR) at the receiver. This is typically caused by atmospheric conditions in long, line-of-site high-frequency links, as well as in ionospheric communications. Flat fades of up to 20 dB are relatively common in both types of links.

Flywheel effect A tuned circuit has the capability to produce a sinusoidal wave even if only a simple pulse is received. It is able to perform this function because the capacitor and inductor of the tuned tank circuit exchange energy back and forth, creating an oscillation at its own resonant frequency. The sinusoid will not decay in amplitude as long as a pulse is received at the proper time to reinforce and restore the power lost within the tank's resistances.

Forward error correction (FEC) A technique of error correction to catch, locate, and correct transmission errors by sending redundant data to the receiver, along with the data payload. This prevents retransmission from being required when errors are found in the data, and improves BER. Reed-Solomon is one such type of FEC.

Frequency diversity Two separate frequency channels are transmitted in a communications link to provide multipath immunity, with the receiver choosing the frequency channel with the strongest signal.

Frequency tolerance The degree in which a wireless device's carrier frequency is permitted to deviate above and below its center frequency, expressed as a percentage.

Frequency translation Converting an input frequency to either a higher or lower frequency, while still maintaining the original baseband modulation information.

Full duplex A communications method that allows both ends of a link to communicate with each other during the same time period.

Fundamental overload A type of interference, sometimes severe, to a communications device. It is created by the fundamental signal of a transmitter (at some other frequency) reaching a receiver, producing undesired results because the receiver is unable to reject out-of-channel signals.

Gain block A high-gain, stable 50-ohm amplifier usable from low frequency to at least 3 GHz and beyond.

Gain compression The point in an amplifier's or receiver's gain curve in which raising the input power of a signal will not linearly increase the power at the device's output.

GASFET Gallium arsenide field-effect transistor; a transistor made of gallium arsenide that can supply gain, with stability, at UHF and above.

Ground bounce The voltage from emitter to ground of a high-power, high-frequency transistor should ideally be 0 V, but is usually some fraction of a volt because of emitter lead inductance. This small voltage is referred to as ground bounce.

Group delay The comparative delay of specific frequencies over that of other frequencies.

Half-duplex A communications link in which each end can communicate with the other, but only at different time periods.

Harmonic generator frequency interference A type of interference in which the harmonics of the fundamental frequency create interference. These harmonics are generated in any nonlinear elements, such as frequency doublers or mixers.

Harmonic suppression The degree to which harmonics of the fundamental frequency are attenuated before being output from a transmitter.

Heterojunction bipolar transistor (HBT) An active device that can operate up to 30 GHz and above in oscillators and power amplifiers. HBTs have poor noise figures, but good flicker noise characteristics.

Heat spreader A thin copper plate that assists in dissipating a device's heat over a wider area, and is located between the device and its heat sink.

Helical resonators A filter that is constructed of a helically wound resonant transmission line surrounded by a conductive shield. Helical filters are used between 50 and 500 MHz, are highly selective, and have a flat response when placed in a series of four or more.

High-electron mobility transistor (HEMT) A new active device that has the capability to operate up to 200 GHz, and has a very similar physical structure to that of a GaAs FET. HEMTs are mainly employed as the active element of a low-noise amplifier (LNA) for microwave receivers in direct broadcast satellite (DBS) television, radio telescopes, and in terrestrial and space telecommunications applications.

HEXFET A high-gain metal oxide semiconductor field-effect transistor found in switching power supplies and audio amplifiers.

House wiring The 120-V main wiring obtainable in a residence or office building. The green wire is at ground (the round plug on the socket or plug), the white wire is the neutral (the wide slot), and the red or black wire is the hot lead (the narrow slot).

Hum A 60-, 120-, or 180-Hz low-frequency interference riding on the RF or baseband signal of interest, or on the DC supply voltage. Hum is normally created by an unshielded power transformer or poor filtering of the power supply.

Hysteresis loss AC transformer power losses caused by friction of the shifting magnetic domains within the core material, dissipated as heat. Air-core coils have virtually no hysteresis loss.

IEEE 802.11 wireless standard One of the most important wireless local-area network (WLAN) specifications today. It is the dominant specification adopted by wireless manufacturers in order to build compliant radios that will interoperate—even when multiple vendors' radios are involved. Noncompliant wireless equipment is still relatively common for specialized applications, such as low-cost WLAN systems and wireless video, and for higher bit rate applications than 802.11 allows. 802.11 gives the *media access control* (MAC) and *physical layer* (Phy) specifications only, and does not indicate how these specifications can be resolved (this is completely up to the designer). The specs concentrate on the 2.4-GHz unlicensed band, with data rates of 1 and 2 Mbps using either FHSS or DSSS, and 11 Mbps utilizing DSSS. The permitted total bandwidth is 83 MHz, between 2.4 GHz and 2.483 GHz. DSSS uses DQPSK (2 Mbps) and DBPSK (1 Mbps) modulations in three 20-MHz bandwidths, while FHSS employs 2- to 4-level GFSK in three hopping patterns with 1-MHz bandwidths in 79 frequency slots (with a hop occurring at a minimum of 2.5 hops each second). An antenna gain of no more than 6 dBi is permitted in the United States, with a maximum EIRP of 1 W. In a departure from CDMA, 802.11 DSSS radios all use the same code. The MAC layer of the 802.11 protocol specifies CSMA/CA (see *carrier sense multiple access / collision avoidance*) as the protocol to share the airwaves.

IF notch filter A tunable wavetrap utilized within a communication receiver's intermediate frequency stages.

Incidental AM (IAM) Undesired amplitude modulation of an FM signal.

Incidental FM (IFM) Undesired frequency modulation of an AM signal.

Instability Oscillation of an amplifier at some specific frequency or input/output impedance. An unconditionally stable amplifier will not oscillate at any RF frequency or input/output impedance.

Intermodulation distortion (IMD) dynamic range The difference between the noise floor and the power of two equal-in-amplitude input signals that cause third-order products to be 3 dB above the noise floor. Measured in decibels.

I/Q signals *I* is a signal with 0 degrees phase shift, while *Q* is a signal with 90 degrees phase shift. *Q* stands for *quadrature,* while *I* stands for *in phase.*

Isolator A device that permits a signal to pass in one direction, but not in the reverse direction. Used to prevent reflected waves or externally generated frequencies from entering the final stages of a transmitter. It can be mounted in a waveguide or on microstrip, and can be constructed of a three-port circulator in which one of the circulator's ports is terminated.

Jitter The rapid fluctuation of a signal in frequency or amplitude. It can be created by phase changes in a digital signal produced by phase noise in one or more of the LOs, causing inaccurate or ambiguous demodulation of a received symbol due to timing errors.

Linear mixing Linear mixing creates one frequency that rides on another. Unlike nonlinear mixing, no extra frequencies are created.

Linearizer An electronic device adopted to reduce spectral regrowth in the power amplifiers (PAs) of a transmitter. Most linearizers work by predistorting the signal before it actually reaches the PA, which counteracts the nonlinear distortion introduced by the PA itself. This will negate most, but not all, of the distortion in the PA, and with less power back-off required.

Loading coils An inductor will electrically increase the length of an antenna to make it appear longer than it really is. This will permit a lower frequency to function with a short antenna. However, the antenna will have a narrower bandwidth than if it were at its own natural resonant frequency.

Local area network (LAN) A common collection of computers that are connected over a small area, either by wire or radio, to share files or peripheral equipment.

Mains voltage Voltage supplied by interior or exterior building power outlets. In the United States, this voltage is at 120/240 V, 117/234 V, or 110/220 V.

Mean time between failure (MTBF) The projected average time in hours before a device is predicted to fail during normal operation.

Microphonics Vibration of an electronic circuit that creates noise amplitude modulation of a receiver or transmitter. This effect is exacerbated by loose components in oscillators, such as crystals or capacitors.

Microprocessor A small computer on a single integrated chip.

Mil One thousandth of an inch.

MMIC Monolithic microwave integrated circuit; a high-frequency integrated circuit for microwave communications. An MMIC can be an amplifier, a mixer, or a switch, but the term mainly refers to a Class A RF gain block with matched 50-ohm input and output. The amplifiers are available with high and low gains (33 to 8 dB), low noise (1.7 dB), high- and low-frequency bandwidths (8 GHz to a few kHz), high and low P1dB's (+19.5 dB to +1.5 dBm), and high and low bias currents (60 mA to 12 mA).

Modulus In digital counters, the number of counting states before the counter begins to repeat itself.

Monolithic Formed from a single substrate material, with etching performed on or in the material. A typical integrated circuit is a monolithic component.

Monolithic crystal filter A miniature, low-cost multielectrode crystal filter good from 5 to 500 MHz.

Monotonic An attribute of a filter's or VCO's response. For a filter we would see a steady, increasing attenuation versus frequency, with no slope reversals; for a VCO a rising tuning voltage would produce a steady increase in frequency, with no frequency reversals.

Monte Carlo analysis A computer simulation of a circuit run n times, with random component values, over a set range. For instance, if an LC circuit is simulated in Spice, and each component has a tolerance of ±10 percent, then Spice will choose random values of ±10 percent from each component's ideal value and run some operator chosen number of simulations (n) to discover how typical component tolerances will affect the circuit's output.

Motorboating Audio amplifier self-oscillations that are caused by a low voltage supply, such as a dying battery.

Multiplexers A device with a single output but many inputs by which, by employing control signals, any desired input can be steered to the single output.

Near-field probes Probes to detect cable and printed circuit board RF emissions.

Netlist A text list of components that make up a circuit, along with how they are connected within the circuit. It is basically a text description of the circuit to be simulated, with each node given a number or name.

Noise figure (NF) The NF is the amount of noise, in dB, that is added by the receiver. This noise is contributed mainly by the receiver's first filter and amplifier stage.

Noise floor The amplitude, in dBm, of all the combined thermal noise of a wireless device. Below this level a modulated signal cannot easily be detected, with the noise floor limiting the maximum amplification of any such low-amplitude signal. It is also a direct function of the bandwidth employed to take the noise measurement itself.

Notch filter A bandstop filter with a very sharp bandwidth of high attenuation. Prevents a specific undesired band of frequencies from passing through a circuit.

Nyquist theory States that the sampling rate must be at least double the highest frequency component of the received signal being processed in order to properly retrieve the information embodied within the signal.

Octave An octave is double a certain frequency, or half a certain frequency. For instance, if a 3-MHz signal is used as an example, then one octave above 3 MHz would be 6 MHz (2×3 MHz), while two octaves above 3 MHz would be 12 MHz ($2 \times 2 \times 3$ MHz, or 2×6 MHz), three octaves would be 24 MHz (or $2 \times 2 \times 2 \times 3$ MHz, or 2×12 MHz).

Occupied bandwidth Generally the bandwidth of a digital signal, in hertz, that contains 99 percent of the signal's power. Other values besides 99 percent can, and are, utilized.

Original equipment manufacturer (OEM) The original replacement parts manufacturer for a circuit, system, or device.

Orthomode transducer (OMT) Functions as a waveguide diplexer at microwave frequencies. An OMT comprises two output rectangular waveguides and a single input circular waveguide. When both a horizontally and vertically polarized wave are sent into the circular waveguide input (which supports both polarizations with very little loss), these two polarizations hit a junction. One offshoot of this junction has a horizontal rectangular waveguide, while the other has a vertical rectangular waveguide. Thus dual polarized signals are separated (or combined) into or out of a transceiver. However, most OMTs have limited bandwidth capability.

Pad attenuator A passive circuit that decreases the amplitude of a signal, while maintaining the proper input and output impedance for the system, over a wide band of frequencies.

Parametric instability Instability of an amplifier produced by the collector-to-base nonlinear capacitance causing low-frequency modulation of the transistor's output. Normally only a problem at high frequencies, causing the outbreak of oscillations within the amplifier. Can be alleviated by high-value bypass capacitors.

Parts per million (ppm) Typically used as a measurement of the frequency accuracy and stability of an oscillator. As an example, a particular LO of a frequency source is specified as accurate to within ±0.1 ppm, with a center frequency of 155.00 MHz. However, since it is rated as accurate to within ±0.1 ppm, then we know that the oscillator's frequency may vary by a maximum of ±15.50 Hz from the rated frequency, or $0.1/10^6 \times 155$ MHz $= 15.5$ Hz.

Peak envelope power (PEP) The average peak power of a linear signal with 100% modulation applied to the carrier.

Peak limiter A circuit used to confine a signal's maximum output amplitude.

Phase noise Real-world local oscillators are not perfect single CW frequency sources, but possess phase noise. Phase noise is similar to a modulated spectrum created by a virtual noise source that is phase modulating the desired signal. Any modulation scheme that uses phase variations to communicate, especially modulation schemes such as QAM-16 and above, are adversely affected by the presence of this phase noise.

Phy layer The Phy, or physical, layer defines the modulation and signaling attributes for wireless data transmission.

Port The location within a circuit or device where a signal can be inserted or extracted, either in a physical device or in a software simulation.

Power spectral density (PSD) The measure of the RF power within a given bandwidth.

Processing gain In direct-sequence spread spectrum (DSSS) communications, a modulated signal is sent into a correlator that spreads the signal over a much wider bandwidth than is actually required for the information being carried. The signal is then received, and the information is despread. This action creates *processing gain*. Processing gain in DSSS systems is obtained by the multiplication of data by the PN code, since it spreads the information out over a low spectral power density (due to the much higher rate of the PN code over that of the information). The processing gain can simply be viewed as the difference in amplitude between the unspread and the spread signal. Because the transmitter spreads the intelligence out before it transmits it, and the receiver despreads it (converts it back to a narrowband signal), any interference at the receiver is converted to wideband, low-amplitude signals with a very low power spectral density. This means that the higher the processing gain, the more an interferer's ability to damage or destroy DSSS communications is reduced.

Quadrature phase The condition when two sine waves that are 90 degrees out of phase.

Radials A simulated earth ground. It is a ground plane or counterpoise, consisting of four or more quarter-wavelength wires laid on or in the earth, that is employed when insufficient conditions exist for grounding a quarter wavelength vertical antenna.

Radiation resistance The RF resistance, in ohms, of an antenna at its feedpoint.

Radio-frequency interference (RFI) Interference created by a generator of RF, such as a wireless transmitter, a computer, or even the LO of a radio receiver, to any other electronic device.

Receiver signal strength meter A relative- or real-power indicator of the strength of the RF signal reaching a receiver. The signal strength is gleaned from the AGC or RSSI circuits.

Reflected impedance A transformer has no true inherent impedance of its own, but merely reflects the impedance of the opposite winding. In transformer action, the impedance of the secondary load is reflected back into the primary winding. This impedance can be calculated by the turns ratio of the transformer times the impedance of the secondary load.

Regenerative feedback Any feedback that is at zero degrees, or in phase, from a device's output back into its input. It can create increased amplification or oscillations. It is also referred to as *positive feedback*.

Repeater A device that is capable of retransmitting an RF signal as it is received from a transmitter. It is employed to extend the range of a low-powered transmitter, or to lengthen the line of site of a low-altitude antenna.

Resonator A component that vibrates or oscillates only at its natural resonant frequency, and will not efficiently resonate at any other frequency, thus acting as a high-Q resonant circuit. A crystal or ceramic filter is a resonator.

Return loss The measurement, in decibels, of the difference between the RF power sent toward an input of a circuit, and the power of the returned (reflected) RF signal power. Most circuits, such as filters, amplifiers, and attenuators, can easily be designed with a return loss of 10 dB or higher. With this value of return loss, only one-tenth of the power incident to the circuit's input will *not* be passed on to the circuit's load, but will actually be reflected back toward the source.

RF The section of the electromagnetic spectrum between 50 kHz and 3 GHz.

RF integrated circuits (RFICs) Integrated circuits designed, optimized, and constructed strictly for RF use. Most of the *LC* passive components must be minimized within the chip's substrate, so many will be located off chip because of the large footprints required.

RF leakage The amount of electromagnetic energy, in decibels, that escapes from a cable, connector, component, or circuit.

Ringing Damped sine wave oscillations that take place within a resonant tank circuit when hit by a pulse of energy.

***S/N* ratio (SNR)** The ratio of the signal voltage or power to the noise voltage or power. Expressed in decibels.

Saturation The condition when a transistor conducts as hard as it can, with $V_{CE} \approx 0$ V. However, $V_{CE(\mathrm{SAT})}$ never truly reaches 0 V because of the transistor's natural internal resistances, and thus V_{CE} may be up to 2 V.

Selective calling Communication on a channel to only a specific receiver or receivers by a distinct tone-coded signal sent to the receiver(s) from a transmitter. This will then break the squelch of only the desired radio(s).

Selective fading Multipath fading that occurs only at certain frequencies.

Selectivity The capability of a receiver to select a single signal out of many, while eliminating all others. Thus the receiver's IF bandwidth must be narrow enough, and with sharp enough skirts, to filter out undesired frequencies, but wide enough to not attenuate the sidebands of the desired signals.

Selectivity factor A measurement of a filter's skirt steepness, or its bandwidth in Hz at the 3-dB-down point divided by the bandwidth at the 60-dB-down point. A perfect brickwall filter would be equal to 1.

Self-resonance Condition when a passive component becomes parallel or series resonant because of its distributed capacitive and inductive reactances at a specific frequency.

Self-quieting Muting of a received external signal when an FM receiver captures a harmonic, or other spurious signal, of one of its own local oscillators. This effect can be reduced in the design phase by checking that no harmonics or spurious mixer products fall within the receiver's IF, and that all stages are properly shielded and laid out.

Sensitivity The capability of a wireless receiver to recover low-amplitude signals. This is a function of the gain and internal noise of the receiver, and is normally expressed as the minimum signal level that will be 10 dB above the noise of the receiver. Sensitivity is influenced by the gain of the IF amplifiers and the noise generated in the front-end filter, amplifier, and mixer sections. However, IF gain will be able to amplify only the desired signal, as well as any noise present, to the appropriate level required of the detector; and will have no positive effect on the actual SNR level. This means that any noise contributed by the first stage of a receiver will have a huge effect on the actual sensitivity of the receiver, since that front-end noise will be pasted over the signal's own noise level. The rest of the receiver will also be almost completely isolated by any further effects of internal noise contribution by the gain of this first stage, because the incoming signal will be much higher in amplitude after this first amplification. Thus, the receiver will be much less affected by added internal noise increases. This is why the NF of a receiver's front-end must be minimized and its gain maximized, which maximizes the SNR, for maximum sensitivity.

Shadowing Blocking of certain areas within a wireless coverage area from receiving a broadcast or signal by foliage or structures in the transmission line of sight.

Shielding A part that is made to protect a sensitive circuit or component from stray electromagnetic fields. Copper is a very good RF shield, while soft iron will function at decreased capability. At low frequency, however, copper is a poor *magnetic* shield, while soft iron is much better in this application. Nevertheless, the lower the frequency, the thicker the shield must be to be effective.

Sideband cutting If a receiver's IF filters are too selective they can attenuate the desired sidebands of a signal, distorting the higher audio frequencies, or increasing the BER in a digital system, after demodulation.

Simplex operation In wireless communications, the successive communication between two transceivers on the same frequency, but not at the same time, without the use of a repeater.

Skin effect At high frequencies a conductor's effective AC resistance increases, causing the RF currents to travel closer to the surface. This skin effect will also create a decrease in the Q of an inductor.

Space diversity Condition in which two antennas are separated by distance in order to mitigate multipath effects. One antenna will receive the faded signal, while the other will hopefully receive a clear signal. The antenna with the strongest signal is chosen for amplification by the digital receiver.

Spectrum The complete range of electromagnetic waves from sound waves to x-rays. The section of the electromagnetic spectrum of interest in wireless communications is between 20 kHz (VLF) and 300 GHz (EHF).

Spread The tolerance variation of a device's characteristics. For instance, a JFET may be rated as having an I_{DSS} of 12 mA, but may really have an I_{DSS} of anywhere between 9 and 15 mA.

Spurious emissions Simply referred to as "spurs," these are interfering frequency emissions that are inside or outside of the passband of a transmitter or receiver.

String A number of series-connected amplifiers or frequency multipliers.

Subharmonics The lower-frequency multiples of a frequency multiplier that are still present at the multiplier's output, along with the higher desired multiplied frequency.

Susceptibility How easily an electronic device is adversely affected by another device's electromagnetic fields.

Telemetry The remote wireless transmission of measurement signals, such as temperature, position, pressure, frequency, and speed.

Television interference (TVI) Hampering of a television's received signal by a producer of RF, such as a transmitter, computer, or the LO of a receiver.

TEM (transverse electromagnetic) mode propagation The propagation mode for free space and coaxial transmission lines. In this mode, the electric field and the magnetic field are propagating at right angles to each other (mutual orthogonality) through space or in an unbalanced transmission line.

Termination A load, at the system's own characteristic impedance (50 ohms), utilized to prevent reflections from a cable's or circuit's open end. This termination is used to prevent damage to or unstable operation of a circuit, and can consist of an antenna or resistor.

Thermal runaway A series of events that occurs to a bipolar junction transistor that results in its destruction. Thermal runaway is caused by poor biasing, which allows increasing internal heat within the BJT to create a rise in current through the device, producing further heat until, if unchecked, the BJT is destroyed.

Transformer An electronic component that is capable of magnetically coupling power in its primary winding into its secondary winding and can, if the windings are unlike, transform the voltage, current, and impedance to a different value. Indeed, a transformer has no true inherent impedance of its own, but merely reflects the impedance of the opposite winding. In transformer action, the impedance of the secondary load is reflected back into the primary winding. This impedance can be calculated by the turns ratio of the transformer times the impedance of the secondary load.

Transient intermodulation Distortion in an amplifier created by its inability to react properly to an input signal's rapid amplitude variations.

Transient voltage suppressor (TVS) Device designed to send a damaging transient to ground by tripping when a certain maximum voltage is reached, such as zener diodes and other such components.

Transition region The region between cutoff and saturation in an active device.

Transmitter noise Noise created and transmitted by a wireless transmitter.

Trimmer A small, adjustable, and fine-tunable capacitor, resistor, or inductor.

Twisted pair A transmission line with a characteristic impedance of 100 ohms, and constructed of two insulated shielded or unshielded wires that are twisted together. Utilized for short runs in low-frequency applications, twisted pair is able to decrease low-frequency magnetic hum pickup and RF-producing radiated fields.

Varistor A part that protects circuits against rapid transients, and consists of metal oxide varistor (MOV) and the zinc oxide resistor (ZOR) devices. These function by decreasing their varistor's resistance when a spike of voltage reaches a certain input level, shunting the transient to ground.

Vector network analyzer (VNA) An instrument that can sweep a chosen bandwidth and then output a screen display of all four *S* parameters, as well as phase, delay, Smith chart representations, log and linear magnitudes, VSWR, etc. A VNA can also output any frequency within its design bandwidth functioning as an RF generator. Many are capable of sweeping the input power of an amplifier to test for P1dB, and operate up to 6 GHz. More exotic VNA versions can reach 110 GHz. Two-port calibration of the unit with high-quality shorts, opens, and a termination of known impedance characteristics is, however, vital for accurate VNA measurements. VNAs are invaluable for testing filters, amplifiers, attenuators, passive components, etc.

Vias Small grommet-like conductors that connect one side of a board to another side; or to another layer. *Through-hole vias* connect tracks between opposite sides of a double-sided (tracks on both sides) PCB, or completely through a multilayer PCB from top to bottom. *Blind* vias connect one top-side layer of a multilayer board to a hidden lower layer.

Voltage-controlled crystal oscillator (VCXO) A VCO with superior frequency stability, but a very small tuning range.

Wavetrap A bandstop filter designed to attenuate a certain frequency in a receiver or transmitter.

Wireless local area network (WLAN) A group of devices connected wirelessly so as to permit mobile, or untethered, connections to each other, the Internet, or shared peripherals (see *IEEE 802.11* and *Bluetooth*).

Bibliography

Books

Abrie, Pieter L. D., *Design of RF and Microwave Amplifiers and Oscillators,* Artech House, 0-89006-797-X, 1999.

Bowick, Chris, *RF Circuit Design, Newnes,* ISBN 0-7506-9946-9, 1982.

Carr, Joseph J., *Microwave and Wireless Communications Technology,* Newnes Publishing, ISBN 0-7506-9707-5, 1997.

Carr, Joseph J., *Secrets of RF Circuit Design,* 2d ed., McGraw-Hill, New York, ISBN 0-07-011673-3, 1997.

DeMaw, Doug, *Practical RF Design Manual,* MFJ Publishing, ISBN 1-891237-00-4, 1997.

Dungan, Frank, *Electronic Communications Systems,* Breton Publishers, ISBN 0-534-07698-X, 1987.

Dye, Norm, and Helge Granberg, *Radio Frequency Transistors,* Butterworth-Heinemann, ISBN 0-7506-9059-3, 1993.

Electronic Tables and Formulas, Howard W. Sams, ISBN 0-672-21532-2, 1983.

Evans, Alvis J., et al., *Basic Electronics Technology,* The Learning Center, 1985.

Feher, Kamila, *Wireless Digital Communications,* Prentice-Hall, ISBN 0-13-098617-8, 1995.

Frenzel, Louis, *Communication Electronics,* McGraw-Hill, New York, ISBN 0-02-801842-7, 1995.

Fry, Jim, *Electronic Circuits,* Heath Publishing, 0-87119-010-9, 1985.

Gottlieb, Irving M., *Simplified Practical Filter Design,* Tab Publishing, ISBN 0-8306-8355-0, 1990.

Gottlieb, Irving M., *Practical RF Power Design Techniques,* McGraw-Hill, New York, ISBN 0-8306-4129-7, 1993.

Gottlieb, Irving M., *Practical Oscillator Handbook,* Newnes, ISBN 0-7506-6312-3, 1997.

Grob, Bernard, *Electronic Circuits and Applications,* McGraw-Hill, New York, ISBN 0-07-024931-8, 1982.

Grob, Bernard, *Basic Electronics,* McGraw-Hill, New York, ISBN 0-07-024928-8, 1984.

Hagen, Jon B., *Radio Frequency Electronics,* Cambridge University Press, ISBN 0521-55356-3, 1996.

Hall, M. P. M., L. W. Barclay, and M. T. Hewitt, *Propagation of Radiowaves,* Institution of Electrical Engineers, London, ISBN 0852968191, 1996.

Hayward, Wes, and Doug DeMaw, *Solid State Design,* ARRL, 0-87259-040-2, 1995.

Hayward, Wes, *Radio Frequency Design,* ARRL, ISBN 0-87259-492-0, 1996.

Heathkit, *Electronic Communications,* Heath Publishing, ISBN 595-2443-01, 1984.

Heathkit, *Semiconductor Devices,* Heath Publishing, ISBN 595-2875-05, 1984.

Hickman, Ian, *Practical Radio-Frequency Handbook,* 2d ed., Newnes, ISBN 0-7506-34472, 1997.

Kaufman, Milton, *Radio Operator's License Q&A Manual,* Hayden Books, ISBN 0-8104-0666-7, 1987.

Larson, L. E. (ed.), *RF and Microwave Circuit Design for Wireless Communications,* Artech House, ISBN 0-89006-818-6, 1996.

Lenk, John, *Lenk's RF Handbook,* McGraw-Hill, New York, ISBN 0-8306-4560-8, 1993.

Maas, Stephen A., *RF and Microwave Circuit Design Cookbook,* Artech House, ISBN 0-89006-973-S, 1998.

Madhu, Swaminathan, *Electronics: Circuits and Systems,* Howard W. Sams, ISBN 672-21984-0, 1986.

Margolis, Art, *Computer Technician's Handbook,* McGraw-Hill, New York, ISBN 0-8306-3279-4, 1990.

Marston, R. M., *Passive and Discrete Circuits Pocket Book,* Newnes, ISBN 0-7506-0857-9, 1997.

Matthaei, G., L. Young, and E. M. T. Jones, *Microwave Filters, Impedance-Matching Networks, and Coupling Structures,* Artech House, ISBN 0-89006-099-1, 1980.
Matthys, Robert J., *Crystal Oscillator Circuits,* Kreiger Publishing, ISBN 0-89464-552-8, 1992.
Metzger, Daniel L., *Electronics Pocket Handbook,* Daniel L. Metzger, Prentice-Hall, ISBN 0-13-251835-X, 1982.
Motorola RF Application Reports, Motorola Literature Distribution, catalog number HB215/D, 1995.
Orr, William, *Radio Handbook,* Howard W. Sams, ISBN 0-672-22424-0, 1995.
Pease, Robert A., *Troubleshooting Analog Circuits,* Butterworth-Heinemann, ISBN 0-7506-9499-8, 1993.
Pozar, David M., *Microwave Engineering,* 2d ed., John Wiley and Sons, New York, ISBN 0-471-17096-8, 1998.
Rhea, Randall W., *HF Filter Design and Computer Simulation,* McGraw-Hill, New York, ISBN 0-07-052055-0, 1995.
Rhea, Randall W., *Oscillator Design and Computer Simulation,* 2d ed., McGraw-Hill, New York, ISBN 0-07-052415-7, 1995.
Rohde, Ulrich, et al., *Communications Receivers,* 2d ed., McGraw-Hill, New York, ISBN 0-07-053-608-2, 1997.
Sargent, Jerry E., and Charles A. Harper, *Hybrid Microelectronics Handbook,* 2d ed., McGraw-Hill, New York, ISBN 0-07-026691-3, 1995.
Sayre, Cotter W., *Complete RF Technician's Handbook,* 2d ed., Prompt Publications, ISBN 0-7906-1147-3, 1998.
Schetgen, Robert (ed.), *The ARRL Handbook for Radio Amateurs,* ARRL, ISBN 0-87259-172-7, 1995.
Schrader, Robert, *Electronic Communication,* Glencoe, ISBN 0-07-057157-0, 1991.
Smith, Bradford L., and Michel-Henri Carpentier (eds.), *Microwave Engineering Handbook,* Van Nostrand Reinhold, ISBN 0-442-13588-0, 1992.
Thomas, Jeffery L., and Francis M. Edgington, *Digital Basics for Cable Television Systems,* Prentice Hall, 0-13-743915-6, 1999.
Tocci, Ronald, *Fundamentals of Electronic Devices,* ISBN 0-675-09887-4, 1982.
Turner, Rufus, *The Illustrated Dictionary of Electronics,* Tab Books, ISBN 0-8306-1366-8, 1982.
U.S. Navy, *Basic Electronics,* Dover Publications, ISBN 0-486-21076-6, 1973.
Vizmuller, Peter, *RF Design Guide, Systems, Circuits and Equations,* Artech House, ISBN 0-89006-754-6, 1995.
Ward, Al, et al., *AARL UHF/Microwave Experimenter's Manual,* ARRL, 0-87259-312-6, 1997.
Williams, Tim, *The Circuit Designer's Companion,* Newnes, ISBN 0-7506-1756-X, 1991.

Papers, Magazines, and Application Notes

"1800 to 1900 MHZ Amplifiers using the HBFP-0405 and HBFP-0420 Low Noise Silicon Bipolar Transistors," Agilent, 1999.
"3-Terminal Adjustable Regulator," National Semiconductor, 1998.
"3-Volt Low Noise Amplifier for 0.8-6 GHz Applications," Hewlett-Packard (Agilent), 1999.
"A 5.0 GHz Bipolar Active Mixer," App. Note S010, Hewlett-Packard (Agilent).
Andren, Carl, "A Comparison of Frequency Hopping and Direct Sequence Spread Spectrum Modulation for IEEE 802-11 Applications at 2.4 GHz," Harris Semiconductor, April 11, 1997.
"A Low Distortion PIN Diode Switch using Surface Mount Devices," HP App. Note 1049.
"A Low-Cost Surface Mount PIN Diode Pi Attenuator," App. Note 1048, Hewlett-Packard (Agilent), 1997.
Abidi, Asad A., "Direct-Conversion Ratio Transceiver for Digital Communications," IEEE, Aug. 29, 1995.
"Active GaAs FET Mixers, using the ATF-10136, ATF-13736, and ATF-13484," Hewlett-Packard (Agilent), 1992.
"An Analysis and Performance Evaluation of a Passive Filter Design Technique for Charge Pump PLL's," App. Note AN-1001, National Semiconductor Corp., 1996.
"Application of Microwave GaAs FETs," CEL AU82901-1-982.
"Applications for the HSMP-3890 Surface Mount Switching PIN diode," App. Note 1072, Hewlett-Packard.
"Base Station Antennas for Digital Cellular Systems," Decibel Products, Inc., 1998.
"Basic MODAMP MMIC Circuit Techniques," App. Note S001, Hewlett-Packard (Agilent), 1993.

Basraoui, Mahmoud, and S. N. Prasan, "Wideband, Planar, Log-Periodic Balun," Bradley University.
Bateman, Andrew, "Theory and Implementation of 16-QAM Radio Modems," Wireless Systems International Limited, 1998.
"Biasing ERA Amplifiers," Mini-Circuits, 1998.
"Biasing MSA Series RF Integrated Circuits," App. Note 5003, Hewlett-Packard (Agilent), 1997.
Blevins, Bruce A., "Small Satellite Antennas," PSL, Bruce A. Blevins, 1998.
Brannon, Brad, "Basics of Designing a Digital Receiver," Analog Devices.
"Capacitive Tap Matching," TSV Engineering, 1999.
Carioca, Cezar A., Paulo H. DeCarvalho, and Umberto Abdalla, Jr., "Computer-Aided Design of Diode Frequency Multipliers," *Applied Microwave & Wireless Magazine,* April 1999.
"Cascaded Input IP3 Calculation," RF Micro Devices.
"Constellation Displays," App. Note LAB303A, LeCroy Corp.
"CTS Crystal Application Notes," CTS Corp.
Cushing, Rick, "Replacing or Integrating PLL's with DDS Solutions," Analog Devices, 1999.
"Difficult Amplifier Specifications and Tradeoffs," Miteq, Inc.
"Digital Modulation in Communications Systems—an Introduction," App. Note 1298, Hewlett-Packard (Agilent), 1997.
"Digital Radio Theory and Measurements," App. Note 355A, Hewlett-Packard (Agilent), 1992.
"Digital Wireless System EVM Measurement," *Wireless Conference,* 1996.
Dittmer, Tim W., "Advances in Digitally Modulated RF Systems," Harris Corp., 1997.
"Eagleware" Help Files, Eagleware Corp., 1999.
"Electronic Prototyping: Tips and Pitfalls," University of Nebraska–Lincoln Electronics Shop Web page, 1994.
Evans, John, "Microwave Circuit Board Design and Manufacturing Considerations," Novacom Microwave.
Fakatselis, John, "Processing Gain in Spread Spectrum Signals," Harris Semiconductor.
Gilbert, Barrie, and Eamon Nash, "Controlling RF Power Transmission using a Demodulating Logarithmic Amplifier," Analog Devices, 1998.
"Glossary of Specifications and Terms," Modco, Inc., 1998.
"High Speed Design Techniques," Analog Devices, ISBN 0-916550-17-6, 1996.
"High-Frequency Transistor Primer," Part 1, "Silicon Bipolar Electrical Characteristics," Hewlett-Packard, 1998.
"High Frequency Transistor Primer," Part II, "Noise and *S*-Parameter Characterization," Hewlett-Packard, 1993.
Howland, Rob, "Impairments `R' Us," *Communications System Design Magazine,* 1999.
"IAM-8 Series Active Mixers," App. Note S013, Hewlett-Packard, 1993.
"ICs Simplify the Design of Digital Communications Links," RFMD, 1994.
"INA Series RFIC Amplifiers," App. Note S112, Hewlett-Packard (Agilent), 1997.
"Introduction to Single Chip Microwave PLL's," App. Note AN-885, National Semiconductor Corp., 1995.
Jeganathan, K., "Design of a Simple Tunable/Switchable Bandpass Filter," *Applied Microwave & Wireless Magazine,* March 2000.
"Kinds of Diodes," Hewlett-Packard, 1997.
Kraemer, Bruce, "Digital Modulation: Today and Tomorrow," *Portable Design Magazine,* December 1997.
Kraemer, Bruce, "Nonlinearity Effects in Wireless," *Portable Design Magazine,* May 1997.
Lazarus, Mitchell, "Keeping the FCC Happy—A Reference Article," *Wireless Design and Development Magazine,* November 1999.
"LMX2306/LMX2316/LMX2326," data sheet, National Semiconductor Corp., 1999.
"LNA/Mixer ICs Ease Wireless Receiver Design," RFMD, 1993.
"Low Cost Integrated Solution for Analog Cellular RF Blocks," RFMD, 1995.
"Low Noise Amplifiers for 1600 MHz and 1900 MHz Low Current Self-Biased Applications using the ATF-35143 Low Noise PHEMT," Hewlett-Packard (Agilent), 1999.
"M/Filter Types," Eagleware, Inc., Chap. 9, 1998.
Mannion, Patrick, "Direct Conversion Prepares for Cellular Prime Time," *Electronic Design,* November 1999.
"MAX2620," data sheet, Maxim, May 1998.
McLarnon, Barry, "VHF/UHF Microwave Radio Propagation: A Primer for Digital Experimenters," 1988.
Mercer, Sean, "Minimizing RF PCB electromagnetic emissions," RFDESIGN.com, Jan. 1999.

"Microwave Alternatives, Competition and Markets," M-Pulse, 1998.
"Microwave Transistor Bias Considerations," App. Note 944-1, Hewlett-Packard, 1993.
"Mismatch Conversions," RMFD, 1999.
"Modulating the Alphabet Soup," Analog Devices, 1999.
"Network Analyzer Measurements: Filter and Amplifier Examples," App. Note 1287-4, Hewlett-Packard.
"OCXO's: Oven-Controlled Crystal Oscillators," Wenzel Associates, Inc.
Olney, Barry, "EMC Design for High Speed PCB's," *VeriBest,* 1996.
"Other Crystal Oscillator Types," Wenzel Associates, Inc.
"PC Board Layout Tips," Sawtek, Inc., 1998.
"Power Conversion Tables," RMFD, 1999.
"Practical RF Design," James M. Bryant, 1994.
"PTI Filter Guide," PTI, Inc., 1997.
"Quadrature Modulator/Demodulator RF 2703," RFMD.
"Representative Active Resonator Parameters," PTI, Inc., 1997.
"RF Connector Interface Styles and Applications," Johnson Components, 1997.
"RFIC Components for Cordless Phones," RFMD, 1994.
Robertson, Dave, "Selecting Mixed Signal Components for Digital Communication Systems—Sharing the Channel," Analog Devices, Inc., 1999.
"S Parameter Techniques," Hewlett-Packard (Agilent), App. Note 95-1, 1996.
"Selecting a Wireless LAN Technology," Proxim, 1997.
"Silicon Bipolar MMIC 56GHz Active Double Balanced Mixer/IF Amp," Hewlett-Packard (Agilent), 1997.
Skolnick, David, and Noam Levine, "Why Use DSP?," Analog Devices, 1999.
"Solectek's Wireless Multipoint Media Access Control Protocol," Solectek, Inc.
"Solid-State Phase-Locked Microwave Signal Sources," CTI, Inc., 1997.
"Surface Mount PIN Diodes," Hewlett-Packard, 1999.
"Switching Diode Frequency Doublers," Charles Wenzel, 1999.
"TA 0019," RFMD, 1999.
"TCXO's—Temperature Compensated Crystal Oscillators," Wenzel Associates, Inc., 1999.
"TechTip 3," Rogers Corp., 1998.
"TechTip 5," Rogers Corp., 1998.
"TechTip 8," Rogers Corp., 1998.
"The IEEE 802.11 Wireless LAN Standard," WLANA (The Wireless LAN Alliance).
"The Quest for the Ideal RF Amplifier," Amplifier Research, Inc., 1999.
"Two-Diode Odd-Order Frequency Multipliers," Wenzel Associates.
"UHF and Microwave Designer's Handbook," Watkins-Johnson Company, 1998.
Umstattd, Ruth, "Operating and Evaluating Quadrature Modulators for Personal Communication Systems," National Semiconductor, 1993.
"Using the ATF-10236 in Low Noise Amplifier Applications in the UHF through 1.7 GHz Frequency Range," Agilent, 1999.
"Using the MSA-0520 and MSA-1023 Medium-Power MODAMP Silicon MMIC Amplifiers," App. Note S007, Hewlett-Packard (Agilent), 1993.
"Using Vector Modulation Analysis in the Integration, Troubleshooting and Design of Digital RF Communications Systems," product note, HP 89400-8, Hewlett-Packard, 1994.
"VCXO's—Voltage Controlled Crystal Oscillators," Wenzel Associates.
"VHF Quadrature Modulator," RFMD.
Vig, John R., "Introduction to Quartz Frequency Standards," U.S. Army Communications Electronics Command.
Vogel, Mark O., "Key RF Upconverter Parameters for Optimizing Your Cable Modem System," 3Com Corp.
Vogel, Mark O., "Measuring the Power of a Digital Signal," 3Com Corp., 1999.
"Wireless Analog Design Solutions," 8th ed., MAXIM.
"XOs—Non-Compensated Crystal Oscillators," Wenzel Associates, Inc.

Index

ABOUT THE AUTHOR

Cotter W. Sayre is a wireless RF hardware engineer with 3COM Corporation's Advanced Development Group and the author of an earlier bestselling book for RF technicians.

DISK WARRANTY

This software is protected by both United States copyright law and international copyright treaty provision. You must treat this software just like a book, except that you may copy it into a computer in order to be used and you may make archival copies of the software for the sole purpose of backing up our software and protecting your investment from loss.

By saying "just like a book," McGraw-Hill means, for example, that this software may be used by any number of people and may be freely moved from one computer location to another, so long as there is no possibility of its being used at one location or on one computer while it also is being used at another. Just as a book cannot be read by two different people in two different places at the same time, neither can the software be used by two different people in two different places at the same time (unless, of course, McGraw-Hill's copyright is being violated).

LIMITED WARRANTY

McGraw-Hill takes great care to provide you with top-quality software, thoroughly checked to prevent virus infections. McGraw-Hill warrants the physical diskette(s) contained herein to be free of defects in materials and workmanship for a period of sixty days from the purchase date. If McGraw-Hill receives written notification within the warranty period of defects in materials or workmanship, and such notification is determined by McGraw-Hill to be correct, McGraw-Hill will replace the defective diskette(s). Send requests to:

McGraw-Hill
Customer Services
P.O. Box 545
Blacklick, OH 43004-0545

The entire and exclusive liability and remedy for breach of this Limited Warranty shall be limited to replacement of defective diskette(s) and shall not include or extend to any claim for or right to cover any other damages, including but not limited to, loss of profit, data, or use of the software, or special, incidental, or consequential damages or other similar claims, even if McGraw-Hill has been specifically advised of the possibility of such damages. In no event will McGraw-Hill's liability for any damages to you or any other person ever exceed the lower of suggested list price or actual price paid for the license to use the software, regardless of any form of the claim.

McGRAW-HILL SPECIFICALLY DISCLAIMS ALL OTHER WARRANTIES, EXPRESS OR IMPLIED, INCLUDING, BUT NOT LIMITED TO, ANY IMPLIED WARRANTY OF MERCHANTABILITY OR FITNESS FOR A PARTICULAR PURPOSE.

Specifically, McGraw-Hill makes no representation or warranty that the software is fit for any particular purpose and any implied warranty of merchantability is limited to the sixty-day duration of the Limited Warranty covering the physical diskette(s) only (and not the software) and is otherwise expressly and specifically disclaimed.

This limited warranty gives you specific legal rights; you may have others which may vary from state to state. Some states do not allow the exclusion of incidental or consequential damages, or the limitation on how long an implied warranty lasts, so some of the above may not apply to you.